Lecture Notes in Geoinformation and Cartography

Series Editors

William Cartwright, Melbourne, Australia
Georg Gartner, Wien, Austria
Liqiu Meng, München, Germany
Michael P. Peterson, Omaha, USA

Jan Brus • Alena Vondrakova • Vit Vozenilek
Editors

Modern Trends in Cartography

Selected Papers of CARTOCON 2014

Springer

Editors
Jan Brus
Alena Vondrakova
Vit Vozenilek
Department of Geoinformatics
Palacký University
Olomouc
Czech Republic

The publication has been completed within the project CZ.1.07/2.4.00/31.0010 Supporting the creation of a national network of new generation of Cartography – NeoCartoLink, which is co-financed from European Social Fund and State financial resources of the Czech Republic.

ISSN 1863-2246   ISSN 1863-2351 (electronic)
ISBN 978-3-319-38440-5   ISBN 978-3-319-07926-4 (eBook)
DOI 10.1007/978-3-319-07926-4
Springer Cham Heidelberg New York Dordrecht London

Editorial

In the past, the maps played a deciding role in the exploration of the world. They were primarily used to chart unknown territories. A completely new phase in mapping the unknown started with computer development. Today, cartographic production, visualization, representation and geographic data publication are rapidly developing areas within the GIScience. The fast exchange of information and knowledge are the essential conditions for successful and effective research and practical applications in this domain. For successful research development, it is necessary to not only follow trends in cartography but also try to adapt new trends and technologies from other areas.

Many examples of new and innovative maps are being produced outside of the traditional cartographer's or map producer's offices. The contemporary map makers may not have come from traditional mapping backgrounds and are frequently using open data and open source mapping tools. This situation blurs boundaries between map producers and map consumers. The availability of data and tools allows them to make their own maps in easy and fast way.

It is hardly to say what current trends in cartography are. There are many challenging cartographic studies and outstanding map prototypes which might establish new fields of cartography or open new perspectives of our domain. Cognitive research, usability studies, web technologies, real-time mapping and virtual visualization are some of them.

Mentioned cartographic areas are also the main topics of many conferences which have the main aim to link research, education and application experts in cartography and GIS&T into one large platform. Such a right place for exchange and sharing of knowledge and skills was also the CARTOCON2014 conference, which took place in Olomouc, Czech Republic, in February 2014. It was a great honour for organizers to hold such a significant meeting of experts from all over the world. Invitations have been accepted by dozens of distinguished guests and many of them also participated in this publication. Whole conference proved that it is very important to share knowledge and experience with other professionals, students, public administrative professionals and overall application in the fields of cartography and geoinformatics. Only an effective dialogue between academics, researchers and representatives of the application area can lead to the efficient development of our branches.

The book content consists of four parts. Each of them represents the main CARTOCON2014 conference topics. The first part *New Approaches in Map and Atlas Making* collects studies about innovative ways in map production and atlases compilation. Following part of the book *Progress in Web Cartography* brings examples and tools for web map presentation. The third part *Advanced Methods in Map Use* includes achievement of eye-tracking research and users' issues. The final part *Cartography in Practice and Research* is a clear evidence that cartography and maps played the significant role in many geosciences and many branches of the society. Each paper is original and brings new ideas to cartography.

We have done our best to enable participants to gain as much new knowledge and professional contacts as possible. This book is a compilation of the best and most interesting contributions of CARTOCON2014 conference. The editors hope that this book should provide the reader an overview of current trends in cartography and enable to acquire new knowledge leading to the further development of world cartography.

<table>
<tr><td>Olomouc, Czech Republic</td><td>Jan Brus</td></tr>
<tr><td>Olomouc, Czech Republic</td><td>Alena Vondrakova</td></tr>
<tr><td>Olomouc, Czech Republic</td><td>Vit Vozenilek</td></tr>
</table>

Acknowledgements

The publication has been completed within the project CZ.1.07/2.4.00/31.0010 supporting the creation of a national network of new generation of Cartography—NeoCartoLink, which is co-financed from European Social Fund and State financial resources of the Czech Republic.

About the Series

The Lecture Notes in Geoinformation and Cartography series provides a contemporary view of current research and development in Geoinformation and Cartography, including GIS and Geographic Information Science. Publications with associated electronic media examine areas of development and current technology. Editors from multiple continents, in association with national and international organizations and societies bring together the most comprehensive forum for Geoinformation and Cartography.

The scope of Lecture Notes in Geoinformation and Cartography spans the range of interdisciplinary topics in a variety of research and application fields. The type of material published traditionally includes:

- proceedings that are peer-reviewed and published in association with a conference;
- post-proceedings consisting of thoroughly revised final papers; and
- research monographs that may be based on individual research projects.

The Lecture Notes in Geoinformation and Cartography series also includes various other publications, including:

- tutorials or collections of lectures for advanced courses;
- contemporary surveys that offer an objective summary of a current topic of interest; and
- emerging areas of research directed at a broad community of practitioners.

Contents

Contents

Part II   Progress in Web Cartography

Integrating User and Usability Research in Web-Mapping Application Design

Part IV Cartography in Practice and Research

Part I

New Approaches in Map and Atlas Making

Aspects of the Thematic Atlas Compilation

Vit Vozenilek

Introduction

Atlases are, probably, the best known and the most flexible of popular cartographic products. Atlases are used to address different issues and to target different audiences. Historically, atlases have played different roles—from instruments of power, in the renaissance to current decision and planning support tools (da Silva Ramos and Cartwright 2006). Atlases are used for general reference, education, research and business. As they evolved, atlases were produced in different ways, from the initial manual compilation to current computer-generated processing. Atlases have experienced many changes in the way they are conceived, produced, disseminated and used.

Many researchers have reviewed the definition of what an atlas is. However, although many definitions of atlases have been made, thematic atlases encompass such a wide range of features and technologies, it is necessary to extend these by defining their content and different thematic atlas types. Also, the recent advances in atlases compilation have raised the question of: are thematic atlases a job of cartographers yet? This question is inevitable as, in some ways, atlases cannot just be comprised by map makers but thematicians (experts in atlas content), as well. If there was a dividing line separating the roles of cartography and science of content in atlas compilation, one could say that it is becoming fuzzier.

V. Vozenilek (✉)
Department of Geoinformatics, Palacký University, Olomouc, 17. listopadu 50, Olomouc, Czech Republic
e-mail: vit.vozenilek@upol.cz

© Springer International Publishing Switzerland 2015
J. Brus et al. (eds.), *Modern Trends in Cartography*, Lecture Notes in
Geoinformation and Cartography, DOI 10.1007/978-3-319-07926-4_1

Books Called Atlases

Atlases can be understood as a logically organised collection of maps with a specific purpose and compiled in the form of a book, which usually includes tables, graphs, figures and text. Generally, atlases are likely the first cartographic products that people use because they are introduced to pupils in early at school. Atlases can be considered the most widely known cartographic products (Kraak 2001) and a top of cartography (Kraak and Ormeling 1996; Kraak 2001; Vozenilek 2005).

The books published as atlases (mentioning the word 'atlas' in the title of the book) can be divided into four groups:

A—an atlas as non-cartographic product ('non-atlas')
Such atlas as a type of book (literary work) is introduced and used across all branches of science. Here, however, defining the term atlas is missing. Only da Silva Ramos and Cartwright (2006) states that the word 'atlas' can also be used to describe a collection of information that covers a field of knowledge, for example, an Atlas of Anatomy or History. Nevertheless there is no definition of atlas in medicine (Atlas of Anatomy), in botany (Atlas of Mushrooms) neither in military industry (Atlas of Firearms). The apposite appellations for the products of this concept are 'lexicon', 'encyclopaedia' or 'catalogue'.
B—an atlas as a collection of maps ('primitive atlas')
A large number of map series generated as a representative selection of museum exhibitions or archival collections are often referred to as atlas. This approach, which initially complied with the concept of the atlas, is currently unsatisfactory. Such map series do not form a thematically or regionally organized whole and the maps are not drawn in a unified map language. The apposite appellation for the products of this concept is 'the map collection'.
C—an atlas as a cut up map ('false atlas')
The large format maps compiled in large or medium scale, the scope of which is beyond the capabilities of printing on one map sheet, are printed in the sheet layout in the page size of the book. These map sheets arranged in a book form and accompanied by registers are referred to as an atlas, although it is still a single map (e. g. road atlases, atlases of orthophotomaps). The apposite appellation for the products of this concept is 'a map' or 'maps' (e. g. road map, orthophotomap).
D—an atlas as a systematically organised maps compiled in a unified map language
    ('true atlas')
Current cartography understands *'an atlas as a set of targeted compiled maps systematically organized according to the thematic content, the spatial extent and temporal viewpoint and assembled in a unified map language'* (the author's definition). Only such books may be called 'atlases'.

The use of the word atlas to describe a collection of maps was introduced by the famous cartographer Gerhardus Mercator in the sixteenth century (Thrower 1972). Mercator used the term 'atlas' for a book of maps that it is in use today. The term atlas as an indication of the literary forms of professional publications has been introduced to denote 'books with maps'. In the fields of geoscience, the term atlas should be used exclusively for group D ('true atlases') of above-mentioned classification.

Traditional atlases, such as books, are bound publications and, therefore, have a fixed linear structure. Topics are developed linearly throughout the publication. The maps are developed according to a fixed format, limited to the size of the page. Alonso (1968) notes that atlas cannot be considered to be a book. He stress-es that atlas is not any book consisting mainly of maps (as laymen assume). Technically to the geographer, no collection of maps deserves the name unless it is comprehensive in its field, systematically arranged, authoritatively edited and presented in a unified format. Atlases are collections of maps of regions, countries, continents, or the world. Maps in atlases are accurate only to a degree and can be used for general information only.

Maps have existed throughout the human history and their evolution can be reviewed from different viewpoints. According to Keates (1989) the use of maps in atlases is very flexible. Many of them provide useful inputs for further research in information-spatial disposition (Tucek et al. 2009; Dvorsky et al. 2009). Atlases can be classified by several aspects. Ormeling (1995) classified traditional atlases in regarding to their contents with respect to: geographical atlases, historical, national/ regional, topographic and thematic atlases. Considering communication purposes, the following types of atlases that could be identified are educational, navigational, physical planning, reference, and management/monitoring. Further-more, Borchert (1999) stated that different categories of atlases can be distinguished according to format, geographical coverage, thematic content, information level, purpose, pub-lisher, quality and price. Vozenilek in Mikulík et al. (2008) distinguishes atlases according to territorial extent (national, regional, municipal), user groups (school, public, scientific) and thematic aiming (geological, climatic, forestry, economic etc.).

In addition to traditional book atlases, modern cartography develops digital atlases that offer different representations as the user may experiment with colours, map types, and classification parameters. Both traditional and digital atlases also complement each other with supplementary data, which adds a new element to both, providing modified structural possibilities. Peterson (1998) made research in aspects of digital atlases—their evolution, interactivity, formats, tools etc.

Prerequisites for the Compilation of Thematic Atlas

Each compiled thematic atlas is determined by expertise and marketing limits. The intent of the atlas also takes into account the group of potential users, their literacy and experience in the theme. In order to take into account all the facts, the cartographer compiles a cartographic project of the atlas in cooperation with the thematician.

Table 1 User groups and examples of themes of atlas content for the thematic atlases

User group	Theme of atlas content
Schools (children, pupils, students etc.)	Climate
Public and interest groups (tourists, fishermen, etc.)	Geology
Researchers (teachers, universities, government, etc.)	Soils
	Forestry
	Public health
	Religion
	Transport
	Trade
	War
	Elections

The cartographic project of the atlas is a specific conceptual document in which the cartographer defines the key cartographic information for the preparation of an atlas, especially name and thematic aiming of the atlas, map scales, map projections, atlas and map layouts, draft of map contents, methods of data processing, draft of atlas symbology, organizational and economic topics. The cartographic project includes the text part and attachments, in which the authors present a first look at the atlas layout and symbology. Processing of the cartographic project for scientific or commercial thematic atlas always has the character of scientific and technological work. It consists of system analyses, practical experiments and visualization of atlas elements.

Thematic atlases encompass maps integrated in their content by a common theme. They are compiled for various user groups in various themes (Table 1).

Atlas makes sense to compile if there are met following prerequisites:

- … there is the content

It is crucial in order to that science, in whose 'responsibility' the atlas thematically belongs to, has plenty of territorial scientific knowledge somehow temporally unified (depending on the nature of the phenomenon, preferably to the selected year).

- … there are serious interest and demand for the atlas compilation

It is expected that community of the mentioned science will welcome and support the atlas and its use. It also includes the ability to 'redeem the atlas' at both the stage production, and sales (to sell out full edition of the atlas soon). The using of atlas means not only its applying in the relevant field but also in related fields (e.g. climatic atlas in agriculture) and the governmental management (e.g. phenological atlas in determining agricultural subsidies). The indispensable argument is also that science promotion and its publicity lurk behind the objective needs, which must always be the central motif.

- … there is a cartographic subject which is capable to compile the atlas

The atlas compilation has to be undertaken by a sufficiently competent authority. The competence of the authority means:

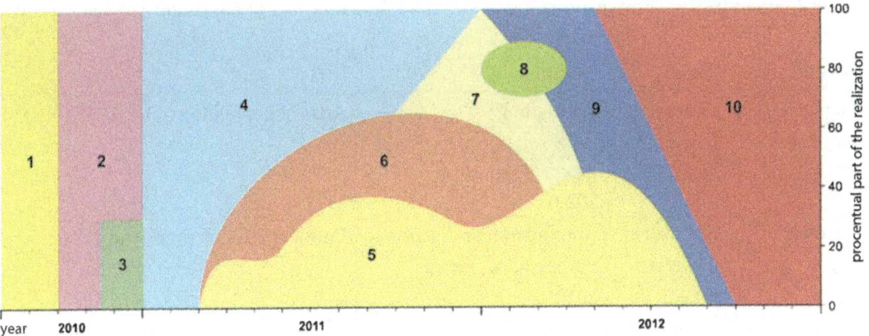

Fig. 1 Timeline for compilation of the Atlas of special educational centres (SECs) in the Czech Republic: *1*—preparation of data collection, database and web form; *2*—pilot data collection; *3*—content optimization for data collection, editing web form; *4*—data collection about services of SECs; *5*—data collection about infrastructure of SECs; *6*—sample preparation of maps; *7*—preparation of atlas model; *8*—consultations among cartographers and the target group; *9*—data processing of data collection; *10*—preparation for atlas production

- it is able to compile the cartographic atlas project
- it is able to manage a large team of authors consisting of experts in atlas content and cartographers
- it is able to provide a technological solution for the atlas and its publishing
- it has experience with previous atlas projects
- … there is a guarantee of sustainability atlas

It means that the authority will assume full responsibility for the atlas topicality in the sense of the content (actual data supplementing) and technology (transferring to new storage media). The question is what sources of funding will cover the sustainability. It usually has been provided by guaranteeing authority.

- … there are available time and money to compile the atlas

The above mentioned prerequisites have to be preserved throughout the compilation process of the atlas, i. e. from the planning stage to the distribution stage of the atlas, which might last 2 years, but even 7 years. The financial assuring of the project includes the costs of human resources (thematicians, cartographers, ICT specialists, GIS experts, economists, lawyers, graphic designers, reviewers, translators, managers, etc.), the cost of data sources (purchase of existing data, conversion of existing data, acquisition of new data) and the cost on the production of the atlas (the atlas book—typographical, printing and bookbinding, the digital atlas—storage, security, availability, transmission rate) (Fig. 1).

Thematic, Territorial and Temporal Aspects

Generally, Keates (1989) highlighted that one could discern atlases by its scale, topic, and target audience. Just as every thematic map a thematic atlas is specified by theme, space/territory and time. Each of these specifications determines the

Table 2 Examples of combination of a theme, space and time as a fixed, controlled and observed aspect

Specification	Aspect	Example	Atlas
Theme	Observed	Population	Atlas of population of USA 1990–2020
Space	Fixed	USA	
Time	Controlled	1990 (2000, 2010, 2020)	
Theme	Controlled	Administrative division	Political atlas of Europe in 2010
Space	Observed	European countries	
Time	Fixed	2010	
Theme	Fixed	Nesting of birds	Atlas of nesting of birds in 2000–2010
Space	Observed	Habitat	
Time	Controlled	Years	

concept of individual maps, and thus the type of thematic atlas. These specifications are established for each atlas as fixed, controlled or observed aspect:

- fixed aspect does not change in all maps in the atlas,
- controlled aspect comes values according to the exact sequences,
- observed aspect is systematically measured and forms the main content of the map in the atlas.

Combining a theme, space and time as a fixed, controlled and observed aspect forms the basic types of atlases (Table 2).

Although the term 'atlas' is frequently associated with the concepts of 'world' and 'small scale', there are atlases with large-scale plans (city street guides), special-purpose atlases (such as road atlases for motorists), and special-subject atlases (such as an atlas of agriculture) (Keates 1989).

Each synthetic thematic map can be a part or even basis of a thematic atlas, particularly when they are the result of extensive geographic research, such Vondraková et al. (2013) and Burian et al. (2013). The atlas resources or stimulus can also be the materials from Earth remote sensing and thematic satellite maps that have already become the full cartographic products (Vozenilek et al. 2012; Bělka and Vozenilek 2013).

Relationships in the Atlas

Atlases are used to collect and disseminate geographical knowledge about the earth. To advance the usage of atlases, development on its methodological base for relationship within both atlas and its maps is a hot topic.

Thematic atlases are comprehensive cartographic products whose content follows the structure and logic of the central theme. The maps in the thematic atlas are organized gradually from simple analytical maps of the theme components. The chapters correspond to these components and lead to the complex and synthetic maps. By arranging chapters and maps each thematic atlas follows a line explanation of basic research issues that initiated the formulation of the atlas objectives. The content of the thematic atlas is arranged like a story book—from simple to complex, from basic information to the culmination as the main message. This is mostly contained in the maps typology and regionalization.

The basic structure of the central theme of the atlas can be described by systems theory, which is in a range of disciplines to current research paradigm (Bossel 2007). A multitude of complex spatial systems pursuing their own agenda shapes the dynamics of the world (Getis et al. 2010). Better understanding of their actions and interactions is crucial, and can be achieved by a profound knowledge of systems and their properties, and their representation in models allowing simulation of probable behavior. Atlases are the perfect tool for it. Drawing on extensive geographical research in modeling and simulation of a wide range of spatial systems—from natural to social systems—atlases represents specific ways of the fundamental concepts and approaches for understanding and modeling the complex systems shaping the dynamics of our world.

All relationships which are essential for a systemic expression of the central atlas theme must be preserved when they are visualized. The basic tool for the expression of these relationships is a map language. The relationships in the theme are reflected in the relationships in geovisualization methods that convey all relationships into a map symbol. The relationships are also maintained in a graphics design of the atlas.

The relationship in a map means the relationship between symbols within map language. There are many relationships originating from a nature of cartographic information conveying—attributes of a real phenomenon are represented by symbol variables (Bertin 1983). Basic understanding is that properties of spatial phenomena are expressed by variables of map symbols. Due to this the similar phenomena (objects or processes) are expressed by similar symbols (in colour, size, pattern etc.). The level of similarity can be difficult to define because most pairs of the object have both similar and different characteristics. The relationships concern a map, map series, atlas structure, atlas design, text, graphs, tables, figures, indexes etc. Each map language relationship represents links between real world and map because it expresses relationships in real system (identical attribute phenomena) with relationships in map language (identical values in symbol variables). The relationships are immersed in shape, size, colour, pattern, thickness and other symbol variables. The relationships in a particular map correspond with groups in map legend.

The relationships related to map language are in:

- map series
The requirements on map language are too general and do not differ from one map to another. When producing a single map, the scope of cartographic creativity in a

map key is wide and is limited only by rules of map language. In a cartographic atlas project, there are map series showing either times changes of investigated processes (e. g. average air temperature) or comparison set of phenomena at several territories (e. g. geology in national parks). There are additional requirements on a key map which has to be the same for all maps in the series. A unique key of map series cannot differ from the single map keys. The only dissimilarity in the map keys is the invisible differences in symbol variables (size, colour etc.) which can make better readability in the single maps in the series. In many cases, a lot of requirements make narrower way for the map key compilation. In some cases, the strong requirements can cause changes in the structure of the central theme, i. e. wider intervals, higher hierarchical levels, more complex groups etc.

- atlas structure and design

These relationships have both graphic and semantic form. The graphic relationships are represented by using of unique symbols or some their elements (shape, colour, fragment etc.). Chapters, titles, subtitles, figure outlines, registers and other atlas components are coloured according to colour scheme. The semantic relationship means standardized terminological, stylistic and typographic framework of text.

- between maps vs. graphs, tables and figures

These relationships follow the graphic and semantic regulations from a map key. Map symbols and their graphic variables are used in graphs elements (coloured lines, symbols in headings etc.). These relationships have influence on other graph elements which do not link with map key. For example, background of graphs is coloured according to atlas colour scheme.

The research of relationships in atlas can be one of topical streams in current cartography (MacEachren et al. 1997). For a number of research methods, eye-tracking technology appears to be promising. It has been successfully applied in the study of cognitive aspects of geovisualisation (Popelka and Vozenilek 2013; Brychtova et al. 2012).

Conclusions

A scientific atlas is a top, extensive and professionally demanding cartographic work that represents a certain theme (Kraak and Ormeling 1996). The scientific atlases function as a means of guaranteeing the development of science and culture throughout human society. It documents not only the state of the presented topic, but also the level of development of the state. In all developed countries is the quality and thematic diversity of scientific atlases always great. The diversity of the presented themes not only documents the development of individual areas of society interests, but they especially provide an essential source of information for international comparisons (Vozenilek 2009; Martha 2013; Köbben 2013).

(continued)

> Conceptual work within the preparation of thematic atlas is one of the crucial factors that decide about the success of the atlas project. The cartographers must not underestimate cooperation with thematicians, they must clarify the aim and purpose of the atlas, they have to define the aspects of the atlas and carefully follow preserving the relationships within the atlas. All these must be done while compiling a cartographic atlas project, thus before his own atlas compilation. This means—think before you draw.

Acknowledgments The author gratefully acknowledges the support by the Operational Program Education for Competitiveness—European Social Fund (project CZ.1.07/2.3.00/20.0170 of the Ministry of Education, Youth and Sports of the Czech Republic).

References

Alonso PG (1968) The first atlases. Can Cartogr 5:108–121

Bělka L, Vozenilek V (2013) Prototypes of orthoimage maps as tools for geophysical application. Pure Appl Geophys 170:745–954

Bertin J (1983) Semiology of graphics: diagrams, networks, maps. University of Wisconsin Press, Madison (originally in French: Semiologie Graphique, 1967)

Borchert A (1999) Multimedia atlas concepts, Multimedia cartography. Springer, Berlin, pp 75–86

Bossel H (2007) Systems and models: complexity, dynamics, evolution, sustainability. Books on Demand, Norderstedt

Brychtova A, Popelka S, Vozenilek V (2012) The analysis of eye movement as a tool for evaluation of maps and graphical outputs from GIS. In: Proceedings of 19th international conference on geography and geoinformatics: challenge for practise and education, Brno, 2011, pp 154–162

Burian J, Brus J, Vozenilek V (2013) Development of Olomouc city in 1930–2009: based on analysis of functional areas. J Maps 9:64–67

da Silva Ramos C, Cartwright W (2006) Atlases from paper to digital medium. In: Stefanakis E, Peterson MP, Armenakis C, Delis V (eds) Geographic hypermedia. Lecture Notes in Geoinformation and Cartography. Springer, Berlin, pp 97–119

Dvorsky J, Snasel V, Vozenilek V (2009) Map similarity testing using matrix decomposition. In: Proceedings of international conference on intelligent networking and collaborative systems (INCOS 2009), Barcelona, pp 290–294

Getis A, Getis J, Bjelland M, Fellmann JD (2010) Introduction to geography, 13th edn. McGraw-Hill, New York

Keates JS (1989) Cartographic design and production, 2nd edn. Longman Scientific & Technical, Essex

Köbben B (2013) Towards a national atlas of the Netherlands as part of the national spatial data infrastructure. Cartogr J 50:225–231

Kraak M-J (2001) Web maps and atlases. In: Brown A, Kraak M-J (eds) Web cartography: developments and prospects. Taylor and Francis, London, pp 135–140

Kraak M-J, Ormeling F (1996) Cartography: visualization of spatial data. Longman, Essex

MacEachren AM, Polsky C, Haug D et al (1997) Visualizing spatial relationships among health, environmental, and demographic statistics: interface design issues. In: Proceedings of the 18th international cartographic conference, Stockholm, 23–27 June 1997

Martha S (2013) The impact of Indonesian policy on the use of high resolution imagery for updating national geospatial information. In: "Sharing Knowledge" joint ICA symposium. Technische Universitat Dresden, Dresden, 23 August

Mikulík O, Vozenilek V, Vaishar A et al (2008) Studium rozvoje regionu založené na vizualizaci geoinformačních databází. Palacky University Olomouc, Olomouc

Ormeling F (1995) Atlas information systems. In: Proceedings of the 17th international cartographic conference, Barcelona, pp 2127–2133

Peterson MP (1998) That interactive thing you do. Cartogr Perspect 29:3–4

Popelka S, Vozenilek V (2013) Specifying of requirements for spatio-temporal data in map by eye-tracking and space-time-cube. In: Processing of international conference on graphic and image processing (ICGIP 2012), Proceedings of SPIE, vol 8768

Thrower NJW (1972) Maps & man: an examination of cartography in relation to culture and civilization. Prentice-Hall, Englewood Cliffs

Tucek P, Paszto V, Vozenilek V (2009) Regular use of entropy for studying dissimilar geographical phenomena. Geografie 114:117–129

Vondrakova A, Vavra A, Vozenilek V (2013) Climatic regions of the Czech Republic. J Maps 9:425–430

Vozenilek V (2005) Cartography for GIS – geovisualization and map communication. Palacky University Olomouc, Olomouc

Vozenilek V (2009) Artificial intelligence and GIS: mutual meeting and passing. In: Proceedings of international conference on intelligent networking and collaborative systems (INCOS 2009), Barcelona, pp 279–284

Vozenilek V, Kudelka M, Horak Z, Snasel V (2012) Orthophoto feature extraction and clustering. Neural Netw World 22:103–121

Mapping Disorder: An Exploratory Study

David Fairbairn

Introduction

The nature of geographical reality is such that disorder, complexity and dynamism are inherent properties of all geospatial datasets reflecting the environment. Both in the natural sphere and in anthropogenic environments, chaotic phenomena are evident in the form of fuzzy boundaries, indistinct objects, moving features, uncertain classifications, dynamic processes, and complex systems. Examples range from the diversity of tropical rainforest habitats with uncertain boundaries and complex classification schemes, to the manifestation of diurnal commuter patterns exhibiting complex networks showing a dynamic human activity with significant regular and irregular patterns of change. The requirement to address the possible representation of such phenomena in map form is part of the process of mapping—defined here as the abstraction of geographical reality using cartographic transformation.

The functions, tools and techniques available for cartographic transformation have, throughout history, concentrated on using static, single-view, two-dimensional graphics to communicate a distillation of reality captured as a 'snapshot'. Such maps are created to give an ordered insight into the complexity and unpredictability of reality.

In general terms, this paper suggests that a new paradigm of cartographic representation is required to address the task of moving away from such standard 'snapshot' maps to representations which reflect the disorder of spatial reality. This is a major ambition, so this paper specifically attempts to contribute, in a more limited and preliminary manner, to an investigation of one example of disorder—in this case in topographic landscapes. It involves the assessment of disorder; its

D. Fairbairn (✉)
School of Civil Engineering and Geosciences, Newcastle University, Newcastle upon Tyne
NE1 7RU, UK
e-mail: david.fairbairn@newcastle.ac.uk

© Springer International Publishing Switzerland 2015
J. Brus et al. (eds.), *Modern Trends in Cartography*, Lecture Notes in
Geoinformation and Cartography, DOI 10.1007/978-3-319-07926-4_2

quantification, description and comparison; its characterisation; and its subsequent representation. Contemporary technologies are opening up significant possibilities in undertaking such tasks, most notably in representation and visualisation. It is such opportunities which may well lead to the new paradigm sought, involving use of animation, hyperlinked documents, interactivity with displays, and editing capability, perhaps in a multi-user environment, possibly connected and web-enabled, certainly multi-sourced and easy to distribute.

At this stage, however, the examples chosen are familiar to the conventional topographic cartographer, with the assessment and handling of spatial data in landscapes being the focus for an initial study of disorder and its effect on mapping.

Handling Disorder in Landscapes

Landscapes, and the processes which create and influence them, can often be dissected, uneven and disturbed. Such environments can be completely natural, for example in peri-glacial areas, or can be anthropogenically influenced, for example in areas which have sustained mining activity over many years. The quantification of such disorder can be initiated by collecting spatial data using direct measurements, by survey and by remote sensing, and subsequently characterising the environments using standard metrics such as diversity, complexity and order indices.

The indices of disorder then need to be translated into graphical representations of space: this will involve a further stage, exploring and extending the range of tools, techniques and technologies available in contemporary geo-visualisation, including methods for communicating multi-variate, multi-dimensional, multi-temporal, chaotic, disordered data.

Sporadic efforts have been made to undertake the first stages of quantifying spatial disorder, most relying on the development of specific calculable indices, such as entropy and diversity indicators. Both the terminology used and the applicability of these indices are contentious, but a consistent and acceptable approach to identifying, recording and measuring disorder is essential to fully utilise this approach.

Example studies following this particular *modus operandi* have been undertaken in the field of **landscape ecology**, a discipline which has led research into quantifying diversity, complexity and order, and applying these to aesthetic and cultural readings of environments (Arnheim 1972; Lewis 1982). The application and analysis of metrical indices, including those mentioned above, is considered more recently by Ode et al. (2010) and Zurlini et al. (2012). Such indices are also applicable in **archaeology**, where notable developments in high resolution remotely-sensed surface data collection, notably by LiDAR, allow detailed measurement of landscape disturbance and perturbation resulting from human activities (e.g. mining, agriculture, military) (Doneus and Kühtreiber 2013; Kovacs et al. 2012). Translated into map representations, and integrated with documentary and interpretative sources, the relationship between anthropogenic impacts and quantitatively-derived landscape characteristics can be systematically modelled. Pfeifer et al. (2011) have taken similar approaches, using LiDAR data in the study of **geomorphology**, whilst the related field of geomorphometry—applying numerical methods and deriving similar indices of topographic structure—can exemplify issues related to characterising disorder (Hengl and Reuter 2008).

The subsequent cartographic representation of disordered raw data and such derived indices is problematic. Both the recording of current situations and the creation of predictive or explanatory models require effective map representations of actual or potential order, disorder, complexity, and change. The representation of such metrics on maps has not been explored in depth by the cartographic community, despite initial attempts at examining the use of maps to actually derive the metrics themselves (Fairbairn 2006, 2011).

Further examples of mapping required in dynamic environments (e.g. periglacial zones, coastal margins and river channels) can be envisaged, along with the need to represent uncertain human behaviour in cartographic products detailing geographic phenomena which are not primarily terrain-oriented, such as transport networks, crime occurrences, migration patterns, and epidemiology.

Archaeological Disturbance in the Landscape

The specific landscape milieu initially considered in this paper is recent archaeology of sites with varying characteristics, and the methodology will primarily seek to empirically quantify the measured variability in order and disorder over such areas. The plan is to examine a number of indices which have relation to disorder (e.g. entropy, diversity indicators, information metrics, concentration and fragmentation measures, randomness measures) and to determine their applicability in characterising those areas which are dominated by natural landscape, as well as those where anthropogenic influences have altered the environment, historically.

This paper introduces a study, therefore, which asks: To what extent do remnant mining sites reflect the human activity which is manifest in them? What landscape parameters can be used to characterise the nature of such sites? To what extent can indices of landscape ecology, geomorphometry, and landscape archaeology, be used to describe and quantify order, disorder, complexity, diversity and change? And what are the possibilities of optimising the cartographic representation of such sites?

The study has identified an area of landscape disturbance as a result of human activities (mining of lead ore and other heavy metals). Such disturbance is variable across the small area chosen, as a result of location (notably the distribution pattern of minerals which have been exploited), the methods of mining and extraction, and the development of transport networks to export the products from the site. It is recognised also that the extent of the site, the scale of sampling of spatial data, the level of resolution of the data (notably the digital terrain model), and temporal variability, may each affect the data collection, data processing, data analysis and data representation tasks. For the site, a range of possible metrics related to disorder will be examined, eventually to establish a relationship between anthropogenic mining impacts and landscape characteristics, and also to further analyse the representation possibilities.

Data Preparation

The site chosen is an area of relict mining activity, active from the fourteenth century, but primarily during the second half of the nineteenth century (mining has now ceased in this area). The geology of the location is relatively straightforward, consisting of sedimentary layers of the Carboniferous era, including some with significant mineral resources, both in the rock and in mineralised veins. The main mineral output was of barium and lead ores, along with some commercially viable iron ore deposits. The workings were in the form of medium-sized quarries in the exposed sandstone layer to the north, with small open pits from earlier centuries for the more southerly measures containing barytocalcite and lead ore: these were subsequently mined using vertical shafts and drift mining (adits) in the later nineteenth century. A large amount of waste resulted from these operations and spoil heaps typify the landscape. This site is near the settlement of Blagill, close to Alston, Cumbria, in northern England.

The assessment and quantification of the disorder in this landscape relies on a suitable-resolution digital surface model: airborne LiDAR survey data was processed to provide a one-metre planimetric resolution gridded dataset. This data did not have vegetation or buildings removed from the surface, but the actual area examined has no trees or buildings covering the site, and the current land use is uniform upland sheep grazing pasture. The initial investigation in this paper uses this digital surface model (DSM) to assess, quantify, and characterise the nature of the terrain.

Figure 1a shows the shaded relief image of the surface model, with added geological mapping (zones and fault lines) and a generalised 10 m contour map. The grid coordinates of these maps are in metres, projected to the British National Grid. It can be seen that the zones chosen for analysis have variability in their geology, their height and their surface characteristics. Detailed analysis of the rock types, their formation and their exploitation is given in Clarke (2008), which indicates that the alternating layers of the Stainmore formation comprise mudstone, sandstone and limestone, yielding rich ore deposits, whilst the Firestone zone consists of relatively mineral-poor uniform sandstone. It is suggested here that variation in mining activity results from the variable geology, and the remaining evidence of that activity has affected the configuration of the current landscape, a configuration that exhibits distinct differences in complexity and hence

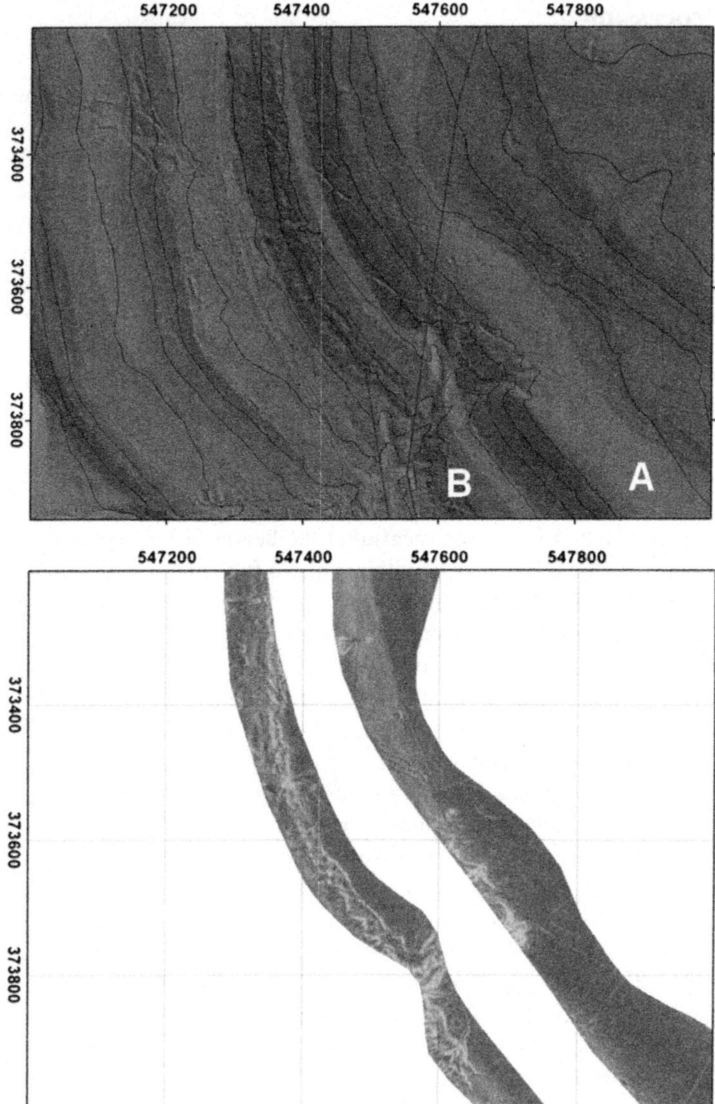

Fig. 1 (from top—west-up) (**a**) LiDAR derived DSM, geological mapping and contours; (**b**) Slope map of study area (steeper slopes in lighter greyscale)

demonstrates contrasting degrees of disorder. Zone A, within the geological zone mapped as Firestone sandstone, can be compared with Zone B, part of the Stainmore formation with its mixed, but mineral rich geology.

Data Processing

The DSM was examined using a range of software, in order to prepare and modify the data, and analyse its properties. A primary parameter which can be obtained from the surface model is a slope map, which can visually be used to detect zones of differing levels of dissection. The slope map for the relevant zones is shown in Fig. 1b. As can be seen, the zone of sandstone (Zone A), which has been subject to less mining activity and spoil heap creation, has fewer steep areas and a more uniform surface. The mean slope in Zone A is 10.11°, whilst in Zone B it is 14.69°. The comparative values of average slope may well be the simplest and most suitable metric to establish variable disorder in this landscape.

However, further confirmation was sought of the distinction between the two zones, by examining several more landscape indices. The Terrain Ruggedness Index (TRI) was initially established for large-area landscape characterisation to assist in wildlife management (Reilly et al. 1999). It was calculated and mapped in this exercise using the Raster Calculator within ArcGIS, the output demonstrating the differences which result from measuring the height differences between adjacent pixels in the DSM. By extension, this is also a function of slope but quantifies, more directly, dissection of the terrain surface and the degree of difference in height of all eight neighbouring cells to the target pixel.

The mean value of the TRI for Zone A was measured at 19.31 whilst Zone B had a higher mean TRI value of 22.55. There are, in fact, many different indices available for characterising surface roughness (see, for example, http://gis4geomorphology.com/roughness-topographic-position/). A further index applied to this dataset is sourced from terrain analysis work presented by Hobson in 1972, and coded as a Python script for incorporation into ArcGIS by Sappington (2008). This index is more comprehensive than the TRI metric, in that it takes account of aspect in addition to slope—clearly, consideration of variable orientation of equal slope values around a point, for example, should yield improved and more faithful measures of dissection. The resultant index (called vector ruggedness measure, VRM, by the script author) was assessed for Zones A and B: once again, an overall mean figure for VRM shows variability, with Zone A calculated at 0.0013 and Zone B at 0.0044.

The measurements taken so far indicate that it is possible to develop realistic measures of terrain variability from LiDAR-derived digital surface models, at sufficiently large scale. The scale must be set to consider the impact, in this small area, of relatively minor features—small spoil heaps, depressions indicating capped pit-shafts, and surface features such as tracks and specially-dug drainage channels.

The figures show that a comparison can be made between nearby zones with differing landscape use histories, and it may be possible to develop models of landscape form and genesis which can be transferable across regional and national landscape characterisation studies. In this case study, a distinction has been drawn, using simple indices of disorder, between an area relatively untouched by human activity, and one which has been comprehensively altered by anthropogenic mining practices.

Additional metrics were examined using alternative software for terrain data handling. For example, a 'patch richness density' index (PRD) was calculated using the Fragstats program, resulting in a metric for a landscape (or any categorised polygonal dataset) which is higher the more individual patches of a class which exist in the image (McGarigal and Cushman 2005). The PRD metric presents the number of distinct patches per 100 ha. Thus, a dissected landscape, classed for example into 32 categories of height (i.e. a layer-tinted terrain model of a complex area) will have smaller individual and more numerous adjacent hypsometric layer tint zones, and a higher density of separate individual patches, compared to a uniform sloped terrain which will have only as many patches as there are classes. In this terrain, for example, Zone A has a PRD of 307.07, whilst Zone B, with its more complex landscape has a PRD of 340.23.

Analysis of the terrain was also undertaken using the Landserf terrain data handling software package. The fractal dimension of the surface in each zone was calculated (Zone A, 2.12; Zone B, 2.20) confirming the higher disorder in Zone B. Feature extraction and landscape feature detection is effectively undertaken in Landserf, with elements such as pits, ridges, channels, passes, flat surfaces etc. being identified and visualised. Graphical output from this routine indicates that Zone A has a lower density of structural features (most of the detected ridges are, in fact, walls and field boundaries rather than mining artefacts), whilst a greater proportion of the pixels in Zone B can be categorised as forming channels and ridges.

In Zone A, the Ridges and Channels form only 3.9 % of the DSM cells, whilst in Zone B they constitute 16.6 %. The planar areas form 96 % of Zone A, but 81 % of Zone B, which has many more peaks identifiable.

The work undertaken so far has demonstrated that terrain surfaces can be captured effectively at appropriate scale and resolution for investigating their structure. Disorder in the terrain can also be quantified, either absolutely (for specific measures to be stored) or comparatively (to detect areas of relative disorder). Furthermore, background information about the nature of the terrain, its formation and its modification, can be used to confirm the disorder inherent in differing landscapes and land uses.

Representation

Once terrain disorder has been identified and quantified, the task of mapping it to reflect the variability and complexity must be faced. It was suggested in the Introduction that new methods of cartographic representation must be sought and established for the most efficient mapping of disorder. Historical and contemporary maps of the area studied in this paper are illustrated and considered here.

The early topographic mapping shown in Fig. 2a, with data (including contour values) collected solely by field observation, demonstrates the use of two-dimensional mimetic symbols to represent breaks of slope, patches of spoil,

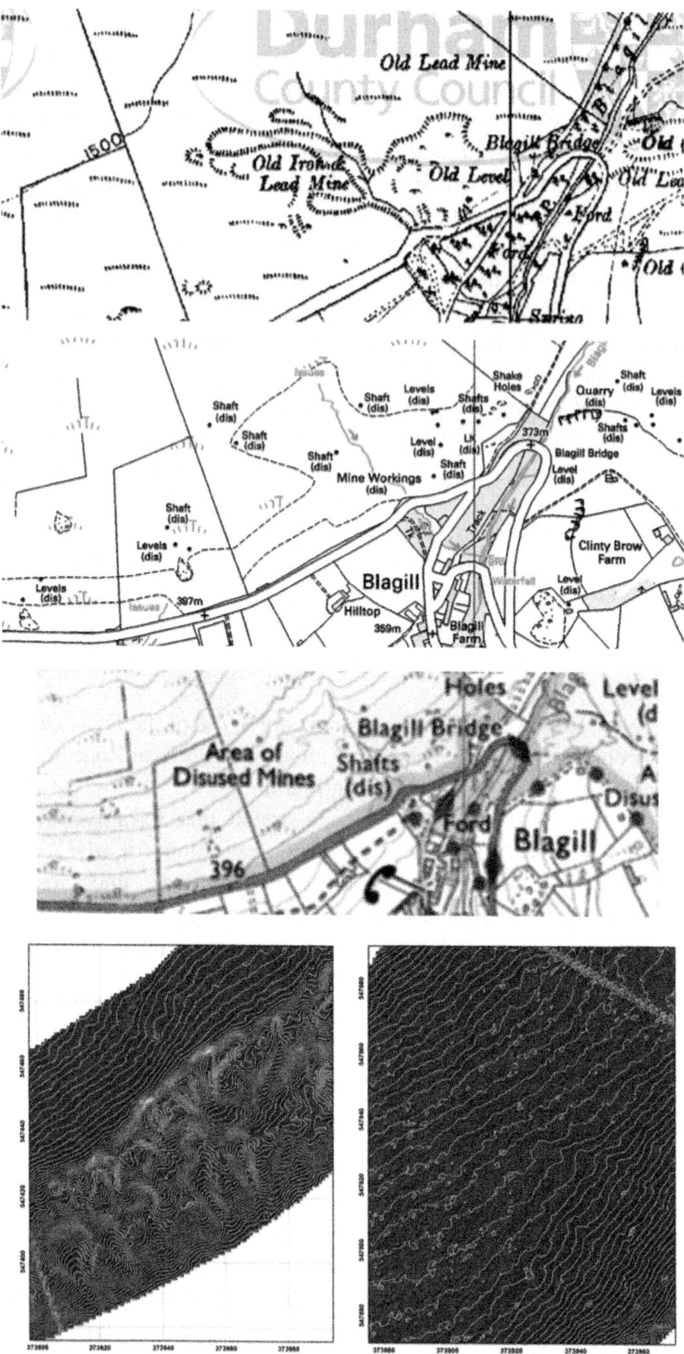

Fig. 2 (from top) (a) Early 20th century OS mapping, original 1:10,560 (source: Durham County Council); (b) Contemporary OS data mapped at 1:10,000 scale (source: OS data disseminated through the Digimap service, University of Edinburgh, Crown Copyright); (c) Contemporary OS data mapped at 1:25,000 scale (source: OS data disseminated through the Digimap service,

and natural rock features. In addition a significant amount of information about land use and feature attributes is conveyed by text. Such mapping has some success at indicating the nature of disordered terrain, although in this case the use of a similar design for symbols representing rough moorland rather confuses the terrain portrayal.

Contemporary Ordnance Survey data is supplied in digital form and can therefore be portrayed at varying scales, and indeed with user-defined symbolisation. The default portrayal shown in Fig. 2b reveals that the combination of text and mimetic symbolisation has been maintained on maps of this area captured photogrammetrically and by GNSS survey update to archival material. The main differences between Fig. 2a, b are the reduction in sketching of spoil heap features—the remaining cliff lines and rock faces on Fig. 2b are mainly showing natural features—and the concentration on point features. The areal depiction of the area of mining is shown by a generalised pecked line surrounding the zone of interest—this mainly outlines Zone B in the study described above.

Figure 2c shows the influence of scale on landscape portrayal. It shows the raster-scanned 1:25,000 mapping of the region which has been the focus of studies here. In this case, the contour pattern does not fully reveal the dissected landscape, and it is primarily text which offers, in a descriptive manner, the major clue to the nature of the terrain.

It is clear that these map representations, like most topographic map products, have had to sacrifice dimensionality, by graphically portraying the third dimension—a major factor in determining the disorder of a landscape—using two-dimensional symbolisation. Techniques of symbolising the third dimension have been developed and applied by cartographers for centuries. The contour line has proven a most effective device for quantitatively communicating terrain data, although an understanding of the whole terrain requires that contour lines be read as a pattern. Further quasi-two-dimensional symbolisation can try to pictographically portray terrain variability, the most obvious example being hill shading. Comparison of Fig. 2d, e using contour lines (1 m interval) combined with shaded relief of the raster DSM to highlight terrain characteristics, re-iterates the differences between Zones B and A, quantified above, and also shows the effectiveness of such methods of representation in portraying disorder.

It is concluded here that map representation, which involves abstracting characteristics and properties of the real world to cartographically transform spatial data into a graphical product, inevitably sacrifices dimensionality. The representation of three-dimensional surfaces using two-dimensional symbols is an obvious example. The mapping of disorder requires a serious attempt at developing cartographic symbols and map representations which can optimise the portrayal of multi-

Fig. 2 (continued) University of Edinburgh, Crown Copyright); (**d**) Extract from contour (1 m interval) and shaded relief of the DSM in Zone B; (**e**) Extract from contour (1 m interval) and shaded relief of the DSM in zone A

dimensional phenomena, such that the complexity of the real world can be most efficiently and effectively portrayed.

> **Conclusion**
>
> In addition, the multi-variate, multi-temporal and chaotic nature of geographic reality means that all the contemporary tools at the cartographer's disposal—including animation, imagery, multiple views, generalisation routines, display platforms, interactivity, and other technologies—will be required to address the representation of disorder. This study has embarked on a consideration of the cartography of disorder by examining one mappable phenomenon, landscape terrain. It has been shown that terrain can be characterised according to its measured disorder, but the representation of that disorder in cartographic terms requires the development of further techniques of representation.

References

Arnheim R (1972) Order and complexity in landscape design in towards a psychology of art. University of California Press, Berkeley, pp 123–135

Clarke S (2008) Geology of NY74NE, NW and NY75NE, SW and SE, Alston, Cumbria. British Geological Survey Open Report, OR/07/032, 51 pp

Doneus M, Kühtreiber T (2013) Airborne laser scanning and archaeological interpretation (Chap. 3). In: Opitz R, Cowley D (eds) Interpreting archaeological topography. Oxbow Books, Oxford, pp 32–50

Fairbairn D (2006) Measuring map complexity. Cartogr J 43:223–237

Fairbairn D (2011) Using entropy to assess the efficiency of terrain representation. In: Proceedings of 25th international cartographic conference, Paris, Paper CO-398

Hengl T, Reuter HI (eds) (2008) Geomorphometry: concepts, software, applications. Developments in soil science, vol 33. Elsevier, 772 pp

Kovacs K, Hanke K, Lenzi K, Possenti E, Brogiolo G (2012) Utilization of airborne LiDAR datasets in GIS environment for prospection of archaeological sites in high Alpine areas. Archeologia e Calcolatori 23:151–164

Lewis P (1982) Axioms for reading the landscape. In: Schlereth T (ed) Material culture studies in America. American Association for State and Local History, Nashville, pp 175–182

McGarigal K, Cushman S (2005) The gradient concept of landscape structure. In: Wiens J, Moss M (eds) Issues and perspectives in landscape ecology. Cambridge University Press, Cambridge, pp 112–119

Ode Å, Hagerhall C, Sang N (2010) Analysing visual landscape complexity. Landscape Res 35:111–131

Pfeifer N, Roncat A, Stötter J, Becht M (2011) Laser scanning applications in geomorphology. Zeitschrift für Geomorphologie 55:2

Reilly S, DeGloria S, Elliot R (1999) A terrain ruggedness index that quantifies topographic heterogeneity. InterMountain J Sci 5:23–27

Sappington M (2008) Vector ruggedness measure. Python script: http://arcscripts.esri.com/details.asp?dbid=15423

Zurlini G, Petrosillo I, Jones K, Zaccarelli N (2012) Highlighting order and disorder in social-ecological landscapes to foster adaptive capacity and sustainability. Landscape Ecol. doi:10.1007/s10980-012-9763-y

The Next Generation of Atlas User Interfaces: A User Study with "Digital Natives"

Raimund Schnürer, René Sieber, and Arzu Çöltekin

Introduction

Digital atlases have been successfully created and edited for over 20 years, but essential information about the usability of atlas interfaces is still largely missing. Similarly to other digital products and technologies, the usability requirements of digital atlases might differ among older and younger users. This distinction based on age and behavioral differences in the use of digital technologies has been classified into "digital natives" and "digital immigrants" (Prensky 2001). More precisely, digital natives are considered as those born on 1980 or later, digital immigrants as those born between 1955–1979, and those born before 1955 are called "silver surfers" (Cody et al. 1999; Prensky 2001). While some critical opinions are voiced against this categorization (Bennett et al. 2008), it is commonly theorized that digital immigrants and silver surfers possess thinking and acting patterns which may differ from each other as well as from the digital natives (e.g., Black 2010; Thompson 2013). This reasoning comes from the fact that digital natives grew up with a ubiquity of digital desktops, mobile devices such as smartphones or tablets, and not least the Internet.

Digital natives are a major target user group for digital atlases among current population. Therefore, we contend that the graphical user interfaces (GUI) of digital atlases need to be optimized specifically for this particular group to enable them to explore and visualize the thematic content of atlases. Based on this reasoning, we take the digital Atlas of Switzerland as an example and explore the *preferences* and *performances* of a sample of digital natives in a classroom environment with five

R. Schnürer • R. Sieber (✉)
Institute of Cartography and Geoinformation, ETH Zurich, Zurich, Switzerland
e-mail: schnuerer@karto.baug.ethz.ch; rene.sieber@karto.baug.ethz.ch

A. Çöltekin
Department of Geography, University of Zurich, Zurich, Switzerland
e-mail: arzu.coltekin@geo.uzh.ch

© Springer International Publishing Switzerland 2015
J. Brus et al. (eds.), *Modern Trends in Cartography*, Lecture Notes in Geoinformation and Cartography, DOI 10.1007/978-3-319-07926-4_3

alternative GUI designs. With the digital Atlas of Switzerland, user experiments have already been conducted with silver surfers, and internal feedback was obtained for this age group. However, atlases are often used in schools, as they constitute an important part of geography education (Häberling and Hurni 2013). As entire books are written on the subject (e.g., Palfrey and Gasser 2011), understanding the needs of digital natives in terms of their interface expectations, design requirements and needs may be a worthwhile endeavor both from fundamental science perspective (observing their spatial behavior) and from an applied science perspective (for obtaining design guidelines). From a design perspective, the goal would be to find what works best, and if possible, why—and so that, the principles applied in one GUI design can be possibly transferred to other GUI designs for digital atlases.

In sum, our study was motivated by the reasoning that (a) atlases are extremely information-rich, therefore the GUIs need careful consideration, and (b) the digital atlas GUIs need to be designed also for the younger generations. This could mean not only avoiding the "old-fashioned" style, but also supporting interaction patterns they are familiar with. To enable the exploration of thematic content for this user group, novel concepts, visualization means and GUIs must be considered. We cannot assume that the youth have a similar expertise level as silver surfers; yet, digital natives may have other strengths. Motivated by this reasoning, we evaluate five alternative visualizations, each tested with five typical 'atlas tasks' to answer the questions, "which layout design facilitates the best exploration and why?", "which layout design is preferred by the participants and why?" and "what can we learn from this experiment in terms of atlas GUI designs?".

Related Work

User interface issues linked to geographic visualizations have been studied from various aspects (e.g., layout design, interaction design, integrating multivariate information to map displays, visualizations themselves as interfaces) as the discipline made the transition from static maps towards interactive, on-demand geovisualization services (e.g., Howard and MacEachren 1996; MacEachren et al. 1997; Çöltekin 2002). More than a decade ago, MacEachren and Kraak (2001) have acknowledged user interface design as a research challenge in geovisualization. At the same time, Cartwright et al. (2001) have detailed the issues and challenges *about* the interface design for geographic visualizations based on a set of priorities for interface design that were defined by the members of International Cartographic Association's Commission on Visualization and Virtual Environments. Cartwright et al. (2001) list a number of challenges that are relevant for designing interfaces in particular for geographic visualizations. They ask "Can geovisualization products offer too much information?" (Cartwright et al. 2001), implicitly making a reference to *information overload* (Ruff 2002), a concept that has strong relevance to information-rich geographic visualization environments, such as digital atlases (Polys et al. 2007).

Atlases are particularly complex geovisualization environments, delivering information on a large collection of themes (MacEachren et al. 2008; O'Dea et al. 2011). As such, providing direct interaction options for all the underlying information at once is out of question, and even organizing the interface categorically in a way that allows access to all the main categories from the main interface can easily cause information overload (Kramers 2008; Bhowmick et al. 2008). In such information-rich environments, various design considerations should be taken into account such as the layout design (e.g., where to place which control, in what size, color, orientation or visual hierarchy based on semantic importance, labeling), organization of the display elements (e.g., what to group together, whether to follow conventions or not when utilizing the "screen estate"), and interaction design (Galitz 2007; Çöltekin et al. 2009).

An interaction design principle that is originally proposed for information visualization displays which can be translated to GUI design is the three-level "Shneiderman's information seeking mantra": overview—zoom-and-filter—details on demand (Shneiderman 1996). While "Shneiderman's mantra" is a rather reasonable starting point in organizing the elements of interaction, it appears that evaluating or validating any design often (if not always) requires a user study. Acknowledgement of the needs to place the user in the center of the design process led to the term "user-centered design" (UCD) which is commonly used today (e.g., Fuhrmann and Pike 2005; Holtzblatt et al. 2009). UCD is a formalized approach which ideally involves user feedback in all stages of iterative design: At the early stages a feedback from the user should be obtained, integrated in the next stage of design, and new user tests should be conducted to further improve the design until the 'right' solution is reached (e.g., Nivala et al. 2007). However, it should be noted that user-centered approaches appeared in cartography and geovisualization designs even before the digital and interactive maps have become common (Bartz 1970; Eastman 1977 via Olson 1979). More recently, we see prototypical examples in which users can personalize the display or the GUI as they wish (Balciunas 2013) or utilize Web 2.0 approaches such as map recommendations, user comments, tag clouds and RSS feeds for online atlases (Özerdem et al. 2013).

The closest work to our project in the literature appears to be by Kramers (2008), as this study features an evaluation of an atlas interface. Kramers (2008) highlights the importance of understanding the users, individual and group differences among users, their 'map use' behavior, the task and the context of use. Presenting how user-centered design improved the user performance over time with the Atlas of Canada, Kramers (2008) focuses on the evolution of a particular design. Contrary to this, in our project we evaluate five alternatives with a specific age group. In this project, we draw concepts from earlier literature as well as from modern interface examples that may be well-known by digital natives. Using five alternative GUI designs in which we allow limited interaction, we experiment with layout density (linked to information overload concept), visually grouping the elements based on their function (or semantic closeness) on the interface and some standard interaction paradigms (linked to information seeking mantra).

Experiment

Materials

We created five "mock-up" atlas GUI layout designs varying in layout density and tool arrangement (Fig. 1). As Fig. 1 shows, the layouts include GUI concepts that are used in web atlases and popular websites such as Google Maps or YouTube.

Layout 1: Minimalistic style

Layout 2: Statistical Atlas style

Layout 3: Tablet style

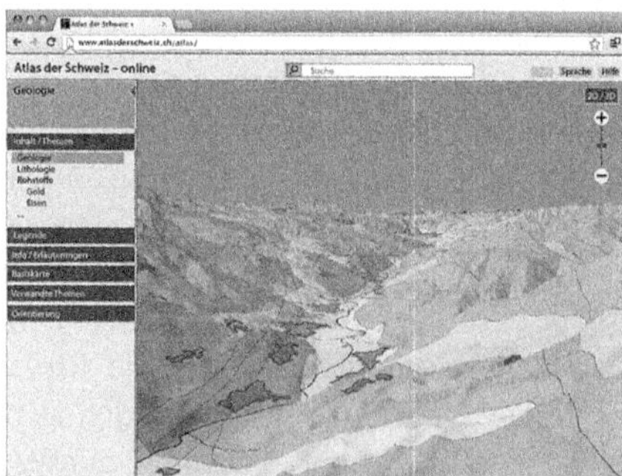

Layout 4: Google Maps style

Layout 5: YouTube style

Fig. 1 Five clickable atlas interfaces as stimuli (*independent variables*) of the study

To control the interactivity, only specific elements were clickable in the layouts at a time. Layouts were kept in grey-scale to avoid any bias that may come from color interactions.

Participants and Procedure

Students at secondary school level (n = 110, age 14–15 years, 61 female, 49 male) were asked to complete predefined tasks using the prepared layouts in the user study. The study was designed as a classroom experiment on desktop computers within a standard 45-min lesson. Students had to fill out a digital questionnaire delivered by a survey software (SelectSurvey.NET[1]) running in a web browser.

The experimental procedure was divided into four parts in which participants: (A) delivered background information about themselves, (B) voted on useful atlas tools and functions, (C) solved tasks in the five GUI layouts and judged their attractiveness and usability, and (D) assessed the "look and feel" of some specific atlas functions. Since the study focused primarily on solving tasks using different layouts, part (C) will be presented and discussed in more detail in the following sections of this paper.

The study started with a short introduction including general information about atlases and instructions on the survey procedure. After this, each student launched the digital survey individually and provided socio-metric and background information (age, gender, formation level, frequency of computer/tablet usage, use of computer games, and use of Internet maps) in part A of the online form. Part B dealt with the usefulness of 17 atlas functions. Using a Likert scale ranging from 1 (not useful) to 7 (very useful), students were asked to judge functions such as zooming, panning, printing, and querying. These feature questions were mainly asked to familiarize the participants with typical atlas functions.

Part C—assessing different GUI layouts—was the core part of this study. *Before* solving tasks, students were asked to rate (without any additional information) the visual attractiveness of our five layouts on a 7-step Likert scale. Following that, students conducted five tasks in all five mock-up interfaces (i.e., the experiment was designed as a within-subject study in which we obtained repeated measures). The tasks represented typical use cases for thematic navigation, spatial orientation and information queries in atlases:

1) Switch the map topic from "geology" to "raw materials",
2) Resolve the meaning of map colors (gray shades) by means of the legend,
3) Query the name of the settlement and the underlying geological structure in the center of the map,
4) Access any additional media (e.g. images, text) of the map, and
5) Find out where the displayed map region is located within Switzerland.

---

[1] http://selectsurvey.net/.

The GUI layouts appeared in random order to avoid learning effects (however, the task order was not manipulated). As indicators for the students' performance, the *number of mouse clicks* (how many clicks does it take to solve a task?) as well as the *mouse-click position* (do they click in the right places?) and the *required time to solve the tasks* (how long does it take for them to solve the tasks?) were recorded on each design. A *maximum limit* of 10 clicks and 20 seconds per task and layout type was implemented and participants were informed about these limits. After succeeding in or failing in a task, an immediate response was given by means of an alert. Hypertext Markup Language (HTML) image maps and jQuery[2] were used to provide layout interactivity on the client-side. A node.js[3] application stored the test data in a SQLite[4] database on the server-side.

After having dealt with the tasks, the students were asked to rate the five GUI layouts in terms of attractiveness and usability for a second time. In addition, they had to choose their favorite layout. By asking them to rate the attractiveness of the design before and after executing the tasks, we were able to determine the influence of the task solving process on the GUI attractiveness rating.

The final part (Part D) was optional and not part of the core study, three topics on specific atlas functions were inspected: (a) the placement and behavior of an information panel, (b) the labeling of tool buttons, and (c) the use and behavior of tool panels. These topics were aimed at revealing the preferences of digital natives on the accessibility of atlas tools and thematic information. The study closed with a text field for qualitative remarks and suggestions on how to improve the atlas concept (and the survey). The entire study was conducted in German.

Results

We carried out a statistical analysis based on collected data, i.e., the responses to questions, mouse click positions and recorded task completion times. Performance (effectiveness and efficiency) and attractiveness measures are derived from these data and their mean scores were compared between the five GUI layouts. Individual differences were identified by pairwise significance tests at a 95 % confidence level (i.e., p-value < 0.05). To explore the clicking activity per task and layout, maps of the first and other clicks as well as density surface maps were created with the Heatmap plugin[5] of Quantum GIS. We calculated success rates per task using SQL statements in the database, and used these as reference values for comparison.

Most tasks were completed successfully using Layout 4 (Table 1); that is, in 93 % of cases neither the time limit of 20 s nor the limit of 10 clicks was reached.

---

[2] http://jquery.com/.

[3] http://nodejs.org/.

[4] http://www.sqlite.org/.

[5] http://www.qgis.org/de/docs/user_manual/plugins/plugins_heatmap.html.

Table 1 Performance metrics (effectiveness and efficiency) for tested layouts

Stimuli	Successfully completed tasks	Time spent on a task on average (95 % confidence interval)	Normalized number of mouse clicks needed for a task on average (95 % confidence interval)
Layout 1	66 %	8.93 s (±0.68 s)	2.51 (±0.20)
Layout 2	72 %	9.21 s (±0.68 s)	3.31 (±0.24)
Layout 3	90 %	5.37 s (±0.51 s)	1.85 (±0.18)
Layout 4	93 %	4.52 s (±0.41 s)	1.87 (±0.21)
Layout 5	78 %	7.87 s (±0.47 s)	2.53 (±0.22)

Layout 4 was also most efficient in terms of required time (Table 1). On average, tasks could be solved twice as fast with Layout 4 as with Layout 2. Since some tasks involved more consecutive actions than others, the number of clicks was normalized for a meaningful comparison by the number of screen changes. For example, students had to click at least twice to solve the first task in Layout 1, whereas in Layout 5 one click would have been sufficient; so we divided the number of clicks by 2 (i.e., the number of screen changes) for Layout 1. The normalized number of clicks is considered as a measure for efficiency, as it indicates the participants' effort (Tamir et al. 2008) to solve a task. A value of 1 would be optimal. However participants needed for instance 2.5 times more clicks per task on average in Layout 1 and 5 before they managed to find the right GUI element (Table 1).

A statistically significant difference in the average time spent on the tasks was found between Layouts 3/4 and Layouts 1/2/5, because their confidence intervals at a 95 % level do not overlap. A pairwise analysis of variances resulted in significantly different average times between Layout 3 and 4 (p-value = 0.01) as well as Layout 1 and 5 (p-value = 0.03), however not between Layout 1 and 2 (p-value = 0.37). Regarding the normalized number of clicks needed to solve a task on average, 95 % confidence intervals do not overlap for Layouts 3/4, Layouts 1/5 and Layout 2. A pairwise analysis of variances did not reveal a significant difference concerning the normalized number of clicks between Layout 3 and 4 (p-value = 0.88) as well as Layout 1 and 5 (p-value = 0.91).

A qualitative visual analysis of click maps suggested that the students were strongly focused on the given tasks, as they were mainly clicking on the task-relevant GUI items. Only a small number of clicks appeared on browser elements like the URL bar, the refresh or the close button. To depict the students' first intuition to solve the task, we separated their first clicks from subsequent clicks (Fig. 2, top). By this, we could identify whether students followed our intention or whether they thought of alternative ways to solve a task. In most cases, first clicks occurred at just a few functions. If the first click did not lead to the right result, following clicks were more scattered. In the latter case, students lost time when functions were not grouped.

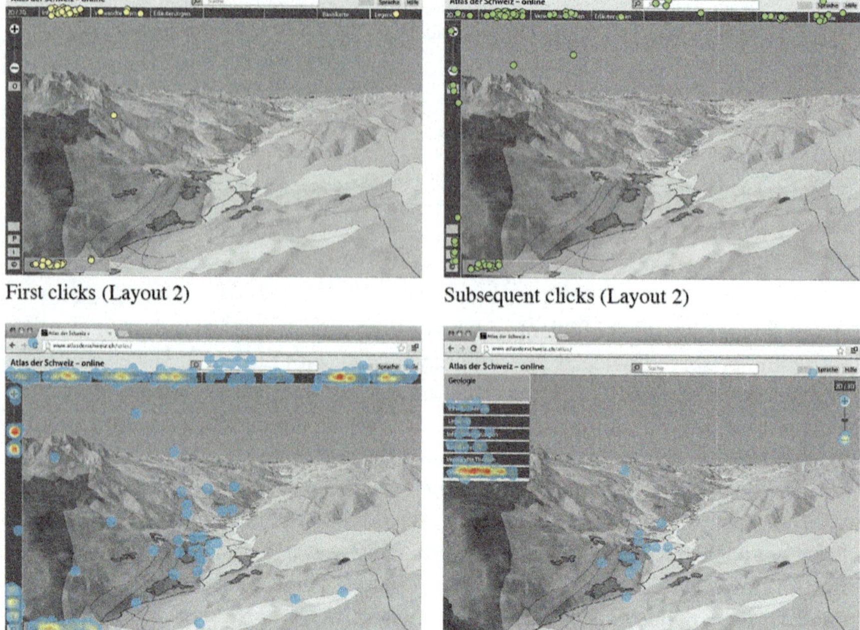

First clicks (Layout 2)   Subsequent clicks (Layout 2)

Scattered clicks (Layout 2)   Clustered clicks (Layout 3)

Fig. 2 *Top:* Separation of first clicks (*left*) from further clicks (*right*). Participants were asked to change the map theme in Layout 2. *Bottom:* Different small mouse-click hotspots (marked with blue–yellow–red colors, red being the most dense) in Layout 2 (*left*) and one large hotspot in Layout 3 (*right*) become apparent on a surface density map. Students were asked to get an overview of the given map extent

Further exploration of click maps and surface density maps revealed different interaction patterns for each layout. In the following section, the most remarkable findings are summarized for each task.

Aggregated for all layouts, 90 % of answers were correct for the *first task*. To be able to solve this task, participants had to change the map theme between geology and raw materials in a list. The list was already open in Layout 4 and 5 or appeared when clicking on a button labeled 'Content/Theme' in Layout 2 and 3. In Layout 1, the list opened when clicking on the map title. Interestingly, 29 % of students, who were faced with Layout 2 for the first time (so being unbiased by the other layouts), expected the map title to be interactive (Fig. 2, top). Approximately one fifth (21 %) of the students clicked at least once into the search box in one of the layouts for this task, however this functionality was not implemented.

For the *second task*, the participants had to discover the meaning of thematic colors (gray shades) on the map and this was possible only by clicking on a button labeled 'Legend' or in the legend itself. This task was completed successfully in

81 % of cases overall. Approximately half of the students (52 %) clicked at least once on the theme menu or on the map title to find the legend.

To finish the *third task*, the students were asked to click on the center of the map in all layouts to query feature information. In the end, 75 % of answers were correct for this task and about 61 % of these correct answers were given in less than two seconds. Comparing the latter value with those of other tasks (Task 1: 6 %; Task 2: 13 %; Task 4: 14 %; Task 5: 36 %), it seems that the third task was very easy for some participants, whereas others found it quite difficult. From the tested GUIs, Layout 1 with its minimalistic style showed the best success rate for this task (86 %). Success rates of the other layouts ranged between 70 and 75 %.

76 % of participants managed to access the required additional map information for the *fourth task*. Most difficulties arose in Layout 5 where only 35 % of students noticed the information text that was implicitly given in a content panel. In other layouts, map media could be accessed by clicking on an information/explanation menu button which seems to have made the task easier.

The *fifth task* resulted in an overall success rate of 77 %. Participants were asked to get an overview of the displayed map extent. In Layouts 1, 3, 4 and 5, this could be achieved by clicking on a menu button with the caption 'Orientation'. In Layout 2, however, this functionality was represented by a button labeled with the letter 'O'. Only one participant found the correct answer with the first click. 30 % of clicks were finally correct in Layout 2 while other clicks were quite disseminated (Fig. 2, bottom). Roughly one quarter (24 %) of students wanted to zoom out for an overview in Layout 2 alternatively.

Besides these five tasks that were related to performance, digital natives' preferences concerning the overall GUI attractiveness and usability were evaluated before and after task solving (Fig. 3). Ratings given on a Likert scale (1–7) indicated distinct differences between the five layout styles.

Two groups of layouts could be discerned: Layout 3, 4 and 5 were preferred more than Layout 1 and 2 at a 95 % confidence level, both before and after task completion. While Layouts 3 and 4 were rated initially on the same level as Layout 5, they improve after having dealt with the tasks. Layout 2 is judged worse after task solving. However, the statistical t-test does not indicate significance for Layout 1 ($p = 0.694$) and Layout 5 ($p = 0.364$). Even in the question about the students'

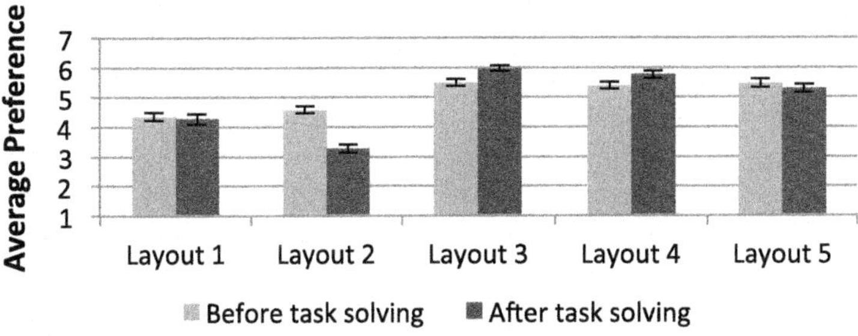

Fig. 3 Rating on attractiveness (preference) of the five GUI layouts

favorite layout, the votes for Layout 3 (30.3 %) and Layout 4 (32.1 %) turned out to be the highest, while the others ranked far behind (Layout 1: 11.9 %; Layout 2: 7.3 %; Layout 5: 18.4 %).

In summary, Layouts 3 and 4 did not only outperform the other GUI layouts in terms of efficiency and effectiveness, but also in terms of visual attractiveness (preference or satisfaction regarding the visual design).

Discussion

Although the five tested atlas GUI designs included exactly the same functionality and they were presented in grey-scale, it seems that students rated some of the layouts as more attractive than others. While this is not always the case in user studies (Hegarty et al. 2009), the subjects' intuition seems to be perfectly aligned with their performances in this case; those layouts which were rated as most attractive were also those where task performance was best. This impression was reinforced by the fact that students retained their favorite layouts—*Layout 3* and *Layout 4*—before solving the tasks and also after solving the tasks. These layouts were characterized by a compact GUI design where functions were grouped and easily recognizable through labels. Since these two layouts (3 and 4) had a quite similar appearance (Fig. 1), the possibility of an indirect learning effect for these two layouts cannot be excluded despite the stimuli randomization. This possible bias can be addressed in a future study when the two layouts are evaluated in an inter-subject test. However, at this point, one should note that while the two interfaces were similar, they are not identical, and especially given that the students stated their preferences before solving tasks, we can assume that the applied design concept indeed may have a favorable configuration compared to the others. Another possible reason for the popularity and success of Layout 4 may be a familiarity bias as it was modeled after Google Maps (which is likely well-tested, as well as used by the participants before).

Contrarily, *Layout 2* showed a poor performance and effectiveness. Besides a more scattered arrangement of functions, ambiguities appeared in this layout between the map title, theme menu and legend. Digital natives seem to have assumed the map title was an interactive area and expected the theme menu or legend to open when clicked. To evade this ambiguity, the map title should either be clearly discernible as such or implement the functionality of the legend or theme menu. Another drawback of Layout 2 was that students did not grasp the meaning of icons as fast as with the labeled functions of other layouts. Tooltips and better-designed icons could have helped the participants identify these functions.

Layout 1 was more efficient than Layout 2; however, tasks were solved less successfully. An explanation for this might be that participants had to click twice at four tasks in Layout 1, so probably some participants proceeded to the next question after the first screen changed although the task has not yet been completed.

Interestingly, a large map panel—as given in Layout 1—did not increase the attractiveness at first sight.

Layout 5 obtained average results altogether. Although all functions could be accessed immediately with only one click in all five tasks, Layout 5 did not outperform the others. While efficiency and effectiveness in atlases do not have to be necessarily the most crucial criteria (often, goals are open-ended for learning and exploration), they implicitly influence user satisfaction. In combination with an appealing design like in Layout 5, it is more likely that the users enjoy using an atlas and revisit the application from time to time.

A persistent affinity to favorite layouts could be noticed although students were not very experienced with digital maps and atlases. Results of all layouts would probably improve when alternative solutions were allowed, e.g., entering the map theme in the search box for the first task or zooming out to get an overview for the fifth task. In general, it can be stated that the implicit hypothesis of the study is true: There are significant differences between the five layout types concerning visual preference and simple task-solving performance.

Conclusion and Outlook

The user interface is *the* gateway to explore maps and media contained in a digital atlas. Therefore, much attention should be paid to the optimal interplay of functionality, graphical design and user behavior. User studies are a crucial means to reveal the intuitive usability of an application, especially with test subjects such as digital natives. The results of this study demonstrate that students strongly prefer atlas GUIs with a medium layout density, conforming to the well-known "inverted U-curve" in performance studies (e.g., Eppler and Mengis 2004). Too little information (minimalistic layouts) and people cannot find what they are looking for, too much information (highly dense layouts) and there is too much to process, thus it gets difficult. These medium density layouts not only have significantly lower average task completion times, number of clicks and a higher percentage of completed tasks, but they also received the best scores of attractiveness concerning the general GUI impression at a first glimpse. This rating preference has been further accentuated after solving the given tasks.

Some general and practical advice for atlas user interfaces could be derived from the study:

- Seemingly, desktop (in terms of screen size) atlas GUIs should have a medium tool density in terms of covered screen area (map space vs. interface space), and proportion of visible tools.
- Based on this experiment, it appears to be better to avoid screens with only a few visible atlas GUI elements; the search-and-find process decreases performance.

(continued)

- Important and often used GUI elements should be immediately accessible; the hierarchical structure of the GUI elements thus has to be determined by a use-oriented priority list.
- Functional grouping within the layout is highly recommended: Tools should be clustered and not distributed.
- GUI elements should be user tested for ambiguity (which needs to be minimized); the use and effect of a tool should be easily recognizable.

These findings might appear as common sense propositions, however, with this study we confirm the principles of user-oriented GUI design (Nielsen 1999) specifically for atlas GUIs. Together with additional information on GUI design and user profiling, other atlas authors might benefit from the methodology of this study to improve their digital atlas interfaces. Most importantly, this study demonstrates again how much a design team can benefit from running a user study, despite the tedious work involved in conducting one.

Follow-up studies related to our project should consider more complex tasks (e.g., interrelating map features spatially and/or temporally), other atlas user groups (e.g., silver surfers), or interfaces including more functionality (e.g., analysis tools). To give more realistic feedback to the participants when performing an action, not only atlas mock-ups but rather prototypically implemented GUIs have to be tested. An iterative testing process within the design cycle should give further validity to our findings as well as to others pursuing similar endeavors.

As mobile devices become a common means for visualization and information retrieval in everyday life, business and education, atlases should also consider tablets as communication devices. The design of an atlas GUI for tablet devices has to conform specific mobile usability concepts (which highly differ from a stationary 'regular' sized display), including gesture-driven commands (Nielsen and Budiu 2012).

To summarize, the investigation of atlas GUI designs is fruitful for both atlas authors and atlas users. In every stage of the development process, it may reveal positive effects and deficits on the graphical and technical level of an atlas application. Involving digital natives with their intuitive sense for technology into usability studies from the beginning seems to be an expedient way to provide carefully designed atlas interfaces in the future.

References

Balciunas A (2013) User-driven usability assessment of internet maps. In: Buchroithner MF (ed) Proceedings of the 26th international cartographic conference, Dresden

Bartz B (1970) Experimental use of the search task in an analysis of type legibility in cartography. Cartogr J 7(2):103–112. doi:10.1179/caj.1970.7.2.103

Bennett S et al (2008) The 'digital natives' debate: a critical review of the evidence. Br J Educ Technol 39(5):775–786. doi:10.1111/j.1467-8535.2007.00793.x

Bhowmick T et al (2008) Distributed usability evaluation of the Pennsylvania cancer atlas. Int J Health Geogr 7:36. doi:10.1186/1476-072X-7-36

Black A (2010) Gen Y: who they are and how they learn. Educ Horiz 88(2):92–101

Cartwright W et al (2001) Geospatial information visualization user interface issues. Cartogr Geogr Inf Sci 28(1):45–60. doi:10.1559/152304001782173961

Cody MJ et al (1999) Silver surfers: training and evaluating internet use among older adult learners. Commun Educ 48(4):269–286. doi:10.1080/03634529909379178

Çöltekin A (2002) An analysis of VRML-based 3D interfaces for online GISs: current limitations and solutions. Finnish J Surveying Sci 20(1/2):80–91

Çöltekin A et al (2009) Evaluating the effectiveness of interactive map interface designs: a case study integrating usability metrics with eye-movement analysis. Cartogr Geogr Inf Sci 36 (1):5–17. doi:10.1559/152304009787340197

Eastman JR (1977) Map complexity: an information approach. M.Sc. thesis, Queen's University, Kingston, ON

Eppler MJ, Mengis J (2004) Side-effects of the E-society: the causes of information overload and possible countermeasures. In: Isaías P, Kommers P, McPherson M (eds) Proceedings of IADIS international conference e-society, vol II, Ávila. IADIS Press, pp 1119–1124

Fuhrmann S, Pike W (2005) User-centered design of collaborative geovisualization tools. In: Dykes J, MacEahren AM, Kraak M-J (eds) Exploring geovisualization. Elsevier, Oxford, pp 591–609

Galitz WO (2007) The essential guide to user interface design: an introduction to GUI design principles and techniques, 3rd edn. Wiley, Indianapolis

Häberling C, Hurni L (2013) The web-based "Swiss World Atlas Interactive": first evaluation of user experiences in modern geography education. In: Buchroithner MF (ed) Proceedings of the 26th international cartographic conference, Dresden

Hegarty M et al (2009) Naïve cartography: how intuitions about display configuration can hurt performance. Cartographica: The International Journal for Geographic Information and Geovisualization 44(3):171–186. doi:10.3138/carto.44.3.171

Holtzblatt K et al (2009) Driving user centered design into IT organizations. In: CHI '09 extended abstracts on human factors in computing systems, Boston, 4–9 April 2009. ACM, New York, pp 2727–2730

Howard DL, MacEachren AM (1996) Interface design for geographic visualization: tools for representing reliability. Cartogr Geogr Inf Syst 2(23):59–77. doi:10.1559/ 152304096782562109

Kramers ER (2008) Interaction with maps on the internet – a user centred design approach for the atlas of Canada. Cartogr J 45(2):98–107. doi:10.1179/174327708X305094

MacEachren AM, Kraak MJ (2001) Research challenges in geovisualization. Cartogr Geogr Inf Sci 28(1):3–12. doi:10.1559/152304001782173970

MacEachren AM et al (1997) Visualizing spatial relationships among health, environmental, and demographic statistics: interface design issues. In: Ottoson L (ed) Proceedings of the 18th international cartographic conference, Stockholm. Gävle Offset, Gävle, pp 880–887

MacEachren AM et al (2008) Design and implementation of a model, web-based, GIS-enabled cancer atlas. Cartogr J 45(4):246–260. doi:10.1179/174327708X347755

Nielsen J (1999) Designing web usability: the practice of simplicity. New Riders, Indianapolis

Nielsen J, Budiu R (2012) Mobile usability. New Riders, Indianapolis

Nivala AM et al (2007) Usability methods' familiarity among map application developers. Int J Hum Comput Stud 6(9):784–795. doi:10.1016/j.ijhcs.2007.04.002

O'Dea E et al (2011) Coastal web atlas features. In: Wright DJ et al (eds) Coastal informatics: web atlas design and implementation. IGI Global, Hershey, pp 12–32

Olson JM (1979) Cognitive cartographic experimentation. Cartographica: The International Journal for Geographic Information and Geovisualization 16(1):34–44. doi:10.3138/R342-258H-5K6N-4351

Özerdem E et al (2013) Evaluating the suitability of Web 2.0 technologies for online atlas access interfaces. In: Buchroithner MF (ed) Proceedings of the 26th international cartographic conference, Dresden

Palfrey J, Gasser U (2011) Born digital: understanding the first generation of digital natives. Basic Books, New York

Polys NF et al (2007) Effects of information layout, screen size, and field of view on user performance in information-rich virtual environments. Comput Animat Virtual Worlds 18 (1):19–38. doi:10.1002/cav.159

Prensky M (2001) Digital natives, digital immigrants Part 1. Horizon 9(5):1–6. doi:10.1108/10748120110424816

Ruff J (2002) Information overload: causes, symptoms and solutions. Harvard Graduate School of Education. http://lila.pz.harvard.edu/_upload/lib/InfoOverloadBrief.pdf. Accessed 21 Oct 2012

Shneiderman B (1996) The eyes have it: a task by data type taxonomy for information visualizations. In: Spencer Sipple R (ed) Proceedings of the IEEE symposium on visual languages, Boulder. IEEE Computer Society Press, Los Alamitos, pp 336–343

Tamir D et al (2008) An effort and time based measure of usability. In: Proceedings of the 6th international workshop on Software quality, international conference on software engineering, Leipzig, 10–18 May 2008. ACM, New York, pp 47–52

Thompson P (2013) The digital natives as learners: technology use patterns and approaches to learning. Comput Educ 65:12–33. doi:10.1016/j.compedu.2012.12.022

The State of Official Statistical Mapping in Switzerland (and Other European Countries)

Thomas Schulz

Statistical Mapping

A Definition: Maps can serve three main purposes in the value chain of official statistics: as a means of the (a) preparation of surveys, (b) daily analysis of new data, and (c) documentation of statistical results (Bollmann and Müller 2002; Witt 1970; Dickinson 1963). Their traditional function as graphical memory for spatial data has strongly declined over the last decades, as we all know. As a matter of fact, it has never played a role in official statistics. If maps were used by statistical institutions, they have at all times only promoted a small part of the statistical material at all which are rather kept in large tables and databases. Thematic maps in statistics usually concentrate on those variables suited for telling us stories—remarkable stories that aid consumers in gaining fundamental knowledge about the most important structures and changes in our society and economy.

Historically, the first category—maps for the preparation of surveys—was of great importance. (Applied) topographical maps helped National Statistical Institutes (NSIs) to delimit census blocks and sample areas before data could be collected. With the new census approach of the twenty-first century, where most statistical data are linked to coordinates and retrieved from existing registers or are based on sample surveys, fix reference areas dissolve. These "internal" maps are no longer needed. It can be assumed that 90 % of all maps used and produced by (NSIs) today belong to the third category—maps for dissemination. Hence, the most important role of maps in statistics is visibly their role as a concise messenger of regional statistical data for the broad public as well as decision makers (Fig. 1).

T. Schulz (✉)
Swiss Federal Statistical Office, Cartography Unit ThemaKart, Espace de l'Europe 10, 2010 Neuchâtel, Switzerland
e-mail: thomas.schulz@bfs.admin.ch

© Springer International Publishing Switzerland 2015
J. Brus et al. (eds.), *Modern Trends in Cartography*, Lecture Notes in Geoinformation and Cartography, DOI 10.1007/978-3-319-07926-4_4

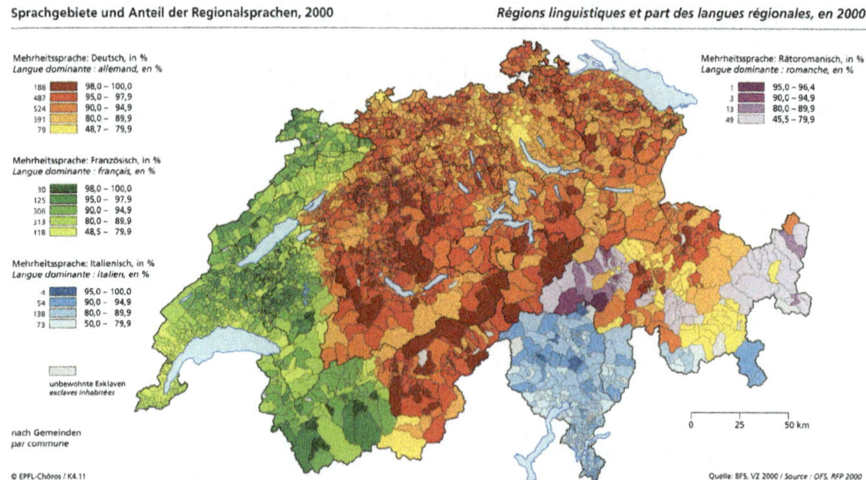

Fig. 1 Classical statistical map from a modern printed atlas (*Atlas of Spatial Change in Switzerland*—Swiss census atlas 2000)

A *statistical map* can, according to Schulz (2013), be defined today as a "map, which portrays by cartographic means of representation current societal and esp. socio-economic facts in mainly analytical manner. In its core sense, it represents, by using primary sources and applying statistical methods, all spatial data collected by official statistical institutions through special surveys or censuses for a broad public insight".

The use of maps in the statistical dissemination process: When maps appear in statistical publications, their significance goes beyond the pure localization of values for users or the "decoration and easing of publications", which are otherwise—without illustrations—not easy to understand (Witt 1964). In practice, maps are considered in statistical publications for manifold reasons (Schulz 2009):

- Maps *locate* results and show the spatial distribution of a phenomenon.
- Maps enable *comparisons* and answer multiple questions: between different areas in one map, between different topics in different maps, between variables for the same area in one map or between time periods.
- Maps *confirm* and validate statistical findings.
- Maps, atlases and geographic information systems *store* spatial knowledge.
- Maps *support* textual and tabular information that are only difficult to explain.
- Maps have a synoptic character and *summarize* large amounts of information, esp. numerical information. They reduce complexity.
- Maps *convey* a concept or an idea and tell stories. They are a democratic means, easy to understand and can help to better communicate the whole statistical agenda and thus contribute to the overall acceptance of statistics.

- Maps are colourful and have a general positive image. They *appeal* to the user's curiosity and attract attention to a publication.

Official Statistical Mapping in Switzerland

Products and Institutions

Swiss Federal Statistical Office: The use of maps and other means of visualisation in official publications by the Federal Statistical Office (FSO) has been on the rise since the late 1980s, when printed publications like the yearbook became colourful (again). While the office published about 150 thematic maps in the early 1990s and about 500 maps in the year 2000 annually, the use of maps has accelerated impressively during the past 5 year. This development was nourished by an equally rising demand for maps from customers and the implementation of new mapping technologies, e.g. the introduction of a central regional data base and a modern atlas platform (Schulz and Ullrich 2009). In 2012, the FSO, based in Neuchâtel, issued in total 16,898 thematic maps (Fig. 2). Most of them, to be statistically exact: 11,386 maps, are available in interactive online atlases, which offer a variety of additional features, e.g. data queries, thematic searches, charts, animations, or links to meta-data and definitions on the Internet. Another 4,200 maps (around 25 %) are available as individual applications, directly integrated in the official webpage. And about 1,300 maps were produced and published as static maps in PDF format,

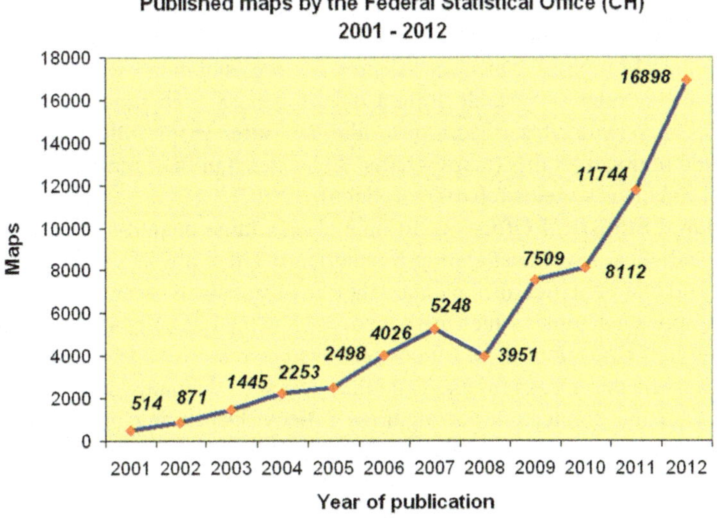

Fig. 2 Published maps (all media formats) by the FSO between 2001 and 2012

to be integrated in the various thematic publications or to be made available through special sources, like the Statistical Encyclopedia.

All maps of the FSO are published either bilingual (German and French) or monolingual in the four official Swiss languages (German, French, Italian, Rhaeto-Romanic) and English. And of course, all maps are free of charge, in coherence with policies by Eurostat and other NSIs to freely disseminate statistical information. Maps are made available in almost any type of media, including new dissemination channels like mobile devices. The maps can cover all administrative geographical levels, grids and special analytical regions for Switzerland, as well as major regions for Europe or thematic country maps of the World.

FSO's Cartography Unit and its copyright sign "© BFS, ThemaKart" are meanwhile well known by many customers as a reliable brand and source for statistical maps of Switzerland and Europe. The unit is today the second largest official map distributor in Switzerland. This implies also a good deal of public relations work and customer service engagement for about 500 enquiries p.a. apart from the core cartographic business. One example of an important PR contribution are live-mapping activities on Sundays, when federal elections or popular votes, which are conducted four times a year, are mapped continuously and interactively throughout Sunday afternoons as new data becomes available. At 8 p.m. ready-to-print maps are provided to several important Swiss newspapers (Schulz 2011). Other major activities include the provision of a large number of geometric bases and maps of spatial divisions to customers, which are updated several times a year, when changes to basic spatial units or their relations occur.

In addition to thematic maps, the unit also provides other regional by-products from the atlas data base, such as interactive regional profiles for cantons, communes or European countries. For many years, these popular portraits were usually put together manually on the FSO website every year from plain tables. Such key data sets can now be extracted and updated almost automatically from the data base (Platform Stat@tlas), taking into account also all spatial changes which happened over the past 12 months. Thus, for the user, the regional data will either be shown visually in the atlas or in table format on the general web pages, and it will be provided as download services—just like all other maps. Other products are automated regional reports (maps, charts, tables, text) in PDF format or interactive regional indicator systems (charts), cf. Schulz (2012).

Regional Statistical Offices contribute also in large quantity to the number of official statistical maps available in Switzerland. There are 25 cantonal statistical offices, and 6 larger cities also maintain their own statistical offices, collecting local data and additional topics which are not covered by the FSO. Looking back 20 years ago, only half a dozen of these offices produced or disseminated thematic maps at all. Today, and this speaks for itself and the success of thematic mapping, maps can be found in the publications of all these offices. Depending on the size of the territory they cover, often they also show several spatial divisions. Twelve cantons also had some kind of a statistical atlas on the Internet by this year. Regional statistical offices cooperate amongst each other as well as with the FSO in regards of technologies (e.g. atlas platforms, data bases) and map know-how.

Other institutions: Apart from the FSO and regional statistical offices, further government offices collect also statistical data according to the Swiss Statistical Law, e.g. the Swiss Tax Administration, the Federal Health Agency or the Federal Office for Environments. All official institutions together form the National Statistical System (NSS). While most of the institutions transfer their collected data to the FSO for publication, larger thematic offices also publish their own regional data through statistical maps and even atlases. In addition, statistical maps (mainly with data from the FSO, which is freely available) can also be found in larger atlases issued by universities, like the Atlas of Switzerland.

Users

Official statistical target groups can be divided into four categories (OECD 2006): Data *miners* (specialists, experts, statisticians), *farmers* (professionals, administration, consultants, businesses), *tourists* (occasional users, media, politicians, occasionally also administration), and *consumers* (passive consumers, media consumers). The urge to communicate and deliver information is inherent to maps and atlases, including, as a matter of course, statistical maps. Therefore, these should naturally be applied where they reach effectively and cost-sensitively the largest possible audience, including stakeholders, e.g. statisticians or politicians. This is usually the case with "farmers" and "tourists". Individual specialists (data miners), such as university researchers, query and analyse statistical data independently and usually create and tailor their own analytical maps to their special needs with the tools they possess. They are therefore not the main target group for official statistical mapping products. Maps and atlases in statistics rather have a "portal function". They allow the average user to discover spatial facts and relations, and they provide access to the underlying, more detailed data (Ormeling 2009).

Statistical maps and atlases are being used today by an extraordinarily diverse and broad audience, which comprises almost all societal and professional levels. The majority (84 %) of users in Switzerland, as found out by a customer survey at the FSO in 2006, come from a professional background (education, media, politics, and science). But the share of private users is constantly increasing (Fig. 3). To further differentiate user groups, the author had the chance to access more than 800 customer mails that were received by FSO's Cartography Unit during the years 2003–2013 and separate the individual professions of their senders.

Table 1 shows the results of this analysis. This data for Switzerland can be compared to other European countries. The leading and most active user group of maps from the Federal Statistical Office are university students with a share of almost 25 %. If school students, teachers and scientists are added to this number, almost 40 % of the audience for statistical maps and atlases can be found in the educational sector. One can even add the 7.4 % publishing companies to this already impressive number. The large majority of thematic maps asked for by

Fig. 3 Institutional backgrounds of users of statistical maps

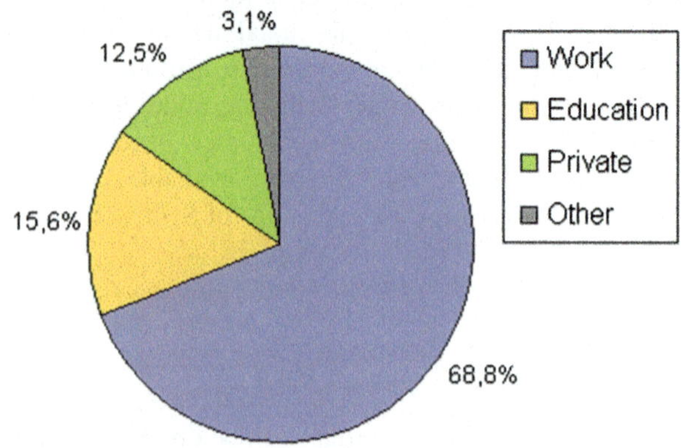

Table 1 User groups of statistical maps and atlases in the professional and educational sector[a]

User group by profession	Share
University students	23.8 %
Journalists	12.0 %
Administration employees	7.5 %
Publishing companies employees	7.4 %
Teachers	7.1 %
Scientists: geographers	5.5 %
NGOs and other organization employees	5.2 %
Politicians	5.0 %
Statisticians	4.4 %
Students, pupils	3.6 %
Cartographers	2.8 %
Employers, Businesses	2.5 %
Planners, Engineers	2.2 %
Private businesses employees	1.9 %
Scientists: others	1.1 %
Others	7.7 %

[a]Based on a survey of 873 client contacts at FSO's Cartography Unit (ThemaKart) during the years 2003–2013

publishers are integrated thereafter in school books, text books or other teaching material. Another large user group is formed by professions that are officially engaged with questions around the status of the administration or the society: administration employees, statisticians and politicians (together 17 %). Not unexpectedly large is also the interest from the media. Journals, newspapers and TV broadcasters love statistical charts and maps to illustrate their stories, e.g. after election days, demographic problems or economic crises.

Contents

A general overview of the current contents in statistical maps can best be attained from the table of contents and an analysis of large statistical atlases. A clear and unique system in the topical structure exists for all statistical atlases, independent of temporal or regional aspects, as proven by Schulz (2013). Due to reasons of data protection or survey methods, which in some cases lead to a deficiency in highly resolute spatial data, only about half of the 30 general statistical topics regularly appear in such atlases. As classical censuses are on the retreat, more data from sample surveys or registers find their way into publications and replace former census data—a fact that at the moment narrows the variety of topics in atlases.

The FSO currently knows 21 themes (or products) to cover and classify the wide range of socio-economic data from all aspects of society. Maps are used for almost all of the 21 products, where regional data is available. Yet, some statistical topics stand out in numbers and importance. These are topics, like demography or politics, where maps have in time become indispensable elements to explain the results. In total, seven topics dominate official Swiss statistical maps (as shown in Table 2) with a

Table 2 The 21 official statistical topics and the number of maps for each topic shown in the Statistical Atlas of Switzerland (January 2014)

No.	Theme	Maps	Share
01	Population	695	27.2 %
02	Territory and Environment	217	8.5 %
03	Work and Employment	22	0.9 %
04	National Economy	7	0.3 %
05	Prices	–	–
06	Industry and Services	156	6.1 %
07	Agriculture and Forestry	140	5.4 %
08	Energy	–	–
09	Construction and Housing	94	3.6 %
10	Tourism	175	6.8 %
11	Mobility and Transport	33	1.3 %
12	Money, Banks, Insurances	–	–
13	Social Security	33	1.2 %
14	Health	93	3.6 %
15	Education and Science	12	0.5 %
16	Culture, Media, Information Society and Sports	104	4.1 %
17	Politics	230	9.0 %
18	Public Administration and Finances	27	1.1 %
19	Crime and Legal System	36	1.4 %
20	Social and Economic Condition of the Population	88	3.5 %
21	Sustainable Development and Regional Disparities	39	1.5 %
	Spatial Divisions for Switzerland	68	2.7 %
	Historical Statistics	181	7.1 %
	International Statistics	100	3.9 %

Link to the atlas: www.atlas.bfs.admin.ch or www.statatlas-switzerland.admin.ch

share of 70.1 % (share of 5 % or more for each individual topic): Population—Territory and Environment—Industry and Services—Agriculture and Forestry—Tourism—Politics—Historical Statistics (long-term times series).

Due to the use of own primary data sources (around 85 % of all data used in official statistical maps come from the NSS) statistical maps and atlases reach a degree of topicality unlike any other thematic atlas type. Around two third of the maps are no older than 2 years, one tenth no older than 1 year (Schulz 2013).

Cartographic Representations

Not without good reason, many authors use the term "choropleth map" synonymously with "statistical map" (cf. Hake et al. 2002 or Bill and Zehner 2001). The overwhelming availability of elementary analytical data in the public statistical domain correlates unambiguously with the cartographic representation methods applied within their products. There is a direct link between the 98 % simple key figures used as data bases and the 98 % analytical representations that result in these cartographic works (Schulz 2013). Only three representational methods dominate statistical maps and atlases today: *area cartograms, proportional symbol cartograms* and *combinations* of both. Other methods (incl. "statistical methods" as named by Spiess et al. 2010) remain clearly at the margin—independent of the regional or temporal context. Schulz (2013) analysed app. 6,300 maps in 20 statistical atlases. In total, the above mentioned representation methods count for 89 % of all maps found in these atlases. Another 5.5 % are qualitative maps. The statistical atlases taken into account by this study come from regions all over the world and from different historical periods, going even back to the nineteenth century. Thus, the results can be generalised and are not only valid for Switzerland or Europe. These results correspond also to other studies conducted by e.g. Freitag (1988), van Elzakker et al. (2005), or Dickmann and Zehner (2001). Van Elzakker et al. (2005) analysed statistical maps and atlases that were published on the Internet by NSIs around the year 2002. The most common maps found on these websites were choropleth maps and proportional symbol maps of all kinds.

The use of seemingly endless rows of simple choropleth maps in statistical publications, often criticised and disliked by cartographic scientists and professionals, has structural and provable reasons linked to the data bases and user demands. It can only to a very limited percentage be derived from the ignorance of other semiotic means and methods by the respective authors and institutions. Maps in statistical publications do primarily not serve scientific purposes, as mentioned before. They are made for the layperson and the occasional visitor. Their intentional simplicity in design and content makes statistical maps so successful, as web statistics of several NSIs show. The Statistical Atlas of Switzerland, for example, is accessed by more than 2,000 users every day, and the numbers are growing, especially for very simple choropleth maps. Already Arnberger (1977) expressed, what is still true today: "The majority of thematic maps are justifiably of analytical nature", which is often the "best and only possible way of expression."

Reference Areas

Reference areas—variously shaped polygons in maps to which thematic data is attributed—are an important part of the base map in any thematic map—and probably the most important feature in statistical maps that otherwise do not show many situational elements (apart from borders and bodies of water). Statistical maps are characterised by two kinds of reference areas. The most common class and almost a "natural" basis are administrative divisions. Rarely, non-administrative divisions (geometric, geographic or value area surfaces) can be found.

It is a fact in geographic science: existing administrative areas are very heterogeneous in size and structure and therefore not suited for geographical analysis. Yet, official statistics, due to their existing survey methods and their political mandate,[1] are generally unable to go beyond the representation of administrative boundaries. Table 3 shows that 96.2 % of all maps in statistical atlases still refer to this area type or its aggregates. Also 96 % of the maps published by the Swiss FSO are based on administrative divisions. Although often asked for, there is no visible trend towards the use of more geographic or geometric units.

Table 4 shows the average distribution of the various spatial divisions in statistical maps. Almost 35 % of the maps correspond to the NUTS 3 level (cantons in Switzerland, districts in Germany, or counties in the United Kingdom). A similar share goes to maps based on communes. Higher or lower hierarchical levels are either not suited for maps, or data is not available for them.

Table 3 Reference units in statistical maps and atlases[a]

Type of reference area	Share
Administrative	96.2 %
Non-administrative	
Geometric	3.2 %
Geographical	0.2 %
Statistical reliefs	0.4 %

[a]Based on the analysis of 20 atlases and 6,500 maps (Schulz 2013)

---

[1] Statistical laws or decrees often explicitly demand data by the NSIs on administrative levels, such as states, cantons, districts, municipalities or communes.

Table 4 Level of detail for regional data on administrative bases[a]

Level of detail according to NUTS[b]	Share
NUTS 0	1.4 %
NUTS 1	12.1 %
NUTS 2	10.0 %
NUTS 3	34.9 %
LAU 1 + 2	37.8 %
No administrative basis	3.8 %

[a]Based on the analysis of 20 atlases and 6,500 maps (Schulz 2013)

[b]*Nomenclature des Unités Territoriales Statistiques*: still based on administrative units from individual countries, this nomenclature was developed by Eurostat since the 1980s in order to identify units that are approximately equal in size and homogeneous for the comparison of data from all EU member states, candidate states and EFTA countries (Eurostat 2007)

Statistical Atlases

General Development Trends

Statistical atlases are the most comprehensive, integral and of course attractive way to disseminate regional statistical data. Like the Statistical Yearbook for data on the national level, they tell vivid "stories" about our society and important regional structures and developments of socio-economic phenomena and thus go far beyond what mere collections of individual maps can show. Atlases have been widely used already in the nineteenth century by statistical offices. Works of this "classical period" comprise the *Census Atlas* and the *Graphical-statistical Atlas* (Central Europe). After a long break and decline in map production, mainly caused by WWI and WWII, statistical atlases have re-emerged in large numbers during the latter part of the twentieth century. The increasing number of surveys and the availability of electronically stored data on ever more detailed regional levels have subsequently led to a new boom, and many NSIs have again seen their potential. Statistical atlases can commonly be grouped into three main sub-groups (Schulz 2013): (a) polythematic atlases (general statistical atlases showing all topics), (b) monothematic or special-topic atlases (showing only topics like demography, agriculture or economy in one atlas) (c) provisional editions (early atlases, which still show physical topics in their content and are strongly linked to other editions) (Fig. 4).

Statistical Atlases of Switzerland

The Swiss Federal Statistical Office (FSO) published its first atlas, the *Graphical-statistical Atlas of Switzerland*, in 1897, followed by a larger and even more

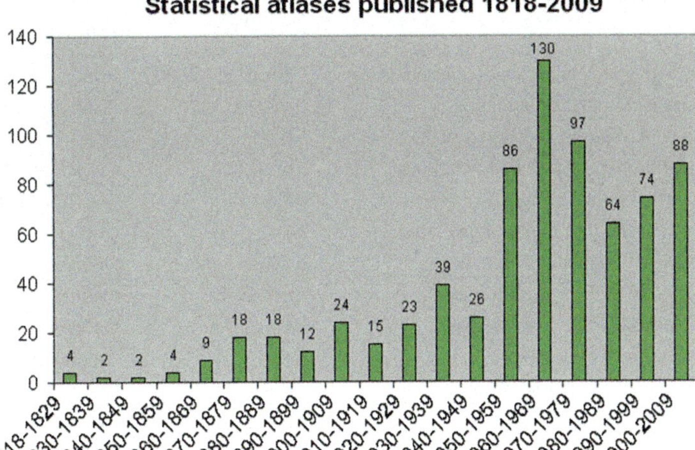

Fig. 4 The development and dynamics of the publication of statistical atlases during the last 190 years since the first publication of a statistical atlas in 1818 (France)

successful second edition in 1914 that has been the last statistical atlas for more than 70 years (!). Only since the 1980 population census a comprehensive printed census atlas—the *Structural Atlas of Switzerland*—has been issued again every 10 years to document regional results of this survey. The 1990 and 2000 censuses have also seen the publication of mono-thematic atlases, e.g. on gender and equality issues or the new phenomenon of the ageing society. Since 2003, the FSO offered a popular annual *Statistical Atlas of Switzerland* on DVD, which accompanied the statistical yearbook with current maps and figures until 2013 when it was replaced by a mobile version (iPad). It also went Online in 2009 and has been continuously updated on the Web ever since as new data arrives. It was supplemented in 2010 by the *Political Atlas of Switzerland* (showing all results of federal elections and popular votes) and in 2013 by a *Statistical Atlas of Swiss Cities* (Table 5).

The Statistical Atlas of Switzerland is the core cartographic product issued by the institution and has been built up intensively over the past 5 years. The number of maps is growing by the hundreds every year. From "only" 76 maps in its first edition (2009), it has increased to currently 2,256 maps (January 2014). All maps are being updated continuously, sometimes more than once a week, together with the issuing of new press releases. To contribute to its already large distribution, the atlas is available in various formats, e.g. Online, on DVD, on USB devices, on mobile devices, such as the Apple iPad, and partially in Print.

The new draft of contents for the web atlas considered for the first time the full spectrum of all 21 statistical products (themes) in an atlas and thus offers an up-to-date and captivating overview. These "thematic" chapters are supplemented by an *International*, a *Historical* (showing statistical time series dating back until the first

Table 5 Modern statistical atlases published by the Swiss FSO

Atlas	Year	Format
Structural Atlas of Switzerland (census atlases)	1985, 1997, 2006	Print
Swiss Atlas of Population and Age	1998	Print
Swiss Atlas of Women and Equality	2001	Print
Statistical Atlas of Switzerland (yearbook)	2003–2007	CD-ROM/DVD
Political Atlas of Switzerland	2004	CD-ROM/DVD
Swiss Atlas of the Population above 50	2005	Online
Swiss Atlas of Women and Equality	2005	Online
Statistical Atlas of European Regions[a]	2006–2009	Online/DVD
Atlas of National Council Elections[a]	2007, 2011	Online/DVD
Statistical Atlas of Switzerland[a]	2009–	Online/DVD/USB/pad
Political Atlas of Switzerland[a]	2010–	Online/DVD/USB/pad
Statistical Atlas of Swiss Cities[a]	2013–	Online/DVD

Links to all atlases: www.atlas.bfs.admin.ch or www.statistics.admin.ch > REGIONAL
[a]Atlases based on an Atlas Content Management System (Platform Statatlas)

census in 1850) and a chapter on *Spatial Nomenclatures* in Switzerland (24 chapters in total). In this last chapter, users can view the various geographical levels (and their changes in time) used as reference areas for statistical analyses. An implemented service offers downloads for these overview maps and the nomenclatures behind. With four navigation steps, the atlas contains currently approximately 580 themes. Data in the atlas comes from primary statistical sources, mainly from within FSO, but also from Eurostat and for some projects from other federal and cantonal institutions. Data is usually made available on the lowest possible regional level. In general, the political communes (municipalities) form the basis for most maps; however, for single subjects this can vary strongly, as some data are available only on district, cantonal or even national level (e.g. health or income data). Analytical and regional-political divisions are also used for various maps. For a smaller part of the themes, e.g. data from the big censuses or land use statistics, also grid maps and other geometric polygons are conceivable and are being integrated into the atlas. The atlas, which is available in German and French, is—like all statistical publications—free of charge for the individual user. Address: www.statatlas-switzerland.admin.ch (Fig. 5).

Statistical Atlases of Other European Countries

In his "Census Mapping Survey" Nag (1984) gives an exhaustive and contemporary overview of the use of maps and atlases in statistical offices just before the start of the digital era. Twenty-one years later, van Elzakker et al. (2005) studied the cartographic offer again and looked at the distribution of maps by 126 NSIs worldwide—esp. at online maps. According to their findings, the websites of

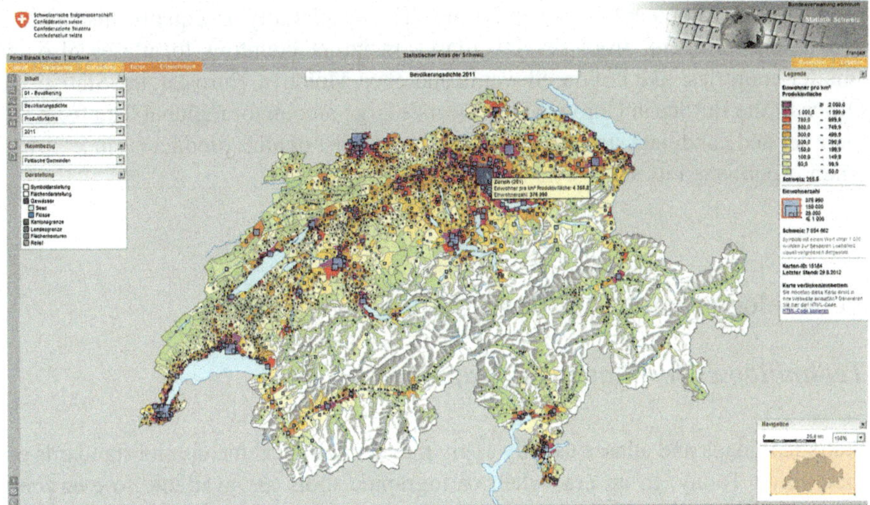

Fig. 5 Map example from the *Statistical Atlas of Switzerland* (population density, 2011)

Table 6 European countries with a current statistical atlas on the national level (January 2014)

Country		Country		Country		Country	
	Albania	X	Finland		Luxembourg		San Marino
	Andorra	X	France	X	Macedonia		Serbia
X	Austria	X	Germany		Malta		Slovakia
	Belarus		Greece	X	Moldova	X	Slovenia
X	Belgium		Hungary		Monaco	X	Spain
	Bosnia and H.		Iceland		Montenegro	X	Sweden
	Bulgaria		Ireland	X	Netherlands	X	Switzerland
	Croatia	X	Italy		Norway		Turkey
	Cyprus		Kosovo		Poland		Ukraine
X	Czech Republic		Latvia	X	Portugal	X	United Kingdom
	Denmark		Liechtenstein		Romania		Vatican City
X	Estonia	X	Lithuania		Russia		

already 25 out of 126 offices possessed a mapping interface and some even an atlas application (in 2002). After certain pioneers, like the *Atlas of Canada* (digital: 1987/Online: 1999) and the *Statistical Atlas of the United States*, also several European NSIs issued web atlases with statistical data shortly after the beginning of the new millennium (Germany, Switzerland, France, Netherlands, United Kingdom, Sweden etc.). Table 6 shows the current offer of statistical atlases by 47 - European countries (study conducted by the author in January 2014). Today, more than one third of all countries call a Statistical Atlas on the national level their own (18 nations), and one to two countries can be added to this list every year, which is a promising development in a short time. The availability is equally distributed across the continent, from several atlases in larger countries to atlases also in smaller countries, like Latvia, Slovenia and even Moldova. Eurostat, the Statistical Office of the European Union, who collects a large amount of regional data from all member and candidate states, can be added to the list of atlas producers since 2012, when it published its successful *Statistical Atlas*.

Technologies and the Role of the Cartographer

Technologies for Statistical Mapping

Statistical maps and atlases usually apply the technological means and standards of their time. Today, in an era, when cartographic tools are available to everyone and every institution at often low costs or even for free, a plethora of different software packages is used by NSIs to create their maps. Apart from some input standards (.csv, .xls, .sdmx) and output standards (Flash, SVG, HTML5) no uniform application exists that is considered by all the different European countries. Applications for statistical mapping can be grouped into four main categories: (1) There are classic *Geographic Information Systems* (GIS), which offer a broad range of analytical functions and more and more convenient map components. (2) There are *statistical software packages* (like SAS) with rather rudimentary map tools—often suited only for quick sketches. (3) There are freely available *Internet tools* that allow for the creation of maps without software skills (e.g. Google). (4) And there are high-end *Cartographic Information Systems* (CIS) on the market that aim directly at the professional map and atlas maker in statistical offices.

As often in cartography (and life), there is no one tool that is suited for every situation. Usually, two or three solutions of the above mentioned categories co-exist in statistical offices to fulfil all needs. The decision to adopt a tool depends on many factors, including the available infrastructure and data bases, the desired output format (print, online, mobile, etc.), the range of possible use cases within offices, the number of staff involved in the preparation of maps or atlases, and last but not least on the professional background of the person who is responsible for the map production. While cartographers will rather chose CIS solutions, geographers will use a GIS; statisticians will use statistical software they have grown up with, and finally other staff that is not acquainted with any of these solutions will search for easy-to-use free Internet software. In any case, it can be concluded: while the production of thematic maps in statistical offices is now much cheaper and faster than in pre-digital times, there are two negative effects as well: Firstly, costs for software introductions and the shift to a new technology are today higher than 10 or 20 years ago, caused by the sheer number of interfaces with an increasingly complex surrounding in statistical information systems. And secondly, all the freely available tools do still not automatically guarantee that the resulting maps are well designed and communicate their messages accurately. In fact, the increase in tools and the quantity of maps has lead to often poor designs and an increase in errors occurring in statistical maps in many institutions.

Statistical atlas platforms: When NSIs generate statistical atlases or produce large quantities of maps every year that are continuously, all of the above mentioned software solutions show severe shortcomings, as they are usually aimed at the non-sustainable production of individual maps or smaller series. In this context, a number of offices begin to use cartographic Content Management Systems (CMS) in the form of atlas platforms—just as they use other CMS to create standard publications and statistical websites. To enhance its in-house production of thematic maps and make new multimedia atlases available faster, the Swiss FSO started with the development of a new atlas platform in 2006 (platform Statatlas). Its Web Content Management System today helps to integrate a very large number of (non-cartographic) editorial staff with different tasks and backgrounds into one easy-to-understand production environment through a generic web interface. It also delivers various services for customers and colleagues alike, allows the exchange of dissemination data with partners in other institutions or the fast production of new and other language editions.

The system is currently employed by around 80 internal and external users and has raised the map production at FSO to a new level—in quantity and quality. Besides the efficient handling of large spatial data and the hybrid output for all different kinds of media at the same time, every production step—and possible error—is now being logged in the spatial data base and can later be identified, traced back and corrected. Furthermore, all cartographic metadata is stored for the first time in one data base, e.g. map titles, legends, descriptions, classes, colour schemes and of course translations, so that they can be reused for any new products and co-operations coming up on the horizon. The platform is described in detail by Schulz and Ullrich (2009) and Schulz (2012).

The (New) Role of the Cartographer

With changing technologies, also the role of cartographers within statistical institutions has altered dramatically. While only 20 or 30 years ago cartographers still manually designed a few dozen thematic maps per annum, state-of-the-art systems—as described before—allow for a map output of thousands of maps a year. At the Swiss FSO the same three cartographers have been responsible for the 514 maps generated in 2001 as well as the 11,744 maps in 2011. It is obvious, that cartographers—if they are still employed in statistical offices—do not "touch" or verify any individual map any more. Instead of the former craftsmen and the artist the cartographers in NSIs today have adopted the role of translators, visual consultants and software engineers. As translators, they transfer statistical numbers and textual information into the graphic language which is unknown to statisticians. As consultants, they advise colleagues to certain representation methods or their bosses to the choice of certain technologies. As software engineers, they develop tools and interfaces that guarantee the functioning of automated mapping processes and constant quality assessment measurements implied.

Cartography itself can not—nor must it—stop these developments. It is her professional obligation to get engaged in the design of these new technologies and algorithms that support high quality map output from systems used by NSIs. Statisticians and the statistical administration will not wait for cartography. They can putatively live without the latter, as they have done for decades before. Yet—and this is where cartography has the whip hand—their customers can not live without good maps and will continue to demand for more visual statistical output.

Currently, professional cartography in many NSIs is at the crossroads. It can either take the lead and step forward with all knowledge when new systems are to be developed and new contents are to be mapped. Or it can stand aside and let others do the developments and products—and finally its job. A good example for this mixed situation is Eurostat. In 2012, the European Statistical Office launched two cartographic products at the same time: (1) a high-quality *Statistical Atlas* that has been produced by editors who selected indicators through a didactic approach and chose relevant map parameters for each representation. (2) A statistical dashboard, *Regional Statistics Illustrated*, has been set-up on the same site. This application also shows maps—at random parameters—for more or less the same data—with the difference that these have been produced automatically by a statistical software package and data base that come together with tables and charts as well. Without any doubt, these two systems (or approaches), paid for twice by the tax payer, will not co-exist forever. So, one can only call upon cartographers who have access to statistics: get involved in these decision making processes, show what professional cartography can offer in the statistical dissemination process. But get involved *now*, before others will take these decisions!

Current Technological Trends and Developments

The use of Atlas Content Management Systems and platforms has well paid off for institutions and customers alike, where they are used. They enabled the offices to produce maps and atlases in large numbers and various output formats while equally guaranteeing the quality of all process steps. Nevertheless, the world doesn't stand still. These platforms will continue to exist for another 5 or 10 years, and many offices will yet commence boarding this comfortable train and begin with data-driven productions of their map output. But in an extremely dynamic environment, characterised by Web 2.0, Web 3.0, fast changing formats and ever faster changing customer interests, any serious prognosis who would attempt to predict the state of statistical mapping in 10 years from now on will fail. As in the general dissemination policies of NSIs, future requirements for mapping products and technologies can only be described with all the uncertainties of forecasts one has to accept. Yet, seven to eight trends emerge at the horizon:

- *Fully automated production:* Automation will continue at high speed—in the interest of clients (quantity, topicality) and institutions (reduction of costs, quality). Editing systems, in which authors individually manipulate and generate maps, will largely be replaced by data-driven systems that automatically and randomly create maps for users—or users will create their own maps.
- *Integration in statistical systems:* Today's software solutions within NSIs are large, heavy, complex, based on statistical standards and more and more integrated. Cartographic aspects and components are of very marginal interest for those systems. Thus, official statistics will not subsidise specific cartographic solutions any more that are not in compliance with their interfaces.
- *Regional data base:* Statistical atlas platforms and their cartographic data bases will either develop into the central regional data bases in the offices (as in Switzerland)—or the will hand over this role to other statistical data bases and continue to exist as cartographic plug-ins (user interfaces) to other systems.
- *Portal function:* Statistical atlases, with their visual simplicity, can provide the central access point to all regional statistical data and give users orientation and guidance within the vast offer of statistical data (Ormeling 2009).
- *Graphical User Interfaces:* The optimization of user interfaces for maps and atlases will steadily continue. There will be more data analysis functions, comparative functions, GIS functions, charts, opportunities for users to import their own data, but also didactic features like story-telling components that summarize the main findings in statistical results in an informative way (Ormeling 2009; Kriz and Pucher 2010; Hurni et al. 2013).
- *New representations:* The list of cartographic methods used in statistical applications will grow. The number of maps that contain more complex symbols, but also the number of value area cartograms and so-called "statistical surfaces" in the respective official publications will slowly increase (Burgdorf 2008; Schulz 2011). Very likely, we will also see more three-dimensional maps in statistics. User studies and promising technical developments in this field are currently conducted for instance at the institutes in Dresden and Zurich (Bröhmer et al. 2012; Sieber et al. 2009; Hurni et al. 2013).
- *Responsive design:* Existing applications that consist of rigid designs made only for large screen displays will be replaced by responsive design solutions. That means, the user interface and its components will adapt in content, shape and size to the individual devices of the client at all times—may it be a desktop PC, a notebook, a smart phone, a tablet PC or any other device of the future (Fig. 6).
- *Mobile apps:* "Classical" web mapping tools, as we have experienced them during the last 20 years, are on the retreat. Statistical maps and atlases will increasingly be made available as mobile apps on the statistical websites or through professional app stores—following customer demands.
- *Printed atlases—eBooks—eAtlases—DigiPubs—DigiAtlases:* Printed statistical publications with maps and statistical atlases are successful and will continue to exist, but they will primarily be derived from existing online publications. They will be published as Print, PDF, eBooks or DigiPubs at the same time and thus try to reach as many customers and reading devices as possible.

Fig. 6 iPad version of the Statistical Atlas of Switzerland (iStatatlas)

Conclusions

Statistical maps and atlases, like official statistics itself, live in a highly dynamic social and technical environment today which is changing the parameters almost every day. While their contents and the cartographic representations methods have remained relatively stable during the last 150 years, the applied technologies and especially the role of cartographers within the statistical dissemination process have dramatically changed—and will continue to change swiftly. In order to increase the quantity and topicality of statistical maps further—as demanded by customers and institutions—automated processes will take over most of the map production in 5–10 years time and replace individual map editing as we know it. At the same time, map and atlas interfaces will become responsive and emerge more and more as mobile apps which fully adapt to individual user needs at any time.

Yet, and this is not only a glimmer of hope, statistical maps—together with other visual means of expression—will be ever more demanded during the coming years. As mirrors of our life, as "travel guides" through our societies, they will become indispensable elements in the statistical dissemination process and continue to tell vibrant stories about the state, about citizens and the living together of them.

References

Arnberger E (1977) Thematische Kartographie. Westermann, Braunschweig

Bill R, Zehner ML (2001) Lexikon der Geoinformatik. Wichmann, Heidelberg

Bollmann J, Müller A (2002) Zensus-Kartographie. In: Bollmann J, Koch WG (eds) Lexikon der Kartographie und Geomatik, vol 2. Spektrum, Heidelberg, Berlin, p 448

Bröhmer K, Knust C, Dickmann F, Buchroithner M (2012) Die Nutzung autostereoskopischer Monitore zur kartographischen Visualisierung von Diagrammen. Kartogr Nachrichten 62 (5):227–234

Burgdorf M (2008) Verzerrungen von Raum und Wirklichkeit in der Bevölkerungskartographie. Kartogr Nachrichten 58(5):234–242

Dickinson GC (1963) Statistical mapping and the presentation of statistics. Arnold, London

Dickmann F, Zehner K (2001) Computerkartographie und GIS, 2nd edn. Westermann, Braunschweig

Eurostat (ed) (2007) Regionen in der Europäischen Union. Sytematik der Gebietseinheiten für die Statistik. NUTS 2006/EU27. Eurostat, Luxemburg

Freitag U (1988) Zum Problem der Darstellung geographischer Systemzusammenhänge in Atlaskarten. In: VEB Hermann Haack (ed) Zum Problem der thematischen Weltatlanten. Vorträge zum Kolloquium aus Anlaß der 200-Jahr-Feier des Verlagshauses, Gotha, pp 75–85

Hake G, Grünreich D, Meng L (2002) Kartographie, 8th edn. De Gruyter, Berlin, New York

Hurni L, Bär HR, Haeberling C et al (2013) Atlas information systems – current developments at ETH Zurich. Kartogr Nachrichten 63(spec. edition):148–153

Kriz K, Pucher A (2010) ÖROK-Atlas online – Ein Atlas-Informationssystem von Österreich. Kartogr Nachrichten 60(2):76–81

Nag P (1984) Census mapping survey. Concept Publishing, New Delhi

OECD (ed) (2006) The future dissemination of OECD statistics. A policy proposal. Paris

Ormeling F (2009) Moderne Atlaskartographie im Spiegel von National- und Regionalatlanten – Bestandsaufnahme und Entwicklungslinien. Kartogr Nachrichten 59(1):13–18

Schulz T (2009) Maps. In: UN-ECE (ed) Making data meaningful. A guide to presenting statistics. New York, Geneva, p 30–40.

Schulz T (2011) Interaktiver Wahlatlas zu den Schweizer Nationalratswahlen 2011. Kartogr Nachrichten 61(5):261–265

Schulz T (2012) The statistical atlases of Switzerland and their atlas content management system. In: Jobst M (ed) Service-oriented mapping 2012, Vienna, pp 453–469

Schulz T (2013) The statistical atlas. Studies on classificatory, conceptual, formal, technical and communication aspects. Dissertation, Technical University of Dresden

Schulz T, Ullrich T (2009) Ein Atlas Content Management System für den neuen Statistischen Atlas der Schweiz. Kartogr Nachrichten 59(1):25–36

Sieber R, Jeller P, Hurni L (2009) Statistische Oberflächen in einem interaktiven 3D-Atlas – Strategien und Techniken. Kartogr Nachrichten 59(4):190–196

Spiess E, Hurni L, Werner M et al (2010) Thematische Kartografie. GITTA, Zürich

Van Elzakker C, Ormeling F, Köbben BJ, Cusi D (2005) Dissemination of census and other statistical data through web maps. In: Peterson MP (ed) Interactive and animated cartography. Elsevier, Amsterdam, pp 57–75

Witt W (1964) Deutsche Regional- und Planungsatlanten. Kartogr Nachrichten 14(3):88–96

Witt W (1970) Thematische Kartographie. Methoden und Probleme, Tendenzen und Aufgaben, 2nd ed. Gebr. Jänecke, Hannover

Online Cartographic Atlas Products: Learning from the Past

Alexander Pucher

Introduction

Harley and Woodward (1987) described the term "map" in their preface to 'The History of Cartography' as "graphic representations that facilitate a spatial understanding of things, concepts, conditions, processes, or events in the human world" They probably had no clear idea about how the maps will change over the next quarter of a decade to this day.

Maps are now much more present to the public than they were a few decades ago. The possibilities of the Internet offer completely new approaches to spatial information and its visualization and communication. Fast and geared to the needs of the users, spatial information can be accessed within a few mouse clicks from literally anywhere in the world.

To take this into account, the requirements of modern cartography, their forms of representation but also the usability of "screen maps" had to be re-thought and researched. This process seems not finished yet, as there are many examples of cartographic products on the Internet that still try to "copy" traditional methods into the digital age.

The working group on Cartography and Geoinformation at the University of Vienna, Department of Geography and Regional Research (IfGR) has more than a decade of experience in the development of online applications with spatial content. From pure proof-of-concept implementations of simple online mapping tools in 2000 towards recent information systems in multi-disciplinary environments, these works reflect the changes in the discipline of web cartography. At the same hand, these applications have always been results of the current status of the use and usability research of Internet-based cartography. Significant changes have taken

A. Pucher (✉)
Faculty of Earth Sciences, Geography and Astronomy, Department of Geography and Regional Research, University of Vienna, Universitätsstraße 7, 1010 Wien, Austria
e-mail: alexander.pucher@univie.ac.at

© Springer International Publishing Switzerland 2015
J. Brus et al. (eds.), *Modern Trends in Cartography*, Lecture Notes in
Geoinformation and Cartography, DOI 10.1007/978-3-319-07926-4_5

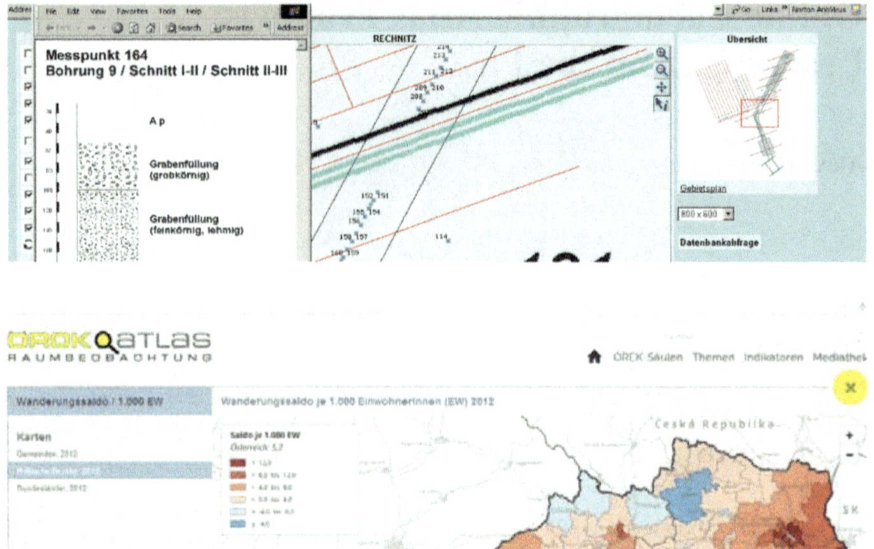

Fig. 1 Examples of Internet-based information systems with spatial content from 2002 (above) and 2013 (below)

place in the transition from technology driven applications towards user-centred products. Recent applications can build on contemporary technical features and utilize the full framework of modern Internet functionalities (Fig. 1).

Apart from that and probably even more important is the positioning of the user within the centre of attention throughout the whole planning and development process. This fact is in such a way important, since cartographic information communication in the Internet has very special prerequisites and characteristics that sets it apart from other online content processing. Online cartographic information must be seen as independent and thus be treated with special methodologies of usability testing.

Over the past years, findings from several online cartographical implementations at the IfGR led to a series of lessons learned, which were considered in recent developments to increase the usability of these latest cartographic information systems.

What's So Special About Online Cartography?

The term "geographic information system" is widely used for all systems that store, process and visualize spatial data. This definition is particularly to be found in English-language literature in which GIS has established itself as a self-evident term, as it is used in Lloyd et al. (2007), Haklay and Zafira (2008) and Skarlatidou et al. (2011). The term 'GIS' stands in these publications in a general context for all mostly Internet-based systems with spatial content. However, a closer examination of this is necessary.

Whereas a GIS is an environment for capture, extraction, modelling, manipulation and analysis of geographic information (Bill 1999), the main focus of a cartographic oriented information system is targeted on a high-quality cartographic visualization of the underlying data.

In 1996, Kelnhofer defined "interactive cartographic information systems" as the development of all cartographic products in which not only the design of the information processing by the user, but also the cartographically processed information on the screen is an essential aspect (Kelnhofer 1996). Cartographic information systems are thus in opposition to geographic information systems, as the visualized information must be optimized in a cartographic process based on the output scale, as he stated in 2000: "Die dargebotene Information muss hierbei nach kartographischen Regeln bearbeitet und gestaltet werden, was zwangsläufig zu einer maßstabsbezogen generalisierten Kartengraphik und zur graphischen Konfliktfreiheit des Präsentationsergebnisses führt" (Kelnhofer et al. 2000). In this context, Hennig (2004) used the term "interactive screen map" in order to clarify a definition of this form of cartographic products and put it in the "gap between the static information transferred by paper maps and the extensive and complex processing possibilities of GIS maps."

In addition to the aspects mentioned above, a cartographic information system relates generally to a particular geographic area and takes an additional "narrative component" on a particular topic (Ormeling 1995) into consideration. In this context, Cartwright used the concept of "narratives" as a leitmotif of a cartographic visualization in the Internet (Cartwright et al. 1999). Thus, greater emphasis is placed on the design of the visualization of spatial information. It offers both high quality cartographic quality, as well as a user friendly interface.

The above points show the differences between geographic and cartographic information systems. Due to different methodological approaches and objectives, as well as a focus on different user groups it is necessary to consider cartographic information systems as an independent application type and focus on relevant properties of such systems, which must be considered within the framework of a development process.

As essential and fundamental properties of a cartographic information system can therefore be stated:

- Internet-based information system with focus on the processing of geographic information: In contrast to geographic information systems data has to be preprocessed to be used in cartographic information systems. The processing of geographical data has to be done by methods of generalization, based on the output scale. Thematic content is to be treated accordingly to ensure the "narrative component" of the system.
- Maps-centred system: A cartographic information system represents a map centred system whose functionality is primarily related to interaction and manipulation of the graphical representation of information. Functionalities

that are taken from a GIS can thereby be integrated, but only in an appropriate thematic context.

- Consistent use of object hierarchies and structures: The stringent use of a strict hierarchy of objects is imperative in a cartographic information system to convey data corresponding to the user in a transparent form. To ensure the effective use of the system, a restrictive-flexible approach of the user has proven its positive aspects (Gartner et al. 2005).
- Extensive search mechanisms: A range of search mechanisms are to be offered in a cartographic information system in addition to the aspects mentioned above and should be followed consistently throughout the system. Search operations should be based on the three axes of the information space (space, time, content), as well as combinations of those. Search mechanisms are the user's entry point to the information structure without having a knowledge of the logical and physical data structures itself.

The above mentioned issues require special demands on cartographic information systems that need to be taken into account in the context of a user-centred development of such an application.

Cartographic information systems represent thematically defined systems that are generally freely accessible. The selection of the user target groups, as well as the identification of user needs is an important aspect in the planning of these type of system, as the essential features and properties of such a systems must be taken into consideration. For this reason it has proved to be advisable to perform a pre-evaluation of existing systems at a later development stage, because such evaluation can lead to a bias in the determination of the user's needs. The generally free access to cartographic Internet-based information systems also prevents the clear definition of whether the user is to be considered as a layman or expert. It is therefore recommended to avoid such a separation of the system.

The functional and content specifications of a cartographic information system have the graphic visualization of the available information as maps or other cartographic expressions—as the conceptual centre of the system. However, as opposed to a printed map, due to the lower graphical resolution of the screen, an increased use of interaction functionalities is necessary.

Due to the complexity of Internet-based cartographic information systems, a transparent information architecture has to be implemented. Procedures for the recognition of the systems content and its functionalities must be taken into account, as well as the narrative component in the form of "geographic storytelling". To ensure the greatest possible flexibility to the user, however restrict the degree of motion in the system, proves as an efficient and effective way of working. This "restrictive-flexible" approach to a user interface, which makes system-dependent limits of the application invisible for the user, has proven itself in practice (Gartner et al. 2005).

The map itself is to be regarded as the primary interface of a cartographic information system to underlying geographical as well as thematic information. Accordingly, a fact-matched navigation is required. In addition to common interaction functionalities to change the map's content and scale, suitable tools for the exploration of thematic and geographical facts have to be implemented.

The visual design of a cartographic information system has to take care of all above mentioned properties and prerequisites of such applications. This applies to the cartographic representation as well as the visual implementation of all system functions. Appropriate measures in the sense of the use of "metaphors" from existing systems for the recognition of functionality shows a high efficacy.

Considerations on Testing Methods for Cartographic Information Systems

The use of individual usability methods in the special discipline of cartography and geographic information is described in numerous publications. Roth (2011) examined particularly graphical user interfaces of web-based information systems with geographic/cartographic focus. A significant increase in research activity in recent years can be seen in the use of "eye-movement-tracking" methods in cartography. Pretorius et al. showed 2005 the potential of this method by recording fixations and eye movements of the user. Çöltekin et al. (2009) used the method to evaluate the effectiveness of interactive cartographic systems. Heil (2010) describes the potential of the method for the use in cartography as follows: "The eye tracking data helped in understanding the cognitive processes of the participants during system usage."

In addition to the exploration of individual methods and their specific application in the evaluation of cartographic products, a tendency of using multiple methods can be seen. Robinson et al. (2005) used this approach for the development and evaluation of interactive systems in epidemiological studies. A comprehensive assessment of the potential of the combination of several methods was published by Bleisch (2011). She emphasized that: "...the weaknesses of one method are compensated for with strengths from other methods while ideally even having strengths in similar areas thus increasing the validity of the research."

A similar holistic approach to the implementation of user-centred development as well as the use of well known, proven methods usability testing is followed by the Commission on Use and User Issues of the International Cartographic Association (ICA). A key objective of the work of this commission is the process of adapting the principles of user-centred development for the use in the planning, design and implementation of cartographic products.

However, the variety of available procedures and methods for testing the usability has hardly expanded in recent years. In addition to both qualitative and quantity methods that require the active participation of the users, procedures which can be performed without the direct participation of the users involvement can be seen as innovative. These forms of test methods, such as the logging and analysis of user interaction with an interactive system, as described by Atterer et al. (2006) as

"Implicit Interaction Logging" are strongly linked to the use of Internet-based systems. Existing methods are well known in its basic design and its implementation and are only adapted to be used with Internet-based cartographic systems. Various publications explain this in terms of increasing the usability quality of the Internet through user evaluation (Schweibenz and Thissen 2002). Pearrow (2006) extends the "user-centred" idea to the fact that finally the technology has to serve the user: "Accessibility and usability form the foundation of human-enabling technology; that is, technology that serves us, rather than technology that we serve".

In general, the determination of the user's needs of a cartographic information system can be increased and thus made much more efficient, more productive and more sustainable by bearing in mind the following considerations:

- Informative individual interviews instead of anonymous survey by means of questionnaires.
- Qualitative findings instead of quantitative results, to obtain "personal" information rather than statistical values.
- Interviews should preferably be carried out without any knowledge of existing systems by the interviewed persons, to prevent bias.

In the scope of the work on Internet-based cartographic information systems, the best results were gained through qualitative interviews with representatives of potential user groups. The actual discussions could contribute to obtain a "feeling" for the typical workflows of users and their needs on the developed system. The talks also aimed to exactly define the main target audience of the system to concentrate the further development process upon these results. Typical modi operandi could be recognized and further utilized to enhance the user-centred idea within the systems.

In the context of user-centred development of Internet-based information systems, special emphasis should be pointed out on the scope and quality of the methods for testing the usability. This situation arises from the multiple special features and the corresponding requirements provided to the users of these systems. A permanent evaluation, accompanying each stage of the development is inevitable.

Qualitative methods have thereby proven a higher information level than quantitative methods, particularly in the identification of user needs and accompanying prototype evaluations. Due to the generally high degree of specialization of a cartographic information system and the fact that users usually have a thematically high, but cartographically low expertise, personal interviews with qualitative orientation showed as the optimal method. With regard to the evaluation of graphical components of the system, such as the maps and the visual implementation of functionalities, "Implicit Interaction Logging" as well as "eye-tacking" methods are efficient solutions.

Lessons Learned from Previous Developments

Methods for testing the usability must be adequate for the evaluation of different aspects of the broad field of user behaviour. This idea was expressed by Wodtke (2003) and Rubin and Chisnell (2008) and further taken into account in the DIN EN ISO 9241 (2011): "Ergonomics of Human System Interaction".

Rubin and Chisnell (2008) describes a product as usable if the lowest possible level of frustration occurs in its use. He specifies the usability as a product by the following parameters:

- Usefulness
- Efficiency
- Effectiveness
- Learnability
- Satisfaction

What lessons could be obtained through the development and the evaluation of Internet-based cartographic information systems and what kind of conclusions could be drawn for following developments?

Knowledge gained from the development and the evaluation of several cartographic information systems were obtained as lessons learned and were taken into account in further developments.

Usefulness

In more recent projects, the determination of user groups, user needs and content requirements was undertaken in a much more structured form than in previous implementations. The thematic focus as well as the presentation of the topics was much more oriented on the basis of the results of the qualitative interviews, the user access on specific questions and issues was paid closer attention. Thus, a significantly higher usefulness of the system was obtained, which was also confirmed in several evaluations concerning this aspect. The holistic approach of the data content integration was regarded by the users as outstanding, in particular the ability to use the cartographic information system in the usual workflows.

Efficiency

The efficiency of a system, which can be determined as the ratio between benefits and costs could be considerably improved by an overall system performance increase in recent applications. Nevertheless the fact that newer systems are even more map-centred than earlier ones, the creation and presentation of the graphical content was significantly faster and more efficient than in previous cartographic information systems. This is possible due to a consistent use of object hierarchies and object structures that allow an optimized information architecture. The added value of linking thematic with geographic information was explicitly stressed. The possibility of the integration of space and time as additional substantial factors of scientific exploration was highly appreciated.

Effectiveness

The effectiveness indicates the degree of the achievement of objectives. Also in this area, recent systems are to be understood as optimized compared to earlier applications, since user needs and fixed objectives were maintained in a much more consistent way. This is due to the fact of the increased and more targeted use of test methods in the development process of recent products. Findings of these tests have been included directly in the development process and therefore serve as controlling factors of each subsequent development phase. The optimized effectiveness of newer systems was verified in post-release evaluations, which emphasized the logical match of available functionality with the content exploration within particular work processes.

Learnability

Poor accessibility to the information due to suboptimal navigation and functionality was often criticized as a weakness of earlier systems. The evaluations carried out during the planning and development stages of recent systems gave clearly more important insights for optimizing the interfaces and navigation and resulted in a much more user-centred system and thus in an optimized accessibility of information. This is confirmed by the evaluated fact that both the amount of functions used, as well as the extent of the retrieved thematic and cartographic information in recent applications are significantly higher than in previous ones.

Satisfaction

The qualitatively evaluated satisfaction in recent developments is considered higher than in previous ones. A wrong or fuzzy approach to the system, a bad preprocessing of the integrated information as well as a poor technical implementation of the system leads to a generally low user satisfaction. The main advantages of recent systems are the scope of, as well as the approach to the integrated content, the added value of cartographic visualization as well as the ease of use of the systems.

> **Conclusion**
> The perennial work on Internet-based cartographic information systems at the working group on Cartography and Geoinformation at the University of Vienna, Department of Geography and Regional Research delivered important insights on the improvement of these special applications. It could be shown that the optimization of Internet-based cartographic information systems is possible through knowledge gained from user-centred development. This fact shows direct potential to optimize the usability of subsequent developments in the areas of usefulness, efficiency, effectiveness, learnability and satisfaction. However, as spatial information and cartographic representations enter steadily increasing areas of the information technology and thereby our common way of communication, even more attention must be paid to satisfy the user's needs.

References

Atterer R, Wnuk M, Schmidt A (2006) Knowing the user's every move – user activity tracking for website usability evaluation and implicit interaction. In: Proceedings of the 15th International World Wide Web Conference (WWW 2006), pp 203–212, Edinburgh

Bill R (1999) Grundlagen der Geoinformationssysteme. Wichmann Verlag, Karlsruhe

Bleisch S (2011) Evaluating the appropriateness of visually combining abstract quantitative data representations with 3D desktop virtual environments using mixed methods. PhD. City University London, London

Cartwright W, Peterson M, Gartner G (ed) (1999) Multimedia cartography. 1. Ausgabe. Springer, Berlin

Çöltekin A, Heil B, Garlandini S, Fabrikant S (2009) Evaluating the effectiveness of interactive map interface designs: a case study integrating usability metrics with eye-movement analysis. Cartogr Geogr Inf Sci 36(1):5–17

Gartner G, Spanring C, Kriz K, Pucher A (2005) The concept of "restrictive flexibility" in the "ÖROK Atlas Online". In: Proceedings of the 22nd international cartographic conference (ICC 2005), pp 105–110, La Coruna

Haklay M, Zafira A (2008) Usability engineering for GIS: learning from a screenshot. Cartogr J 45 (2):129–138 (Maney Publishing, London)

Heil B (2010) Usability testing of interactive maps by using eye tracking a case study. In: Haklay M, Morley J, Rahemtulla H (eds) Proceedings of the GIS Research UK 18th annual conference GISRUK 2010, pp 213–218. University College London, London

Hennig S (2004) Das kartographische Informationssystem "MaaT" Interaktive Bildschirmkarte – Kartographische Nutzeroberfläche und Digitaler Kartenplan. In: Strobl J, Blaschke T, Griesebner G (eds) Angewandte Geoinformatik 2004, Beiträge zum 16. AGIT-Symposium, Salzburg, pp 217–226

International Organization for Standardization (ISO) 9241 (2011) http://www.iso.org/iso/iso_catalogue/catalogue_ics/catalogue_detail_ics.htm?csnumber=16873. Last seen 20.01.2014

Kelnhofer F (1996) Geographische und/oder Kartographische Informations systeme. In: Kartographie im Umbruch – neue Herausforderungen, neue Technologien. Beiträge zum Kartographiekongress Interlaken 96, 45. Deutscher Kartographentag. pp 9–26

Kelnhofer F, Pammer A, Schimon G (2000) "GeoInfo-Austria" – Interaktives Multimediales Kartographisches Informationssystem von Österreich. In: Beiträge zum 5. Symposium zur Rolle der Informationstechnologie in der und für die Raumplanung. CORP 2000. TU Wien, pp 69–74

Lloyd D, Dykes J, Radburn R (2007) Understanding geovisualization users and their requirements; a user-centred approach. In: Winstanley AC (ed) GIS Research UK 15th annual conference, Maynooth

Ormeling F (1995) Atlas information systems. In: Proceedings of the 17th international cartographic conference (ICC 1995), pp 2127–2133, Barcelona

Pearrow M (2006) Web usability handbook, 2nd edn. Thomson

Robinson AC, Chen J, Lengerich EJ, Meyer EG, Maceachren AM (2005) Combining usability techniques to design geovisualization tools for epidemiology. Cartogr Geogr Inf Sci 32 (4):243–255

Roth ER (2011) A comparison of methods for evaluating cartographic interfaces. In: Proceedings of the 25th international cartographic conference, Paris. http://icaci.org/files/documents/ICC_proceedings/ICC2011/Poster%20Presentations%20PDF/POSTERS%20SESSION%201/P-019.pdf. Last seen 20.01.14

Rubin J, Chisnell D (2008) Handbook of usability testing: how to plan, design, and conduct effective tests, 2nd edn. Wiley

Schweibenz W, Thissen F (2002) Qualität im Web: Benutzerfreundliche Webseiten durch Usability-Evaluation, 1st edn. Springer, Berlin

Skarlatidou A, Wardlaw J, Haklay M, Cheng T (2011) Understanding the influence of specific Web GIS attributes in the formation of non-experts. In: Proceedings of the 25th international cartographic conference, Paris

Wodtke C (2003) Information architecture, blueprints for the Web. New Riders, Indiana

Non-technological Aspects in Atlas Cartography Based on the Czech Approach

Alena Vondrakova

Introduction

Maps are produced for different target groups with different purposes and present many different topics. They are created primarily by thematic experts in collaboration with cartographers. However, with widespread access to geographic information systems in the form of simple programs and on-line applications in which it is possible to create maps without a technical background, maps are largely produced by thematic experts without cartographic education or even by the general public (Voženílek 2004). Maps are very various, not only because of the potential availability of technologies. The differences are due to the specifics of the presented issue, the chosen methods of cartographic representation, professional education of authors, graphic design, but mainly because of adaptation to the main purpose of the map and the requirements and preferences of the target group.

In most cases, the non-technological aspects of map-making are influenced by the subjective perception of cartographic works by individual users. Their importance is so fundamental that it is undoubtedly important to deal with them. Research on non-technological aspects of map-making using expert system, modern scientific methods and using available technology, brings a lot of knowledge and recommendations for cartographic production process and different procedures based on scientific knowledge in the field of cartography. On the basis of these facts, the paper focuses on the definition of non-technological aspects of map making, analyse and draw appropriate conclusions and recommendations for creating cartographic works.

The most complex process of map-making is usually associated with implementation of state map series, large mapping projects and almost always the complex

A. Vondrakova (✉)
Faculty of Science, Department of Geoinformatics, Palacký University, Olomouc, 17. listopadu 50, Olomouc, Czech Republic
e-mail: alena.vondrakova@upol.cz

© Springer International Publishing Switzerland 2015
J. Brus et al. (eds.), *Modern Trends in Cartography*, Lecture Notes in Geoinformation and Cartography, DOI 10.1007/978-3-319-07926-4_6

process of map-making is mentioned in atlas cartography. Atlas creation it is not only a random collection of separate cartographic works, cartographic atlas represent a set of maps that are associated with certain themes and concepts. It can be said, that cartographic atlas "tell the story" about presented topic. Therefore, for its uniqueness and value, atlases have been appreciated at any historical stage. The importance of single non-technological aspects of map making varies in different areas of cartography (Vondráková 2012). With regard to the use of technologies varies for example the importance of hardware and software aspects (different for the web or printed maps), as well as digital and paper maps are based on different colour models and they require different organizational arrangements, etc.

Atlas Cartography in the Czech Republic

Atlas cartography in the Czech Republic has a long history of over 250 years. Klečková (1999) introduces the first atlas in our country *Atlas Silesiae*, which was created in 1752 as a result of the Müller mapping (extensive mapping in the Czech lands). However a long time before the first Czech cartographic atlas, individual maps of Bohemia, Moravia and Silesia occurred in well-known foreign atlases. Publishing house "Matice česká" has released the first Czech geographic atlas in 1842 and in subsequent years there were produces other atlases based on the same cartographic content, such as *Small Geographic Atlas* (Malý zeměpisný atlas—Merklas, 1843 and 1853; André, 1858), *Small Hand Atlas* (Malý příruční atlas—Matice česká, 1846; Kronberger, 1853) or *Old World Atlas* (Atlas starého světa—Merklas, 1850; André, 1853). The tradition of Czech atlas produced for the general public and extends to approximately half of the nineteenth century.

The following period was marked by historical events—the world wars, regional power struggles, etc. Although historical basis certainly has an impact on the current mapping creation, the research was focused on the cartographic atlas works published in the Czech Republic after 1993, when the independent state—the Czech Republic—was established.

After 1989, when the state monopoly in the field of cartography began to crumble, cartographic works from the past were still used. It means that there was not revolutionary or rapid change in cartographic production, such a uniform system of school cartographic atlases. For example the Atlas of Czechoslovakia was replaced by Atlas of ČSFR (Czechoslovak Federative Republic), and subsequently by the School atlas of the Czech Republic—but mostly the only change in maps was caused just by renaming some places and with implementation of necessary borders modifications. This situation was very similar also in other cartographic production, such as road maps, tourist maps and other traditionally published cartographic works. Commercialization of cartography in the Czech Republic fully developed after the formation of a democratic state in 1993 and the development of new cartographic departments and the creation of new cartographic works carried out over the whole of the 1990s.

The most important part of the Czech atlas production is definitely formed by national atlases, so it is important to mention older works. Among the Czechoslovakia national atlases include: Atlas of the Czechoslovak Republic (Atlas republiky Československé RČS—1935), Atlas of the Czechoslovak Socialist Republic (Atlas Československé socialistické republiky ČSSR—1966) and Atlas of the Slovak Socialist Republic (Atlas Slovenské socialistické republiky—1980). So far the only atlas, which is known as a national atlas in the history of independent Czech Republic is the Landscape Atlas of the Czech Republic (Atlas krajiny), which was released in 2009. All these atlases presents different issues and national maps.

As mentioned, after 1993 the cartographic production was based on cartographic works revised from the previous period and later, mostly at the turn of the millennium, in the Czech Republic began the mass production of foreign publications translated into the Czech language. Foreign works translated into Czech and presented as geographical atlases are mostly popular encyclopedias, or they are not primarily cartographic publications, but they contain some maps, and therefore in the Catalogue of the Czech Republic they are identified as cartographic documents. Many foreign works is also a creation for small children, but there are also genuine Czech products. Perhaps the most controversial issue in the Czech atlas cartography is taking over foreign works presented as "School atlas", such as atlases from Ikar (2003) and Svojtka & Co. (2004 and 2009), which were translated from "Student atlas", "Kingfisher children's atlas" and "World atlas". These works, of course, do not follow the content ingrained in the Czech school atlases. They are not adapted to the Czech curriculum for primary and secondary school (now the School educational programs), and at the time of publishing they had not clause of the Ministry of Education, Youth and Sports of the Czech Republic. But because they were presented as "School atlases" many users have been misled.

Statistics in Table 1 show the information obtained from the Union Catalogue of the Czech Republic and from the Catalog of the Czech National Library. In any work there is not detailed expert evaluation of all these cartographic works and because of processing such a vast number of publications and technical and timing reasons, such an analysis is basically impossible. Nevertheless, we can distinguish traditional cartographic works and atlases that are unique in its concept or theme. Such works are gaining awards from the Cartographic Society of the Czech Republic in the competition Map of the year. This award was given for example to the School Atlas of the Today World in digital form (Školní atlas dnešního světa, TERRA-KLUB, 2011) or Climate Atlas of Czechia (Atlas podnebí Česka, 2007).

Production of atlases is significantly affected by the production of one road-atlas, which was published for various companies (with different ISBN) in years 2007 and 2009. The most important producers of Czech cartographic atlases are Kartografie PRAHA and SHOCart publishing houses. Roughly a fifth of total amount is represented by small producers or publishers who publish cartographic atlases irregularly and in a limited number of titles.

Atlas production in the Czech Republic is comparable to the world's best cartographic production. This is evidenced by a number of international awards, such as honorable award of the Commission of the International Cartographic

Table 1 Number of different titles—cartographic atlases published in the Czech Republic	Publishing year	Total sum of titles
	1993	5
	1994	12
	1995	14
	1996	22
	1997	34
	1998	47
	1999	22
	2000	16
	2001	15
	2002	21
	2003	26
	2004	48
	2005	55
	2006	63
	2007	322
	2008	63
	2009	402
	2010	39
	2011	55
	2012	24

Exhibition at XXI. ICC in Durban in 2003 for the *Atlas of Prague – Prague Integrated Transport*, a silver plaque for *Bike-Atlas Czechia* from the International Cartographic Exhibition at XXIII. ICC in Moscow or the most recent first place in the Atlases of the International Cartographic Exhibition during the XXV. ICC in Paris for *Landscape Atlas of Czechia*.

General Aspects of Map Production

Aspect of map production can be defined as a general way of solving specific steps in the process of cartographic work realization, may be external or internal and includes a variety of influences and factors. Every person can imagine something else, but in complex, all aspects of map production should describe the whole process of map-making. These aspects can be also understood as "complex factors" od "characteristics" – this is an English (semantic) wording problem, which has no definite answer. The author of the presented research chose after long discussion the term "aspect of map production".

Analysis and definition of the aspects of map production belongs to the meta-cartography, which is understood as the science of cartography, the subject of its knowledge, its methods and expression and is engaged in the scientific exploration of the fundamental theoretical problems, by definition, terminology and classification issues (Murdych and Novák 1988).

Generally aspects of map production represent influence and perspective, covering a wide area in the cartographic process and defending a large number of specific steps of map-making (Voženílek 2004). Their definition is always based on the specific purpose of the analysis process of map creation. The various aspects of map production include a range of influences affecting the cartographic production and are also largely affected by regions (continents, countries).

Aspects of map production can be divided according to many various characteristic, but the basic distribution is to technological and non-technological aspects. The concept of technological aspects is quite often used in publications in various fields of science and technology; however there is not any uniform definition. Generally, technological aspects are described by the basic technical specifications and workflows in various fields of human activity. They describes facilities, materials, structures, strategies, tools and their uses, methods and other issues dealing with technical resources. In the context of the research, technological aspects are regarded as the processes or elements for which the importance of technology (technologies) prevails over the others. Restrictions on non-technological (sometimes also called as non-technical) aspects are composed primarily of a range of knowledge and skills of professionals and people who are involved in the process of cartographic production, not the possibilities of the technique used.

In the research there was created the following list of technological aspects of map production and its subsequent detailed definition as a result of a detailed study of Morrison (1974), Robinson et al. (1995), Kraak and Ormeling (1996), Kaňok (1992), Reichenbacher (2001), Veverka (2001), Voženílek (2004), Pravda and Kusendová (2007) and others. Technological aspects of map production, defined in the research, include: safety aspect, data aspect, geoinformatic aspect, hardware aspect, mathematical aspect, mapping aspect, software aspect, standardization aspect, visualization aspect and the aspect of technical production. It is possible to include all these technological aspects of map production in the general division stages of the map creation (Voženílek 2004).

Mentions dealing with non-technological aspects are usually used in technical fields where technical nature specified and there is a reference to any other "non-technological" or "non-technical" factors. Among those are usually classified aspects of economic, organizational (or the organizational aspect is called as a concept aspect), as well as aspects related to various issues in the field of human resources. According to the scope are also mentioned aspects of the political, historical and more. Non-technological aspects are described in detail below.

Non-technological Aspects of Map Production

The list of non-technological aspects of map production is based on the basic procedures for creating cartographic work. Among the non-technological aspects of map-making are arranged (in the alphabetical order): aesthetic aspect, conceptual aspect, economic aspect, ethical aspect, geoinformatic aspect, historical aspect, legal aspect, methodological aspect, organizational aspect, political aspect, psychological aspect, sociological aspect, user aspect and visual aspect (Vondráková 2013).

The list of non-technological aspects of map production has been debated for quite a long time to be completed in the view of the majority expert opinion of experts. Aspects of map production can be understood in different ways, as well as the methods of cartographic representation are presented differently by different authors. It is obvious that the defined non-technological aspects of map production can be presented in the larger units. All aspects are classified into three basic groups of non-technological aspects of map production: social, professional and user aspects. Individual groups are not disjoint and some aspects are also included in multiple groups. The designated groups are: **social aspects** (political, ethical, legislative, economic, organizational), **technical aspects** (geoinformatic, ethical, historical, legal, conceptual, methodological, visualization, organization) and **user aspects** (sociological, historical, conceptual, psychological, aesthetic organizational and user).

Social aspects represent an important part of the process of map production, because without their presence there would not be possible to initiate the implementation of map production. Professional aspects include everything that is dealing directly with the cartographic production. User aspects are undoubtedly the most discussed area that is focused by many publications. They also include knowledge from other disciplines—sociology, psychology, art and more.

Definition and Properties of Individual Non-technological Aspects

Aesthetic aspect of cartographic products affects the overall look and user-friendliness of the cartographic work. It includes art design of cartographic works, overall design, methods of cartographic visualization, etc. In terms of user perception this aspect can affect the emotions of the user and his future preferences.

Analysis of the aesthetic aspect has led to the following information: aesthetic perception is a subjective assessment of a user and cannot be generalized as to generally applicable patterns; results of aesthetic evaluation is possible to objectify on the basis of statistical analyzes using evaluation methods of cognitive cartography (preferably a combination of research methods); assessment of user emotions can be realized by using eye-tracking technology in combination with medical devices, however, these studies are not yet very widespread in cartography;

aesthetics (graphic) rendering of maps significantly affects map reading by user, especially time-consuming search for the required information and the accuracy of the information received.

In atlas cartography is very significant influence of specific needs of the target user group (school atlases for various levels of primary and secondary schools, scientific thematic atlases for professionals, etc.). In school atlases for younger students is elected more artistic style, while the scientific thematic atlases often use austere expression without any graphic design process. Atlases also should always be presented in consistent graphically and aesthetics design.

Aesthetics of maps is needed to evaluate by representative group of respondents from the target group of users because evaluation from different groups of respondents may be completely irrelevant. To evaluate the aesthetics of maps there must be used a combination of research methods and a multidisciplinary approach involving knowledge of psychology, art, cartography, and other related fields.

Conceptual aspect of cartographic products is mainly related to the content of map and is directed to a method of presenting information and results of the analysis and synthesis of input data in the map. In connection with the creation of thematic maps, there are many information concepts: communication, systemic, cognitive, mathematical and cartographic modeling, language, etc. A more general approach to the conceptual aspect is the distribution of cartographic works according to their concept to groups of analytical, synthetic and complex maps.

Analysis of the conceptual aspect and the evaluation of case study led to the following information: in the largest quantities there are produced analytical maps, in a lesser extent are produced complex maps and in the least number are produced synthetic maps; there is no automatic tools for map synthesis, because the process always need an expert opinion on the issue being addressed; map synthesis has an irreplaceable role in atlas cartography and should be provided within the professional treatment of the topic. This aspect is more important for the atlas cartography than, for example, the creation of monothematic editions (tourist maps, cycling maps, etc.). It is also always important to pay attention to the object and purpose of maps.

Economic aspect is an important part of the process of cartographic production. Determining is a demand and purchasing power of potential users of cartographic products. Ensuring economic requirements is the determinant of the success of the cartographic project implementation, so there must be given great attention. Data relevant to the economic plan are usually obtained from research among users of maps, market analysis, analysis of the experience of publishing cartographic products and other similar procedures. Atlas cartography is economically one of the most expensive areas of map making.

Detailed analysis of the economic aspect and the evaluation of case study has led to the following information: economic arrangement is an existential condition for the implementation of the project in cartographic production; basic requirement for exact economic planning is target market and users exploration; there are many cartographic publishing houses, which focus only on custom manufacturing or only on the creation of free products; users perceive the price of cartographic works as adequate, but atlas cartography they perceive as slightly more expensive; the most important parameters when buying cartographic work are the topic, the quality of cartographic treatment, previous experience and recommendations, quality of workmanship, price and design and the least important are advertising products (presented by map users in on-line questionnaire); users are willing to pay a higher amount for scientific atlases; users consider the offer of cartographic atlases only slightly above average, while the producers of cartographic products believe that they fully reflect user requirements; data sets costs and commercial software for GIS and graphic design producers consider as rather expensive, the price of hardware and storage is adequate according to them.

Specificity of atlas cartography is higher economic requirements than in the case of any other cartographic products. In the field of school atlases (with the guidance of the Ministry of Education, Youth and Sports of the Czech Republic), the market is relatively stable and in the case of scientific thematic atlases, the majority of projects is currently directly or indirectly financed from the budget of the state or from grants under auspices of the European Union (specific situation in the Czech cartography, not general worldwide situation).

Ethical aspect includes an assessment of the values and principles that helps to create human behavior. In the context of map-making this aspect mostly deals with the evaluation of individual processes in the cartographic production: from the data gain, through the use of program resources and symbol sets, to the realization of a particular topic in the form of map. Ethical aspects can be closely related to the legislative aspects, such as abuse of know-how or copyright infringement. Ethical rules are valid even beyond the legislative measures.

Analysis of the ethical aspect led to the following information: copyright status and legal ethics in cartography is not uniform; there are "codes of ethics" and it is always necessary to become familiar with the rules applicable to the area in which cartographic creation is realized.

In the atlas production an ethical aspect has not specifics that would not apply to other cartographic works. Any violation of ethical principles in the case of cartographic atlas, however, may occur stronger or more extensive. It is always necessary to avoid creating any map in controversial circumstances in relation to the ethical-moral principles of society.

Geoinformatic aspect in the concept of non-technological aspects represents particularly the knowledge and skills of geographic information systems—using a variety of geospatial tools, for example for the realization of analysis and synthesis of data processed, leading to a previously unknown information, which are then presented through the cartographic work.

Detailed analysis of the geoinformatic aspect and the evaluation of case study has led to the following information: geoinformatic aspect involves the use of knowledge and skills of geographic information systems, but very important factor is the education of specialist who works with the software; this aspect stands on the border of technological and non-technological aspects; the importance of geoinformatic aspect significantly increases with the content intensity of cartographic works and the manner of its preparation and cartographic production.

Software (geographic information system) must be chosen with regard to the requirements of the final cartographic works (e.g. vector or raster processing, demands on computing capacity). It is necessary to subordinate the functionality of the software to the needs of resulting cartographic work and its visualization methods, not the other way around. Specifics of atlas cartography are like in other aspects—increased performance requirements.

Historical aspect involves influencing of contemporary cartographic production by old traditions and practices, such as map key established by usage (for example in Czech forestry maps) or the selected methods of cartographic representation (for example climate zones). For most of cartographic works that are currently published, the historical aspect has minor importance. An important factor is the cartographic education provided in primary school and cartographic education of cartographers, which is also subject dealing with historical context.

Analysis of the historic aspect and the evaluation of case study has led to the following information: historical aspect is perceived subjectively and does not affect users and creators in the same way; aspect is seen more as a marginal for users and more as significant for creators and producers of cartographic works; the most important for all groups of respondents is the tradition of cartographic methods used in past decades and education and training in schools; it is not necessary needed to confirm the assumptions that experts predict, that in the future the development of new technologies must necessarily mean the transition from printed atlas creation to the creation of digital atlases.

Specifics of atlas production are significant in Czech cartography, for example in the field of school atlases the previous user experience and a tradition are an important factor in their production. In the case of thematic scientific works have often affect the newly created works such as the fact that in the past were designed themes satin processing implemented (newly created work may include a comparison, methods of cartographic visualization and data processing can be inspired by the theme of the original processing) or not (satin work is newly created, it is necessary to design a completely new concept).

Legislative aspect is the collection of all legislative measures that affect individual processes within the cartographic production. It is all about protecting intellectual property rights (copyright), proprietary rights to the cartographic works and their components. Aspect involves a method of acquisition and data handling software and hardware equipment, the presentation of data outputs and cartographic products, their distribution and last but not least their protection.

Legislative aspect significantly affects Czech cartographic production in many respects; it is mainly a copyright legal issue, such as the availability of data and the possibility of their use, the possibility of cartographic work protection, etc.

Recommendations for cartographers are never do not disclose or reuse the work of another author in the original or modified form, if not obtained written consent (it means all products of cartography and geoinformatics—maps, figures, tables, data link layer, or any adaptation). It is also important to study in detail the licensing and condition terms of use of each used work. Although the use of some cartographic works is enabled by licenses such as Creative Commons, it still must follow the rules for further processing, reproduction and dissemination. It some cases can be for example a map server used in the form of interactive windows with a specific location such on the user's personal pages, but must not be used on sites with commercial content.

Methodological aspect in the process of map creation involves various methods used for data collection, data management, and cartographic methods of processing and cartographic reproduction, including the methods of cartographic research. It is therefore an aspect that involves some treatment processes of the map production. Aspect includes the issue of mathematical and logical processing and evaluation of information with regard to the geographical detail, custom applications, content completeness of the resulting map, the structure of the linkages and the suitability of the graphic map symbols with respect to the properties of real objects and also requirements of the target user group.

Methodological aspects of map production process involves processing the creation of cartographic works from the perspective of data collection methods, approaches and forms of implementation of data management, use of methods of cartographic visualization and others. Analysis of the aspect led to the following information: frequent form of processing methodological aspect in the process of making maps represent specific certified methods; methodological aspects of cartographic processing is closely linked to the main aim of the map, its purpose and characteristics and requirements of the target user group; there are tools for automatic or semiautomatic creation of cartographic outputs (including cartographic atlases), which include the methodology inside, so the methodological aspects of using these tools is difficult suggestible.

In terms of methodological aspect atlas cartography is specific because the entire cartographic work must be processed in a uniform approach, from working with data, analysis and implementation of other operations, to the choice of methods of cartographic visualization, map symbology and the final composition. Modern issued Czech cartographic atlases, belonging to the categories of school and scientific atlases, usually meet this condition and are really compact.

Organizational aspect of map production deals with the main idea or intention and with a plan for the implementation of particular cartographic project and its realization. The form of the arrangements must be clarified before the start of the process of map production and should identify the key steps or tasks of individual processes. Organization and implementation of cartographic project must be uniform for the entire body of work; it should be clear and should include ensuring data sources, technical equipment, suitable processing methods, aesthetic requirements, and it should also take marketing and user aspects into account.

Analysis of the organizational aspect has led to the following information: correct and appropriate organizational arrangement is a prerequisite of any cartographic work; each cartographic work that is unique, it also has specific requirements on the organization of the project; within the organization and implementation of cartographic project also a number of other influences plays a role, such as market factors, financial costs or users demand.

The concept of the organization and implementation of cartographic project must be specified before the start of the process of map-making and should determine all the steps and processes, including the responsibility of the individual guarantors of the processing phase. With special emphasis it is needed to apply this recommendation for atlas cartographic production.

Political aspect includes the impact of state information policy, for example in the form of state data providing (charging data sets, distribution, update, etc.), as well as government incentives and especially projects that are directly or indirectly funded by the state (such as projects under the Grant Agency of the Czech Republic, within which there have been realized many cartographic researches and carto-graphic atlases).

Analysis of the political aspect and consultations with cartographic producers has led to the following information: development of infrastructure for spatial information in the Czech Republic is very closely related to the state information policy and transnational activities; political aspect is understood from many per-spectives, such as relations with the ruling political parties, influence of policy decisions, legislative influences to the grant state policy, tax policy and others; assessment of the political aspect is very difficult, there is given the large number of different possible concepts with regard to the personality opinions of various experts; producers marked as the most important factor in cartographic production the grant policy of the Czech Republic, grant EU policy and providing of data by the state. It is also needed to say, that due to the operation of the grant policy (and in general funding from the state budget) could be realized many scientific carto-graphic atlases in the Czech Republic.

Psychological aspect is a very important area which can affect other non-technological aspects and it is mainly user related; it can affect users' opinion on a particular presented phenomenon and it may equally influence the understand-ing and perception of a geospace. The psychological aspect is a complex of specific user aspects that are further investigated by using scientific research tools. It is an area of subjective perception of cartographic products based on the color schemes, map design and a map composition, cartographic methods and their application, specific adaptation to user needs etc.

Within the psychological aspect much attention is paid to cognitive cartography and researches, which use modern technologies. Knowledge of psychology pro-vides to cartographers many important clues in the process of creating maps. Psychological perception map is very closely related to the user aspects, the difference is in the definition of scientific issues.

Analysis of the psychological aspect led to the following information: it is appropriate to use advanced technology to evaluate the psychological perception of cartographic works; it requires an interdisciplinary approach for the evaluation of cartographic works through qualitative and quantitative evaluation methods.

Sociological aspect includes the influence of society on the process of carto-graphic production. The society was tilted differently to the cartographic produc-tion in different historical periods; nowadays cartography becomes a regular part of life and regular part of professional and public education, which places additional demands on the processing of different themes and an increasingly more detailed and higher quality of cartographic production. Society's impact on the perception of cartographic products is also closely related to school education, where pupils are sufficiently or insufficiently encouraged to use maps, their understanding and even possibly the basics of map creation. There is also a link to the political aspect.

Analysis of the sociological aspect led to the following information: sociological aspect is related to the conduct of the society, its values and attitudes; within cartographic research sociological approaches are used mostly for characteristics of respondents in the user groups.

It is necessary to promote the general cartography and cartographic production, so that the society was positively inclined to maps. It is needed to create cartographic works that are beneficial and they represent a high quality.

User aspect is determined by the user of cartographic products and it presents one of the most important influences in the process of map creation (in the user centered designed maps). The user needs and preferences need to take special attention and should be the determining factor in creating the concept of cartographic works in general and they have to be taken in account through the whole process of map creation. There are generally rated aspects of cartographic products that are related to the needs of the map users (e.g. optical aspects), and these can be involved in a user testing for further evaluation. There is also a narrow connection with psychological aspect.

Users are one of the primary determinants of any cartographic work (e.g. van Elzakker and Wealands 1999; Morville 2004). User aspects of map-making are understood in different concepts, an example is an adaptation of map-making to the needs of the target user group and user perception of maps.

Analysis of the user aspect led to the following information: user needs may differ with regard to their intellectual and physical condition; some Czech publishing houses focuses on the specific needs of users, such as people with minor eye defects, but maps for people with other disabilities (e.g. mental) is missing; adaptation of cartographic production to the target group of users has a significant impact on the processing technology; specifics of atlas cartography at the user aspect is closely related to factors of production.

The view of users and creators to the offer of cartographic production with respect to user demand varies considerably. Producers believe that the cartographic creation is adapted to specific user needs, but still there is a need to focus more on specific user needs (carried out by market research purposes).

Visualization aspect involves creating a complete map, its design, composition with all additional and superstructure elements and all particular changes in the map field. Within the visualization aspect belong the selection of the methods of cartographic representation, using of map symbology, map composition, its design and appearance. To the visualization factors also include the form of presentation of the resulting map.

Cartographic visualization is one of the most important processes in the formation of cartographic works; in real, visualization is the main aim and the means of cartography. Specific areas of cartographic visualization (e.g. semiotics) address many of the studies and visualization aspect is closely related to map design.

Atlas cartography does not have any special requirements to the visualization aspect. As with the other aspects, however, need to maintain a consistent appearance throughout the process of atlas work, both in terms of design, and with regard to the chosen methods of cartographic representation and the choice of parameters.

Evaluation of the case study led to the following information: when choosing methods of cartographic visualization and individual parameters it is necessary to take into account the user needs; it is needed to look for new possibilities of application and adaptation of cartographic visualization for the specific needs of visualization (fuzzy approach, thematic maps, etc.); cartographic visualization must be subordinated to the requirements of the resulting cartographic work and should not be limited by the functionality of the software used, if there is another possible solution.

Conclusion

Many publications of Czech and foreign authors deal with comprehensive description of cartographic production, but most of these publications are the teaching materials or the description of the cartographic production as a whole. The concept of the technological and non-technological aspects of map making in mentioned in many works, but had been never completely developed. General level of work is a detailed definition of non-technological aspects of map production, complemented by specific case studies that touch each non-technological aspect. These studies are extensive and are described separately in other contributions; this paper therefore presents only outputs leading to knowledge of non-technological aspects of map production. Case study findings and the resulting recommendations were always focused on the general mapping creation, highlighting specifics of atlas cartography.

Some of the case studies were rather general; some may show options of relevant cartographic research answering the specific questions of modern cartography. Although the theme and focus of the work, as well as its individual parts, were consulted with numerous experts among cartographers, cartographic producers or teachers, the results cannot be presented as generally accepted consensus. The work is largely influenced by the subjective views and experiences of the author, which is involved in the creation of maps (e.g. Vondráková et al. 2013; Tuček et al. 2013; Michalík et al. 2013; Hájková et al. 2012).

Theoretical foundations of cartographic production are very similar, despite different approaches. But the specific perception and influence of non-technological aspects are significantly affected by territoriality. It is the influence of culture, education, politics and more. Significant is for example the method of financing cartographic production in some countries of the world and providing funding for research in this area.

Acknowledgments The author gratefully acknowledges the Department of Geoinformatics, Palacký University, Olomouc, where the most of dissertation thesis "Non-technological Aspects of Map-production in Atlas Cartography" was implemented, and also the author gratefully acknowledges the support by the Operational Program Education for Competitiveness—European Social Fund (project CZ.1.07/2.3.00/20.0170 and CZ.1.07/2.4.00/31.0010 of the Ministry of Education, Youth and Sports of the Czech Republic).

References

Hájková L, Voženílek V, Tolasz R et al (2012) Atlas fenologických poměrů Česka/Atlas of phonological conditions in Czechia. Czech Hydrometeorological Institute, Praha, 311 p

Kaňok J (1992) Kvantitativní metody v geografii (Quantitative methods in geography). Ethics, Ostrava, 233 p

Klečková K (1999) Školní zeměpisný atlas (School geographical atlas). Diploma thesis, VUT FAST Brno Ústav geodézie, Brno, 114 p

Kraak MJ, Ormeling FJ (1996) Cartography: visualization of spatial data. Longman, Harlow, 222 p

Michalík J, Voženílek V, Brychtová A, Vondráková A (2013) Atlas činnosti speciálně pedagogických center (Atlas of special educational centres in the Czech Republic). Palacký University in Olomouc, Olomouc, 144 p

Morrison JL (1974) Cartography and geographic information science. In: Changing philosophical-technical aspects of thematic cartography. Cartography and Geographic Information Society, Washington, pp 5–14

Morville P (2004) User experience design. Semantics Studios, on-line

Murdych Z, Novák V (1988) Kartografie a topografie (Cartography and topography). Státní pedagogické nakladatelství, Praha, 318 p

Pravda J, Kusendová D (2007) Aplikovaná kartografia (Applicated cartography). Geografika, Bratislava, p 224 p

Robinson AH et al (1995) Elements of cartography, 6th edn. Wiley, New York, 674 pp

Reichenbacher T (2001) Adaptive concepts for a mobile cartography. J Geogr Sci 11:43–53

Tuček P, Caha J, Janoška Z, Vondráková A, Samec P, Bojko J, Voženílek V (2013) Forest vulnerability zones in the Czech Republic. J Maps, ISSN, pp 1744–5647. doi:10.1080/17445647.2013.866911

Van Elzakker C, Wealands K (1999) Use and users of multimedia cartography. In: Cartwright W, Peterson MP, Gartner G (eds) Multimedia cartography. Springer, Berlin, 343 p

Veverka B (2001) Topografická a tematická kartografie (Topographic and thematic cartography). ČVUT, Praha, p 220 p

Vondráková A (2012) Význam vybraných netechnologických aspektů mapové tvorby v atlasové kartografii (Importance of selected non-technological aspects of map production in atlas cartography). In: Němcová P (ed) Aktivity v kartografii 2012 – Zborník referátov zo seminára konaného 11. 10. 2012. Kartografická spoločnosť Slovenskej republiky a Geografický ústav Slovenskej Akademie vied, Bratislava, 188 p

Vondráková A (2013) Netechnologické aspekty mapové tvorby v atlasové kartografii (Non-technological aspects of map production in atlas cartography). Dissertation thesis, Palacký University in Olomouc, Olomouc, 154 p

Vondráková A, Vávra A, Voženílek V (2013) Climatic regions of the Czech Republic based on the climatological conditions in 1961–2000. J Maps 1744–5647. doi:10.1080/17445647.2013.800827

Voženílek V (2004) Aplikovaná kartografie I. tematické mapy (Applicated cartography I., thematic maps). Palacký University, Olomouc 187 p

The Evolution of Digital Cartographic Databases (State Topographic Maps) from the Beginnings to Cartography 2.0: The Hungarian Experience

László Zentai

Introduction

The history of Hungarian digital maps can be understood only in conjunction with other disciplines (especially with the development of information technology) and the political changes. It is also important to present which state topographic maps were available at that time for the users and how suitable these maps could fulfil the users' requirements. The effect of the political and economic changes at the end of 1980s is most relevant concerning the situation in Hungary (the one-party communist system was replaced by a multiparty democratic social order). This period was important both for the mapmakers and for the map users as the classification of the state topographic maps was ended and the digital technology totally replaced the traditional map productions techniques.

The hitherto secretive role and activities of the Army Cartographic Institute became open. The Institute took the name of a nineteenth century Hungarian cartographer, *Ágoston Tóth* (although the name was changed again after some years), and even entered the market with some non-military maps.

The situation was similar in most of the so-called socialist countries of Central and Eastern European, but the transition period was different from country to country.

L. Zentai (✉)
Department of Cartography and Geoinformatics, Eötvös Loránd University, Budapest, Hungary
e-mail: lzentai@caesar.elte.hu

© Springer International Publishing Switzerland 2015
J. Brus et al. (eds.), *Modern Trends in Cartography*, Lecture Notes in Geoinformation and Cartography, DOI 10.1007/978-3-319-07926-4_7

State Topographic Maps of Hungary

There are two types of state topographic maps: civil and military. In many countries only one system exists (either military or civil). Nowadays in many countries these attributes have no relevance: these terms mostly refer to scales or scale ranges. Since Hungary became independent after WW I, the base maps of the military state topographic series was 1:25,000. Between the two world wars, there was no official military mapping organisation in Hungary due to the peace treaty after WW I, but the one Hungary had was practically a military organisation. After WW II, in the early 1950s, all Central and Eastern European countries became part of the Soviet bloc (the Warsaw Pact, the mutual defence treaty of these countries was formed in 1955). In military mapping, all the countries accepted the Gauss-Krüger system, which was used in military cartography in Hungary till 2004. It was a logical step to change the mapping system for UTM after joining NATO in 1999.

The state organisation of the civil topographic mapping was established after WW II. The first survey was completed between 1952 and 1980 in 1:10,000 scale. This was the largest survey campaign in the Hungarian cartography: more than 4,000 sheets were published in the base map scale. In 1969, the government decided to make the cadastral and civil topographic mapping uniform (projection, geodetic background, sheet numbering). The first sheets of this new topographic survey (EOTR, Unified National Mapping System) were published in 1976. Although the last sheets were published in 2000, the digital transition and the update started around 1995; since then the update has been an ongoing process. The cadastral sheets are available in scales 1:500–1:4,000. The topographic series contains a little bit more than 4,000 sheets in the base scale.

All topographic maps were classified after WW II (the topographic maps were easily available between the two world wars). These strict rules made these civil maps available only for a very limited number of users, but even in the 1980s the classification was changed to internal use only. When the political system changed in Hungary at the end of the 1980s, the access to the civil topographic maps became open. The military topographic maps have been publicly available since 1992 (Papp-Váry 2010).

The First Cartographic Database

It is not easy to determine what the first digital cartographic database in Hungary was. In the beginning of the informatics age both the military and civil organisations of the state cartography were busy with the production and update of new sheets, and the experts were able to deal with IT research and development to a small extent only.

Although it is not a real cartographic database, we have to remember the very first digital product, the digital terrain model (DTM-200) of the *Hungarian Post*

Office Experimental Institute around 1978. At that time, the organisations of the Hungarian cartography did not use computers, but the organisations of the tele-communication wanted to solve their problems by using IT supported planning methods. For the purpose of the Hungarian Post Office, a lower resolution database was suitable, although the resolution was not comparable to a real digital carto-graphic database. The base of the database was the 1:25,000 scale civil topographic maps. As mentioned previously, these maps were classified; therefore, the Hungar-ian Post Office Experimental Institute had to ask the help of the civil state cartog-raphy. With their contribution the institute finally got the permission from the Ministry of Defence.

The equidistance of the 1:25,000 scale maps was 5 m, which would allow the interpolation of heights within less than one metre, but due to the poor digital storage capacity of the time the operators read the heights with one metre precision only.

In this special project, the map sheets were only used for interpolating the heights in grid points, although other factors on a terrain would also influence the wireless telecommunication. However, it was not possible to determine these factors were in proper accuracy from the paper map sheets. For the general calculations of the microwave and very high frequencies the 200-m-grid was appropriate, but it was a compromise between the special needs of the Hungarian Post Office Experimental Institute and the IT opportunities of the time (Fig. 1) (Koós 1996).

The First Digital Map Production

The first GIS software applications became available in Hungary in the middle and at the end of the 1980s. Although this was the last decade of the cold war period, the

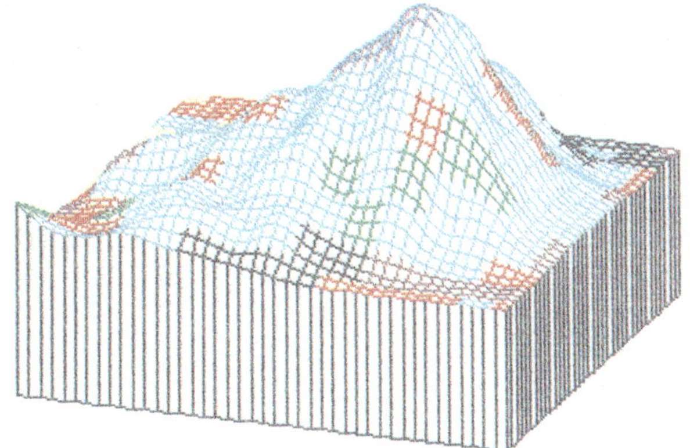

Fig. 1 Digital terrain model based on DTM-200, printed by contemporary plotter

Fig. 2 COMAPO thematic
map (Switzerland)

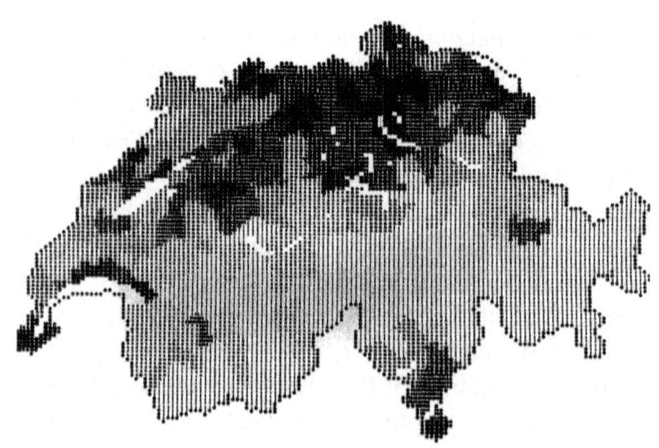

access of high tech devices (including computers) for the Soviet bloc countries was
strictly limited. The most developed Western countries formed the so-called
CoCom (Coordinating Committee for Multilateral Export Controls) to put an
arms and different developed industrial technologies embargo on the Soviet bloc
countries.

One of the very early cartographic/GIS software was the COMAPO system,
which was developed at my department in 1972. The computational capabilities
were good enough to manage the analyses of various statistical data, but the main
drawback was the lack of suitable output devices (printers). The COMAPO application was similar to the well-known SYMAP systems, which used line
(dot-matrix) printers to create thematic maps. Computer output, on monitors and
printers, was limited to typical typewriter characters (letters, numbers and simple
ASCII symbols). These limited symbols could be used to create area patterns on
maps. A modification of the line printer hardware and some programming allowed
the overprinting of characters.

COMAPO was mainly the output system of an application developed by an
institute of the *National Planning Office*. This was a set of methods to print thematic
maps, which were created for analysing and researching regional planning data.
The only method for producing such maps was the traditional (analogue) paper map
production process, which was very time-consuming. Although the digital method
required very expensive infrastructure, they could afford for such equipment on a
state level. As these thematic maps were used for planning purposes and were not
really published in books or atlases, the low quality of the visualization was not a
problem; similar methods were not widely used at that time or later on (Fig. 2).

The Beginning of GIS in Hungary

The first real GIS company was *Geometria*, which was established in 1986. Hungary was still a socialist country and all economic activities were state owned and controlled. Geometria has had a determining role in network information and has been the leading service supplier of geographical information system applications in Hungary since its foundation. The company developed *alfaGrafik* and *topoLogic*, their first basic GIS software, which reached significant success in several professional forums including also the International Cartographic Conference in Budapest in 1989.

The first GIS-based applications were municipality systems, environmental protection systems, town and route designing systems, police and fire brigade supporting systems, regional and city planning projects and utility systems.

The first legal copy of the ESRI PC ArcInfo was sold to the *Budapest Land Administration* in 1988, which was not a simple process due to the restrictions by CoCom.

The National Atlas of Hungary was published in 1989. Although it was made by traditional cartographic technologies, some supplementary sheets were published a few years later to reflect the political and economic changes; these maps presented topics which were not published in the Soviet times (like gipsy population, elections, unemployment rate). These supplementary sheets were probably the first offset printed maps which were produced by GIS software.

Digital State Topographic Maps

The first digital map which covered the whole country was the *DTA-200 (Digital Cartographic Database)*, which was created between 1987 and 1989. DTA-200 is the digital version of the 1:200,000 scale military topographic maps in Gauss-Krüger projection system. Due to the contemporary software and hardware opportunities, the Cartographic Institute of the Ministry of Defence shifted to the 1:500,000 scale version of those maps. Even this map contained so much data that the relief details (contour lines) and forested areas were omitted. Relief details could be replaced by the DTM-200 of the Hungarian Post Office Experimental Institute. The lack of forested areas in the digital map was probably the information that did not reduce considerably the number of users. These paper maps were classified, so the DTA-200 was used only in military environment, and the number of users was quite limited in the beginning.

The total size of the DTA-200 in the Gauss-Krüger projection system was only 7.2 MB, but it was also available in the Unified National Mapping System. The main importance of this work was to provide enough practice and experience to the experts of the military cartography in order to manage a much larger digitizing project.

By the end of 1980s, Geometria created the *National GIS Basic Database (OTAB)*, which was the digital version of the 1:100,000 scale EOTR civil topographic map (all content except the relief). Completing the digital topographic map of the entire country constituted a very important step in the development of GIS in Hungary, since it enabled work on different regional GIS designing and analysing systems. This was also the time of the political changes in Hungary, when a lot of international companies from Western Europe invested here, and most of them were very keen on using GIS.

OTAB has three levels (in fact, they were made in different scales and at different generalization levels):

- detailed level: 1:100,000–1:250,000
- overview level: 1:500,000–1:1,000,000
- presentation level: 1:1,000,000–1:2,000,000

For most of the business users the thematic layers (like transportation, boundaries and hydrology) were most important; it was especially important that Geometria offered update on these thematic layers because the civil state cartography was not prepared for such requests (at least not at that time).

The transition process of Central and Eastern European countries was quite fast especially in the case of the Czech Republic, Hungary and Poland. These countries officially joined NATO in 1999, but due to various partnership programmes these countries were already familiar with the NATO requirements earlier. Hungary offered a former military air base (Taszár) in the Balkan Wars in 1995; this place became the primary staging post for US peacekeeping forces coming and going into the Balkans.

As modern military air navigation is based on digital maps, it was important to the US and NATO forces to have a detailed digital topographic map of Hungary available. When the Cartographic Institute of the Ministry of Defence finished DTA-200 in 1989, they immediately started the preparation of the digitization of the 1:50,000 scale military topographic maps (altogether 319 sheets), the *DTA-50*. The real work started in 1993, when both the hardware options (Laserscan) and the financial background was fixed or at least the project seemed realistic enough to start. The first version of this digital map was nothing else than just a digital version of the paper maps. The 2.0 version of the DTA-50 was released in 1998, which was much closer to a real GIS database. Altogether more than 700 different objects were entered into the database. The most common feature categories were the geodetic points, settlements, facilities, transport, bridges, hydrology, relief, vegetation, boundaries, and scripts.

When the DTA-50 2.0 was released, all updating process of these maps was moved to a complete digital environment. The 1:25,000 scale military topographic maps are still in paper form (more than 1,100 sheets). Although the military cartographers have already enough experience, the process started only in 2004 due to the lack of suitable financial support. This was the time when the military cartography changed to the UTM projection system.

The DTA-200 2.0 was released also in 1998. This was available in the most commonly used file formats (MicroStation/DGN, MapInfo, AutoCad/DWG and DXF). Based on this, the continuously updated databases of DTA-500 and DTA-1000 were also created in 1:500,000 and in 1:1,000,000 scales (Zentai 2012).

Cadastral Maps

The very first IT developments in this sector were started at the *Budapest Geodetic and* Cartographic *Enterprise* (BGTV) in 1974. Although they started the first test of digitizing large scale cadastral maps in 1975, the real digitizing process was started only after 1990 due to the very large number of these sheets, when both the technical and the political circumstances made this process feasible.

The following requirements were taken into consideration:

- the process should be uniform and authoritative ("state sealed");
- it should fulfil the national and European demands concerning the technical, economic, business and legal demands;
- be conform to the national and international standards (precision, data format);
- allow the users the conversion of map data between different projections and legends.

The estimated time of the digitization process (about 60,000 map sheets) was 15 years. In 1996 the national data standards for cadastral maps were established (DAT) and the *National Cadastral Programme* Non-Profit Company (NKP Kht.) was formed. This organization controlled and supervised the process and, with the contribution of the local land offices, they managed the final acceptance. The law on land surveys and cartography stated that these digital cadastral maps should be the base of municipality GIS.

The DAT-Standard gives prescription of the following main groups of information:

- cadastral and real estate data,
- natural and man-made features.

Prescriptions are formulated as adequate to resolution available in the scale range 1:1,000–1:4,000. The data handling unit is the settlement.

The sheets of the residential areas were digitized by 2002. After Hungary joined the European Union, the process of digitizing the cadastral maps of rural areas had to be speeded up. The European Union specified that this database should be ready by the end of 2005.

Other Digital Databases

- Budapest-2000 and Budapest-4000 digital cadastral maps were created in 1993 in 1:2,000 and in 1:4,000 scales both as blocks and line segmented maps (including underground features).
- Gazetteer of the Hungarian Geographic Names (FNT) was created from the large scale topographic maps including 39 different name types including the following names: cities, districts, smaller rural areas, protected areas, hydrographic and relief features, man-made objects. Names are linked to their location by geographic coordinates. The second version of this database is still under construction and will be expanded with the names of the 1:10,000 scale topographic maps.
- One of the most frequently used databases is the administrative borders of Hungary (region, county, municipality, settlement level). This database is available in various scales; the accuracy of the geographic coordinates is 1 m in the largest scale. FNT is part of the spatial data infrastructure of Hungary.
- Street maps of Hungary (DSM) were based on the DTA-50. All relevant map features were replaced by the GPS-based fieldwork-checked versions. The most important element of the DSM is the address (street numbers) and the transportation database, which includes all state and municipal road sections. The database has been updated four times a year.
- In the frame of the European Union integration programme of the Department of Lands and Mapping, different projects were launched by the Institute of Geodesy, Cartography and Remote Sensing in 2000. The archives contain about 7,000 aerial photos at scale 1:30,000, which are available in analogue and digital forms. These photographs have been continuously updated (Winkler 2004).

Legal Issues

A law on land surveys and cartography was issued in 1996. The law set out the duties of the State and regulated land surveys and cartographic activities in order to create a framework for such activities and to provide a professional and efficient service. The responsible bodies, in addition to the Ministries of Agriculture and of Defence, are the district land offices, county land offices, military cartographic institutes and the central land survey organ with national competencies.

This law prescribes to use the state base map for any GI database in Hungary as mandatory. The law divides the state obligation task of the geodetic control networks, of the state surveying base maps and the state topographic maps of 1:10,000 scale to the civilian Mapping Agency, and the state topographic maps of the scales smaller than 1:10,000 to the Hungarian Defence Forces.

One of the most important items of this law is to describe the duties of the state concerning the mapping and the clear definition of the civil and military cartography. Although the responsibilities of these two areas are different, the experts of these areas have similar experience in this area and the law has not prevented the co-operation of cartographers between these two sectors.

The law was re-issued in 2012 aligned to the rapid development of the information technology.

Cartography 2.0: Global Cartographic Services

The user demands for digital cartographic databases completely changed when global and national map services started. The very first global map service was the Xerox MapViewer in 1994, but only very limited and relatively small-scale data were available (continent and country borders and hydrography). In 1996, MapQuest offered street maps and route finding (but the most detailed maps were available only in US areas).

Google Maps, introduced in 2005, transformed the online map into an extension of a search engine thereby making it possible to search for features on the map. This service has the largest effect: since then, most of the similar services are using the same tile-based map data, and the user interface of the global cartographic services resemble to Google Maps/Earth. Image tiling has been used since the early days of the World Wide Web to speed the delivery of graphics. It was a little bit unusual to insist on this technology at that time, but taking into consideration all factors this was the most effective solution (Peterson 2011).

Google Maps is regularly mentioned as one of the first and most prominent appearances of web 2.0 (although the term itself has already been mentioned some years before) together with Wikipedia and social networking sites. An even more characteristic example of web 2.0 and cartography 2.0 is the Open Street Map (OSM) project, which was inspired by the success of Wikipedia. The crowdsourced data of OSM fits very well to the web 2.0 concept.

Web 2.0 is a platform; this is the most common short definition. Web 2.0 is the second phase of development of the web, including its architecture and its applications. The companies that succeed in the Web 2.0 era are those that understand the rules of this platform, rather than trying to go back to the rules of the desktop PC era. Every significant Web 2.0 application to date has been backed by a specialized database.

Web 2.0 is first of all a business and the service providers are looking for more income, and they are not afraid of using new, easily marketable terms, but the term is really short, unusual and understandable for everybody (Zentai 2007).

Most of the national mapping authorities established their on-line map services at the end of the 1990s. Unfortunately, none of the two state cartographic institutes started a nationwide service in Hungary; only partial information was provided (orthophotos), but not free.

The military cartography tested a web service of the DTA-200, called Hunet-200 in 2002. This required a special application (based on GeoMedia) at that time, because general engines were not available for vector map publication. Several layers were provided, but the relief was excluded as the DTA-200 did not include contour lines. The original paper maps had this information, but when these map sheets were digitized, the military cartography decided to omit this content due to the contemporary hardware and software limitations. The development of the test version was financially supported by the government; however, the service was suspended after a few years (before the beginning of the web 2.0 era).

The civil NMA of Hungary offers raster versions of the state topographic maps in 1:10,000, 1:100,000 and in 1:200,000 scales (although the last two scales legally are not state topographic maps). The DITAB-10 V.0 is a real vector-based GIS database (object-oriented database structure) of the 1:10,000 scale topographic maps, but it has two serious drawbacks:

- not freely available,
- very slow update circle.

The innovations of recent years are also characteristic in the development of maps as mass media. Maps on the Internet or those developed for mobile devices reach a further variety of social classes and more than ever can shape the behaviour and attitudes of individuals or communities, as well as opinion former. Additionally, maps can be characterized in these new use environments by their temporary nature; they vary constantly and can be adapted and removed by the users at any time (Faby and Koch 2010).

Conclusion

Maps have always been models of the real world and never simply a reflection of a perceived space. This paper reviewed the digital cartographic/topographic databases of Hungary, concentrating mostly on the 1980–1990 years. When I studied this area, I faced a characteristic problem: the contemporary literature is void and the contemporary digital files have already been lost or they are available in a format which is not known by the recent applications. One of the main reasons of writing this paper was to save this information for future users. I hope that experts of this area will complete this information and users of other countries will write their summary in order to make the transition process understandable.

One of the most interesting parts of this research is to recognise and identify the relationship of the development of information technologies and their effect on the maturation of the digital cartography. The first step of the creation of digital cartographic databases was the implementation of digital technologies (to be familiar with computer hardware and software). The second and most time-consuming process was the digitization of the existing state topographic (and cadastral) maps. From one aspect, it was much easier to manage the digitization of cadastral maps, because these maps are

(continued)

very simple graphic products consisting of lines and points only. From another aspect, the number of cadastral map sheets is so large that this step took quite a long time (depending on the size of the country and the ratio of urban areas, where the cadastral maps have importance). In most countries when these cadastral maps were digitised they were also used to create the first GIS databases.

Topographic maps are much more complicated as graphic products (many different features, symbols, lot of lines and area symbols, most of the line symbols are curves, such as contour lines and water lines, several text elements). This is why digitizing topographic map sheets required much more preparation and experience. Technical development was also necessary, because digitizing tablets were not the most suitable devices on the input side. The on-screen digitizing (using scanners) is a more suitable procedure on the input side, but handling large raster images required more modern computer infrastructure. The first digitised state topographic maps were only the digitized versions of the paper maps, and in the beginning they were not intended to be databases.

The last step of the process was the conversion of the digitized maps into a real GIS database: building topology, linking database elements to map features, connecting different databases, establishing spatial data infrastructure. Meanwhile the updating process of state maps has also changed. The mapping campaigns (surveys) were mostly replaced by continuous updates of the database usually based on regular orthophoto campaigns. The topography as it was known before the digital era has nearly disappeared (the real fieldwork and controlling measurements became obsolete), or at least the NMAs do not want or they are not able to afford this process, but rather offer the users the newest orthophotos and digital terrain models.

Cartography 2.0 can also have its share using web 2.0 features like crowdsourcing (to collect update information provided by the users themselves) and volunteered geographic information. Nevertheless, only few countries of the world can really integrate these opportunities to their spatial data infrastructures. The development is very rapid and the most developed countries will solve this challenge in the near future.

Acknowledgement The present study was sponsored by the Hungarian Scientific Research Fund (OTKA No. K100911).

References

Faby H, Koch A (2010) From maps to neo-cartography. In: 3rd international conference on cartography and GIS, 15–20 June 2010, Nessebar

Koós Á (1996) Digitális terepmodellek a vezeték nélküli összeköttetések tervezésében PKI 1972–1994. Távközlési Könyvkiadó, Budapest

Papp-Váry Á (2010) Két évtizede nem titkosak a térképek. Geodézia és Kartográfia 1:22–25

Peterson M (2011) Twenty years of the world wide web: perspectives on the internet transition in cartography. In: Proceedings of the 25th international cartographic conference, Paris, 3–8 July

Winkler P (2004) The National Orthophoto Program of Hungary completed under strict quality control. In: The international archives of the photogrammetry, remote sensing and spatial information sciences, vol 34, Part XXX

Zentai L (2007) Application of web 2.0 in cartographic education. Is it time for cartography 2.0? In: Proceedings of the 23th international cartographic conference, Moscow, 3–10 August

Zentai L (2012) A digitális térképek Magyarországon az első digitális adatbázisoktól a kilencvenes évek végéig. Remote Sensing Technol GIS Online 1:14–38

Creation of the Accurate Raster Driven Polygonal Environment for the 3D Surface Models Based on the LIDAR Technology

Jan Hovad and Jitka Komarkova

Introduction

Light detection and ranging (LIDAR) is a frequently used technology that provides information about elevation of scanned objects, their position, classification etc. Laser-based ranging instrument measures distance (range). The basic principle of the measurement is simple. The measurement is based on the precise measurement of time. Intense laser pulse is transmitted from instrument to the object; it is reflected from the object and detected by the instrument. Elapsed time period is measured. Range is calculated, based on the knowledge of the speed of light. Alternatively, the laser transmits a continuous beam. In this case, transmitted and received sinusoidal wave patterns are compared to obtain the slat range. LIDAR allows to collect explicit elevation (3D) data at a very high accuracy. Simultaneously, LIDAR scanning produces data in large volumes (Shan and Toth 2008). Authors use raw LIDAR scans as inputs to create final photorealistic 3D polygonal surface models.

Creation of the digital terrain model (DTM) is frequently discussed topic and it can be evaluated in the detail from authors like Kraus and Pfeifer (2001). DTM serves as the ground for the surface objects. Its creation and precision is dependent on the used algorithm. These approaches can be further studied from authors like Hu (2003) and Elmquist (2002). None of the approaches are suitable for the intended result. Authors propose to use slope based polygonal terrain model system.

Vegetation modelling, e.g. placement of trees in the space and identification of their height, belongs to the most important parts of this article. Klimanek (2006), Omasa et al. (2008) deeply dealt with this topic. The main idea is to use the laser

J. Hovad (✉) • J. Komarkova
Faculty of Economics and Administration, University of Pardubice, Studentska 95, 532 10
Pardubice, Czech Republic
e-mail: jan.hovad@upce.cz; jitka.komarkova@upce.cz

© Springer International Publishing Switzerland 2015

J. Brus et al. (eds.), *Modern Trends in Cartography*, Lecture Notes in
Geoinformation and Cartography, DOI 10.1007/978-3-319-07926-4_8

beam which is reflected from the certain height level of an object (i.e. tree) to identify the bottoms and/or tops of the vegetation.

Vegetation analyses and operations based on the Digital Canopy Height Models (DHCM) are the major part of this article. The main aim of the authors is to use LIDAR data to reconstruct large scale areas of vegetation in the 3D by raster driven distribution of surface objects. The output is presented in the photo-realistic model of the surface, which is based on the real-world values derived from the LIDAR data.

To reach the given aim, first the quadrilateral and polygonal model of the terrain is created. Next, filter for the tree extraction from the raw LIDAR point cloud is created using Python. Filtered data representing real trees are then transformed into the 8 or 16 bit black–white raster. This structure is used to place the individual objects to the correct locations in the terrain. The same approach is utilized for the cloud distribution.

A very small abstraction of factors in the recreation of the faint vegetation details in the large sized areas provides the possibility to perform photorealistic spatial, or view-shed based, analyses. Proposed approach of this text works with the idea to derive the most important attributes of small objects into the black–white raster structure. Pixel values are suitable to lead the distribution of 3D polygonal models in the area in the form of proxy objects. Every tree has the appropriate height and also a proximate position.

Authors utilize slope based polygonal terrain model system, which was developed by them to match the resolution of the quadrilateral grid with the terrain adaptively.

The most hardware demanding task is to recreate polygonal models for the vegetation and the environment globally. However, this task is the key issue for preserving the faint details of surface objects. Linking particular digital models together (e.g. terrain, buildings, vegetation, clouds, etc.) and visualizing them in the form of a photorealistic and scientific output (real properties of objects are preserved) is a multidisciplinary approach which involves different scientific fields.

Used Data, Software and Hardware

There were two data inputs available. The *first one* is the detailed surface scan downloaded from the OpenTopography.org, which contains the point clouds of the Pennsylvania forests. Scanned values are in the both cases stored as XYZ coordinates. This data set was used during proposal of the whole procedure of creation of vegetation model (vegetation reconstruction).

The *second input* is the LIDAR scan for the area size roughly 10×20 km situated in the Czech Republic, Europe. Data set is scanned by aircraft Turbolet L-410 FG. Two types of LIDAR models are created as an output. The first type is the Digital Surface Model 1st Generation (DSM 1G), which includes all objects on the surface (terrain + vegetation, buildings...). The second type is the Digital Model of the Relief 5th Generation (DMR 5G) (Belka 2012). This data set is used for the final 3D photorealistic visualization of a larger area of interest.

The infra-red NOAA imagery from its archive is used to illustrate utilization of the proposed procedure in the case of clouds.

Hardware requirements correspond to the big data needs. AMD Opteron and Intel i7 2600K CPU, with minimum of 16 GB memory, fast SATA III SSD (500 MB/s) and GeForce 460 GTX. Some parallelizable tasks can be distributed across the distributed architecture, e.g. Apache® Hadoop®.

Mainly a 64 bit software is used to fully utilize RAM capacity. Fully sufficient for the GIS operations is the ArcGIS 10 SP3 or a free version of the SagaGIS. Advantage of the ArcGIS is that it allows implementing Python made additions directly in the map environment. Graphical tasks are completed in the Autodesk 3D Studio Max 2012, which allows object oriented scripting and processing of wide range of data inputs. Any similar application can be chosen upon the needs of the set analysis.

Methods and Procedures

This chapter characterizes details of the individual steps needed to reach the given set of the aims. The proposed procedure consists of the following steps:

- Creation of the 3D polygonal digital terrain model from LIDAR data
- LIDAR data filtering and trees recognition, the following layers are created:

  - Top cover layer (ground and tops of the trees)
  - Treetops layer
  - Inner vegetation layer

- Raster transformations: creation of black–white rasters from filtered LIDAR data to describe location and height of trees in terrain

  - Raster describing real heights of trees (in the given length unit)
  - Recalculated raster of normalized trees heights
  - Raster describing positions of each tree

- Raster driven digital cloud model: creation of black–white rasters from NOAA images to describe location and size of clouds
- Creation of output 3D photorealistic model of surface

Creation of the 3D Terrain Model

A 3D terrain model (the ground for the vegetation and other surface objects) and its creation is based on the LIDAR scanned values. Raw LIDAR data are transferred into the quadrilateral polygonal form. This form is easy to edit, subdivide and has a nice, clean and predictable edge flow.

Laser scan is carefully checked and corrected—registered, merged, de-noised etc. It is simplified by chosen interpolation technique into the form of quadrilateral set of grids (C++/Python). Each grid has different resolution to lower hardware requirements. Next, the raw LIDAR scan is analysed from the point of view of the slope factor and is used further to classify interpolated grids into groups. Classified grids are processed by 3D scripting language to form 3D polygonal digital terrain model. The procedure was described in detail by Hovad et al. (2013).

Resulting DTM is used in the end of the whole procedure together with other partial models (vegetation and clouds) to form a resulting photo-realistic scientific model of the surface based on the real-world values scanned by LIDAR.

Data Filtering and the Tree Recognition

Authors use simple method to extract the needed information from the LIDAR data. Basic vegetation and ground recognition is based on the scattering or gradual reflection of the LIDAR laser ray from the targeted objects. The ray, which is wholly returned from the first reflection, is classified as a ground. In some cases, the ray is partially reflected and returned more than once. In these occurrences, objects are classified as vegetation because the ray propagates down through the treetops, which increases the reflected counter. The last reflected pulse is with a high probability the ground as well. There are enough points available within the LIDAR data to calculate minimums or maximums for the area of interest to increase precision and exclude those rays that do not reach the ground entirely.

It is possible to filter raw LIDAR data to select only the points that are reflected first and leave behind the rest of the point cloud, which includes points scattered through the treetops. The scattered points represent the *inner layer*. The result of the selection (firstly reflected points) creates a *cover layer*—the top layer describing ground and the highest occurrences of treetops. As it was mentioned before, the irregularly distributed points form an *inner vegetation layer* which is later used to create the base for trees. As the next step, this irregular point cloud is transformed into the point grid by means of the maximum function that aggregates points in the given area together. An appropriate spatial resolution of the raster must be chosen at this moment too. One meter resolution is used in this case. Resulting grid introduces some uncertainty. This uncertainty is a sparse in the given square area of each grid cell. In this case, each tree has a 1 meter spatial tolerance of a measuring accuracy. This spread is sufficient in the case of reconstructing large vegetation areas like forests. Also, it reduces the point cloud and simplifies the next data processing (see Fig. 1a).

The point grid cover layer includes treetops along with terrain. Data have a form of a matrix, which can be filtered by a customized filter to obtain *treetops layer*. Simultaneously, vertical or/and horizontal elevation is computed as a difference between two adjacent points (see Fig. 1b).

The above mentioned filter is quite simple, written by authors in the Python 2.7. The code is object oriented and easy to read because of the Python features. The main idea is to use Hash-table (dictionary—key value pair) to store and process the

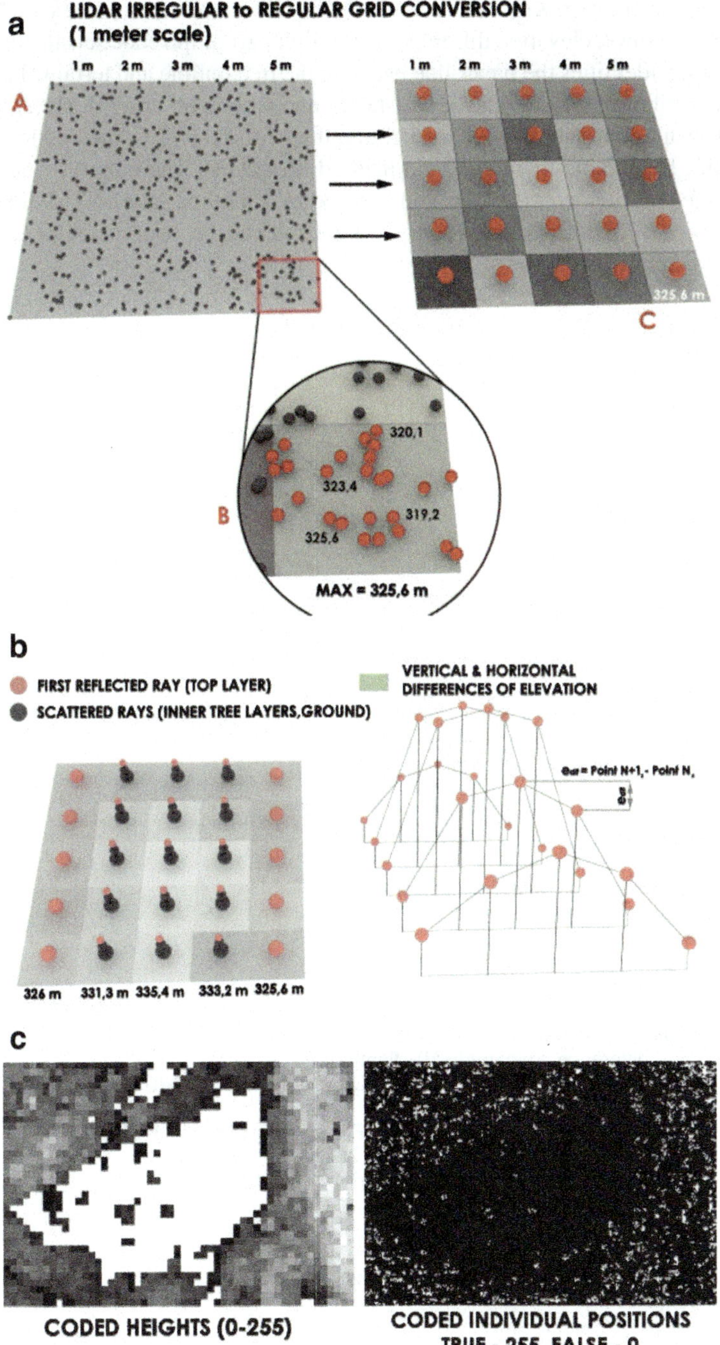

Fig. 1 (**a**) Transformation of irregular LIDAR point cloud into the point grid (spatial resolution: 1 m), (**b**) Cover layer and computation of vertical and horizontal elevation, (**c**) Raster describing normalized heights (on the *left*) and raster describing positions of particular trees—both rasters are used as the masks to control placement and height of proxy shapes

point grid cover layer. Key values are the X/Y (2D) coordinates; values are the lists (array of Z values, elevation difference). The first part of the code sets the threshold value which identifies the break-line between the tree outline and terrain. Tree class has two attributes: a raster_dict (Hash-table) and an output_array (List). A .csv file is used as an input data file and loaded into the Hash-table by a method loadPoints (filePath). Hash-table is filtered by a method filterOut(). The output string is built during the iteration over the keys and is saved by a method saveXYZ(filePath) as a filtered output. This approach is very simple and can be customized or extended. The source code of the script follows.

```python
THRESHOLD = -5.0
DELIMITER = " "

class Trees():
  ## ATTRIBUTES
  raster_dict = {}
  output_array = []
  ## METHODS
  def __init__(self):
    print("Object created\n")

  def loadPoints(self,filePath="D:\\input.csv"):
    f = open(filePath, 'r')
    for line in f:
      line_ar = line.split(";")
      if self.raster_dict.has_key(line_ar[0]) == False:
        self.raster_dict[line_ar[0]] = [[],[]]
      self.raster_dict[line_ar[0]][0].append(line_ar[1])
      self.raster_dict[line_ar[0]][1].append(line_ar[2])
    f.close()

  def filterOut(self):
    c = 0
    for key, value in self.raster_dict.iteritems():
      for index, ar in enumerate(value):
        if index == 1:
          print("VERTICAL LINE: "+str(c))
          c+=1
          i = 0
          while i < len(ar)-1:
            difference = (float(ar[i+1]) - float(ar[i]))
            if difference > abs(THRESHOLD) or difference < THRESHOLD:
              pLine =
str(key)+DELIMITER+str(value[0][i])+DELIMITER+str(ar[i]).rstrip("\n")+DELIMITER+str(round(
difference,3))+"\n"
              self.output_array.append(pLine)
              print pLine
            i+=1

  def saveXYZ(self,fPath = "D:\\filteredXYZ.xyz"):
    f = open(fPath,'w')
    f.writelines(self.output_array)
    f.close()
```

Raster Transformations

Filtered LIDAR point cloud is used again as an input to create black–white rasters which can be used as master objects to lead the object (trees) distribution throughout the terrain. It means one raster will represent height of trees; the second one will describe their location.

The next steps can be done in two slightly different ways—according to the required accuracy. Filtered treetops can be further aggregated into the raster of specified spatial resolution to find out the maximum reflection. Or, in the second case, the points can be used in the form of the raster directly to minimize uncertainty. Following steps work with the average diameter of trees in the given area of interest. Authors extract the maximum value of every raster cell from the minimum value of every corresponding raster cell from the inner layer. Inner layer contains the inner ray reflections and the last one is, with the high probability, the ground. It is enough to find one minimum from 20 points on the given cell to get the ground value. This subtraction provides the *real tree heights* in the given units.

The next step is to recalculate the tree height values into the given bit depth of the black–white raster. This step can be done in the Python or the FreeMat (imwrite method). Authors use 8 bit precision, which increases the rounding of the tree heights. Demonstration is thus much more apparent. Subtracted values are computed and normalized into the interval of 0–255 by Eq. (1).

$$HeightRaster = \frac{TreeHeightMatrix * 255}{Max(TreeHeightMatrix)} \tag{1}$$

This operation provides the final *raster with normalized heights*, which is ready to *control the height* of objects (trees in this case) on the terrain.

The similar operation can be used to create the *raster of planting positions of each tree*. This is only an approximate evaluation. Authors are not able to evaluate the bending or precise profile of individual trees but they take into account that every maximum has a planting position at the same X/Y coordinate (simply said, every tree is straight). This is done by raster division (Eq. 2).

$$PositionRaster = \frac{LowResolutionHeights}{HighResolutionHeights} \tag{2}$$

Pixel values with high deviations after the division are filtered out. This leads to the preserving only the matching pixel pairs.

Position and height rasters (see Fig. 1c) created in the previous steps can be directly used to control the distribution of any object. The distribution can be demonstrated by placing the box shapes as proxy shapes on the top of the terrain model. Proxy shapes are placed to the cells according to the raster of planting positions. Their heights are controlled by the height raster mask. In the next step, the proxy shapes can be replaced by appropriate graphical models of trees. This more complex visualisation is shown in the last part of this chapter.

Raster Driven Digital Cloud Model

There are many more raster datasets publicly available on the Internet. One of them is the NOAA satellite archive. Images are available to be downloaded in the different wavelengths and time intervals. The Infra-Red (IR) images can be used in the same way as the normalized heights and positions of the trees, created in the previous part. In this case, the only difference is, that the distributed objects are not trees but cloud particles. This raster dataset does not provide extended information about the elevation of a cloud layer, density or cloud types. However, it is sufficient enough to display the weather conditions that closely represent the real state in the area of interest.

Firstly, a 3D projected texture map is created. Next, the same texture is used as a master object, which controls the distribution of basic 3D primitives (boxes)—proxy shapes, based on the raster value. This extruded structure works as a volume for the cloud particles. Authors use the built in module inside of the 3D Studio Max 2012 called Particle Flow but any other alternative can be used in the similar way. In ParticleFlow module, the particle is positioned into the chosen object and its normal vector is turned directly into the camera of the spectator. This allows correct Sun ray redirection. The object shape is set to the simple 1 polygon plane with procedural material. The material is composed from the black–white mask and gradient built smoke structure, which gives the good impression of the photorealistic look of the clouds. The test object used to generate the particles is made out of the deformed sphere. This system can be used and implemented along the created vegetation system, 3D terrain and partial outputs of other branches (architecture, engineering). The final composition which includes the cloud model is shown in the following chapter.

Final Composition and Outputs

The proposed solution describes how to create photorealistic 3D polygonal vegetation model based on the raw LIDAR point clouds. The same procedure is demonstrated on creation of digital clouds model. Both of these processes utilize raster encoded values to control the object properties (namely height) and its placement in the given area of interest. The next picture (see Fig. 2) shows the resulting photorealistic outputs. All outputs are completely Computer Generated Imageries (CGI) taken from the random cameras placed in the 3D model. The setup of the model is focused into the parametrical and procedural way, which means, that the model can be immediately changed by a couple of values. All the objects have the coordinate system WGS 1984 preserved and their properties are very close to the real-world values that were acquired during the scan period (8 bit limits). The leading rasters can be recreated anytime in the future, which automatically affects the object redistribution.

Fig. 2 Example of resulting photorealistic 3D model of surface including vegetation

Discussion and Conclusion

This chapter summarizes the results and outlines possible ways forward in the future research.

Discussion

Authors successfully completed all stated goals. However, some tasks can be adjusted and improved. The biggest issue is the implementation of the particle systems. Authors designed a modelling approach to process LIDAR point clouds and transform the LIDAR point clouds into the form of the interpolated, quadrilateral and slope based polygonal planes that can be used for very large areas of interest. Creation of the particle system in the current state is not possible for very large areas (e.g. local regions) in sufficient quality and reasonable time. Another issue is the setup of the cloud base plane elevation. This is currently placed randomly. Authors do not have an access to publicly available source of such detailed meteorological data that could be interpolated in the required quality for the whole area. These facts slightly limit the procedural usage of the particle systems.

Conclusion

Authors describe a newly proposed method, which encapsulates outputs from the GIS analyses, LIDAR data processing, attribute data acquisition and terrain model creation to utilize all these steps in parametrical and procedural raster driven abstraction of the LIDAR point cloud data. The output is in the form of the photorealistic scientific visualisation. The resulting 3D model can be widely used for 3D GIS analyses, visualizations for public contracts, planning and construction of new communications (roads, bridges, etc.),

(continued)

simulations, 3D printing, GPS navigation and generating individual static/ dynamic views from any location. The result connects GIS with other sciences like architecture, agriculture, forestry or road design and can be further used as a GIS server with large dimensioned 3D terrain model.

3D environment creates an imaginary bridge which allows to interconnect GIS outputs with other disciplines like architecture or transportation because of the support of wide-range 3D file formats.

Acknowledgments This work was supported by the project No. CZ.1.07/2.2.00/28.032 Innovation and support of doctoral study program (INDOP), financed from EU and Czech Republic funds.

References

Belka L (2012) Airborne laser scanning and production of the new elevation model in the Czech Republic. Vojenský geografický obzor 55(1):19–25

Elmquist M (2002) Ground surface estimation from airborne laser scanning data with morphological methods. Photogramm Eng Remote Sensing 73(2):175–185

Hovad J et al (2013) Slope based grid creation using interpolation of LIDAR data sets. In: Cordeiro J et al (eds) Proceedings of the 8th international joint conference on software technologies (ICSOFT), Reykjavík, 29–31 July 2013. SciTePress, Porto, pp 227–232

Hu Y (2003) Automated extraction of digital terrain models. Ph.D. thesis, University of Calgary, Calgary

Kraus K, Pfeifer N (2001) Advanced DTM generation from LIDAR data. Int Arch Photogramm Remote Sensing 34(3):23–30

Klimanek M (2006) Optimization of digital terrain model for its application in forestry. J For Sci 52(5):233–241

Omasa K et al (2008) Three-dimensional modeling of an urban park and trees by combined airborne and portable on-ground scanning LIDAR remote sensing. Environ Model Assess 13 (4):473–481

Shan J, Toth CK (2008) Topographic laser ranging and scanning: principles and processing. CRC, Boca Raton

A Framework for Color Design of Digital Maps: An Example of Noise Maps

Beate Weninger

Introduction

Colors have been used in cartography "to label (color as noun), to measure (color as quantity), to represent or imitate reality (color as representation), and to enliven or decorate (color as beauty)" (Tufte 1990, p. 81). In thematic mapping, such as traffic noise maps, color especially is used to represent quantity. Colors are therefore arranged in a color scheme to represent a range of ordered values. Used properly colors then have the power to reveal the structure within the data, but they can also contribute to misinterpretation if used carelessly.

Principles for so-called sequential and diverging color schemes that are used to represent ordered values have been introduced (Harrower and Brewer 2003) and put into practice in *ColorBrewer*, a library for color schemes. *ColorBrewer* laid a theoretical and practical basis for color design in cartography and also indicates the suitability of schemes for users with color vision deficiencies (CVD). Although color design has been addressed in depth and cartographers have described its complexity, examples on the internet show that guidelines are still neglected, especially by untrained map makers. Rainbow color schemes, e.g., are still used in maps and scientific visualizations although their unsuitability was pointed out (Light and Bartlein 2004). Also, the difference between qualitative and sequential schemes has been stressed and the importance of a perceptual order of the color patches has been highlighted (Brewer 1994). The reason why guidelines are ignored is sometimes simply because map-makers, often non-cartographers, do not know *ColorBrewer* or similar libraries and standard software does not give cues to suitable schemes based an data characteristics. However, for some applications,

B. Weninger (✉)
Lab for Geoinformatics and Geovisualization, HafenCity University Hamburg, Hamburg, Germany
e-mail: beate.weninger@hcu-hamburg.de

© Springer International Publishing Switzerland 2015
J. Brus et al. (eds.), *Modern Trends in Cartography*, Lecture Notes in Geoinformation and Cartography, DOI 10.1007/978-3-319-07926-4_9

like noise mapping, there might not be a suitable, ready-made color scheme. Reasons for this are, e.g.:

- Ready-made palettes are finite (Wijffelaars et al. 2008) and color specifications in *ColorBrewer* should never be treated as ironclad guarantees since color reproduction, whether on screen or in print, is an inexact science (Harrower and Brewer 2003, p. 33).
- *ColorBrewer*, e.g., offers sequential schemes for up to nine classes, but due to a low contrast caused by the high number of groups some colors are quite hard to distinguish, which is especially a problem if the enabling of a correct identification of classes is the objective.
- Nature of data varies strongly for different applications (Harrower and Brewer 2003) and can be ordered (ordinal, interval, ratio scale), with a neutral point, or qualitatively (nominal). Transformed scales, as in our use case logarithmic scales, are also ordered, but it should be considered that, in this case, arithmetic operations are not valid and that higher values contribute more to a mean value. To our knowledge, no guidelines or recommendations are available how this characteristic can be reflected by color schemes.
- Spatial distribution of data can be regular vs. irregular, ordered vs. unordered, or coarse vs. finely detailed. This can lead to different perceptual effects: "The more complex the spatial pattern of the maps, the harder it will be to distinguish slightly different colors" (Harrower and Brewer 2003, p. 32). Schemes with lightness steps only, hence, are not appropriate for more than five or six classes.

With the increasing number of devices, and therefore variety of screen sizes, resolutions, use cases and contexts, color design becomes even more complex. The major challenge of color design for online maps is that output devices present colors differently. Consequently colors have to be optimized to suit a variety of devices, such as computer screens, tablets and smart phones.

Due to this high number of criteria influencing the suitability of schemes, guidelines are highly dependent on use cases, user tasks, respective data and aims that are achieved. Starting with fundamentals, such as color perception and scheme types we thus introduce a framework that concentrates guidelines and combines them with criteria that facilitate the decision on a suitable color scheme. As a proof of concept, we give an outline of the guideline's relevance for the design of noise maps.

Fundamentals of Color Design

Color Perception

How color is perceived depends on many things: "All colors, whether they are seen as direct or reflected light, are unstable. Every change in light or medium has the potential to change the way a color is perceived. [...] Not only are colors themselves unstable, ideas about colors are unstable as well" (Holtzschue 2011, p. 11). Not only physiology, also psychology and cultural background effect how people see and interpret color. The subsequent paragraphs outline the most important principles.

Color Contrast

Color contrast is a result of the perceived difference of at least two colors. The *World Wide Web Consortium* (W3C) defined straightforward recommendations on color contrast in their "Web Content Accessibility Guidelines". Comparable rules are not available for maps because they do not just consist of background and text that have to be balanced. It is necessary to consider the different kinds of contrast.

Classical contrasts are the ones between two colors: light and dark, warm and cool or complementary colors (red/green, yellow/purple, blue/orange). Special contrasts are quality and quantity contrast (Itten 1987). The first occurs when colors of different saturation or purity are combined. The latter describes the effect when contrastive saturations are used for objects of different sizes. Itten (1987) suggests to use colors only in harmonic proportions otherwise it results in an *effect of depth* and warm or saturated colors appear further in the front than cool or unsaturated colors. Especially saturated yellow, orange and red have the characteristic to be salient, therefore they should only be used for values that are meant to be highlighted and in lesser amounts in contrast to blue and green. Figure 1a illustrates, how yellow and orange stand out in contrast to other colors in the map. Hence "the intensity of color which should be used is dependent on the area that that color is to occupy" (Ihaka 2003).

The described contrasts together cause simultaneous-contrast that effects the appearance of colors in dependence on adjacent and background colors. Light colors surrounded by light colors look darker than surrounded by dark colors. An object surrounded by an unsaturated or complementary color appears more saturated than on saturated background (Lübbe 2012). This effect is hard to avoid and predict because the spatial patterns in maps differ significantly.

Effects on Perceived Object Size

Schumann and Müller (2000) describe that the perceived purity of a color is dependent on the object size: A small colored area on black background is perceived more saturated than a bigger area of the same color on black background. In contrast, the color of an object has effects on the perceived object size too because lighter colors stimulate retinal cells more due to higher light reflection (Schumann and Müller 2000, p. 85). Consequently colors can facilitate misinterpretation if the area size needs to be estimated for interpretation.

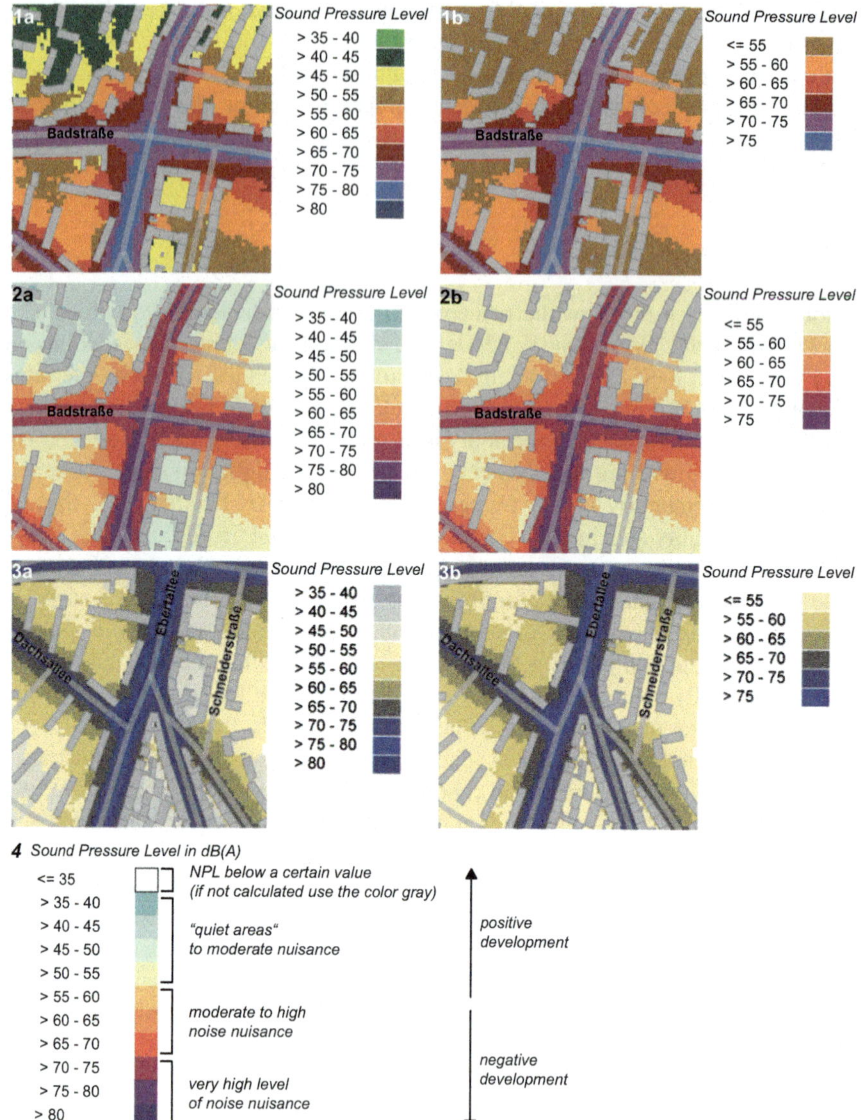

Fig. 1 A map in the color scheme according to DIN 18005-2 with ten classes (**1a**) and six classes compliant to German law (**1b**); the map in the new scheme with ten (**2a**) and six classes (**2b**) and as people with CVD (protanopia) see the new colors (**3a, b**). Please consider that the scheme has *only* been optimized for digital use. A color illustration and color codes can be found on the website www.coloringnoise.com

Color: Highly Associative and Emotional

Color is generally connected with mood and emotions, especially hue has strong effects on associations. This is because experiences of everyday life are connected to sensory impressions. A popular example are the connotations of red and green: While red is used to issue a warning, green is assigned a positive semantic (Fischer 2012). Blue and yellow do not have any comparable effects and are less stimulating than red and green (Goldstein et al. 2008). A study on the effects of color schemes on the estimation of nuisance in maps did not show any effects of blue and green schemes in contrast to schemes in warm colors (Weninger 2013). Only vivid schemes with hue-steps lead to higher estimations.

For online maps, as they are accessible from all over the world, it is advisable to consider the cultural aspect of color too. The color for mourning, for example, is black in European countries and white in India.

In psychology effects of color have been studied in depth, nevertheless the consideration of the emotional and affective components in cartography is a recent research area. The *Society of British Cartographers* highlighted the importance of emotion in one of five principles for good map design as early as 1996: "Engage the emotion to engage the understanding" (Jones 2010, p. 150). Griffin and McQuoid (2012) and Klettner et al. (2013) have elaborated on emotion and Fabrikant et al. (2012) present a framework to study emotional responses on map design, under consideration of different color schemes. In the proposed framework associations, connotations, and emotions are seen as crucial aspects of sound map design.

(Situational) Color Vision Deficiencies

Color vision deficiencies (CVDs) and their practical consequences are well known to the scientific cartographic community. As indicated in *ColorBrewer* not all schemes are suitable for people with color vision deficiencies. Programs such as *Adobe Illustrator*, *Photoshop*, and *Color Oracle* (Jenny and Kelso 2007) have a special proof mode. Nevertheless practical examples show that it is still not general knowledge that especially red and green should not be used in combination because 4 % of users have difficulties to distinguish these two colors or are not able to do so at all. Men are much more likely to be affected because of heredity, 8 % have to live with it (Lübbe 2012).

Strictly speaking CVDs are not a perceptual issue but a physiological one, because they are caused by a misfunction of one cone-type and therefore one of the three primary colors—red, green, or blue—cannot be perceived leading to protanopia, deuteranopia, or tritanopia. However, Schumann and Müller (2000, p. 97) claims that deviations from "normal" color vision are quite common and lead to problems distinguishing colors. Besides people that have been diagnosed with CVD, inherited or caused by a disease, color vision deficiencies can also occur if cone photolabile pigments show anomalies which results in a deformed spectral sensitivity curve (Welsch and Liebmann 2012). Affected people therefore perceive colors slightly differently. Additionally, CVDs appear more often during screen handling for a small field of vision (FOV) for up to 2° even though persons concerned do not have any CVD with a FOV around 10° (Schumann and Müller 2000). Results of an experiment showed that 23 % percent of participants were not able to identify numbers on two Ishihara plates while only 13 % of the participants indicated that they in fact had some type of CVD (Weninger 2013). We explain this as a result of *situational* CVD caused by different devices and thus various patterns of use, screen contrast, visual angles and FOV. The issue of *situational* color vision deficiencies and its consequences on map design, according to our best knowledge, has not been studied in cartography, but suggests important insights to support web mapping for inclusion.

Scheme Types

Brewer (1994) identifies two kinds of schemes besides qualitative schemes: sequential and diverging. Diverging schemes have a positive and a negative end which are divided by a neutral point. Sequential schemes can represent ordinal or numeric data in a perceptual order and consist of one or more hues in different lightness steps. If multiple hues are used in combination it has to be assured, preferably by means of a change in lightness *and* saturation, that they can be brought in a perceptual order. Multi-hue schemes with at least one hue-transition are especially reasonable for more than five to seven classes.

As mentioned before, color is highly associative. Therefore the specification of different kinds of scheme types is important to find a suitable scheme for a use case. We therefore suggest a further subdivision of sequential schemes on the basis of some of Imhof's scheme types (Imhof 1965). This characterization is an important basis to decide on the level of abstraction that is described as one decision criteria (see section "Criteria for Decision Support"). His research into relief presentation involved a comprehensive analysis of color schemes as well. Relief presentation as the representation of classified heights is comparable to the presentation of other thematic contents, like physical distributions and phenomena that are usually represented with a low level of abstraction. Amongst other types Imhof defined the following ones:

- **Contrasting schemes**: Contrasting colors are supposed to enhance distinguishability. The amount of contrast of single-hue schemes is defined by the lightness of the first and darkness of the last class in connection with the number of classes. Many single-hue schemes therefore do not contrast much. In addition, the amount of contrast of multi-hue schemes is determined by factors described in section "Color Contrast". It should be adjusted in line with the color design guidelines and decision criteria, that are illustrated in the subsequent section.

- **"The more of a phenomenon, the darker the color"**: This type is based on Imhof's "the higher, the lighter" example, resulting in a presentation that appears three-dimensional. In thematic cartography this is usually inverted.
- **Modified spectral schemes:** Spectral schemes are popular in visualization—61 % of the papers of the IEEE Visualization Conference from 2001 to 2005 without medical figures made use of the spectral scheme (Borland and Taylor 2007)—nevertheless they are to challenge:

Perceptually [...] this scale does not appear linear. Equal steps in the scale do not correspond to equal steps in color, but look instead like fuzzy bands of color varying in hue, brightness and saturation. When mapped onto scalar data, this colormap readily gives the user the erroneous impression that the data are organized into discrete regions, each represented by one of the rainbow colors. This can lead the user to infer structure which is not present in the data [...] (Bergman et al. 2006, p. 119).

If only parts of the spectral scheme are used and modified—for example saturation and lightness balanced—it can be an interesting option for some use cases with a big data range or a logarithmic scale, as our practical experience showed.

Color Design Aspects for Digital Maps

Color Design Guidelines

Levkowitz (1996, p. 97) defines general guidelines for color design: Colors of a scheme should:

1. "preserve the order of the original values";
2. "convey uniformity among values they are representing, and representative distances between them"; and
3. "create no artificial boundaries that do not exist in the original data"

The first rule is established in cartography and basis for sequential and diverging schemes. The other two rules are reasonable, but it is difficult to put them into practice, especially for non-cartographers. Especially representative distances is an aspect that has not been further defined in cartography, therefore it is the aim to contribute to this discussion (refer to guideline 2). Moreover, Levkowitz's guidelines focus on the data and neglect the complexity of perceptual and user issues that go along with color. Color is a physical stimulus that causes physiological as well as psychological reactions. In respect thereof we describe design guidelines that help to develop effective color schemes:

1. To represent partially or fully ordered data colors should build a **harmonical hierarchy** (Jones 2010, p. 46), i.e., lightness *and* saturation should in- or decrease systematically at one end of the scale. Harmonical hierarchy is achieved in single-hue sequential schemes, or multi-hue sequential schemes

that consist of hues that have a perceivable order, like yellow to dark–red, yellow to purple, or yellow to dark–blue. The addition "harmonical" refers to the necessity to balance the scheme considering the saturation and object size (cf. section "Color Contrast").

2. **Match values or class distances and color distances**: The distance of elements or represented classes is next to ordering (see above) and continuity type one data characteristic (Andrienko and Andrienko 2006). Class ranges can be equidistant, i.e., all classes have the same size, or they can be different according to quantiles or natural breaks. Perceptual uniformity is often described as an aim within cartographic science. In color systems like Munsell (1905) or CIE L*a*b the Euclidean distance between any two colors fits with the perceptual distance. For equidistant classes this approach is appropriate, but not for the representation of classes with different ranges or logarithmic scales. Logarithmic data is special, it appears to be equidistant although classes of the antilogarithms would not be. Equidistant schemes could give a wrong impression and classes with higher values would be underrepresented. This has to be considered in the representation. At this point, however, we cannot give any straightforward recommendations about color distances.

3. **Consistency of colors** on a variety of output devices is a major aim. It is the aim to use colors that can be recognized regardless of object size, adjacent colors, or device in use and enable a correct assignment of the object colors to the color patches in the legend. However, we have described many effects that counteract this objective (cf. section "Color Perception"), therefore we recommend the use of different hues in a harmonical hierarchy and fewer steps in lightness per hue—in contrast to single-hue palettes with a high number of classes—to support the recognition of colors.

4. **Avoid colors that are not suitable for (situational) color vision deficiencies.** To consider (situational) CVDs is crucial to facilitate the accessibility of digital maps (cf. section "(Situational) Color Vision Deficiencies"). Already small improvements of the scheme, without completely changing it, turned out to have positive effects (Kröger et al. 2013). Also a reduction of the amount of classes and the use of colors with strong contrast can work against CVD.

Criteria for Decision Support

Each use case is different and therefore a consideration of the specific objective, data characteristics and context is needed. Consequently we propose criteria that support the decision for or development of a color scheme and help to set priorities. Within the development process several criteria can be considered, but there need to be priorities. Although the underlying color design guidelines are the same for each use case, the aim to achieve a high distinguishability leads to other recommended color schemes than the aim to highlight patterns. At this point we cannot present a complete list, but an initial concept.

The basic question for color design—and of course map design in general—is: *What is the aim of the map-maker? What aspect(s) shall be highlighted in compliance with the aim? And what color choice is evoked by the character of the phenomenon that is presented?* Options are:

1. **The user task:** The tasks users are to complete by means of a map are a crucial decision criterion for map design. Andrienko and Andrienko (2006) give a comprehensive overview of tasks for exploratory data analyses, such as:

    a. In **lookup tasks** users are searching for an individual value in the map, such as a sound pressure level of 65 dB(A), or looking up the value of an attribute in a defined place. To complete these tasks it has to be ensured that colors can be distinguished.

    b. For **pattern identification** patterns in the data based on data-interpretation according to the objective are highlighted:

        i. Highlight a critical value, e.g., by means of a contrast-increase or a hue-step. Resulting schemes are sometimes similar to diverging schemes which makes the exceedance of a value striking (see section "Color Contrast").

        ii. Induce a three-dimensional effect by highlighting, e.g., one end of the scheme by means of color effects (see section "Color Contrast").

2. **The level of abstraction:** This facet determines if the color scheme is consistent with the natural color of the occurrence or an associated color, in line with the characteristics of the sensory perception or if it is an abstract choice. We distinguish:

    a. *Realities and physical distributions and phenomena*: e.g., altitude levels, or population density. Because these phenomena have some kind of visual appearance the representation should be less abstract. Representations are more intuitive if an association between their visual appearance and the representation is facilitated. Therefore, e.g., high densities are usually presented by means of darker color, giving the intuitive impression of *more* color meaning *more* subjects or objects per area.

    b. *Phenomena that have a physical, but not visible presence* in space and are measurable and perceivable, such as sound or air pollution. In this case the association between the phenomenon and the representation is not based on their visible appearance, but on other characteristics based on the user's sensory perception, e.g., annoyance for environmental noise. Consequences for color design in such use cases have been described in section "Color: Highly Associative and Emotional".

    c. *Data that do not have any physical presence* in space, e.g., inflation, national debt, energy consumption, age of the population etc. The representation of these examples is most abstract, they do not cause any sensory perception, unless visualized. Hence the choice of a hue does not have to be intuitive, but it will be helpful to consider certain color connotations.

3. **Engage the emotion:** This objective is the most vague of the mentioned herein. Latest research in cartography argues that the way data is presented effects the emotions of users, such as trust (Skarlatidou et al. 2011). In usability engineering user satisfaction is also considered as one of the three parameters according to ISO 4291-11. We regard this facet as crucial in cartography for the following reasons:

   a. to foster positive effects on look and feel; or
   b. to raise attention and increase the recognition value; or
   c. to foster trust in the information; or
   d. to make it persuasive.

However, at this point we will not present straightforward rules as to emotion can be engaged by means of color, besides the aspects of association and connotation (cf. section "Color: Highly Associative and Emotional"). Further research is needed to evaluate if the results of studies on psychology, online marketing etc. are applicable to map reading and data interpretation as well. Nevertheless we would like to stress the importance of this aspect and that it is worthwhile to spend time on the choice or development of a color scheme.

Applicability to Noise Mapping

Strategic Noise Maps According the Environmental Noise Directive

Noise mapping has been obligatory in Europe since 2002 when the European Union adopted the *Environmental Noise Directive* (END). Noise maps have to be drawn up every 5 years by the member states for major roads, railways, airports, and agglomerations. Color represents the sound pressure level in the logarithmic scale dB(A) as equal loudness contours (also isophones) in 5 dB classes. Although these maps are the basis for informing the public [Directive 2002/49/EC, art. 9(2)] examples throughout Europe lack appropriate and satisfactory cartographic presentation. Especially the color scheme is subject of debate (Alberts and Alferez 2001; Schiewe and Weninger 2013) [Fig. 1 (1a, b)]. It is based on the ISO 1996: 2 of 1987. Although the scheme was left out when the ISO standard was revised in 2007 it is still defined in the German Industrial Standard DIN 18005-2 and used in countries such as Austria, Denmark, France, or Italy (Alberts and Alferez 2001). The problem of the scheme is that the colors cannot be put into a perceptual order because the seven hues vary strongly in lightness and can therefore not be intuitively assigned to dB(A) values. Also the signal color red and saturated yellow and orange are used in the middle of the scheme which overemphasizes these values in contrast to the very high values of the range. Moreover, the scheme includes red and green and is therefore not suitable for people with CVD.

Requirements for the Color Scheme

We recommend strongly to define requirements before the development of a scheme. These should consider the color design guidelines specified in section "Color Design Guidelines" and should give an indication of the criteria for decision support (section "Criteria for Decision Support"). In our use case described herein, prerequisites were the class range and the number of classes determined by EU and national law.

Alberts and Alferez (2001) suggest the following considerations for the END-scheme: The scheme should cover a wide range of noise bands from 40 to more than 80 dB(A); noise bands below 50 dB should be in green colors; the bands over 65 dB(A) in red; it should be suitable for different noise indicators like L_{day}, $L_{evening}$ and L_{max} and noise bands of 5 or 10 dB; colors should differ sufficiently; rather topographic objects should be transparent instead of noise bands; color codes such as HEX and RGB should be provided.

We agree with the considerations above, but we add more detailed requirements based on our color design guidelines. Thus the colors of the scheme should be:

- **Distinguishable** for people with CVD as well as when used for areas of different size, in different scales and a variety of screens;
- **consistent** and therefore facilitate a matching of colors used in the map with colors of the map legend;
- **logically assignable to the characteristics of the noise data**; i.e., that presented noise levels should not be under- or overestimated, hotspots and silent areas should be determined by the users without referring to the map legend and colors should facilitate an association with the categories of noise levels.

According to our criteria for decision making the priorities are therefore to highlight a critical value (1.b.i) and to induce a three-dimensional effect (1.b.ii). The level of abstraction is 2.b., hence the colors should be in association with the annoyance levels.

Designing a Color Scheme

The design was carried out iteratively in three steps. Several empirical user studies have been conducted throughout the development process. The first version of the scheme was designed according to cartographic standards as a sequential scheme with two hue-steps (yellow/orange, orange/red), but it did not satisfy noise experts because the different shades could not be distinguished in cases where ten classes were visualized, which is done frequently in Germany. After a discussion with experts from the *German Standardization Organization* requirements were defined not only with a focus on cartographic rules but also with focus on the specific use case and practical requirements. In the second design, therefore, more hue-steps were introduced to support distinguishability and the association of the colors with the annoyance levels. An experiment, including the developed scheme, proved the hypothesis that class and color distances should be matched (cf. guideline 2) (Weninger 2013). Especially Brewer schemes in orange, green, blue, and red, that appear approximately uniform, lead to an underestimation of the sound pressure level, in contrast to schemes with hue-steps. Consequently, we recommend the use of schemes with hue-steps. Moreover, further discussions of the results showed light colors, representing low noise levels, were too saturated and, therefore, too salient. In the third design [Fig. 1 (2a, b) and (4)] a special focus was put on the gradation of saturation. To achieve that areas representing high values are more salient they have to be more saturated than the other colors. The adjustment of saturation was achieved by re-designing the scheme in the Munsell color space, which defines colors by the parameters hue, saturation and lightness (Munsell 1905). This way an increase of saturation and a decrease of lightness for higher values can be facilitated which results in a systematic color order and thus hierarchy. It also helps to induce a three-dimensional effect to highlight hotspots. The five hues—blue–green, yellow, orange, red, purple—are used, which results in a maximum of three lightness steps per hue. The hue-steps in combination with high contrasts are meant to represent a big data range to indicate the logarithmic scale. They also enhance differentiation and thus counteract effects of simultaneous contrast. Each hue stands for a certain level of noise nuisance to support recognition of levels. In the style of a traffic light scheme, which is seen clearly to communicate the exceedance of a value, a diverging scale was implied, using light yellow as a neutral point between moderate nuisance and higher levels of nuisance. By means of the diverging scheme the positive and negative extreme values are highlighted. A bluish green was used to help people with CVD to discriminate green and red.

Conclusion and Outlook

We gave an overview of the manifold perceptional and psychological effects on colordesign, such as color contrast, effect of depth, effects on the perceived object size, the emotional and associative aspects of color, and (situational) color vision deficiencies. While in recent cartographic work often the technical aspect of color design, for example, automatically generating palettes are the focus (Steinrücken and Plümer 2013) we emphasized perceptional issues in the development of guidelines and decision criteria that especially consider digital maps. In particular different kinds of color contrasts and the emotional aspect have been covered in more detailed.

Our hypothesis is that a big variety of output devices and manifold patterns of use, which are characterized by different lightning conditions, screen sizes and visual angles, effect color perception and thus color design. One of the major aspects here is the consideration of color vision deficiencies, both inherent and situational, in the design of a color scheme.

(continued)

Due to the high complexity and interaction of effects it is almost impossible to come up with straightforward rules or a set of color schemes that fits every use case. Thus, we state that the design of color schemes cannot solely be done on the basis of four design guidelines, but this has to be supported by the use of decision criteria. They are necessary to further define the use case specific objective of the map, user tasks and to adjust the scheme to the respective requirements.

The guidelines for color design—(1) harmonical hierarchy, (2) match value or class distances and color distances, (3) consistency of colors, and (4) avoid colors that are not suitable for (situational) color vision deficiencies—are based on research in color science and cartography. The design of a color scheme to represent sound pressure level in maps showed its relevance and proved there was need to come up with decision criteria to choose a final scheme. The decision criteria are in a preliminary state and have to be refined by considering a variety of use cases. Of particular importance is the consideration of emotional responses and the effects on trust; initial foundations for this have been laid by Skarlatidou et al. (2011) and Fabrikant et al. (2012). Additional research on color distance and the combination of effects of hue, lightness and saturation on the association of colors with respective data values and the interpretation of maps is needed to establish clearer rules. Results of the latest user study on the association of the proposed color scheme showed the essential role of connotations with hue and the amount of red/blue that is decisive if a color is interpreted being warm or cool. Results also gave an indication of the perception of color distances, but systematic experiments are needed to analyze patterns.

As soon as the guidelines and decision criteria have been refined by adapting them to other use cases and considering existing factors for color design as described in Slocum et al. (2009), a comprehensive framework can be presented. It will not only list several aspects, but it will also connect them with straightforward recommendations and illustrations in a decision tree. Then this framework can supplement prevalent mapping tools like *ColorBrewer*.

References

Alberts W, Alferez J R (2001) The use of colours in END noise mapping for major roads. In: Proceedings of Euronoise, Prague 2012

Andrienko N, Andrienko G (2006) Exploratory analysis of spatial and temporal data: a systematic approach. Springer, Berlin, Heidelberg, New York

Bergman LD, Rogowitz BE, Treinish L (2006) A rule-based tool for assisting colormap selection. In: Proceedings of the IEEE visualization conference, pp 118–125

Borland D, Taylor RM (2007) Rainbow color map (still) considered harmful. Comput Graph Appl 27:14–17

Brewer CA (1994) Color use guidelines for mapping and visualization. In: MacEachren AM, Taylor DRF (eds) Visualization in modern cartography. Elsevier Science, Tarrytown

Fabrikant SI, Christophe S, Papastefanou G, Maggi S (2012) Emotional response to map design aesthetics. In Proceedings of GIScience 2012, seventh international conference on geographic information science

Fischer S (2012) Ursprung der emotionalen Semantik von kongruenten Farben und Tönen, Assoziationen und Emotionen Erwachsener sowie Blickpräferenzen in der frühen Kindheit. Ph.D. thesis Universität Bielefeld, Bielefeld

Goldstein EB, Irtel H, Plata G (2008) Wahrnehmungspsychologie: Der Grundkurs. Springer, Berlin, Heidelberg, New York

Griffin AI, McQuoid J (2012) At the intersection of maps and emotion: the challenge of spatially representing experience. Kartographische Nachrichten 6:291–299

Harrower M, Brewer CA (2003) ColorBrewer.org: an online tool for selecting colour schemes for maps. Cartogr J 1:27–37

Holtzschue L (2011) Understanding color: an introduction for designers. Wiley, Hoboken

Ihaka R (2003) Colour for presentation graphics. In: Hornik K, Leisch F, Zeileis A (eds) Proceedings of the 3rd international workshop on distributed statistical computing (DSC)

Imhof E (1965) Kartographische Geländedarstellung. De Gruyter, Berlin

Itten J (1987) Kunst der Farbe. Ravensburger Buchverlag

Jenny B, Kelso NV (2007) Color design for the color vision impaired. Cartogr Perspect 58:61–67

Jones CE (2010) Practical cartography. In: Haklay M (ed) Interacting with geospatial technologies. Wiley, Chichester

Klettner S, Huang H, Schmidt M, Gartner G (2013) Crowdsourcing affective responses to space. Kartographische Nachrichten 2(3):66–73

Kröger J, Schiewe J, Weninger B (2013) Analysis and improvement of the OpenStreetMap street color scheme for users with color vision deficiencies. In Proceedings of the 26th international cartographic conference, Dresden

Levkowitz H (1996) Perceptual steps along color scales. Int J Imaging Syst Technol 7:97–101

Light A, Bartlein PJ (2004) The end of the rainbow? Color schemes for improved data graphics. EOS Trans Am Geophys Union 85:385–391

Lübbe E (2012) Farbempfindung, Farbbeschreibung und Farbmessung: Eine Formel für die Farbsättigung. Springer, Berlin, Heidelberg, New York

Munsell AH (1905) A color notation. G. H. Ellis Company

Schiewe J, Weninger B (2013) Visual encoding of acoustic parameters framework and application to noise mapping. Cartogr J 50(4):332–344

Schumann H, Müller W (2000) Visualisierung: Grundlagen und Allgemeine Methoden. Springer, Berlin, Heidelberg

Skarlatidou A, Wardlaw J, Haklay M, Cheng T (2011) Understanding the influence of specific Web GIS attributes in the formation of non-experts trust perceptions. In: Ruas A (ed) Lecture Notes in Geoinformation and Cartography. Springer, Berlin Heidelberg

Slocum TA, McMaster RB, Kessler FC, Howard HH (2009) Thematic cartography and geovisualization. Pearson, Upper Saddle River

Steinrücken J, Plümer L (2013) Identification of optimal colours for maps from the web. Cartogr J 50:19–32

Tufte ER (1990) Envisioning information. Graphics Press, Michigan

Welsch N, Liebmann CC (2012) Farben: Natur, Technik, Kunst. Spektrum Akademischer Verlag, Heidelberg

Weninger B (2013) The effects of colour on the interpretation of official noise maps. In Proceedings of the 26th international cartographic conference, Dresden

Wijffelaars M, Vliegen R, Van Wijk JJ, Van Der Linden EJ (2008) Generating color palettes using intuitive parameters. Comput Graph Forum 3:743–750

Digital Aeronautical Charts: Survey of 64 Czech Air Force Pilots

Jakub Vilser, Jana Merickova, and Scott Bell

Introduction

An air force is an essential component of a modern army. Like all branches of the military, in order to effectively carry out its tasks there must be an excellent support. Production of high quality cartographic material for the air force is one such essential task; such maps are normally provided by a geographic service. Like other branches, work of this service over time undergoes changes that result from new policies and requirements and changing technology. In this research, we focused on the Czech Army and its Air Force. We conducted a survey of **Czech Air Force (CZAF)** pilots regarding their experience with the digital geographic data. The Czech Army **Military Geography and Hydrometeorology Office in Dobruška (MGHO)** provides NATO certified aeronautical charts for the CZAF aircraft both in paper and digital form. Due to the demanding nature of this job it is essential to follow requirements and needs of the target users. A user and user aspects of the use represent the motivation for this research. Functional map research is based on the clear assumption that a map is made for a particular purpose. It is useful and important to find out whether and to what extent, a particular map meets that purpose (Elzakker and Wealands 2007). In our case, this means verifying the functionality of both digital and paper geographical

J. Vilser (✉)
Department of Aircraft and Rocket Technology, University of Defence, Brno, Czech Republic
e-mail: jakubvilser@gmail.com

J. Merickova
Department of Geoinformatics, Palacký University, Olomouc, Czech Republic
e-mail: merickovajana@gmail.com

S. Bell
Department of Geography and Planning, University of Saskatchewan, Saskatoon, Canada
e-mail: scott.bell@usask.ca

© Springer International Publishing Switzerland 2015
J. Brus et al. (eds.), *Modern Trends in Cartography*, Lecture Notes in
Geoinformation and Cartography, DOI 10.1007/978-3-319-07926-4_10

products for the CZAF. In the research we have also focused on the applicability and use of a smart legend for the CZAF.

Background and Relevance

Following are described and explained concepts and contexts used in the questionnaire and its evaluation.

Geographic Service of the Czech Army

The Geographic Service of the Czech Army is a part of the MGHO Dobruška. The origin of the Service is associated with the creation of the independent Czechoslovak Republic. In 1918, a department within the Ministry of National Defence was created. The department was responsible for topographic services of national defence. Thanks to its long history, the Service has much experience and expertise. It is also a member of the NATO's Partnership for Peace program. Among other things, the JOG map issued by the Geographic Service won the Map of the Year award in 2002.

The Geographic Service provides high quality NATO certified aeronautical charts for the Czech Air Force. Maps are updated annually and on demand. The aeronautical charts issued and distributed by the MGHO are:

JOINT OPERATIONS GRAPHIC 1:250,000 SERIES 1501—AIR (JOG 1:250,000)
The map is designed for unified planning and control of the joint land and air operations of the NATO forces.

TRANSIT FLYING CHART (LOW LEVEL) 1:250,000 (TFC 250,000)
The map is intended for flight planning, air operations control, and crew pre-flight preparation. It allows radio and comparative navigation during the flight.

LOW FLYING CHART CZE 1:500,000 (LFC 1:500,000)
The map is intended for flight planning, air operations control, and crew pre-flight preparation. It allows radio and comparative navigation during the flight. It provides the information necessary to navigate and maintain a safe course while flying at low and medium altitudes.

AERONAUTICAL CHART CZECH REPUBLIC 1:500,000 (LOM ČR 1:500,000)
The map is intended as a basis for flight planning and air traffic control over the Czech Republic territory, radio, and comparative navigation, as well as orientation in challenging meteorological conditions and when flying at supersonic speeds.

TACTICAL PILOTAGE CHART 1:500,000 (TPC 1:500,000)

The map is intended for flight planning, air traffic control, crew pre-flight preparation, radio, and comparative navigation during the flight. It provides information needed for fast visual and radar navigation and for maintaining a safe course while flying at low and medium altitudes.

OPERATIONAL NAVIGATION CHART 1:1,000,000 (ONC 1:1,000,000)

The map is intended for flight planning and air traffic control, as well as crew preflight preparation, comparative, and radio navigation during the flight. It provides information needed for fast visual and radar navigation and for maintaining a safe course while flying at low and medium altitudes.

In addition, the MGHO distributes other map data such as DTED,[1] RETMs[2] and more.

Czech Air Force (CZAF)

The CZAF is an important component of the Czech Army; as such, it is necessary to meet the political-military ambitions of the Czech Republic. The main task of the CZAF is to defend the airspace of the state as a part of the NATO Integrated Air and Missile Defense (NATINAMDS). The tasks in peacetime include search and rescue service (SAR), air ambulance, transport of government officials, and tasks of the Integrated Rescue System (IRS). In the case of armed conflict, the CZAF is designed to gain air superiority, close air support of ground forces, reconnaissance, and providing air transport and supply troops. The CZAF is also deployed in international operations, for example, protecting the airspace of allied countries, performing ISAF operations in Afghanistan, and providing an advisory team for the training of the Afghan Air Force.

The CZAF currently includes three air bases, one airport management unit, and two air defence regiments. The air fleet includes fighters (JAS-39 Gripen, L-159 ALCA, L-39 Albatros), transport airplanes (A-319CJ, C295M, CL-601, L-410, Yak-40), and helicopters (Mi-8, Mi-17/171, Mi-24/35, W3A Sokol) (Armada 2014).

Electronic Flight Bag (EFB)

Digital aeronautical charts and other navigation data are displayed on a device called the Electronic Flight Bag (EFB). Simply stated, it is an electronic version of a

---

[1] Digital Terrain Elevation Data.

[2] Raster Equivalents of Topographic Maps.

pilot's flight bag. It carries the documentation pilots must have with them during a flight, such as flight manuals, operation manuals, and navigational plates. This size of this bag can range from a navigation briefcase used in transport airplanes to a small publications bag used in fighter aircraft. Some fighter pilots use the G-suit pockets to hold many of the publications—but for various reasons this not the best place for them for instance, in support of a safe ejection (Fitzsimmons 2002). The EFB can be a part of aircraft navigation equipment or a portable electronic device (PED) such as a tablet. EFBs are coming into the flight deck and bringing along with them a wide range of issues and opportunities. In order to understand the full impact of EFB, it is important to study how crews might use these devices and their opinions of them (Chandra and Yeh 2003).

Smart Legend

Cartographic legends are traditionally intended to explain the content of a map. It provides a summary of symbols, codes, names of variables, and other information from the map (Sieber et al. 2005). However, if we assign an interactive function to the map legend, through which a user can control features to be displayed, the map's capacity to include information of different types grows. This smart function optimally, clearly, and comprehensively describes the content of the map. As reported by Dykes et al. (2010) creating a legend in cartography generally relies on a subjective assessment of a maps use and the experience, knowledge, and preferences of the user group. In other words, cartographers must rely to their good judgment.

In the classical approach when working with a map, some initial understanding of the map elements is expected. This approach has been used for decades and seems to be encumbered with certain conventions. A large portion of the carto-graphic community are unlikely to accept a legend created and edited by the user according to his needs and knowledge; furthermore his professional experience will not be considered, it is only his cartographic expertise (or lack thereof) that is important. In 1997, Kraak et al. describe the possibility, and the certain potential in user interaction with a map, through its legend. In his article he suggests inserting a legend in order to prevent dividing the attention of the user. He also suggests the possibility of changing the map background in response to the time of day. Midtbø (2003) describes legends used to choose map content by a click of the mouse. In addition to these methods Peterson (1999) used the term "active legend" as an interface between the user and interactive map animations. As part of his research he examines the use and design of active legends for interactive cartographic animation. By using an active legend he displays animations on the map. Worm (2001) showed that the map legend could be moved around the map, as well as in the map. It uses reverse connection between the map and the legend; a click in the map leads to highlighting of certain elements in the legend.

Methods

Participants

64 pilots from 3 CZAF bases completed an anonymous survey assessing their experience with paper and digital aeronautical charts. All pilots had made at least one flight during the last 6 months prior to the survey date. The final survey data included 64 pilots ranging in age from 26 to 52 years. Gender has not been questioned because of very low number of female pilots in the CZAF; it would also reduce anonymity. All respondents were offered the opportunity to receive the results of the survey as well as answers to their questions in December 2013.

Survey Content

The 28-item questionnaire was designed to gather demographic data as well as information about the flight background, types of aircraft flown, flight experience, etc. The main section was designed to collect information regarding their experience with and attitudes of digital aeronautical charts, system of updates, and the smart legend.

Procedure

Participants received questionnaires from their commanders. Some of them received the questionnaire in paper form, some in electronic form. The time limit for the completing and submitting the questionnaire was 14 days. Participation was voluntary, but a minimum of 14 participants from each air base was ordered. Later, completed questionnaires were analyzed and processed by the authors.

Results

Pilot Demographics

64 pilots of the CZAF were included in the final database. The flight backgrounds were varied (see Table 1). The flight experiences range from 200 to 8,100 h of flight time.

Table 1 Pilots' flight background

Flight background	%
Fighter aircraft	28
Helicopters	50
Transport airplanes	22

Table 2 What types of aeronautical charts do you know?

Name	Yes (%)	No (%)	No answer (%)
LOM 500	95	2	3
TFCL 250	63	34	3
JOG 250	56	41	3
LFC 500	56	41	3
TPC 500	34	63	3
ONC 1mil	19	78	3

Aeronautical Charts: Knowledge and Training

Pilots' knowledge of aeronautical charts was somewhat surprising. In a list of the charts produced by the MGHO, pilots have marked those that they know. The best-known chart is the LOM 500. It has been ticked in 95 % of cases. This map is a part of the basic equipment of the CZAF pilots. The least known chart is the ONC 1 mil with only 19 % of the positive responses. Two respondents skipped the question. In some cases, a negative response may be caused by the chart name unfamiliarity. For the results see Table 2.

Close cooperation with geographers is important for a variety of tasks of the Air Force. But the question of the frequency of aeronautical chart training with geographers showed that 58 % of the questioned pilots have probably never had such training. 26 % of the pilots had some training in the last 2 years and 16 % of them had it sometimes more than 2 years ago.

75 % of the pilots responded that they are interested in such training. When asked how often, 42 % of them think it should be once a year or more often. 14 % of them think it could be less than once a year or just when needed. 19 % do not specify the frequency.

Steps should be taken to change this situation. Training in aeronautical charts should be done at least annually and whenever significant changes occur. Pilots can receive training from professional teachers and at the same time can be involved in self-learning through online courses. The next step would be to design a training method that would make it easy for pilots to increase their knowledge about maps. The method should cater to the particular needs of the pilots.

There are several factors that contribute to pilots not possessing sufficient knowledge about maps. First, their work in the cockpit is demanding leaving little time or willingness to study maps. Secondly, pilots are accustomed to the aeronautical charts that they are currently using and therefore they are not willing to experiment with the proposed maps. Finally, pilots only work with the types of maps associated with their needs and are not aware of the range of maps that are available. Finally, not all of maps distributed by the Geographic Services are useful to them. However, having a basic knowledge of the different kinds of available maps could prove beneficial during certain situations.

Paper vs. Digital Map, Experience with the Digital Map

98 % of the respondents have had personal experience with flight navigation using digital maps. This implies that the subsequent conversion from paper to digital forms of cartographic products will be possible. 91 % of the pilots currently fly an aircraft equipped with a device able to display a digital map.

An important question was what form of map suits them best for their job. Only 2 % of the pilots prefer paper, 14 % of them prefer a digital map, and 84 % of them choose a combination of the two forms.

Most the pilots prefer a combination. It means both forms of maps must be prepared during the pre-flight preparation (and thus more work). The explanation lies in the fact that one form of map backs up the other. The pilots would rather spend a few minutes longer preparing than to fail a task in the event of a map problem. These results also lead us to believe that we are still not ready to fully replace paper charts with electronic. It is therefore desirable to create such digital material that will be sufficient enough so that the pilot does not have to use the paper. During the planning process on the ground it is easier for pilots to create an overview of long distance routes on paper charts. Interestingly there was a subset of pilots, like all people, who are interested in innovation and moving to new technology. Perhaps this represents a map use situation where complete and rapid conversion to a new technology will not happen as we might hope.

An important and related question was whether the pilots think that the digital map can fully replace the paper one. 27 % of the respondents believe that it can, while 70 % of them believe that it cannot and 3 % do not answer.

In terms of the category of pilot, fighter pilots are most responsive to change and are more open-minded than other pilots. The pilots of transport planes have rather conservative attitudes and mostly believe that the paper map will never disappear. However, a different nature of work and a different amount of space in cockpit of each aircraft type may play a role in these responses.

70 % of the pilots believe that the digital map cannot replace the paper one, but they are not sceptical about its use. The digital map is regarded as a tool that facilitates their work. Still, there is a threat of failure of electronics and many pilots consider the paper map as a necessary backup.

On the other hand, there are some cases when the pilots consider the digital maps necessary for the performance of certain tasks (see the chart below). 42 % of the pilots think that digital maps are necessary for tactical tasks, such as NVG operations, low altitude operations, joint or combined air operations, air to ground operations, and air reconnaissance. 30 % of the pilots mention a necessity of the digital maps for instrument flights (IFR) on civil ATS routes, especially abroad. This type of flying has its specifics and is based on navigation system guidance, such as satellite navigation. Some of these devices (FMS equipped) are able to navigate the aircraft using navigation schemes, to lead the aircraft using data from international navigational databases, to display and navigate the aircraft on arrival, approach, departure aerodrome charts, and also to guide the aircraft on the airport map during taxiing. Some of these devices are able to combine the navigational data with map layers and therefore the pilot can see the current geographical position of the aircraft. (However, the map layers are not as important for this type of flying.) 13 % of the pilots think the digital map is necessary for every flight. 8 % of them mention cases where there is no time to prepare and also some special tasks, such as SAR missions, photogrammetric scanning, etc. 33 % of the respondents claim that the digital map is not necessary for any flight or did not answer.

Whether or not digital maps are important in cockpits may reflect the level of safety and reliability of EFBs. As was mentioned above, pilots do not yet trust the new technology. Another explanation is that pilots are simply not experienced enough to work with such maps yet. The fact that digital map could be more distracting factor from the routine must not be overlooked. On the contrary, tactical flights require greater geographical awareness, therefore having with digital maps could make such a task easier. Digital maps could minimize crew error and some users are well aware of this point.

Other questions reflect the current situation of digital maps in the CZAF aircraft. 56 % of the questioned pilots claim that the digital maps on their aircraft are sufficient for the tasks carried out. 31 % of the pilots are dissatisfied with the current equipment. 13 % respondents claim that they do not have such a device on board or do not answer.

The most common causes of dissatisfaction are the system in place for map updates, lack of data availability, and the geographer—pilot relationship (41 %); hardware and software problems, and problems with system and user interface (31 %). Furthermore, pilots would welcome much more configurable system, e.g. user points, zones, layers (11 %). They also complained that they often don't get maps in the correct data formats, e.g. vector maps (9 %).

System of Map Data Updates

What is already clear from the above is that one of the main problems is the map data update system. We examined what kind of experience the pilots have with this system. 8 % of them have positive and 33 % of them have negative experience with it. 59 % of the pilots have no experience or did not answer.

23 % of the respondents think that there is a need to unify and centralize the whole system. 14 % of them reported an insufficient frequency of updates, 5 % of them mentioned the lack of correct data format or reliance on the manufacturer and 3 % of the pilots mentioned poor availability of the actual foreign countries charts.

A large number of "none, or no answer" responses is likely due to the fact that pilots often rely on technical staff (on contract with the manufacturer) for updates to the system. 33 % of the negative answers point to an inefficient system of updates and call for centralization, unification, and streamlining. Part of this process might involve a closer inspection of the relationship between pilots and those people responsible for making updates.

The truth is that while the updating process is very complex pilots do not see the time, energy, and other resources that are involved. In many cases the cartographer is not able to meet their requirements, further diminishing the importance of the updating step, in their opinion. It is not always easy to procure maps of many foreign countries. Thus, the process is prolonged. In addition, some less developed countries do not have geographic services at a level to be able to provide the maps of the required quality.

The update process of the paper and digital charts in aircraft includes an order from the Chief of the General Staff of the Czech Army that a number of paper charts must be held in stock for each military base. Also for each crew, there is an exact amount of background paperwork for navigation preparation and each pilot must always have a current paper chart of the area where he flies. On the other hand, there is no regulation dealing with the updating process of the digital charts, the manufacturer updates some devices. But these are commercial products. For the fulfillment of combat and special missions of Air Force, digital charts issued by the military geography office are needed. Currently, from the perspective of the geographers, their involvement is instigated by individual requests from pilots and engineers.

Mobile EFB Devices

31 % of pilots reported dissatisfaction with the digital maps in the CZAF aircraft. Additionally, the Czech Air Force still has a number of aircraft without such equipment in its fleet. Equipping the aircraft with the mobile EFBs could be a partial solution. These devices could also perform other functions.

56 % of the respondents have some personal experience with an EFB or similar portable device, 38 % of them don't and 6 % do not answer.

77 % of the respondents would appreciate such device onboard their aircraft. 16 % of them have no interest in such tool and 8 % do not answer.

Reported benefits were the following—better UI; faster and easier to update; possible function as an "electronic kneeboard"; good tool for the IFR flights under the civil air traffic control; flight training facilitation; simple plan changes during the flight. Possible problems reported were—a difficult placement in the cockpit;

software would have to be modified according to the needs of each aircraft type; an external GPS module is needed; power supplies differ across aircraft.

Some reported problems are not attributable to the geographic service. The value from this part of our research is that pilots are willing to have electronic device in the cockpit, which opens the door for digital charts. Only 16 % of respondents are not willing to have a mobile EFB device in cockpit. This is a positive result, if we assume that people generally view changes in their work environment negatively. Furthermore, 38 % of respondents do not have experience with such a device, a level of experience we believe is unsatisfactory. From our point of view, every user should have at least some basic experience with such a device. With such experience the adoption of digital maps is more likely. On the other hand, integrating new electronic devices with older flight desks could obstruct access to or use of other equipment. Installation must ensure that the EFB is installed for easy access if used during high workload flight phases. The structural cradle can obstruct visual and physical access to flight controls and displays (Chandra and Yeh 2003).

Pilots' Attitude to the Smart Legend

Pilots' reactions to the smart legend are mostly positive. 84 % of them stated that it would be a helpful tool. 9 % are not interested, and 6 % did not answer.

The most frequently mentioned required functions of the smart legend are the following: aerodrome information, runways, and schemes (56 %), ability to display obstacles and terrain relief (53 %), displaying published airspaces, air zones, significant and radio navigation points, airways, and corridors (50 %). Less frequently mentioned functions are frequencies of radio navigation systems (17 %), user points, tags, layers, tactical situation (13 %) and current use of the airspace (6 %).

It turns out that the concept of smart legend is correct and it knocked to the needs of pilots. Further research in this area will be valuable.

Collaboration and Reported Problems on the Geographical Basis

We asked the pilots about problems with geographical (cartographic) support. Pilots would especially like faster distribution of maps, better communication, and a centralized system of distribution. Some have expressed dissatisfaction with the transition to NATO standards in the map production. The reason for such complaints is a loss of quality in details, especially in the maps used when flying at low altitudes.

Frequently mentioned problem was the inability to edit the digital geographical data. Since the data are available in raster and vector formats, this issue has two dimensions. Only the Geographic Service can edit the raster data; in case of the vector data a user can only edit an extension. However, if the pilot himself edits a map, there is a necessity of approval process, which currently has nothing to do with the Geographic Service.

There was a suggestion to design a user map base: geographic data (populated places)—simplified depending on the scale. The Geographic Service can ensure that only for the Czech territory. At this moment there are four forms. 1:25,000 blocks of buildings, 1:100,000—reduced, 1:250,000 blocks of populated places, 1:250,000 wheels. Digital maps do not contain as much information as the paper ones (e.g. missing names of some populated places). While performing tactical tasks, such information may be missing. However, the truth is that there are problems associated with the description of vector models. For instance the "annotations" are not part of the analytical data. A flight map is not always tailored to suit each individual or flight plan an important question is how pilots would like to implement such a feature. The Geographic Service is not able to modify all maps according to the needs of individual pilots or specific types of flights.

More than once there was a question whether the degradation to NATO standards, in case of some maps, is beneficial for us. Within the Czech Republic we do not need these standards. However, by joining the Alliance the MGHO has committed to fulfil the obligations resulting from the NATO Geographic Policy. That means providing the geographic support to units deployed in international peacekeeping and humanitarian missions in both the development phase and during operation at a mission site. This implies that all members of the Alliance use standardized products in domestic operations, so when a pilot is operating outside their home country they use products with which they are familiar. It would be illogical to use one product on domestic territory and a different product beyond the borders.

Symbols of populated places in the aeronautical charts are often rated very negatively. The shape and size of towns and villages are important for a pilot, such information supports orientation when flying at low altitudes. For example, there are two villages, one of them is vast and more prominent in the landscape than the other, but they have the same number of inhabitants, so they both have the same symbol. Specifically, this occurs in JOG 250A and TFC (L) 250 maps. In such a situation, despite the presence of different symbology for these two different places being present in the Czech data, we have to follow NATO standards. The NATO standard states that a village with less than 2,000 inhabitants is indicated by a point symbol.

Obviously one of the major problems is lack of communication between the geographers and the pilots. One solution is the creation of a website on the Czech Army internal network. All reported problems and deficiencies are answered there and there is also a possibility for further communication.

Results Discussion

A modern Air Force needs modern support. The Geographic Service of the Czech Army provides high quality aeronautical charts for the Czech Air Force aircraft both in paper and digital form. A survey among the CZAF pilots on the experience with digital maps and their attitudes to new technologies revealed several shortcomings and opportunities. The survey also opened the question of whether electronic data can completely replace paper maps and paper documentation.

There is an obvious lack of cooperation between pilots and geographers in the Czech Army. Pilots' knowledge of the aeronautical charts issued by the Czech Army Geographic Service is poor. Only 26 % of the questioned pilots recalled having had at least one aeronautical charts training session in the last 2 years. Most of them (75 %) are interested in such training and the strongest believe (42 %) is that it should be once a year or more often.

Most of the pilots (70 %) believe that the digital map cannot fully replace paper. Most of them (84 %) also claim a combination of paper and digital maps best for their job in the cockpit. This suggests that there is reluctance to abandon what is known and that the quality of available digital data and related equipment is not at a satisfactory level yet. Pilots consider the paper map as a necessary backup. On the other hand, many pilots consider the digital maps necessary for the performance of certain tasks, such as tactical missions (42 %) and instrument flights (30 %). Only 56 % of the respondents claim the satisfaction with the quality level of digital data on their aircraft. Pilots report problems with updates, data availability, or simply the system geographer—pilot (41 %), hardware and software problems with the displaying devices (31 %). Most of the pilots (approx. 59 %) have no experience with the system of digital data updates. This is most likely due to the reliance on technical staff, 14 % reported an insufficient frequency of updates and 23 % believe that there is a need to unify and centralize the system. We believe this represents an important opportunity for further research.

The CZAF fleet still contains a number of aircraft without a device capable of displaying digital geographic data or the equipment does not meet the needs of pilots. One option is the adoption of a portable EFB. An EFB would also serve as a useful complement to existing systems. An EFB can carry a lot of charts, information, and manuals in a small package and thus save space and weight. 77 % of the pilots would appreciate a portable EFB device on the aircraft flown.

The concept of a smart legend is generally well accepted. A smart legend allows a user to control what features are displayed on the map. 84 % of pilots believe it would be a helpful tool. Most requested features include aerodrome information and schemes (56 %), ability to display obstacles and terrain relief (53 %), and displaying published airspaces, air zones, and significant and radio navigation points, airways, and corridors (50 %). The concept of smart legend is evidently promising and further research in this area will be valuable.

This research identified several shortcomings in the geographic support of the Czech Air Force that should be dealt with. Some are serious, some less so. However we cannot always blame the geographic service or the army system. In some cases, pilots have to specify more precisely their requirements, collectively agree on them and then pass to the geographers. This article provides a basis for further research of the identified problems such as system of data updates, EFBs on CZAF aircraft and smart legend.

Conclusions
Pilots have a strong opinion about whether paper maps can be replaced by the digital maps. Most of them believe paper maps cannot be fully replaced by a digital format. Interestingly, the majority of pilots claim that the combination of paper and digital charts is would be most ideal for their work. Digital maps could support a pilot's work more efficiently in certain missions and situations. Nevertheless, digital charts often contain some problems such as the absence of up-to-date information or missing information. For this reason, pilots prefer to keep the paper map as a backup option and pilots believe the paper map cannot be removed from the cockpit even if advanced digital maps are available.

EFBs are already part of CZAF. However further research regarding EFBs along with an analysis of the current state of the CZAF aircraft equipment is needed. The idea of improving digital charts by using smart legend was also well accepted. Our next step will include implementing smart legend into digital charts and testing its efficiency.

References

Armada CR (2014) Velitelstvi vzdusnych sil. http://www.acr.army.cz/struktura/generalni-stab/velitelstvi-vzdusnych-sil-86864. Accessed 25 Mar 2014

Chandra DC, Yeh M (2003) Human factors considerations in the design and evaluation of electronic flight bags (EFBs) version 2. Available via National Technical Information Service, Springfield http://ntl.bts.gov/lib/34000/34200/34292/DOT-VNTSC-FAA-03-07.pdf

Dykes J, Wood J, Slingsby A (2010) Rethinking map legends with visualization. IEEE Trans Vis Comput Graph 16(6):890–899

Elzakker CJM, Wealands K (2007) Use and users of multimedia cartography. In: Cartwight W, Pererson M, Gartner G (eds) Multimedia cartography. Springer, Berlin, Heidelberg, pp 487–504

Fitzsimmons FS (2002) The electronic flight bag: a multi-function tool for the modern cockpit. In: IITA Research Publication 2 information Series, p 66

Kraak MJ, Edsall R, Maceachren AE (1997) Cartographic animation and legends for temporal maps: exploration and or interaction. In: Proceedings of the 18th ICA international cartographic conference, Stockholm, pp 1, 253–260

Midtbø T (2003) Multivariable "seeing maps" trough interactive animations. In: Proceedings of the 21th international cartographic conference. Available via ICACI. http://icaci.org/files/documents/ICC_proceedings/ICC2003/Papers/016.pdf. Accessed 25 Mar 2014

Peterson MP (1999) Active legends for interactive cartographic animation. Int J Geogr Inf Syst 13 (4):375–383

Sieber R, Schmid C, Wiesmann S (2005) Smart legend – Smart atlas! In: Proceedings of the 22nd international cartographic conference, A Coruña

Worm J (2001) Web map design in practice. In: Kraak MJ, Brown A (eds) Web cartography: developments and prospectus. Taylor & Francis, London, pp 87–108

Comparison of Standard- and Proprietary-Based Approaches to Detailed 3D City Mapping

Lukáš Herman, Andrea Kýnová, Jan Russnák, and Tomáš Řezník

Introduction

According to the United Nations, Population Division (2013) most of the world's population lives in urban areas. Urbanized areas with a high population density are abundant in the developed world. Specific abilities for mobility and orientation are required in these urban areas. Modern urban space is not only defined by latitude and longitude, but also highly affected by altitude or elevation. Elevation information does not represent only altitude values, relative heights, or elevation, but also information about the number of floors. Urban areas are therefore often depicted using 3D visualization. Details of 3D models have been significantly enhanced and improved with the development of technology in recent years. Detailed 3D models provide a more realistic impression and higher immersion. Konečný (2011) explains immersion in the scope of 3D virtual environments.

Detailed 3D models on the scale of a city (e.g. Berlin or London) usually contain detailed textures. Spatial information about the interiors is a part of less territorially extensive models (examples of these 3D models are discussed below). Detailed three-dimensional modelling of indoor space is valuable for many applications like navigation, indoor facilities, energy management, and architectural information systems. This paper is devoted only to methods, technologies and applications of 3D models that contain information about the interiors and exteriors of buildings.

L. Herman (✉) • A. Kýnová • J. Russnák • T. Řezník
Faculty of Science, Department of Geography, Masaryk University, Brno, Czech Republic
e-mail: herman.lu@mail.muni.cz

© Springer International Publishing Switzerland 2015
J. Brus et al. (eds.), *Modern Trends in Cartography*, Lecture Notes in
Geoinformation and Cartography, DOI 10.1007/978-3-319-07926-4_11

Existing Detailed 3D City Models

Application of 3D Models

Today, even 3D models are nothing new in GIS, since they are very popular among layman users. Besides providing an attractive view of landscape or city, these models also cover many practical aspects. Decades ago, building a model meant a physical model from paper or balsa. Today, talking about models or 3D models means, for many people, virtual digital models. These models are used across different application areas.

3D models bring, for example, new possibilities to the tourist industry, because they are good for the presentation and promotion of historical sites, castles or monuments (Jedlička et al. 2012; Popelka and Brychtová 2012). However, techniques of 3D city modelling are also used for the reconstruction of already defunct towns or their parts, as in the *Rome Reborn* project, which deals with the reconstruction of Rome from 320 AD (University of Virginia 2010).

Realistic building models have mode forward disciplines such as architecture or the building industry. We may also expect that the 3D model will be a necessary part of the documentation of public utility networks, which may result in the 3D cadastre for estate evidence. Using 3D building models for facility management and BIM (*Building Information Modelling*) is also quite common (Kolbe 2009). A useful application based on the model of a real town is the German energy-GIS project Sun Area. It uses 3D data to evaluate which roofs are well situated for solar energy generation (SUN-AREA 2011). 3D city models are also used for analysing heating and building energy demands (Strzalka et al. 2011).

3D building models are also very useful for navigation in large buildings, such as, for example, indoor navigation for the visually impaired tested at the University of Nevada (Wolterbeek 2012). A very important project using 3D building models is the NATO STANDEX Counter Terrorism Project. This project deals with the remote detection of suicide attacks based on microwave scanning in real time and 3D space (STANDEX 2012). Other utilizations, in addition to those mentioned above, are also possible. Löwner et al. (2013b) provides a general overview of the applications of 3D city models.

Detailed 3D City Models

Detailed 3D models usually have a smaller territorial extent. 3D campus maps are good examples of this. 3D models of campuses may be found all over the world, e.g. at airports in Beijing, London, Prague, Rotterdam; at hospitals in Johannesburg, Panama, Sevilla; at commercial and financial centres such as the World Financial Center in New York and the World Trade Center in Barcelona; at convention and exhibition centres in Hong Kong, Las Vegas, etc. (3D Warehouse 2013).

University campuses are also a very popular subject to model. Most 3D virtual campuses have been published on Google Earth. In 2007 and 2008, the motivation to create 3D models of university buildings on Google Earth was reinforced by Google's announcement of the "Model Your Campus Competition". Detailed models of Zhongnan University of Economics and Law, China; Nicolas Copernicus University, Poland; Stockholm University and KTH, Sweden; and Cardiff University, UK, were among the winners in 2008 (Winners Collection 2008). In Armenakis and Sohn (2009), a 3D model of York University was created.

However, several other approaches to 3D campus modelling have also been published. Over et al. (2010) dealt with the generation of interactive 3D city models based on the OpenStreetMap (OSM) project and presented a virtual complex of the University of Heidelberg online at www.osm-3d.org. A CityGML-based 3D model of the University of Cologne Campus was published in Willmes et al. (2010).

Mapping interiors and navigating inside buildings have become recent topics for Google (GoogleBlog 2011), OpenStreetMap (IndoorOSM 2013) and also other companies and researchers (Řezník et al. 2013; Indoor/Projects 2013). Goetz and Zipf (2012) presented an approach extending an OSM tagging schema to the indoor environment and applied the extension in modelling a selected building of the University of Heidelberg.

In the Czech Republic, several 3D models of campuses were created as part of theses. Slováček (2002) processed a pair of aerial photogrammetric images in ERDAS IMAGINE software to model a building complex of the Technical University of Ostrava. The same technique was used in Olivík (2003) to create a virtual 3D model of the University of West Bohemia. Malý (2009) and Russnák (2012) used a building passport to model the University Campus at Bohunice and the Faculty of Science at Masaryk University. A detailed interactive indoor 3D model of the Faculty of Electrical Engineering, Czech Technical University in Prague, was created by Kratochvíl (2006) and Holý (2008).

Technologies for Detailed 3D City Mapping

3D city models are not only analysed in the context of geoinformation systems. The fields of architecture, engineering, and construction as well as computer graphics provide their own approaches for the representation and exchange of 3D models. Both proprietary technology and widely accepted standards are applied to all the fields mentioned above. Among the standardized solutions are included IFC (*Industry Foundation Class*), which is designed to store FM data; CityGML, which is described in detail below; and INSPIRE Data specification for Buildings, which has a wide range of applications (Herman and Řezník 2013). The proprietary approach is represented by the Building and Technology Passport.

CityGML

CityGML (*City Geography Markup Language*) is an open data model and XML-based format for sets of 3D urban objects. It is implemented as an application schema for the GML3 (*Geography Markup Language version 3*). Both are extendible international standards for spatial data exchange issued by the OGC (*Open Geospatial Consortium*). CityGML is used for the representation, storage, and exchange of virtual 3D city models. Generalization hierarchies among thematic classes, aggregations, and relations between objects are also included in this application schema (Kolbe 2009). CityGML provides a standard model and mechanisms for describing 3D objects with respect to their geometry, topology, semantics and appearance; furthermore, it defines five levels of detail (LoD). Individual levels of detail contain, for example, following elements:

- LoD0—2,5D model of terrain, building footprints,
- LoD1—extruded building footprints,
- LoD2—3D buildings with simplified modelling of roofs,
- LoD3—3D buildings with modelling of exterior details,
- LoD4—3D buildings with modelling of interior details.

CityGML provides much more than simply 3D content for visualization, as it has many diverse applications. CityGML supports class definitions, regulations, and explanations of the semantics for geographic features in 3D city models, including digital terrain models, buildings, vegetation, transportation objects, and city furniture. Practical applications often require the storage and exchange of data which do not belong to any of the predefined classes. CityGML provides two different methods of extension for such cases. The first is the use of generic features and attributes; however, this method reduces semantic interoperability. A concept called *Application Domain Extension* (ADE) is used to eliminate this problem. In contrast with generic objects and attributes, ADE has to be defined within a separate XML Schema definition.

CityGML includes a relatively general data model, whose content can be linked to a topographic database. Handling specific issues that are associated with the 3D modelling of buildings with interiors requires the extension of this structure. And, for this purpose, various specialized ADEs are used (Kolbe 2009; OGC City Geography Markup Language Encoding Standard 2012).

ADEs, which describe in detail the interiors of buildings and technical facilities inside the building, are for example CAFM (*Computer Aided Facility Management*) ADE, GeoBIM ADE and UtilityNetworkADE. Class Buildings is through CAFM ADE extended to other descriptive information used for facility management. This ADE is designed for LoD 4 and adds to existing classes (e.g. *FloorSurface*), abstract subclasses, and attribute data characterizing a manufacturer, condition, a responsible person, or a year of construction (Bleifuss et al. 2009).

GeoBIM ADE combines information contained in the IFC format with CityGML and designs standard GeoBIM. Like the previous ADE, this ADE extends the *Building* class, but this extension is in LoD 3. Semantic classes, such as Room, are extended by a subclass *VisibleElement* and additional elements (*Railing, Beam, Stair* or *Annotation*). The classes *Window, Door* and *RoofSurface* are extended by attributes (De Laat and Van Berlo 2011). UtilityNetworkADE (latest version 0.9.0) defines a topological network model facilitating analyses and simulations on utility networks. Furthermore, it allows the representation of network components such as 3D topographic city objects (Becker et al. 2011).

Building and Technology Passport: Building Information Modelling

This paper presents a pilot study using data from the Building and Technology Passport. Since 2004, the Facility Management Division of the University Campus at Masaryk University has collected digital spatial data on the real-estate property of the University. Recently, a Building Passport has been created for more than 160 buildings owned by the University. However, this number will continue to grow in the context of new construction. The Building Passport also registers assets in buildings used but not owned by the University. Altogether, this means over 250 buildings with over 20,000 rooms (Glos and Souček 2009; Kroutil 2009).

In 2007, shortly after the Building Passport was created, work on the Technology Passport began, this relating to technologies in buildings and rooms including the location of all electronic devices, air-conditioning systems, fire equipment, water distribution systems, and heat distribution systems. The Building and Technology Passports together form BIM, or Building Information Modelling. Using BIM, the Facility Management Division began modelling a new university campus in Brno Bohunice and has gradually been collecting data about other buildings and rooms at different faculties. So far, the Technology Passport has been completed for 40 buildings all over the University. The University Campus, the Faculty of Economics and Administration, the Faculty of Social Studies, and the Faculty of Education have already been finished. In 2013, a test of technology passport data acquisition started at the Faculty of Science (Facility Management Division of the University Campus Bohunice 2013; Kučera et al. 2013).

Spatial data include DWGs (*DraWinG*) of all floors, side-sights or cuts of each university building. DWG blends symbology and attributes together, so, for the GIS solution and web presentation as well, the most important features like ground plans, steps, windows, and load bearing walls are also saved in ESRI geodatabases. Passportization documentation is executed according to methodologies for building and technological passports, which are an integral part of the tender documentation for construction. Methodologies establish and describe the necessary operations in the process of collecting, managing and updating data passports (Facility Management Division of the University Campus Bohunice 2012).

Naturally, the Building Passport is used for building and room plans. However, the main utilization is the connection of spatial data and context, such as in key-room systems, academic computer network information systems, alarm system zones, evacuation fire plans, and interior drafts. This means that these data are used by both the concrete faculty and the all university organizations. Many theses based on the Building Passport have already been published at Masaryk University from the GIS and IT points of view (Glos and Souček 2009; Kroutil 2009). Height values stored in attributes can also be used to create a 3D building model, such as the model of the Faculty of Science at Masaryk University shown in Fig. 2.

Comparison and Conversion

CityGML as a standardized solution is suitable for exchanging 3D data among different organizations and institutions. However, the Building and Technology Passports are optimized for use in the Facility Management Division. In the Building Documentation System, the unified identification of rooms, buildings and sites is realised through hierarchical identifiers. Position code looks for example like: BVB04N01023a, where meaning of each component is:

- BVB—identification of locality,
- 04—number of the building,
- N—type of the storey
- 01—number of the storey,
- 023a—marking of the room.

UUID (*Universally Unique IDentifier*) stored as gml-id attributes are usually used in CityGML and this notation is not hierarchical.

LoD used in CityGML represents another mechanism for the easier selection of elements. The concept of LoD enables features on a given scale or on a given significance to be identified and selected. LoD can differ for various objects (*Roof* as LoD 2, *Door* as LoD 3). In the future, this multi-level approach may also be suitable for the Building Documentation System. The LoD concept is the most successful (and the most cited) part of CityGML.

The Building Documentation System is primarily 2D. Transformation to 3D visualization can be laborious and some inconsistencies can arise during this process. In contrast to the Building Documentation System, CityGML supports full 3D geometry. In addition, it is XML-based and thus supports automated processing, which is better with respect to consistency. On the other hand, the use of ESRI technology in the Building Documentation System allows, among other things, the relatively simple creation of 2D printed plans of buildings and rooms.

Despite the existing differences, it is possible to transform data from building passports to CityGML and vice versa. Conversion from the Building Passport into CityGML is demanding from the conceptual point of view. It is necessary to:

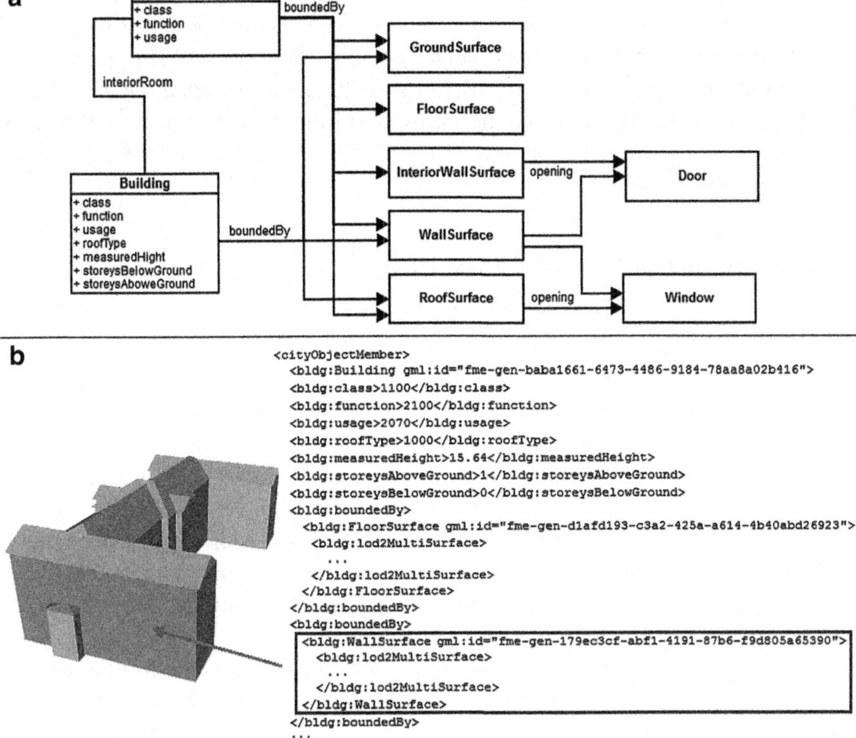

Fig. 1 Conceptual scheme of 3D model of buildings from Faculty of Science campus formed according to CityGML in LoD 4 (**a**) and 3D model of buildings from the Faculty of Science campus in CityGML format in LoD 2 and the record of this data structure in XML (**b**)

- define semantic classes,
- supplement information about level of detail,
- add attribute values or derive them from the geometry.

Individual steps can be performed either automatically or manually. Both approaches were tested and their combination was used to perform the final conversion. The conversion was carried out on two levels: 3D models in LoD 2 (see Fig. 1a) and LoD 4 were made (the data model of this level is shown in Fig. 1b).

ArcGIS 10.0 software was used for manual pre-processing, which was the first step in conversion. The extrusion of vertical surfaces (walls), the base height definition for horizontal surfaces (surfaces of ground, floors and flat roofs) and the completely manual creation (in ArcScene module) of complex gabled roof surfaces were performed in this step. While processing the variant in LoD 4, 3D versions of difference and intersect operations were also used (these algorithms are in ArcGIS parts of 3D Analyst extension).

The second step of the transformation was the conversion of pre-processed Shapefiles themselves. These Shapefiles have polygon geometry with Z (elevation) information (in LoD 2) and Multipatch geometry (in LoD 4). This part of the conversion was implemented in FME software 2012 and included the assignment of individual elements into semantic classes, and the definition and fulfilment of attributes.

Utilization of the Campus Map of the Faculty of Science

Visualization of BIM Through Web and Printed Maps

Data from the Building Passport are widely used for navigation, people and place location, and orientation at various levels of Masaryk University. They are used at the level of the entire university for the searching of persons and premises within the Information System of Masaryk University (IS MU). However, they are also used for specific purposes at the Facility Management Division of the Management of the University Campus, for example to analyse and visualize heating data. The following paragraphs are devoted to the utilization of Building Passport data at the Faculty of Science and the Department of Geography.

The plan of the Department of Geography was made both in the form of a static plan and in the form of a web map. The web map[1] is based on open internet technologies (JavaScript and SVG—*Scalable Vector Graphics*) and uses simple methods of representation with added interactivity. This interactive map appears on the Department's website. The main purpose of the static plan is the possibility of printing, which is offered by the web map. For this reason, the plan optimized for A4 pages is saved in PDF (*Portable Document Format*). However, the static plan does not illustrate all floors of the buildings of the Department of Geography, but only those that provide information required by readers. An emphasis on cognitive style was applied and verified by usability testing (these concepts are explained, for example, in Konečný et al. 2011). The selected floor is viewed from the side, creating the impression of 3D. Each building is shown separately and list of employees can be displayed for individual rooms.

Another variant of the web map is used at faculty level. The webpage of the Faculty of Science uses a simple static campus map[2] in PNG format (*Portable Network Graphics*), which contains the localization of individual departments. Printed campus maps with thematic information are used for various promotional events (e. g. open days or "Night of Scientists" actions), when the Faculty of Science is visited by people who are unfamiliar with the area. Maps for these events include information about buildings and places where the program runs.

---

[1] http://geogr.muni.cz/plan-gu-eng.

[2] http://www.sci.muni.cz/cz/Kontakty/Mapka-arealu-Kotlarska.

3D Visualization

A 3D model can be saved in many formats, only a few of which can be easily opened directly in a web browser, i.e. without the necessity to add a plug-in and/or customize the Web browser in some way. The web presentation of 3D models created from the Building Passport and in the CityGML format requires their transformation into a different form. Both 3D models created in ArcGIS from the Building Passport and CityGML data can be transformed to X3D format.

The CityGML format can be displayed in a web browser using a plug-in, which is an obstacle to use. A number of programming libraries in JavaScript language that display 3D information in the web browser have been developed in recent years. Such a technology is the X3DOM library, which renders data stored in the X3D (*eXtensible 3D*) format. A 3D model made in the ArcScene module can be exported to VRML (*Virtual Reality Modeling Language*), which can be transformed through freely available tools to X3D. This conversion is not difficult, as X3D is based on VRML. The conversion of CityGML is not complicated either. Although both formats are based on XML, CityGML is transformed into X3D using an XSLT (*eXtensible Stylesheet Language Transformations*) stylesheet and XSLT processor. The issue of the XSL transformation of 3D building data is described in detail in Herman and Řezník (2013).

The advantage of X3DOM is the relatively simple construction of 3D scenes and also considerable support in different web browsers, not only on desktop computers, but also on mobile devices. X3DOM is thus useful for developing 3D navigation applications for these devices. It is planned to place the described X3D model on the Department of Geography website or, alternatively, the Faculty of Science website (the first version of this 3D model is shown in Fig. 2).

Future Development

The previous sections showed the various ways of utilizing detailed data about buildings, and the various options available. These data, whose acquisition is not cheap, can be assessed mainly by multiple applications. In addition to the examples mentioned above, these data can also be used for interactive mobile applications for navigation. As mentioned above, CityGML and the Building Passport are conceived quite generally and the indoor navigation issue requires an extension or modification of these general data models. However, creating functional interactive indoor navigation may not require further transformation into a new specific format.

The navigation extension that is designed to work with different types of 3D geometry is IndoorGML. CityGML, IFC, KML (*Keyhole Markup Language*) and 2D digital data represent, for IndoorGML, sources of geometry. IndoorGML will be derived from these datasets and contain mainly topological data supplemented with semantic information. Original technologies are also used for the visualization of

Fig. 2 3D visualization of Faculty of Science campus on Kotlářská street in web browser

buildings, rooms and routes. Synchronization between CityGML and IndoorGML will be implemented through XLink technology (OGC Candidate Standard for Indoor Spatial Information 2013).

In addition, new specialized standards are constantly developed; the continuous adjustment of existing standards is ongoing. Currently, CityGML 2.0 is being revised and discussed and a new version, 3.0, will be prepared. The major changes include the revision of the LoD concept. Definitions of LoDs are currently vague, informal and allow a great deal of ambiguity. Thus, the LoDs in the next version of CityGML will be described by separately specifying:

- semantic and geometric LoDs,
- the definition of separate interior LoDs and LoDs for the building's exterior shell

This division produces four different variants of LoD; each of these options contains five levels of detail. This approach has so far been proposed only for the Building semantic class, since it is complicated; however, it is possible to apply it to other semantic classes. Different testing use-cases bring different requirements. The division into indoor and outdoor LoD is important for combined indoor and outdoor navigation. However, this new concept of LoD brings new challenges as well. It will be composed of many combinations which it will be necessary to quantify and sort (Löwner et al. 2013a). Benner et al. (2013) propose a solution to this task.

Another modification which can improve the use of indoor navigation is support for dynamic (time-dependent) entities. Dynamic entities contain dynamic properties (e.g. energy demand) and moving objects (dynamic geometry). Dynamic features and attributes are not yet supported in CityGML. The utilization of 3D building models in navigation or in simulations has wider consequences including the modeling of the functional dependencies between objects and the functions of objects and also use of metadata. These should include representation of the quality of individual information pieces (metadata at the levels of objects, attributes), lineage, acquisition methods, accuracy, age, and propagation of quality data for aggregations and other types of computations (Löwner et al. 2013a).

While the changing of standardized technologies is a complex process based on a broad consensus and compromises, updates of the Building Documentation System data model are implemented in a simpler and faster way. Regarding the development of the Building Passport, besides updating data due to renovations and other adjustments, updates are also made to acquisition data from buildings under construction (new buildings of the Faculty of Informatics or the Faculty of Arts). It is being planned to create 3D geometry in LoD 2 for each building owned or used by Masaryk University.

Even greater use of data maintained for operating activities, property records, and facility management is assumed (e.g. there are requirements for determining the number of windows for cleaning). This includes the area of space management for maintaining, updating and displaying data relating to the use of Masaryk university's space, where a building or part thereof is used by external entities.

Finally, there should be greater integration with the Information System of Masaryk University (IS MU) and with other existing spatial data (indoor spherical images, thermal images of facades) to provide data from the Building Passport for study purposes and research activities. In this area, an advanced application for navigation will be implemented. This navigation will not simply search for the shortest route, but should also take account of the user's context. Thus, for example, if the building has an access control system, navigation will be addressed with regard to whether the user has access rights to the room. Navigation can also be adapted to the needs of disabled users

Conclusions

This article presents the issues relating to the re-use of a Building Information Model (BIM) in the form of 3D models and visualizations. One of the main motivations for this re-use is the value of data acquisition needed to establish a BIM. Re-use of a BIM in the form of 3D models and visualizations opens the door to various fields of study, such as architecture, engineering, or spatial planning. We may also perceive these models as a foundation stone for a 3D cadastre supporting decision making processes within and beyond public administration. 3D models and visualizations based on spatial data from BIM are also closely connected to existing standardization activities. The implementation specification of the Open Geospatial Consortium (OGC) called CityGML enables the handling of BIM spatial data. Feedback on CityGML was given to the OGC to support progress in the standardization of CityGML. The authors of this paper have recommended correcting

(continued)

inconsistencies in enumerations which occurred in the previous version of the specification.

Transformation from proprietary BIM to CityGML and vice versa was successfully verified. Moreover, 3D models were visualized in an opened way in Web browsers when using X3D technology. As such, 3D models and visualizations created from BIM are useful for creation harmonized background spatial data as in the European Location Framework (see http://www. elfproject.eu for more information).

Acknowledgments This research has been supported by funding from Masaryk University grant No. MUNI/A/0952/2013, entitled 'Analysis, evaluation, and visualization of global environmental changes in the landscape sphere'. The authors would like to thank the Facility Management Division of the University Campus at Masaryk University for consultation on the parts devoted to the Building Documentation System.

References

3D Warehouse (2013) http://sketchup.google.com/3dwarehouse/. Accessed 5 Dec 2013

Armenakis C, Sohn G (2009) iCampus: 3D modeling of York University Campus. In: Proceedings of American Society for photogrammetry and remote sensing 2009 annual conference. American Society for Photogrammetry and Remote Sensing, Baltimore, pp 66–71. ISBN:978-1-615-67322-3

Becker T, Nagel C, Kolbe TH (2011) Integrated 3D modeling of multi-utility networks and their interdependencies for critical infrastructure analysis. In: Kolbe TH, König G, Nagel C (eds) Advances in 3D geo-information sciences. Springer, Berlin, pp 1–20. ISBN 978-3-642-12669-7

Benner J, Geiger A, Gröger G, Häfele KH, Löwner MO (2013) Enhanced LoD concepts for virtual 3D city models. In: ISPRS annals of the photogrammetry remote sensing and spatial information science. Copernicus GmbH, Göttingen, II. No. 2, pp 51–61. ISSN: 2194-9042

Bleifuss R, Donaubauer A, Liebscher J, Seitle M (2009) Entwicklung einer CityGML-Erweiterung für das Facility Management am Beispiel Landeshaupstadt München http://www.gis.bv.tum.de/tum_content/tum_files/Publikationen/studentischeArbeiten/agit_citygml_ade.pdf. Accessed 29 Oct 2013

De Laat R, Van Berlo L (2011) Integration of BIM and GIS: the development of the CityGML GeoBIM extension. In: Kolbe TH, König G, Nagel C (eds) Advances in 3D geo-information sciences. Springer, Berlin, pp 211–225. ISBN 978-3-642-12669-7

Facility Management Division of the University Campus Bohunice (2012) Methodology for building passportization in the scope of the integrated management system (in Czech). Masaryk University, Brno

Facility Management Division of the University Campus Bohunice (2013) Methodology of passportization at Masaryk University (in Czech). Masaryk University, Brno

Glos P, Souček J (2009) Manager, whom nothing escapes (in Czech). In: CAD.cz. http://www.cad.cz/component/content/article/2350.html. Accessed 29 Nov 2013

Goetz M, Zipf A (2012) Extending OpenStreetMap to indoor environments: bringing volunteered geographic information to the next level. In: Zlatanova S, Ledoux H, Fendel EM, Rumor M (eds) Urban and regional data management: UDMS annual 2011. CRC, Leiden, pp 51–62. ISBN 978-0-415-67491-1

GoogleBlog (2011) http://googleblog.blogspot.cz/2011/11/new-frontier-for-google-maps-map ping.html. Accessed 4 Dec 2013

Herman L, Řezník T (2013) Web 3D visualization of noise mapping for extended INSPIRE buildings model. In: Hřebíček J, Schimak G, Kubásek M, Rizzoli AE (eds) Environmental software systems. Fostering information sharing. Springer, Heidelberg, pp 414–424. ISBN: 978-3-642-41150-2. doi:10.1007/978-3-642-41151-9_39

Holý P (2008) Virtual guide through FEE building (in Czech). Bachelor's thesis, Faculty of Electrical Engineering, Czech Technical University, Praha

Indoor/Projects (2013) http://wiki.openstreetmap.org/wiki/Indoor/Projects. Accessed 5 Dec 2013

IndoorOSM (2013) http://wiki.openstreetmap.org/wiki/IndoorOSM. Accessed 2 Dec 2013

Jedlička K, Hájek P, Čerba O (2012) Experience with methods of 3D cartography gained during visualization of detailed geographic data for purposes of documenting cultural heritage (case study at the castle Kozel). In: GeoCart'2012 and ICA regional symposium on cartography for Australasia and Oceania. New Zealand Cartographic Society Inc., Auckland, pp 89–100. ISBN: 978-0-473-22313-7

Kolbe TH (2009) Representing and exchanging 3D city models with CityGML. In: Lee J, Zlatanova S (eds) 3D geo-information sciences. Springer, Heidelberg, pp 15–31. ISBN 978-3-540-87395-2

Konečný M (2011) Cartography: challenges and potentials in virtual geographic environments era. Ann GIS 17(3):135–146. ISSN: 1947-5683. doi:10.1080/19475683.2011.602027 (Taylor&Francis, Honkong)

Konečný M, Kubíček P, Stachoň Z, Šašinka Č (2011) The usability of selected base maps for crises management: users' perspectives. Appl Geomatics 3(4):189–198. ISSN: 1866-9298. doi:10.1007/s12518-011-0053-1 (Springer, Heidelberg)

Kratochvíl P (2006) 3D model of CTU building in Dejvice (in Czech). Bachelor's thesis, Faculty of Electrical Engineering, Czech Technical University, Praha

Kroutil P (2009) Building passport MU in present (in Czech). Zpravodaj UVT MU XIX(5). http://www.ics.muni.cz/zpravodaj/articles/619.html. Accessed 29 Nov 2013

Kučera A, Glos P, Pitner T (2013) Fault detection in building management system networks. In Slanina Z (ed) 12th IFAC conference on programmable devices and embedded systems. International Federation of Automatic Control, Velke Karlovice, pp 416–421. ISBN: 978-3-902823-53-3. doi:10.3182/20130925-3-CZ-3023.00027

Löwner MO, Benner J, Gröger G, Häfele KH (2013a) New concepts for structuring 3D city models – an extended level of detail concept for CityGML buildings. In: Murgante B et al (eds) Computational science and its applications – ICCSA. Springer, Heidelberg, pp 466–480. ISBN: 978-3-642-39645-8

Löwner MO, Casper E, Becker T, Benner J, Gröger G, Gruber U, Häfele KH, Kaden R, Schlüter S (2013b) CityGML 2.0 – ein internationaler Standard für 3D-Stadtmodelle, Teil 2: CityGML in der Praxis. In: Zeitschrift für Geodäsie, Geoinformation und Landmanagement, Augsburg: DVW – Gesellschaft für Geodäsie, Geoinformation und Landmanagement e.V. 138 (2):131–143. ISSN: 1618-8950

Malý M (2009) 3D visualization of selected build-up area (in Czech). Master's thesis, Faculty of Science, Masaryk University, Brno

OGC Candidate Standard for Indoor Spatial Information (2013) http://indoorgml.net/. Accessed 8 Dec 2013

OGC City Geography Markup Language Encoding Standard (2012) http://www.opengeospatial. org/standards/citygml. Accessed 25 Oct 2013

Olivík S (2003) 3D virtual model of UWB Borské pole campus (in Czech). Master's thesis, Faculty of Applied Sciences, University of West Bohemia, Plzeň

Over M, Schilling A, Neubauer S, Zipf A (2010) Generating web-based 3D city models from OpenStreetMap: the current situation in Germany. Comput Environ Urban Syst 34 (6):496–507. ISSN: 0198-9715 (Elsevier, Philadelphia)

Popelka S, Brychtová A (2012) The historical 3D map of lost Olomouc fortress creation. In: Svobodova H (ed) Proceedings of the 19th international conference on geography and geoinformatics: challenge for practise and education. Masaryk University, Brno. ISBN: 978-80-210-5799-9

Řezník T, Horáková B, Szturc R (2013) Advanced methods of cell phone localization for crisis and emergency management applications. Int J Digital Earth. doi:10.1080/17538947.2013.860197

Russnák J (2012) 3D model of the Faculty of Science at the Masaryk University (in Czech). Master's thesis, Faculty of Science, Masaryk University, Brno

Slováček R (2002) Creation of 3D model of buildings from aerial images (in Czech). Master's thesis, Faculty of Mining and Geology, VŠB – Technical University of Ostrava, Ostrava

STANDEX Counter Terrorism Project: 10 Years 10 Stories Anniversary Feature. (2012) http://www.nato-russia-council.info/en/articles/20121123-nrc-10-years-standex/. Accessed 12 Dec 2013

Strzalka A, Bogdahn J, Coors V, Eicker U (2011) 3D city modeling for urban scale heating energy demand forecasting. HVAC&R Res 17(4):526–539. ISSN: 1078-9669. doi:10.1080/10789669. 2011.582920 (Taylor&Francis, Honkong)

SUN-AREA (2011) Nutzen für den Bürger. http://www.sun-area.net/Nutzen-fuer-den-Buerger.93. 0.html. Accessed 1 Dec 2013

United Nations, Population Division: UN Population Division Urban and Rural Population – Population in urban and rural areas 2012 (2013) http://esa.un.org/unup/unup/index_panel1. html. Accessed 10 Dec 2013

University of Virginia: Rome Reborn (2010) http://www.romereborn.virginia.edu/. Accessed 24 Nov 2013

Willmes C, Baaser U, Volland K, Bareth G (2010) Internet based distribution and visualization of a 3D model of the University of Cologne Campus. In: Bandrova T, Konečný M (eds) Proceedings of the digital earth summit. Digital Earth Society, Nessebar. ISBN: 978-9-547-24039-1

Winners Collection (2008) http://sketchup.google.com/3dwarehouse/cldetails? mid=2d5e8dfbfd5d9985bdac737426f6915&prevstart=0&hl=en. Accessed 4 Dec 2013

Wolterbeek M (2012) University of Nevada, Reno scientists design low-cost indoor navigation system for blind http://newsroom.unr.edu/2012/05/18/university-of-nevada-reno-scientists-design-low-cost-indoor-navigation-system-for-blind/. Accessed 24 Nov 2013

Part II
Progress in Web Cartography

Integrating User and Usability Research in Web-Mapping Application Design

David Schobesberger

Introduction

The evolution of digital cartography with all its derivative products creates a vast array of possibilities for cartographic application designers including static and interactive Web maps, multimedia products, geovisualization applications, mobile applications, location-based services and cartographic products which are emerging in the context of Web 2.0, Web 3.0, and Web 4.0. Interactive Web mapping changed people's conception of maps (Peterson 2008). *"The technology-driven evolution from view-only maps to mobile web maps ... has not only extended the typology of cartographic products, but also progressively enriched the map functions"* (Meng and Reichenbacher 2005, p. 5). Web Maps have become capable of delivering sophisticated functionality and of assisting the users with everyday tasks, which lead to a drastic increase of the availability and use of Web maps (Peterson 2003) in many different contexts.

The strong increase in numbers of smart phones and users of wireless broadband (cf. RNCOS E-Services 2011; mobiThinking 2011) presents both a chance and a challenge for cartographic product design. On the one hand, cartographic products are on the verge of becoming ubiquitous in daily use on mobile devices that assist everyday tasks, which means that there is vast potential for new apps and Web applications. On the other hand, the different devices, and the various contexts in which these devices and applications are utilized, demand a reinforced interest in examination and evaluation of user and usability issues in cartographic application design. The philosophy for delivering mapping products has to change from considering map application development as a self-contained exercise, towards a more holistic approach that takes all the evolving technological and social aspects

D. Schobesberger (✉)
Kartograf.at, Cartography, Geoinformation and Consulting, Linz, Austria
e-mail: david.schobesberger@kartograf.at

© Springer International Publishing Switzerland 2015
J. Brus et al. (eds.), *Modern Trends in Cartography*, Lecture Notes in
Geoinformation and Cartography, DOI 10.1007/978-3-319-07926-4_12

into account. This underlines the need for research which focuses on use and usability aspects as well as the methods for evaluating these aspects.

Although the importance of user-centred design is increasingly recognized among cartographic application designers, often the knowledge about how to design and conduct user studies is humble. The field of user research and usability testing is a highly complex one and there is a lack of articles and textbooks about the methodology of user research for cartographic application design (van Elzakker and Wealands 2007). This chapter presents the concept of user-centred design followed by a framework for user and usability research in the Web-mapping domain which has been developed as part of the author's doctorate research (Schobesberger 2012a) and refined by a meta-analysis of published user studies. In the end common usability problems which have been identified across several studies are discussed.

User-Centred Design

User-centred design (UCD) is a design philosophy that originates from the discipline of Human-Computer Interaction (Dix et al. 2005). The main paradigm of UCD is to incorporate the needs, wants and limitations of the end-users in all stages of application design. Through an active involvement of users, an appropriate balancing of functions between users and technology, and iterative design, UCD aims to reach high quality product levels, a high utility and high usability of the end-product, and thus a reduction of support and training costs, higher user satisfaction and a higher productivity (Jokela et al. 2003). Gould and Lewis (1985) were among the first researchers suggesting a UCD approach. Today the ISO 9241-210 standard proposes 6 key principles for user-centred design. These are (Travis 2011):

- Explicit understanding of users, tasks and environments. An understanding of who the users are must be gained in the beginning of design. This knowledge is gathered by directly studying the users' cognitive, behavioural, anthropometric, and attitudinal characteristics and by studying their task requirements as well as the context of use;
- Users are involved throughout design and development;
- Design is driven and refined by user-centred evaluation. Usability testing should be carried out throughout the design process;
- Iterative design. There should be a cycle of design, test and measure, and redesign until usability goals are met;
- Design addresses the whole user experience (UX). This principle addresses perceptual and emotional aspects as well; and
- The design team includes multidisciplinary skills and perspectives.

UCD builds on a number of empirical research methods to collect data about the requirements for an application and for evaluating the usability and utility of prototypes and final products. In the following a framework for a user-centred approach towards the development of Web-mapping applications is presented. Suitable methods for user and usability testing in different development stages have been identified by a thorough literature review and refined by the meta-analysis of 17 case studies in the domains of Web-mapping and mobile mapping.

Meta-Analysis of User Studies

A total of 26 user or usability studies were identified from reviewing academic publications and from a thorough Internet search. User studies were selected for the meta-analysis based on the following criteria:

1. The studies must have been conducted in the domain of Web-mapping.

   It was assumed that the approach to studying users and usability of other cartographic applications or printed maps are different to the approaches for online applications. Exceptions to this were tolerable in the area of mobile mapping applications. Web-based geovisualization tools were exempt from the analysis, because of their completely different presentation approaches and usage goals;

2. The user studies should either have been conducted as part of application development (formative), or in case of summative or comparative evaluations, they must inspect product usability, or compare methods of user and usability research.

   Studies that only compare applications in terms of performance or functionality of existing systems do not provide much input for enhancing the framework for user and usability research. However, summative and/or comparative evaluations can provide input in respect to design guidelines and the application of usability inspection techniques. Studies comparing different methods of data collection were also consulted;

3. The material should include results and must discuss the methodology applied (methods, of data collection, number of participants, data collected).

   The methodology of the study must become evident in the materials. Otherwise the study cannot contribute to the development of the theoretical framework;

4. Material about the studies should be accessible with justifiable effort.

   Studies which have been disseminated more generously were preferable to studies with less information; and

5. Formative studies should have been conducted with prospective users, not with students or scholars as a substitute.

   In academic studies, students or colleagues from the same discipline are often used as a test group. This eliminates the need to recruit a group test subjects, which is often a difficult and/or expensive process. However, subjects might be biased through preoccupation with a subject or the student-teacher relationships. Students as subjects for testing different data collection methodologies were acceptable.

Table 1 User/usability studies in the Web-mapping/mobile mapping domain identified from a review of academic literature and Internet search including assessment of criteria 1–5 (see above)

User study	1	2	3	4	5	Sources
Google Maps vs. MapQuest	✓	✓	✓	✓	✓	Murphy et al. (2005)
Comparative evaluation of 4 major Web-mapping sites	✓	✓	✓	✓	✓	Nivala (2005)
Usability evaluation of 8 comm. Web-mapping sites	✓	✓	✓	✓	✓	Skarlatidou (2005)
Usability Google Maps and Czech Web maps	~	✓	~	✓	~	Voldán (2010)
Czech Regional Authorities Web maps	✓	~	~	~	×	Komarkova et al. (2007)
Maps on official tourism destination Web sites	✓	✓	✓	✓	✓	Richmond and Keller (2003)
Effectiveness of Interactive Map Interface Designs	✓	✓	✓	✓	✓	Coltekin et al. (2009)
ÖROK Atlas online	~	✓	✓	~	✓	Pucher (2008)
Cybercartographic Atlas of Antarctica	✓	✓	✓	~	✓	Pulsifier et al. (2005)
Pennsylvania Cancer Atlas	✓	✓	✓	~	✓	Bhowmick et al. (2008)
The Atlas of Canada	✓	✓	✓	✓	✓	Kramers (2007)
Common GIS usability evaluation	✓	✓	✓	✓	✓	Andrienko et al. (2002)
Eval. of virtual globes & GIS-based Web services	✓	✓	✓	~	~	Vilaca (2008)
Usability of the Lakeshore Nature Preserve Interactive Map	✓	✓	✓	✓	✓	Roth and Harrower (2008)
Natural Hazard Information System for Switzerland	✓	✓	✓	✓	✓	Heil (2009)
Cultural History Information System	✓	✓	✓	✓	✓	Schobesberger (2010)
Touristic Information System for ski-touring	✓	✓	✓	✓	✓	Pucher and Schobesberger (2011)
Water Isotope Information System	✓	✓	✓	✓	✓	Schobesberger (2012b)
Field-based methodology for eval. of mobile applications	✓	✓	✓	✓	✓	van Elzakker et al. (2008)
Comparison of methods for field-based evaluation of mobile devices	✓	✓	✓	✓	✓	Burghardt and Wirth (2011)
Prototype Mobile Navigation Interface for pedestrians	✓	✓	✓	✓	✓	Delikostidis and van Elzakker (2011)
Australian mobile LBS	✓	✓	✓	✓	✓	Wealands et al. (2005)
Topographic maps on mobile devices	✓	✓	✓	✓	~	Nivala et al. (2003)
Usability eval. of mobile navigation and LBS	✓	✓	✓	✓	✓	Chincholle et al. (2002)
Map-based multi-publishing service	✓	✓	✓	~	✓	Flink et al. (2011)
Zooming and panning animations in Web maps	✓	✓	✓	✓	✓	Midtho and Nordvik (2007)

(continued)

Table 1 (continued)

User study	1	2	3	4	5	Sources
Usability evaluation of zoom and pan functions	✔	✔	✔	✔	~	You et al. (2007)
Usability eval. of cartographic animations	✔	✔	~	✔	~	Opach and Nossum (2011)

Studies selected for meta-analysis are highlighted in bold

Of the 26 studies identified, 17 have met the criteria and were to be reviewed and used as cases in the research (cf. Table 1). Although meeting the criteria, a few studies were dismissed because of low actuality or the particular focus of research. Some of the projects followed a user-centred design approach throughout the application design cycle, but most of the reports present singular user or usability research efforts. The user experience (UX) of Web maps was not particularly in the focus of any of the reviewed studies, clearly indicating that the user experience of mapping applications offers potential for further research.

A Framework for User-Centred Web-Map Design and Evaluation

A framework for user-centred Web-map design and evaluation was developed from theory and the existing model by van Elzakker et al. (2008) and refined based on the meta-analysis of user studies. The model of the framework (Fig. 1) features the key elements for user-centred design and evaluation of Web-mapping products in four stages (application goals & requirements, application concepts, application prototypes, final application). Methods that were used successfully in the individual studies for the development of Web-mapping applications are listed between the elements of the model. The methods which are in bold type can be used when following a minimal UCD strategy.

Integrating user and usability research in all the stages of the framework is a time-consuming, expensive and, depending on the methods, complex activity. This is the reason why smaller project teams have difficulties to incorporate a complete user-centred design approach in Web application design. On these grounds, one of the first steps of the integrated framework is to develop and formalize a project-specific strategy for user and usability research activities.

A formal user research plan provides direction and a schedule, helps to deliver results when they are most needed for product development and avoids unnecessary, redundant, or hurried research (Kuniavsky 2003). Moreover, a formalized strategy can be used for communication and coordination between stakeholders and developers. The user/usability research strategy defines which types and methods of user and usability research, in which stages are necessary and feasible. Further additional elements comprise characteristics and resources of the development

Integrated Framework for User-centred Web-map Design and Evaluation

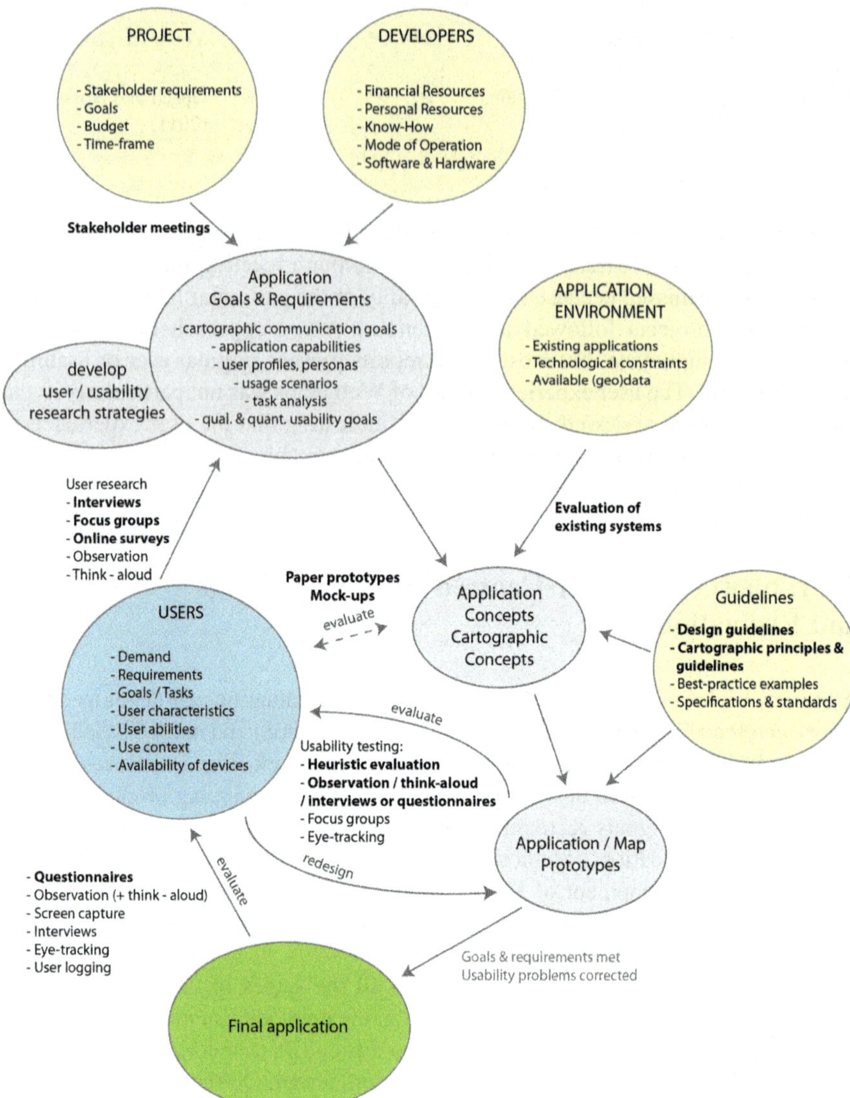

Fig. 1 Integrated framework for the user-centred design and evaluation of Web-mapping applications. Methods which have been commonly used in Web-mapping development throughout the reviewed studies are listed. Methods that can be used when applying a minimal UCD strategy are shown in bold type

team, such as personal and financial resources, know-how and available software and hardware. The application environment, plays an important role within the framework, such that existing applications, technological constraints, and the available (geo)data further influence application design. Guidelines, standards, and best practice examples from cartography, general design and Web design are consulted for both, the building of concepts as well as evaluation processes.

Requirements Stage

The most important stage towards developing a usable and utile application is the thorough analysis of user requirements and application goals. A crucial question which needs to be answered initially is, whether there is any demand for the kind of application planned. Moreover, user research needs to be carried out with prospective end users to gather information on their requirements, goals/tasks, user characteristics, use contexts and which devices they want to use. In this stage methods which are typically applied for the development of Web-mapping applications are interviews, focus groups, online surveys, observation and think-aloud.

Project requirements are often already clearly defined through proposals or contracts. They influence application goals and requirements in terms of the expectations and required outcomes of stakeholders, as well as the scope, budget and time-frame for the project. From the developer's view, further elements that influence application requirements and goals are personal & financial resources, the available know-how (including technological know-how and know-how of user research and usability testing), as well as the available software and usual mode of operation. Smaller development teams will take a more informal approach to application design and user testing, whereas in bigger companies the mode of operation will be more formal. Together all these elements form the application goals and requirements (application capabilities, user profiles, scenarios of use, device requirements, and usability requirements). The know-how and resources of the developers do not only influence the application requirements, but also the development of strategies for user and usability research. The research strategies also have to be adapted according to the user groups and availability of subjects. Furthermore, it is important to select a combination of user and usability research methods which create data needed for facilitating decision making at the stake-holder level.

In a second stage the established application goals and requirements lead into application and cartographic concepts. These concepts present a theoretical model of how the Web-mapping application should work, which functions it should have, how information is going to be presented and how the interaction between users and the application will take place. The cartographic concepts should at least include the types of information shown, in the map, the projection, scale levels, and symbolisation of data (Tyner 2010).

The concepts are further influenced by the environment in which the application will have to sustain (especially by an evaluation of similar existing applications), technological constraints which may impede the development, and the availability of data. Paper-prototypes and mock-ups can be used to test the application concepts with prospective users. Furthermore, the concepts should be developed with the consideration of established design principles, existing specifications and standards and best-practice examples. The application concepts should be thoroughly verified against the previously established goals and requirements. In this stage the expectations and requirements of all groups involved need to be balanced and compromises found between conflicting requirements and goals.

Prototype Stage

After the concepts have been developed and refined the next step involves creating a prototype or a series of prototypes of the application. Prototype development is an iterative process that sometimes involves several rounds of usability testing and refinement. Prototypes can be early paper and pen sketches of the application elements and different configurations, interactive mock-ups (e.g. interactive *PowerPoint* presentations), up to very sophisticated programmed versions of the Web-mapping application. The development team regularly needs to check the prototypes against the established goals and requirements. Lidwell et al. (2003) advise to actively involve members of the target audience in various stages of iterations to support testing and to verify if requirements are met. User testing with the actual prototypes allows problems in the application concepts to be identified, as well as actual and potential usability problems. Prototypes give the development team the opportunity to test and redesign the prototype until the application fulfils the goals and requirements and is error-free in almost all instances. In this stage the testing methods which were used in the reviewed studies were heuristic evaluation, observation, think-aloud, interviews, questionnaires, focus-groups and eye-tracking.

Post-Release Stage

When the latest version of a Web-mapping prototype meets application goals and requirements to a satisfactory level and all major usability problems have been corrected, the application can finally be released to the users. Once an application is released, regular testing of the usability, performance and user satisfaction can help to make sure that smaller bugs are identified and can be corrected in the released application, and that the application stays utile and up-to-date. In this stage questionnaires administered to end users, observation (plus think-aloud), screen capture methods, interviews, eye-tracking, and user-logging were providing good results in the meta-analysis of user-studies. Since the Web and associated technologies are changing at a fast pace, it is important to regularly evaluate user requirements and goals. Major changes in user expectations and requirements give an indication when revisions of the Web-mapping application become necessary.

Common Usability Problems

The meta-analysis of 17 user studies in the domain of Web mapping did uncover usability problems which did reappear in different projects. Therefore it is advisable that the developers of Web-mapping applications devote some time to check if one or more of the following usability problems might occur within their systems. The most common usability problems with desktop Web-mapping applications were:

- Problems with the intuitiveness and grouping of functions and icons;
- Problems with the position, size and function of the search box;
- Missing help, tutorials, and/or tooltips;
- Problems with the legend which was either missing or unreadable;
- Map window sizes were too small;
- Problems with the map quality. Maps were missing detail;
- Maps were overloaded and unreadable;
- System interfaces made use of unclear language (jargon);
- Loading times for maps were too long;
- Minimalist icons (just graphics, with no explanation) were not intuitive enough;
- Users were missing functions for exporting/printing the maps;
- Some interfaces had consistency problems (language, style, fonts);
- Problems with the menu structure; and
- Insufficient feedback on system status was given.

For mobile mapping applications the most common usability problems were:

- Problems with the position and selection of landmarks and with their iconic representation;
- Confusion about the map-orientation when users were standing still or maps were not adapting themselves to the direction; and
- Problems with positioning (GPS, indication of position on the maps).

It is possible that some of these issues could have been avoided if the design of prototypes or applications would have been supported by established design principles, or if heuristic evaluation would have been carried out prior to usability testing with users. Therefore it is advisable to consult principles and heuristics for interface design and for cartographic design when building a new Web-mapping application.

Conclusion

The proposed framework provides a theoretical foundation for conducting user and usability research in various stages of application design. The framework can help cartographic application developers who are not experienced in methods of user and usability research to conduct user studies and to design applications which are utile and usable for the target audience. The

(continued)

framework can help to overcome the typical technology-centeredness of Web-mapping application design and instead to put the prospective users in the focus of design. What is still missing and could become a project for further research is the inclusion of user experience (UX) factors within a UCD framework for Web mapping applications as well as the development of a methodology for testing the user experience of Web maps

This paper presented an excerpt of the doctorate research (Schobesberger 2012a) conducted in the area of user-centred design for Web-mapping applications. For further information, the full dissertation can be downloaded at the Website of the ICA Commission on Use and User Issues (http://www.univie.ac.at/icacomuse).

Acknowledgments The research underlying this paper was partly funded by the University of Vienna Research Grant 2011 and the Austrian Science Fund Research Network (FWF S-98) "The Cultural History of the Western Himalaya from the 8th Century". I am very thankful to my doctorate supervisor Prof. William Cartwright for his invaluable support and assistance.

References

Andrienko N et al (2002) Testing the usability of interactive maps in CommonGIS. Cartogr Geogr Inf Sci 29(4):325–342

Bhowmick T et al (2008) Distributed usability evaluation of the Pennsylvania cancer atlas. Int J Health Geogr 7(36)

Burghardt D, Wirth K (2011) Comparison of evaluation methods for field-based usability studies of mobile map applications. International Cartographic Association, Paris

Chincholle D, Goldstein M, Nyberg M, Eriksson M (2002) Lost or found? A usability evaluation of a mobile navigation and location-based service. In: Paternò F (ed) Mobile HCI 2002. Springer, Berlin, Heidelberg, pp 211–224

Coltekin A, Heil B, Garlandini S, Fabrikant SI (2009) Evaluating the effectiveness of interactive map interface designs: a case study integrating usability metrics with eye-movement analysis. Cartogr Geogr Inf Sci 36(1):5–17

Delikostidis I, van Elzakker CP (2011) Usability testing of a prototype mobile navigation interface for pedestrians. International Cartographic Association, Paris

Dix A, Finlay J, Abowd GD, Russell B (2005) Human–computer interaction. http://cognac.ai.ru.nl/studie/MMI_summary.pdf (online)

Flink H-M et al (2011) Usability evaluation of a map-based multi-publishing service. In: Ruas A (ed) Advances in cartography and GIScience: selection from ICC 2011, Paris, Lecture Notes in Geoinformation and Cartography. Springer, Berlin, Heidelberg, pp 239–257

Gould JD, Lewis C (1985) Designing for usability: key principles and what designers think. Commun ACM 28(3):300–311

Heil B (2009) Design and evaluation of a natural hazard information system. In: Kriz K, Kainz W, Riedl A (eds) Geokommunikation im Umfeld der Geographie – Wiener Schriften zur Geographie und Kartographie. Institut für Geographie und Regionalforschung, Wien, pp 62–70

Jokela T, Iivari N, Matero J, Karukka M (2003) The standard of user-centered design and the standard definition of usability: analyzing ISO 13407 against ISO 9241-11. s.l., ACM

Komarkova J et al (2007) Usability of GeoWeb sits: case study of Czech regional authorities Web sites. In: Abramowicz W (ed) Business information systems. Lecture Notes in Computer Science. Springer, Berlin, Heidelberg, pp 411–423

Kramers RE (2007) The atlas of Canada – user centred development. In: Cartwright W, Peterson M, Gartner G (eds) Multimedia cartography, 2nd edn. Springer, Berlin, Heidelberg, pp 139–160

Kuniavsky M (2003) Observing the user experience – a practitioner's guide to user research. Morgan Kaufmann Publishers, San Francisco

Lidwell W, Holden K, Butler J (2003) Universal principles of design. Rockport Publishers, Beverly

Meng L, Reichenbacher T (2005) Map-based mobile services. In: Meng L, Zipf A, Reichenbacher T (eds) Map-based mobile services: theories, methods and implementations. Springer, Berlin, Heidelberg, New York, pp 1–10

Midtho T, Nordvik T (2007) Effects of animations in zooming and panning operations on Web maps: a Web-based experiment. Cartogr J 44(4):292–303

mobiThinking (2011) Global mobile statistics 2011: all quality mobile marketing research, mobile Web stats, subscribers, ad revenue, usage, trends. . ... http://mobithinking.com/mobile-marketing-tools/latest-mobile-stats. Accessed 29 Nov 2011 (Online)

Murphy R, Rowe C, Hull J (2005) Google maps comparison study. San Jose State University, San Jose

Nivala A-M (2005) User-centred design in the development of a mobile map application. Licentiate thesis, Helsinki University of Technology, Helsinki

Nivala A-M, Sarjakoski L, Jakobsson A, Kaasinen E (2003) Usability evaluation of topographic maps in mobile devices. International Cartographic Association, Durban, pp 1903–1914

Opach T, Nossum A (2011) Evaluating the usability of cartographic animations with eye-movement analysis. International Cartographic Association, Paris

Peterson M (2003) Maps and the Internet: an introduction. In: Peterson M (ed) Maps and the Internet. Elsevier, Oxford, pp 1–16

Peterson M (2008) International perspectives on maps and the Internet: an introduction. In: Peterson M (ed) International perspectives on maps and the Internet. Springer, Berlin, Heidelberg, pp 3–10

Pucher A (2008) Use and users of the ÖROK-atlas online. Cartogr J 45(2):108–116

Pucher A, Schobesberger D (2011) Implicit user logging as a source for enhancing the usability of Web-delivered cartographic applications. International Cartographic Association, Paris

Pulsifier PL, Parush A, Lindgaard G, Taylor F (2005) The development of the cybercartographic atlas of Antarctica. Cybercartography: theory and practice. Elsevier, Amsterdam, pp 461–490

Richmond ER, Keller CP (2003) Internet cartography and official tourism destination Web sites. In: Peterson M (ed) Maps and the Internet. Elsevier, Oxford, pp 77–96

RNCOS E-Services (2011) World GPS market forecast to 2013. www.researchandmarkets.com/reports/836704/world_gps_market_forecast_to_2013.pdf. Accessed 27 Nov 2011 (Online)

Roth RE, Harrower M (2008) Addressing map interface usability: learning from the lakeshore nature preserve interactive map. Cartogr Perspect 60:46–66

Schobesberger D (2010) User-centred design of a Web-based cartographic information system for cultural history. In: Kriz K, Cartwright W, Hurni L (eds) Mapping different geographies. Springer, Berlin, Heidelberg, pp 159–170

Schobesberger D (2012a) Towards a framework for improving the usability of Web-mapping products. Dissertation, Universität Wien: s.n.

Schobesberger D (2012b) WISER user study January 2012 – Internal Report, Vienna: s.n.

Skarlatidou A (2005) The current state of the online mapping industry: a usability study of Multimap's public service and that of its competitors. M.Sc. thesis, University College London, London

Travis D (2011) User focus – ISO 13407 is dead. Long live ISO 9241-210! http://www.userfocus.co.uk/articles/iso-13407-is-dead.html. Accessed 20 Mar 2014 (Online)

Tyner JA (2010) Principles of map design. The Guilford Press, New York

van Elzakker C, Wealands K (2007) Use and users of multimedia cartography. In: Cartwright W, Peterson M, Gartner G (eds) Multimedia cartography. Springer, Berlin, pp 487–504

van Elzakker CP, Delikostidis I, Van Oosterom P (2008) Field-based usability evaluation methodology for mobile geo-applications. Cartogr J 45(2):139–149

Vilaca S (2008) Human computer interaction's role in geographical information systems development. Master thesis, Faculdade de Engenharia da Universidade do Porto, Porto

Voldán P (2010) Usability testing of Web mapping portals. Czech Technical University, Prague

Wealands K, Cartwright WE, Miller S, Benda P (2005) User assessment for developing optimal cartographic representation models within an Australian mobile location-based services travel application. International Cartographic Association, A Coruna

You M, Chen C-W, Liu H, Lin H (2007) A usability evaluation of Web map zoom and pan functions. Int J Des 1(1):15–25

Interlinking Opensource Geo-Spatial Datasets for Optimal Utility in Ranking

D. Bhattacharya, P. Pasquali, J. Komarkova, P. Sedlak, A. Saha, and P. Boccardo

Introduction

There is a huge amount of digital geospatial data available freely over the internet from various agencies, groups, consortiums and companies (OpenStreetMap data; Natural Earth data; GEONAMES; OSM Place-ranks; Ashton). These data can be downloaded raw, as graphical maps and images, as well as geospatially organized database tables. Some examples are Open Street Map (OSM), Natural Earth (NE), GeoNames database, Global Administrative Unit Layers (GAUL), Database of Global Administrative Areas (GADM), Second Administrative Level Boundaries dataset (SALB), not to forget geospatial projects by corporations like Google, Microsoft etc. The utilization of such large sized databases is mostly towards enhanced decision making and generating as much informative a map as possible (OpenStreetMap data; Natural Earth data; GEONAMES; OSM Place-ranks; Ashton). Bodies such as Open Geospatial Consortium (OGC) and

D. Bhattacharya (✉)
Faculty of Engineering, Geomatics Engineering Department, Bulent Ecevit University, Universite Caddesi, Room 201, Zonguldak 67100, Turkey
e-mail: devanjan@gmail.com

P. Pasquali • P. Boccardo
Information Technology for Humanitarian Assistance, Cooperation and Action (ITHACA), Politecnico di Torino, C.so Castelfidardo 30/a, 10138 Torino, Italy

J. Komarkova • P. Sedlak
Faculty of Economics and Administration, Institute of System Engineering and Informatics, University of Pardubice, Studentska 84, 53210 Pardubice, Czech Republic

A. Saha
Fakultě elektrotechniky a informatiky, Univerzity Pardubice, Arnošta z Pardubic 2604, 53002 Pardubice, Czech Republic

© Springer International Publishing Switzerland 2015
J. Brus et al. (eds.), *Modern Trends in Cartography*, Lecture Notes in Geoinformation and Cartography, DOI 10.1007/978-3-319-07926-4_13

other open-source communities are propagating open geospatial data in a big way (OGC 2007; UCGIS 2004). But the availability of numerous and such large sized databases puts a pointer towards comprehensiveness and pervasiveness of each dataset related to any geospatial extent on earth. It has to be determined which dataset gives most elaborate and complete information, or whether an integration of multiple datasets is required (Batini et al. 1986; Beeri et al. 2004, 2005; Chen et al. 2003; Butenuth et al. 2007). One of the essential aspects of digital mapping and online visualization of maps is the prioritized ranking of geolocations with respect to their attributes and this facility is available as rank columns in Natural Earth data tables which need to be merged with other datasets for creating a complete and exhaustive mapping example.

The online mapping systems facilitate many geospatial datasets which are used, created, edited and maintained including the use of GeoNames as the layer for populated places. Herein lies the challenge addressed in this paper which delves in-depth into the completeness of three datasets OSM, NE and GeoNames and whether these can contribute singly or integrally towards better ranking of geolocations. In this section the paper describes the definitions of the important terms that define the current research leading to describe the objective. The next section elaborates on the available literature and research up to now towards merging of geospatial data and what gaps remain those need to be filled. The methodology for filling those gaps is discussed in the subsequent section. The test-bed is also described. The result section puts down the success of the research. Basically this paper deals with the success rate, accuracy and efficiency of joining diverse open-source databases and proves the concept presently using only OSM, NE and GeoNames and adding ranking column correctly to the final database.

Simplistically put ranking of a geolocation (any place on earth depending on scale) is the priority associated with that place to make it visible at a particular zoom level when viewed on a dynamic web map such as OSM or Google Map. When the user zooms on a map then which places and attributes would be visible is decided according to their rankings (Budak et al. 2006; Chang and Park 2006; Safra et al. 2010). Another important field in the database of featured indexed contents is geometry (Safra and Doytsher 2006a). Geometry is the representation of latitude-longitude of a point with the spatial reference system taken into account. It is an important attribute in any geospatial database as it helps to index any given point and search it uniquely. In this paper, we have developed join algorithms under the following assumptions. First, we assume that locations of objects are recorded as points. More complex forms of recording locations (e.g., polygons) have been approximated by points (e.g., by computing the geometry). Second, in each dataset, distinct objects represent distinct real-world entities. This is a realistic assumption for many GIS applications. Finally, we consider join algorithms that use locations of objects, and then move on to matching with names in a range around the location. As is evident, location-based join is a non-trivial task involving extracting elements from multiple datasets. Understanding the factors that determine the quality of location-based join algorithms is a basis for developing join algorithms that use all available properties.

The online mapping systems facilitate many geospatial datasets which are used, created, edited and maintained including the use of GeoNames as the layer for populated places (Chang and Park 2006; Safra et al. 2010; Safra and Doytsher 2006a). Many of them are not classified for importance because of the lack of additional information such as population or administrative level. A way to give an importance scale to the names is by linking the GeoNames to other datasets (OSM, natural earth). OpenStreetMap data provides a limited number of place classifications (such as city, town, village). For the best cartographic results we need classes that are more comprehensive about how they rank cities. "Which of these labels should be visible?" and "how much should this label be emphasized?" are important decisions that need to be made in cartographic design. To do this the present research is to join additional information from Natural Earth, OSM and Geonames and keep the path open for other datasets. Hence the objectives of this study are:

- To develop an algorithm for joining open-source geospatial datasets.
- To find a common link between the datasets considering the parameters of the geospatial databases.
- To establish accurate and efficient results.

This brings to the forefront the fact that each dataset has a unique set of properties and also attributes that are common to others but only a properly studied union of these can yield a completely versatile and comprehensively pervasive body of single point reference repository of open-source geospatial data. The requirement analysis reveals several challenges that have not been addressed in past and recent researches as shown in the next section of literature review that builds up the background of our study.

Background

A careful study of relevant literature brings forward the fact that in this area the major thrust has been on joining multisource geodata for localized areas rather than take-up the challenge of integrating globally used opensource databases and study their strengths when they are combined. So numerous studies have laid down the architecture for joining varied data ranging from remotely-sensed, scanned maps, geographical information system (GIS) layers, global positioning system (GPS) acquired navigation data and more (Butenuth and Heipke 2005; Doytsher 2000). This has brought about a localized effect of study to this domain since the datasets acquired were at best of a very limited extent (Egenhofer et al. 1989; Goesseln and Sester 2004; Rigeaux et al. 2001). The need for integration of heterogeneous data sources arises in many different cases, one example of which is interoperability of information systems (Butenuth and Heipke 2005; Doytsher 2000; Egenhofer et al. 1989), including geographical/geospatial information systems. Another example is mediator systems (Goesseln and Sester 2004; Rigeaux et al. 2001; Sattler et al. 2000) that help in data acceptance and understanding from one system to another or from a subsystem to next. For geo-spatial information systems, the data integration problem has two important sub-problems. In map merging or mosaicing, two digital maps are integrated to produce a new map (Walter and Fritsch 1999; Devogele 2002; Gal et al. 2003; Gravano et al. 2003). Another such area of use is in data fusion, where raster data received from sensors, is processed, by means of image-processing techniques, and then integrated (Laurini 1998; Lemarie and Raynal 1996; Sester et al. 1998). Most of these examples work upon algorithms for discovering corresponding objects and that can be a possible part of the solution to these problems. Whereas in geospatial location based problems such as used in this work, it has been necessary to investigate location-based join of open source geo-spatial datasets. For the current research work datasets from Open Street Map (OSM), Geonames and Natural Earth have been used from which an integrated dataset has been created.

There have been studies (Papakonstantinou et al. 1996; Park 2001; Safra and Doytsher 2006b; Spaccapietra and Parent 1994) where two join approaches, namely, the sequential and the holistic approaches, are presented and compared. The novelty of those works had been in developing, for each of the two approaches, effective join algorithms that use only locations of objects. They showed that the sequential normalized weights method is effective, that is, the result has a high recall precision combination when the "right" order of joins is being applied. In the "right" order, the studies joined first the pair of datasets that have the largest overlap and the fewest errors. For the holistic approach, the studies presented several novel methods. One version of the holistic normalized-weights method provides high recall and precision, under all circumstances (Baltsavias 2004; Spaccapietra et al. 1992; ESRI 2004; Friis-Christensen et al. 2005). Comparing the two approaches, it was found by the studies that the time complexity and the space complexity of the sequential normalized-weights method are lower than those of the holistic normalized weights method (GSDI 2005; Hampe et al. 2004; Lake 2005). The holistic approach, however, is capable of providing higher precision than the sequential approach (at the cost of lower recall). Another advantage of the holistic approach is that each join set is given a confidence value. These studies marked several problems that remained for future work. One problem they state was to optimize the run time of the algorithms. This was particularly important if those algorithms are to be included in real- time applications (Hampe et al. 2004; Lake 2005; Levy 1999). A second problem was of how to utilize most effectively locations that are given as polygons or lines, rather than just points (Lu et al. 2007; Ma et al. 2000; Malhotra 2000). A third research direction is to combine one approach with other approaches, such as the feature-based approach of (Chen et al. 2003; Butenuth et al. 2007; Beeri et al. 2004; Budak et al. 2006), topological similarity (e.g., Batini et al. 1986; Beeri et al. 2005) or ontologies (e.g., Egenhofer et al. 1989; Goesseln and Sester 2004; Rigeaux et al. 2001; Sattler et al. 2000; Walter and Fritsch 1999; Devogele 2002; Gal et al. 2003).

Some studies went forward taking the concept that GIS is a repository for geographic objects (Ashton) and that each object contains information about a real world entity; a real world entity is described uniquely in the system by one object. In the system the objects are organized in datasets, where each dataset contains objects about a certain subject (OGC 2007; UCGIS 2004; Batini et al. 1986; Beeri et al. 2005). The input for fusion process is several datasets from different sources (Batini et al. 1986; Chen et al. 2003). The datasets should overlap on both the subject and the described area. In a preprocessing filtering stage, objects, that does not belong to the intersection between the datasets, are eliminated. The studies assumed however, that the overlap (i.e. the number of world entities that appear in both sets) between the datasets, after the filtering stage, was not complete, i.e. there existed world entities which appeared only in one dataset (Batini et al. 1986; Beeri et al. 2005; Safra and Doytsher 2006a; Butenuth and Heipke 2005). When geographic datasets get integrated, the main task has been to identify when two or more objects, from the different sources, represent the same entity and fuse those objects to a single object. Since each entity represented at most one object in a dataset, a fusion set contained at most one object from each dataset (Batini et al. 1986; Chen et al. 2003). Thus, the matching between objects from two datasets was shown to be 1:1.

The papers (Masuyama 2006; Sagayaraj et al. 2006; Sripada et al. 2004) point out the fact that integration of geo-spatial data from heterogeneous sources has many important applications. One example they showed was combining up-to-date data, maybe from a satellite image, with data from a map that contains verbal descriptions of entities (Zhang et al. 2003; Ziegler and Dittrich 2004). When geographical entities are represented in different sources, and each source stores different properties of the entities, integration makes it possible to obtain all the available information on each entity. Integration is essentially a join of datasets. The main task in a join is to find all sets of corresponding objects, i.e., objects that represent the same real-world entity in distinct sources (Papakonstantinou et al. 1996; Park 2001; Safra and Doytsher 2006b). Over heterogeneous sources, however, finding corresponding objects is difficult, since there are no global identifiers. In principle, both spatial and non-spatial properties may be used, in lieu of global identifiers, for integrating geographical data. However, only location is always available for spatial objects (Sattler et al. 2000; Walter and Fritsch 1999; Devogele 2002). Since in many cases, locations uniquely identify objects in a dataset, location-based join seems to be an easy task. This is not so, however, for several reasons. First, measurements introduce errors, and the errors in different datasets are independent of each other. Second, each organization has its own approach and requirements. Hence, different organizations use different measure- ment techniques and may record spatial properties of entities using a different scale or a different structure (Friis-Christensen et al. 2005; GSDI 2005; Hampe et al. 2004). For example, one organization might represent buildings as points, while another could represent them as polygons. While an estimated point location can be derived from a polygonal shape, it may not agree with a point location in another database. A third reason could be displacements caused by cartographic generalizations. For the above reasons, location-based joins do not provide a precise answer, but rather an approximation. The quality of the approximation is determined by characteristics of the joined datasets, such as sizes of the errors, the density of objects, and the relative overlap. In this paper, we introduce algorithms for location-based join of three or more sources, under the assumptions that locations are given as points and each dataset has at most one object per real- world entity. The rationale underlying all our algorithms is that even in the presence of measurement errors, corresponding objects should have close locations and string matching using names could be done.

The review brought out several challenging research gaps that need to be addressed in this study, such as:

- merging opensource geospatial datasets has not been taken up.
- global datasets need to be utilized for merging such as OSM, NE and GeoNames.
- common attributes identification for reference of databases.
- algorithmic resolution of geo-location matching amongst the databases for higher accuracy.

Methodology

To meet the primary objective of the study which is to insert scale ranking data of populated places into OSM dataset, it is required that the places appearing in NE be matched with OSM places. OSM has a much bigger dataset than NE but NE has the all important attribute of place_rank. Hence both datasets need to be merged. The merging process is helped greatly by taking help from some fields in the GeoNames database, for which the reason and procedure is explained below in Fig. 1. The scale ranking data of populated places can be found in Natural Earth dataset. For ease of computation, it has been decided to begin with the populated places of a single country, Czech Republic (CZ) at present. NE dataset for populated places has 7,312 populated places among which 12 are from CZ.

The work environment has been with Postgis with Postgres on Windows along with Python. The pre-requisites in datasets and run-time environment are that for OSM place-ranks the algorithm joins city rankings from Natural Earth into OSM data with fuzzy matching. The setup computer script initiates by creating a PostGIS-enabled PostgreSQL database. By default this script assumes it is named 'osm'. It imports the Natural Earth cities information included in Fig. 1. Eg: psql -U postgres -f ne_cities.sql -d osm. It imports OSM places with Imposm if not already done. These can be from a full planet dump, a regional extract,or an Overpass API query. It has to be ensured that the Python package 'unidecode' is installed.

Eg:

```
sudo pip install unidecode.
```

Usage:

```
python rank-places.py | psql -U <pg_user> <pg_database>
```

The columns used while joining the OSM and NE datasets are 'way' and 'geom' respectively Fig. 1. Both are geometry type fields, however, they could belong to different spatial reference systems. Hence conversion of the geometry field of one of the datasets is performed based on the spatial reference system of the other. This transformation of data brings the geometry fields of both the datasets to the same

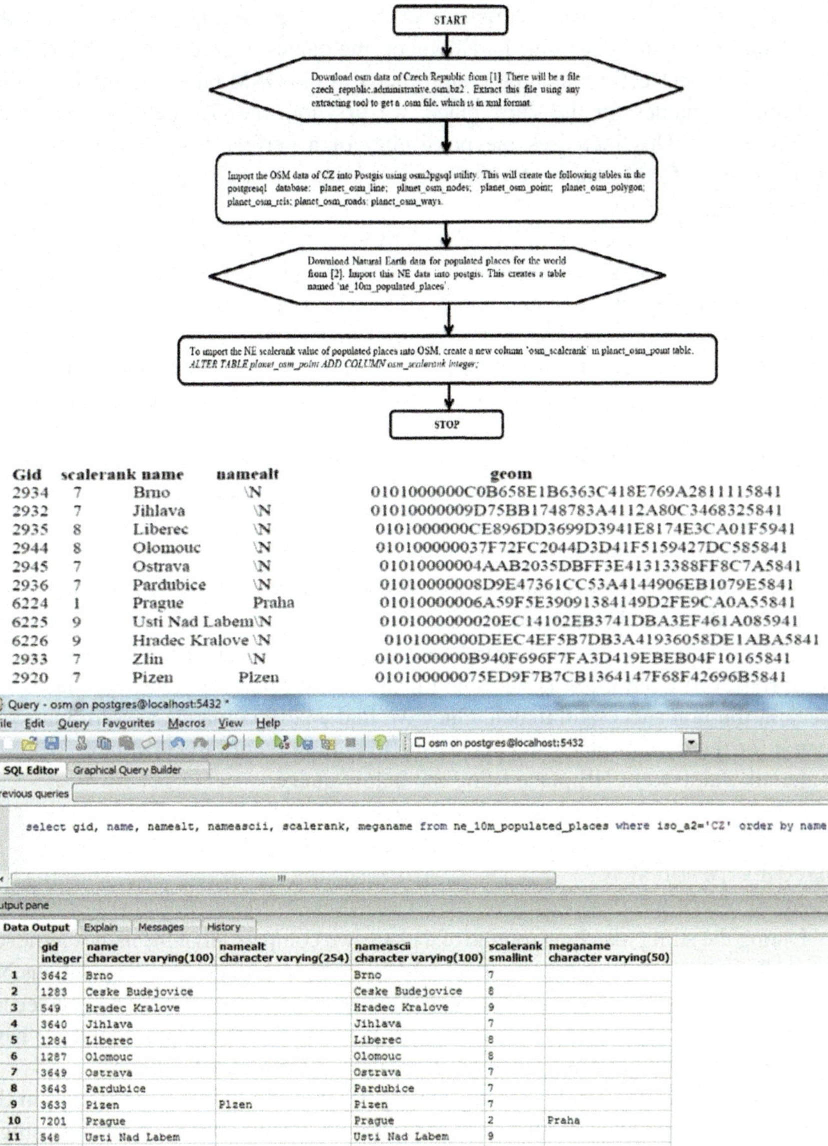

Gid	scalerank	name	namealt	geom
2934	7	Brno	\N	0101000000C0B658E1B6363C418E769A2811115841
2932	7	Jihlava	\N	01010000009D75BB1748783A4112A80C3468325841
2935	8	Liberec	\N	0101000000CE896DD3699D3941E8174E3CA01F5941
2944	8	Olomouc	\N	010100000037F72FC2044D3D41F5159427DC585841
2945	7	Ostrava	\N	01010000004AAB2035DBFF3E41313388FF8C7A5841
2936	7	Pardubice	\N	01010000008D9E47361CC53A4144906EB1079E5841
6224	1	Prague	Praha	01010000006A59F5E39091384149D2FE9CA0A55841
6225	9	Usti Nad Labem	\N	0101000000020EC14102EB3741DBA3EF461A085941
6226	9	Hradec Kralove	\N	0101000000DEEC4EF5B7DB3A41936058DE1ABA5841
2933	7	Zlin	\N	0101000000B940F696F7FA3D419EBEB04F10165841
2920	7	Pizen	Plzen	010100000075ED9F7B7CB1364147F68F42696B5841

Query - osm on postgres@localhost:5432 *
File Edit Query Favourites Macros View Help

osm on postgres@localhost:5432

SQL Editor   Graphical Query Builder

Previous queries

```
select gid, name, namealt, nameascii, scalerank, meganame from ne_10m_populated_places where iso_a2='CZ' order by name;
```

Output pane

Data Output | Explain | Messages | History

	gid integer	name character varying(100)	namealt character varying(254)	nameascii character varying(100)	scalerank smallint	meganame character varying(50)
1	3642	Brno		Brno	7	
2	1283	Ceske Budejovice		Ceske Budejovice	8	
3	549	Hradec Kralove		Hradec Kralove	9	
4	3640	Jihlava		Jihlava	7	
5	1284	Liberec		Liberec	8	
6	1287	Olomouc		Olomouc	8	
7	3649	Ostrava		Ostrava	7	
8	3643	Pardubice		Pardubice	7	
9	3633	Pizen	Plzen	Pizen	7	
10	7201	Prague		Prague	2	Praha
11	548	Usti Nad Labem		Usti Nad Labem	9	
12	3641	Zlin		Zlin	7	

Fig. 1 Starting procedure for joining NE, Geonames and OSM for CZ places, their attributes and tabularization

spatial reference system and hence makes them comparable. Next, to work on joining these datasets, we use a Python script to connect to the postgis database. From the NE table ne_10m_populated_places, fetch all records available for Czech Republic. We fetch only the required columns in the above rows in Fig. 1 so as to avoid unnecessary memory wastage. The important columns are the various fields specifying the primary and alternate names of the place, the geometry and the scalerank of the place (see Fig. 1). However, the names of places in the NE dataset may not have perfectly matching counterparts in the OSM table. To include other variations of names for the same place, we first join the Geonames dataset with Natural Earth. This increases the possibility of a perfect string match of place names with the OSM dataset. To perform this join, we use the 'geonameid' column present in both datasets.

For every record (place) found in NE for Czech Republic, we can scan the entire OSM table to find a match based on the name of the place. This can have a problem that the OSM table is huge and a search based on any non-indexed column will delay the query. To solve this issue, we first query the OSM table for all nearby places of a record fetched from the NE table based on the comparison of the geometry fields of both datasets. To define nearby places, we set a search_range variable to a distance, say for example, search_range = 10,000 m. So, we fetch all places from the OSM table that are within a distance of search_range from the selected place in NE. The OSM table is indexed on its geometry column, hence the query runs faster. We also limit the search of nearby places to 'city', 'town' and 'village' from OSM dataset.

Now, we have a list of nearby places from the OSM table, of which ideally one place should match the selected place from the NE table. We now perform a string matching routine by comparing the selected NE place name with every nearby place retrieved from OSM table, going nearest to farthest from the NE place. We break at the first successful match. Once we find a perfect match, we terminate our search for the current place and update the matching record in OSM table with the scalerank value from NE place using the below query.

```
UPDATE planet_osm_point SET osm_scalerank = 9 where
osm_id = 1601566699;
```

During the string matching routine of a place, we compare all of the names retrieved from NE as well as Geonames dataset with the OSM 'place' column. The various NE columns compared are 'name', 'namealt', 'nameascii', 'meganame' and 'namepar'. The Geonames column used here is 'name'. We continue this process until we have gone through all the places fetched from NE dataset for Czech Republic.

It is possible in a very few cases that a perfect string match is not found in the OSM dataset for a place from NE. In this case we attempt string matching using a fuzzy approach. For this approach we define the percentage of closeness to perfect match. For example, the 2 strings must be 80 % close to perfect match to qualify as the same place. If such a match is found, the scalerank value for the place in OSM is updated with the value from NE. The intermediate utilization of GeoNames to search alternate names of places of NE and OSM and match using partial, complete or fuzzy matching and put it properly in the OSM table column, the query is below. The string matching with GeoNames alt_names and using geoid + geom field of NE and OSM yields much higher accuracy than just 80 % and is displayed in results section next.

```
COPY ne_cities (gid, scalerank, name, namealt, geom)
FROM stdin;
```

Columns of Geonames data: The main 'geoname' table has the following fields: geonameid; name; asciiname; alternatenames (alternatenames, comma separated varchar(5000)); latitude: latitude in decimal degrees (wgs84); longitude: longitude in decimal degrees (wgs84); feature class (OpenStreetMap data; Natural Earth data; GEONAMES); feature code (OpenStreetMap data; Natural Earth data; GEONAMES); country code; admin1; admin2; admin3; admin4; population; elevation; dem: digital elevation model; srtm3; timezone: the timezone id; modification date: date of last modification in yyyy-MM-dd format. The collated dataset for maintaining ranking related experiments has been developed using GeoNode and GeoServer which depicts the names attributes of various locations across the globe. At a certain zoom level there are predetermined attributes that are called from the database table according to the rank specified.

Test Results for the Above Approach for Czech Republic

The columns updated in planet_osm_point table after the dataset joining is shown below (Fig. 2). Before joining, the osm_scalerank column for the below places were 0 (zero). We see that scale ranking of all the places for CZ present in NE dataset have been propagated correctly to OSM dataset. Hence the 12 cities that exist for CZ as common between NE, OSM, GeoNames are found to be correctly matched yielding an accuracy of 95–100 %. The geometry, name and geo-id complete and fuzzy searching and matching around a buffer of 50 km took a minimum of 30 s to maximum 1 min in a commodity computer with 2 GHz, 2 GB memory. The query run to provide the part screenshot in Fig. 2.

```
select st_astext(way) as geom, name, place, population,
osm_scalerank from planet_osm_point where osm_scalerank != 0
order by name;
```

The results accomplished are:

- Linked together open-source geospatial datasets Natural Earth, Geonames and Open Street Map.
- Joined city rankings from Natural Earth into OSM+Geonames data with fuzzy matching.
- Linked datasets, starting from populated places, integrated attributes such as administrative class, population, and also how the name is spelt, local form and English form and so on. Sample datasets used are from Czech country dataset.

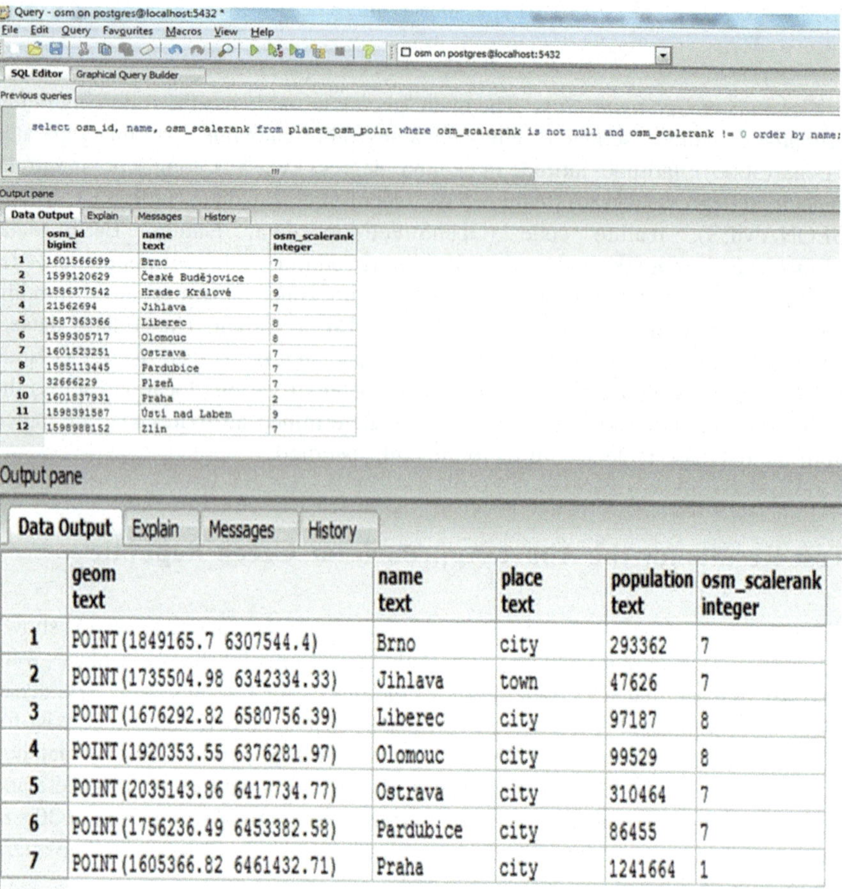


```
select osm_id, name, osm_scalerank from planet_osm_point where osm_scalerank is not null and osm_scalerank != 0 order by name;
```


	osm_id bigint	name text	osm_scalerank integer
1	1601566699	Brno	7
2	1599120629	České Budějovice	8
3	1586377542	Hradec Králové	9
4	21562694	Jihlava	7
5	1587363366	Liberec	8
6	1599305717	Olomouc	8
7	1601523251	Ostrava	7
8	1585113445	Pardubice	7
9	32666229	Plzeň	7
10	1601837931	Praha	2
11	1598391587	Ústí nad Labem	9
12	1598988152	Zlín	7

	geom text	name text	place text	population text	osm_scalerank integer
1	POINT (1849165.7 6307544.4)	Brno	city	293362	7
2	POINT (1735504.98 6342334.33)	Jihlava	town	47626	7
3	POINT (1676292.82 6580756.39)	Liberec	city	97187	8
4	POINT (1920353.55 6376281.97)	Olomouc	city	99529	8
5	POINT (2035143.86 6417734.77)	Ostrava	city	310464	7
6	POINT (1756236.49 6453382.58)	Pardubice	city	86455	7
7	POINT (1605366.82 6461432.71)	Praha	city	1241664	1

Fig. 2 Natural Earth data for Czech Republic cities, Scale-ranks updated in OSM data of Cz based on NE data of Cz, Part screenshot of the combined OSM and NE database with attributes like alt_names from GeoNames, yielding 11/12 places correct match out of 12 CZ cities multiple times

Conclusions, Limitations & Future Scope

The research brings to the forefront an important concept of utilizing multiple web-based geodata sources and the possibility to get a master repository of such databases integrated together. This paper lays down the basics of such joining of opensource data tables that has been achievable by following certain breakthroughs like first combining OSM with GeoNames and then searching NE for place matching. This approach increased accuracy to almost 95–100 % which had not been achieved before. Joining of datasets requires a

(continued)

number of searching and matching operations which grow exponentially when the number of countries increases. Currently the algorithm works one country at a time so it is not that time consuming but when run on global datasets, it is going to take days to complete which is not desirable. Hence faster processing techniques need to be explored and researched. Distributed processing and acquiring facilities for faster processing research of joining multiple datasets so outside channeling of work is the viable option. The Distributed Processing Hadoop Lab is a feasible research facility for undertaking the resolution of above limitations. The utility of the approach lies in identifying places having names in multiple languages, scripts and popular names, official names, and nicknames. So complete matching, partial matching, fuzzy matching have been employed. For example related to a city such as Prague in OSM the properties of the city is bound to be scattered across naming conventions in local language and urban expansions of Prague. In such a scenario getting 100 % match is a commendable achievement and it has been achieved in this paper. No simple or straight forward algorithm could have done that. In a single country like India, where the dialect changes every 200 km, GeoNames contains information of a place in India in a lot of ways and this is true for OSM, GoogleMap and a lot of others. It is true that a lot of places at present are not ranked but that will change very soon with addition of data. We need to improve the ranking constantly. To some extent the paper has achieved cartographic results for the better where more comprehensive information was extracted from geodatasets towards ranking cities. By no means is the task over since to proceed towards the important decisions of "Which of these labels should be visible?" and "how much should this label be emphasized?" only partial breakthroughs have been achieved in this paper. To do further future task is to join additional information. One more advancement envisioned is to place the joined datasets in GeoServer to integrate with the Graphical Interface of World Maps.

Acknowledgments The Ministry of Education, Youth and Sports of the Czech Republic, Project CZ.1.07/2.3.00/30.0021 "Strengthening of Research and Development Teams at the University of Pardubice", financially supported this work.

References

Ashton AJ. Processsing OpenStreetMap data for effective cartography, website: https://www.mapbox.com/blog/processing-osm/. Published 17 Oct 2012. Accessed 2 July 2013. https://www.mapbox.com/blog/2012-08-09-mapbox-streets-design-update/. Accessed 15 Aug 2013

Baltsavias EP (2004) Object extraction and revision by image analysis using existing geodata and knowledge: current status and steps towards operational systems. ISPRS J Photogr Remote Sensing 58(3–4):129–151. http://www.sciencedirect.com/science/article/pii/S0924271603000546

Batini C, Lenzerini M, Navathe SB (1986) A comparative analysis of methodologies for database schema integration. ACM Comput Surv 18(4):323–364. http://pdf.aminer.org/000/338/660/on_the_equivalence_among_data_base_schemata.pdf

Beeri C, Kanza Y, Safra E, Sagiv Y (2004) Object fusion in geographic information systems. In: Proceedings of the 13th international conference on very large data bases, Toronto, ON. http://www.vldb.org/conf/2004/RS21P4.PDF

Beeri C, Doytsher Y, Kanza Y, Safra E, Sagiv Y (2005) Finding corresponding objects when integrating several geo-spatial datasets. In: Proceedings of the 13th ACM international symposium on advances in geographic information systems, Bremen, 4–5 November 2005, pp 87–96. http://dl.acm.org/citation.cfm?id=1097078

Budak A, Sheth A, Ramakrishnan C, Lynn Usery E, Azami M, Kwan MP (2006) Geospatial ontology development and semantic analytics. Trans GIS 10:551–575. http://ncgia.ucsb.edu/projects/nga/docs/ontology.pdf

Butenuth M, Heipke C (2005) Network snakes-supported extraction of field boundaries from imagery. In: Kropatsch WG, Sablatnig R, Hanbury A (eds) 27th DAGM symposium, Wien, Österreich, LNCS, vol 3663. Springer, Berlin, Heidelberg, pp 417–424

Butenuth M, Gösseln G v, Tiedge M, Heipke C, Lipeck U, Sester M (2007) Integration of heterogeneous geospatial data in a federated database. ISPRS J Photogr Remote Sensing 62:328–346 (Elsevier). http://www.sciencedirect.com/science/article/pii/S0924271607000275

Chang Y-S, Park H-D (2006) XML web service-based development model for Internet GIS applications. Int J Geogr Inf Sci 20:371–399. http://www.tandfonline.com/doi/abs/10.1080/13658810600607857#.UzVm6KhdVHQ

Chen CC, Thakkar S, Knoblock C, Shahabi C (2003) Automatically annotating and integrating spatial datasets, advances in spatial and temporal databases. In: Hadzilacos T, Manolopoulos Y, Roddick J, Theodoridis Y (eds) 8th international symposium, SSTD 2003, Santorini Island, Greece, Proceedings, Lecture Notes in Computer Science, 2003. doi:10.1007/978-3-540-45072-6_27. Springer, Berlin, Heidelberg, pp 469–488

Devogele T (2002) A new merging process for data integration based on the discrete Frechet Distance. IAPRS, vol 34. http://link.springer.com/chapter/10.1007%2F978-3-642-56094-1_13#page-1

Doytsher Y (2000) A rubber sheeting algorithm for non-rectangular maps. Comput Geosci 26(9–10):1001–1010

Egenhofer MJ, Frank AU, Jackson JP (1989) A topological data model for spatial databases. Lecture Notes Comput Sci 409:271–286

ESRI (2004) Xml schema of the geodatabase – an ESRI technical paper, ESRI 380 New York St., Redlands, CA, 92373–8100, USA. http://downloads.esri.com/support/whitepapers/ao_/J9620_XML_Schema_of_Geodatabase.pdf

Friis-Christensen A, Nytun JP, Jensen CS, Skogan D (2005) A conceptual schema language for the management of multiple representations of geographic entities. Trans GIS 9:345–380

Gal A, Trombetta A, Anaby-Tavor A, Montesi D (2003) A model for schema integration in heterogeneous databases. IDEAS, pp 2–11. http://ieeexplore.ieee.org/xpls/abs_all.jsp?arnumber=1214906&tag=1

GEONAMES, website http://www.geonames.org/export/codes.html

Goesseln G v, Sester M (2004) Integration of geoscientific data sets and the German digital map using a matching approach. Int Arch Photogr Remote Sensing 35 (Part 4B):1249–1254. http://www.cartesia.org/geodoc/isprs2004/comm4/papers/534.pdf

Gravano L, Ipeirotis PG, Koudas N, Srivastava D (2003) Text joins in an RDBMS for web data integration. In: Proceedings of the 12th international conference on World Wide Web. http://www2.research.att.com/~divesh/papers/giks2003-textjoins.pdf

GSDI (2005) Spatial data infrastructure. Global Spatial Data Infrastructure Association

Hampe M, Sester M, Harrie L (2004) Multiple representation databases to support visualisation on mobile devices. Commission IV, WG IV/2. http://citeseerx.ist.psu.edu/viewdoc/download?doi=10.1.1.184.3303&rep=rep1&type=pdf

Lake R (2005) The application of geography markup language (GML) to the geological sciences. Comput Geosci 31:1081–1094. http://www.sciencedirect.com/science/article/pii/S0098300405001032

Laurini R (1998) Spatial multi-database topological continuity and indexing: a step towards seamless GIS data interoperability. Int J Geogr Inf Sci 12(4):373–402. http://www.tandfonline.com/doi/abs/10.1080/136588198241842#.UzVh96hdVHQ

Lemarie C, Raynal L (1996) Geographic data matching: first investigations for a generic tool. In: Proceedings of GIS/LIS, pp 405–420. http://training.esri.com/bibliography/index.cfm?event=general.recordDetail&ID=10135

Levy A (1999) Logic-based techniques in data integration. Department of Computer Science and Engineering, University of Washington. http://citeseerx.ist.psu.edu/viewdoc/summary?doi=10.1.1.33.4733

Lu C-T, Santos RD, Sripada L, Kou Y (2007) Advances in GML for geospatial applications. GeoInformatica 11:131–157. http://link.springer.com/article/10.1007%2Fs10707-006-0013-9

Ma C, Chou DC, Yen DC (2000) Data warehousing, technology assessment and management. Ind Manage Data Syst 100(3):125–134. http://dx.doi.org/10.1108/02635570010323193

Malhotra Y (2000) Knowledge management for [E-] business performance. Information strategy. Exec J 16(4):5–16. http://www.brint.org/KMEbusiness.pdf

Masuyama A (2006) Methods for detecting apparent differences between spatial tessellations at different time points. Int J Geogr Inf Sci 20:633–648. http://www.tandfonline.com/doi/pdf/10.1080/13658810600661300

Natural Earth data, website: http://www.nacis.org/naturalearth/10m/cultural/ne_10m_populated_places.zip

OGC (2007) Geography Markup Language By Open Geospatial Consortium Inc., https://portal.opengeospatial.org/files/?artifact_id=11339

OpenStreetMap data, website: http://downloads.cloudmade.com/europe/eastern_europe/czech_republic#downloads_breadcrumbs

"OSM Place-ranks", website: https://github.com/mapbox/osm-place-ranks. Accessed: 1 July 2013. http://wiki.openstreetmap.org/wiki/Nominatim/Development_overview. Accessed 12 Aug 2013

Papakonstantinou Y, Abiteboul S, Garcia-Molina H (1996) Object fusion in mediator systems. In: Proceedings of the 22nd international conference on very large databases. http://db.ucsd.edu/pubsFileFolder/147.pdf

Park J (2001) Schema integration methodology and toolkit for heterogeneous and distributed geographic databases. J Korea Ind Inf Syst Soc 51–64. http://citeseerx.ist.psu.edu/viewdoc/download?doi=10.1.1.203.1655&rep=rep1&type=pdf

Rigeaux P, Scholl M, Voisard A (2001) Spatial databases with application to GIS. Morgan Kaufman Publishers. http://store.elsevier.com/Spatial-Databases/Philippe-Rigaux/isbn-9781558605886/

Safra E, Doytsher Y (2006a) Integration of multiple geo-spatial datasets. In: ASPRS annual conference, Reno, Nevada, 1–5 May. http://www.asprs.org/a/publications/proceedings/reno2006/0132.pdf

Safra E, Doytsher Y (2006b) Integrating a sequence of geo-spatial datasets. GISRUK 2006. http://www.asprs.org/a/publications/proceedings/reno2006/0132.pdf

Safra E, Kanza Y, Sagiv Y, Beeri C, Doytsher Y (2010) Location-based algorithms for finding sets of corresponding objects over several geo-spatial data sets. Int J Geogr Inf Sci 24(1):69–106. http://dx.doi.org/10.1080/13658810802275560

Sagayaraj F, Thambidurai P, Bharadwaj BS, Balagei GN, Hemant N (2006) An improved and efficient storage technique for GIS geometric primitives based on minimum bounding rectangles. In: 9th annual international conference, Map India 2006

Sattler K-U, Conrad S, Saake G (2000) Adding conflict resolution features to a query language for database federations, pp 41–52. http://wwwiti.cs.uni-magdeburg.de/iti_db/publikationen/ps/00/SatConSaa00.ps.gz

Sester M, Anders KH, Walter V (1998) Linking objects of different spatial data sets by integration and aggregation. GeoInformatica 2(4):335–358

Spaccapietra S, Parent C (1994) View integration: a step forward in solving structural conflicts. IEEE Trans Knowl Data Eng 6(2):258–274

Spaccapietra S, Parent C, Dupont Y (1992) Model independent assertions for integration of heterogeneous schemas. VLDB J 1(1):81–126

Sripada LN, Lu CT, Wu W (2004) Evaluating GML support for spatial databases. Compsac 2004. In: Proceedings of the 28th annual international computer software and applications conference. http://ieeexplore.ieee.org/xpls/abs_all.jsp?arnumber=1342680&tag=1

UCGIS (2004) University Consortium for Geographic Information Science 2002 Research Agenda. Http://Www.Ucgis.Org/Priorities/Research/2002researchagenda.Htm

Walter V, Fritsch D (1999) Matching spatial data sets: a statistical approach. Int J Geogr Inf Sci 13 (5):445–473. http://www.tandfonline.com/doi/abs/10.1080/136588199241157#.UzViZ6hdVHQ

Zhang C, Peng Z-R, Li W, Day MJ (2003) GML-based interoperable geographical databases. Cartography 32:1–16. http://gis.geog.uconn.edu/personal/paper1/journal%20paper/2%202003%20InteroperableDabase--scanned%20one.pdf

Ziegler P, Dittrich KR (2004) Three decades of data integration – all problems solved? Database Technology Research Group. Winterthurerstrasse 190, Ch-8057, Zurich, Department of Informatics, University of Zurich. http://citeseerx.ist.psu.edu/viewdoc/download?doi=10.1.1.200.546&rep=rep1&type=pdf

Exploring Class Discussions from a Massive Open Online Course (MOOC) on Cartography

Anthony C. Robinson

Introduction

The recent emergence of the Massive Open Online Course (MOOC) has become a transformative force in distance education. MOOCs are designed to provide class experiences that scale elegantly to tens of thousands of students working online (McAuley et al. 2010). MOOCs are normally delivered through dedicated content management platforms that host lecture videos, written/graphical content, assessment tools, and discussion forums. These platforms also capture student interactions and contributions, which form a massive dataset worthy of exploration after the course has ended. In 2013, we designed and delivered a MOOC on the fundamentals of Cartography called Maps and the Geospatial Revolution. This course enrolled over 48,000 students from more than 150 countries for its first 5-week session beginning in July, 2013.

In this paper we highlight the potential for discussion archives and other qualitative contributions from students in courses like Maps and the Geospatial Revolution to reveal interesting patterns about what diverse audiences think about cartography. Students enrolled in our MOOC generated over 95,000 forum posts in more than 13,000 threads. Weekly discussion prompts urged students to focus on geospatial privacy, mapping change, mapping hazards, mapping social media, and telling stories with maps. In addition to those prompted topics, students started discussions on nearly every other aspect of mapping (and the course itself) that one might envision. Since this corpus is impossible for one to readily make sense of, it serves as excellent fodder for text analysis using visual and quantitative methods to uncover key topics and to explore the places that were used to explain those topics.

Here we show how visual methods such as Phrase Nets (van Ham et al. 2009) can be applied to reveal how students say they use maps. We also highlight how

A.C. Robinson (✉)
Department of Geography, The Pennsylvania State University, University Park, PA, USA
e-mail: arobinson@psu.edu

© Springer International Publishing Switzerland 2015
J. Brus et al. (eds.), *Modern Trends in Cartography*, Lecture Notes in
Geoinformation and Cartography, DOI 10.1007/978-3-319-07926-4_14

computational methods such as latent Dirichlet allocation (LDA) (Wallach 2006) can be used to comb thousands of posts to reveal key themes in course discussions. Finally, we use a combined named-entity recognition (NER) and geocoding approach (Karimzadeh et al. 2013) to extract and visualize the places that students mention in their discussion posts. The results of this work provide important insights for cartography educators, as they reveal key motivations for students in a globally-diverse MOOC to use, make, and understand maps. These insights may then feed into future iterations of cartography courses of all sizes and delivery modes.

Background

At a broad level, our research is concerned with the general problem associated with making sense out of large text collections. Visualizing text is possible through a wide range of means, including methods such as tag clouds (Hassan-Montero and Herrero-Solana 2006), which can provide a general overview of the frequency of terms, and more sophisticated methods that use self-organized maps to develop spatialized representations to show the topics that appear in a text collection, and to expose their relative similarity across text datasets (Skupin and Fabrikant 2003). The former approach leverages a visual technique alone, with minimal computational effort required. The latter approach requires sophisticated computational methods to identify and extract patterns, before visualization becomes possible.

Our work leverages data developed by students taking the first MOOC to focus on cartographic design (hereafter referred to as the Maps MOOC). The Maps MOOC featured 5 weeks of lessons on the most general cartographic competencies. Students learned how to recognize spatial thinking, understand basic spatial analyses, and apply core cartographic design principles. The activities of the more than 48,000 students who enrolled to take the class were logged in a variety of ways, and the datasets that result constitute large, complex datasets in the context of distance education. Characterizing what can be uncovered from large textual datasets is a common contemporary problem for information visualization (Dörk et al. 2010) and geographic visualization researchers (MacEachren et al. 2011), and the Maps MOOC offers a very large text dataset in the form of over 95,000 discussion forum posts. An advantage of this dataset for cartographic inquiry is that it's reasonable to expect a large proportion of this discussion to be grounded in discussions about Geography, and therefore lend itself to geovisualization research.

To begin making sense of this large and diverse textual data, we apply the use of three complementary methods in this paper. In the following sections we apply one relatively simple visual method in the form of Phrase Nets to evaluate statements students contributed regarding how they currently use maps. We then explore how the computational approach of topic modeling through latent Dirichlet allocation can be used to mine discussion forum posts to reveal major topics that students discussed. Finally, we make use of a modified named-entity extraction method to identify the placenames that students talked about in discussion assignments and to geocode those places to explore the relevant geography for major discussion themes in the class.

Phrase Nets

The Maps MOOC class experience began with a prompt emailed to students to "pin themselves" on a web map to develop an overall view of the people and countries represented in the class cohort. Students in this early class activity were asked to add a pin to a world map to represent their home, and to provide basic demographic information (age range and gender). Students were also asked to provide a simple, one sentence answer to the prompt, "How do you use maps?" This prompt was intended to elicit a wide range of opinions from around the world regarding how novices view Cartography prior to completing the Maps MOOC. Of the 22,781 pins added to the map, 11,710 had complete data for age, gender, and the map usage question. We focus here only on these complete observations.

To evaluate student responses to the question "How do you use maps?" in the opening map assignment for the course, we turn to a technique developed to support visual analysis of phrases. Phrase nets were conceived by van Ham et al. (2009) to aid exploration and visual analysis of phrases in text. The method allows users to select relations (either syntactic or lexical) and to view what frequently comes before and after those terms. For example, in Fig. 1 (Top), we show how the word "and" is used to link words commonly found in the responses we gathered from students reflecting on how they currently use maps. Links between pairings are shown with lines of varying thickness depending on the degree to which those terms often co-occur. Some of the pairings of interest include *Maps and GIS*, *Travel and Work*, and *Place and Directions*.

Changing the relation used to form the Phrase Net can reveal other interesting patterns in this dataset. If we choose "the" as the relational linkage (Fig. 1, Bottom), one can see that a key phrase in our responses was *Understanding the World*. This was complemented by less frequent pairings such as *Discover the World*, *Explore the World*, and *Navigate the World*.

Expanding the lexical relations in Phrase Nets can further reveal how maps are viewed by students in the Maps MOOC. Figure 2 (Top) shows the resulting visualization for the "of the" relation. This structure pulls out a colloquialism in *Lay of the Land*. The most common phrase here is *Understanding of the World*, with evidence that significant numbers of students also wrote about *Sense of the Place* and *Parts of the City*, the latter of which is one of the few scale and context-specific references that appears in any of the phrase nets we developed.

Since Phrase Nets can also use syntactical relations to arrange text, we can also look at the basic pairings of words separated by spaces alone. Figure 2 (Bottom) shows the extent to which students view technology as a key aspect of their map

                                                                A.C. Robinson

Fig. 1 (*Top*) Terms in student responses about how they use maps that are linked by the word "and." (*Bottom*) Terms in those same responses that are linked by "the"

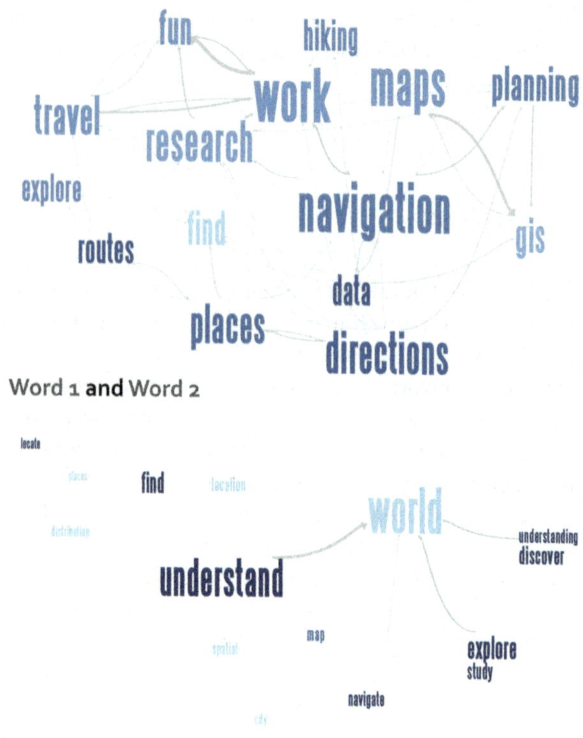

usage, in pairings such as *Google Maps*, *GIS Data*, and *Find Directions*, among others.

These and other examples using our "How do you use Maps?" dataset can be further explored using IBM's ManyEyes Phrase Net tool at http://ibm.co/1f8B2EO.

Fig. 2 (*Top*) Terms in student responses about how they use maps that are linked by the word "and." (*Bottom*) Terms in those same responses that are linked by "the"

Topic Modeling

Topic modeling is one popular computational approach for analyzing text data. Specific techniques for topic modeling include basic probabilistic methods that predict the likelihood that one word follows another (Wallach 2006), and somewhat more sophisticated methods such as latent Dirichlet allocation (LDA) which can model topics independent of word order (Blei et al. 2003). Many options exist today for alternative approaches which advance upon these basic examples, with new combinations and modifications appearing all the time. LDA, however, has remained a popular method for topic modeling, and a large number of tools are available today for researchers to apply which leverage the LDA approach. The Machine learning for language toolkit (MALLET) is one such example that uses LDA to mine topics from text (McCallum 2002). MALLET is built using the Java programming language and provides command line controls for processing large text collections to extract key topics using LDA. Since its first iteration in 2002, MALLET has been improved in several stages, and the tools now include methods for tagging sequences and classifying documents, among others.

To make MALLET easily usable by non-experts, David Newman at the University of California-Irvine created the Topic Modeling Tool (TMT) to provide a graphical user interface to MALLET (http://code.google.com/p/topic-modeling-tool/). We used the TMT in our work to reveal key topics found in discussions in

Mapping Technology

1. gps phone technology access road lost car route directions system
2. gis mapping work geography learn software project tools open tool
3. maps map google paper digital make find love making made
4. http www map link bit story html ly org home

The Course Itself

1. students taking hope study knowledge mooc state research courses experience
2. class post thread coursera interesting ve read discussion forum video
3. map arcgis online add web layer click create layers file
4. time assignment didn work final thought lesson questions question answer
5. time good great idea lot pretty cool thing make ve
6. de en la wikipedia el wiki es los org spain

Geography

1. area city areas live years land town urban small cities
2. map change show level shows image esri images vegetation color
3. interesting place places find lot ve thought amazing sites big
4. important issues space issue spatial human sense community things make
5. population country high countries states years growth life number age
6. point view understand points world things earth book called related
7. water natural north earthquake earthquakes disaster south sea river disasters
8. location information sharing people social privacy share pin feel media
9. people don world long today back place time power ago
10. data information analysis scale based public spatial government specific provide

Fig. 3 Major topics uncovered from discussion forum posts using MALLET

the Maps MOOC. Input data for TMT was comprised of the text from 95,958 discussion forum posts created by students in the class.

Figure 3 shows an overview of the top 20 topics identified by MALLET from our discussion post dataset. We have further categorized the top 20 topics into three key themes that appear to link individual topics; Mapping Technology, the Course Itself, and Geography. Students frequently discussed the impact of mapping and location technology, with a specific interest on changes in the ways maps are made today. One topic shows artifacts of a very common URL shortening service, which was widely used by students who shared the web maps they created with one another.

The second major category of topics concerns the course itself. Students frequently talked about class policies, their goals for taking the MOOC, and aspects of assignments that they struggled to complete or understand. One outlier here is the final listed topic in this category which shows several Spanish words and references a wiki. A large group of Spanish-speaking students took this course, and the Spanish-speaking study group thread in the study group forum was among the most active of all study groups. Students in that thread and others frequently posted links to Wikipedia articles in Spanish to help elaborate concepts from the lectures in the course.

Finally, the largest category of topics concerns Geographically-focused discussions. Students talked about the discoveries they made about populations, land cover change, hazard analysis, and social media—all of which were key themes in

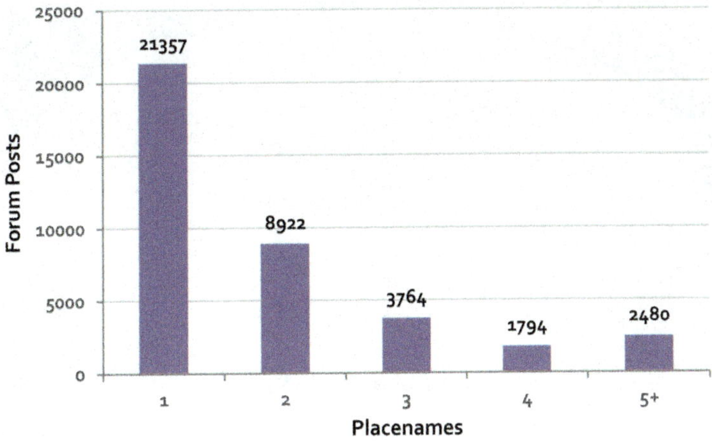

Fig. 4 Chart showing the number of placenames found in MOOC forum posts

their lab assignments for the course. Urban geography was of particular interest, as were topics centered on change of all types. One of the first discussion prompts in the course asked students to consider spatial privacy concerns, which resulted in a few related topics shown here in this category.

Geo-Parsing Forum Posts

We continued our exploration of text generated in the Maps MOOC by applying text mining techniques to identify and geocode the place names mentioned by students in forum posts. This general process is frequently called geo-parsing in contemporary literature (Gelernter and Zhang 2013). To analyze our course forum dataset of over 95,000 posts, we used the GeoTXT.org service. GeoTXT combines named-entity recognition methods with geocoding capabilities in order to extract and locate placenames found in text (Karimzadeh et al. 2013) (Fig. 4).

Processing the forum post dataset with GeoTXT resulted in the extraction of 38,317 places from 95,958 posts. There were 21,357 posts that included at least one place mention. Some posts referenced multiple placenames; an overview of which can be found in Fig. 3.

After extracting and geocoding the placenames mentioned in class discussion forum posts, we sought ways to map these data in order to understand the frequency and distribution of placename mentions. First, we aggregated all placenames found into hexagons (2° wide at the Equator), in order to explore the overall global density of place-oriented discussion in the class. Figure 5 shows the resulting map, which looks much like a population density map of the Earth, with some notable exceptions in China and parts of Africa.

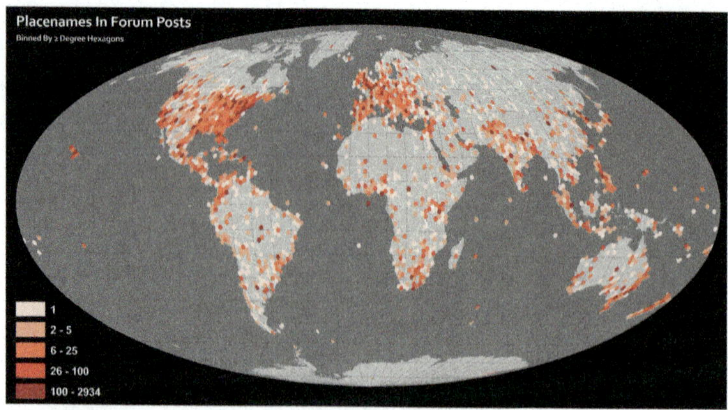

Fig. 5 Map of placenames found in MOOC forum posts, aggregated by 2° wide hexagons

While ordinarily it would be inappropriate to use a non-normalized choropleth map to show these data, in this instance, the raw totals aggregated by country helps show the general pattern of placename mentions in the class forum that is less apparent in the hexagonal binning results. Many of the hex bins are located on the centroids of countries, indicating the mention of a country name in a discussion forum post. Figure 6 shows the raw totals aggregated to countries, revealing a not too surprising trend around highly populated, English-speaking countries. The map does not track overall population trends when one considers Asia and Africa.

These analyses prompted us to think of a way to normalize the data by a reasonable measure. We found overall population normalization was not particularly interesting, as it simply emphasizes very small population countries that received a handful of mentions (Greenland and Antarctica, for example). Instead, we used the "Pin Yourself" mapping activity data from the beginning of the class to count enrollments by country, and then used those enrollment data to normalize the placename mentions in our forum data. The resulting map is shown in Fig. 6. What is clear from this map is that there are some places that are of greater interest in discussion than they were in terms of attracting students, and that this is particularly true in Africa and parts of Asia. Some countries, like Brazil and Australia, were talked about in discussion frequently, and also constituted sizable proportions of the overall student population (both appearing in the top ten placename and student enrollment lists).

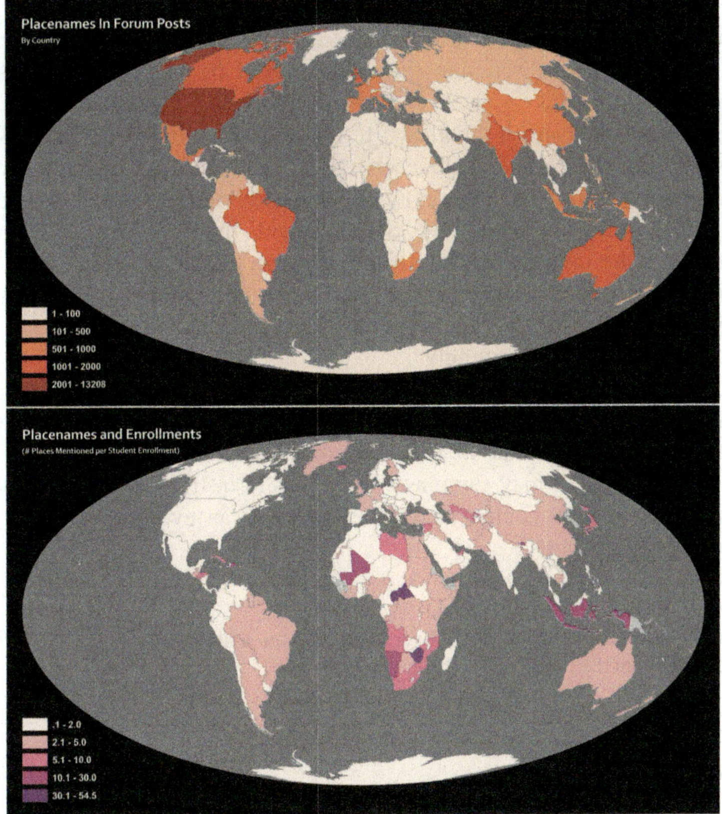

Fig. 6 Map of placenames found in MOOC forum posts; raw counts shown in the *top map*, and counts normalized by the number of students enrolled from each country in the *bottom map*

Conclusions

In this paper we have contributed examples of visual, computational, and combined visual-computational analysis of forum data from a MOOC on Cartography. Our results reveal the ways in which novice students perceive the utility of maps, the key topics students discussed during the course itself, and the Geography embedded in those forum discussions.

There remains much work to do to further explore and analyze these data. Of specific interest to us is the relationship between time and the analyses we have shown here. For example, discussion topics changed over time from week to week in the class, and one wonders whether or not the placename references may vary along with those changing topics as the course progresses. We can also further explore the extent to which where a student comes from has an influence on the places they mention as examples in topic

(continued)

areas like spatial privacy, natural hazards mapping, and social media mapping (all of which were discussion themes prompted by the instructor in this course).

Sentiment analysis is another fruitful potential direction. Students in MOOCs are expected to help each other, given the fact that no single instructor (or team of instructors) could possibly interact with each student individually in a significant way. Some students find this frustrating, while others take the opportunity to self-organize into study groups to help one another. We hypothesize that there are likely geographic differences in students who choose to organize versus those that do not, and that student sentiment may vary accordingly.

Ultimately there remains much left to do with these data beyond the basic analysis we have shown here. A larger future goal should be to support rapid analyses of MOOC data to uncover spatio-temporal patterns of student activity and interest may serve as critical aids to instructors who are trying to adapt their class content and assessments to an ever-changing, global population of students.

Acknowledgements We thank Jan Oliver Wallgrün for assistance in using GeoTxt.org to geo-parse the forum data analyzed in this paper.

References

Blei DM, Ng AY, Jordan MI (2003) Latent Dirichlet allocation. J Mach Learn Res 3:993–1022

Dörk M, Gruen D, Williamson C, Carpendale S (2010) A visual backchannel for large-scale events. IEEE Trans Vis Comput Graph 16:1129–1138

Gelernter J, Zhang W (2013) Cross-lingual geo-parsing for non-structured data. In: 7th ACM SIGSPATIAL workshop on geographic information retrieval, Orlando, FL

Hassan-Montero Y, Herrero-Solana V (2006) Improving tag-clouds as visual information retrieval interfaces. In: International conference on multidisciplinary information sciences and technologies, Merida, pp 1–6

Karimzadeh M, Huang W, Banerjee S, Wallgrün JO, Hardisty F, Pezanowski S, Mitra P, MacEachren AM (2013) GeoTXT: a web API to leverage place references in text. In: 7th ACM SIGSPATIAL workshop on geographic information retrieval, Orlando, FL

MacEachren AM, Jaiswal A, Robinson AC, Pezanowski S, Savelyev A, Mitra P, Zhang X, Blanford J (2011) SensePlace2: Geotwitter analytics support for situation awareness. In: IEEE conference on visual analytics science and technology, Providence, RI

McAuley A, Stewart B, Siemens G, Cormier D (2010) The MOOC model for digital practice

McCallum AK (2002) MALLET: a machine learning for language toolkit

Skupin A, Fabrikant SI (2003) Spatialization methods: a cartographic research agenda for non-geographic information visualization. Cartogr Geogr Inf Sci 30:99–119

van Ham F, Wattenberg M, Viegas FB (2009) Mapping text with phrase nets. IEEE Trans Vis Comput Graph 15:1169–1176

Wallach HM (2006) Topic modeling: beyond bag-of-words. In: 23rd international conference on machine learning, Pittsburgh, PA, pp 977–984

Evaluating Mapping APIs

Michael P. Peterson

Introduction

The use of multi-scale panable (MSP) maps with user-supplied information is based on Web 2.0 mashup technology. Web 2.0 represents a variety of innovative resources, and ways of interacting with, or combining web content that began in about 2004. It includes wikis, such as Wikipedia, blog pages, podcasts, RSS feeds, and AJAX. Social networking sites like Facebook and Google+ are also seen as major Web 2.0 applications.

Mashups combine multiple cloud resources. Central to mashups are Application Programming Interfaces (APIs). These are function libraries that are integrated with objects and are usually based on a language called JavaScript. APIs are the tools that facilitate the melding of data and resources from multiple cloud resources by providing the means to acquire, manipulate and display information. Many different APIs have been written for the user-driven web. The most commonly used is the Google Maps API.

In a strict sense, a map mashup combines data from one website and displays it with a mapping API. The term has come to be used for any mapping of data using an API, including data supplied by the user. The ease of mapping spatial information has resulted in all kinds of different maps, many showing information that has never been mapped before.

Map mashups have had a major influence on how spatial information is presented. One particular advantage of using an API with a major online mapping site is that the maps represent a standard and immediately recognizable representation of the world. Available through mobile phones and computers, these maps represent a standard depiction—in many ways, more standardized than any other

M.P. Peterson (✉)
Department of Geography/Geology, University of Nebraska at Omaha, Omaha, NE 68182, USA
e-mail: mpeterson@unomaha.edu

© Springer International Publishing Switzerland 2015
J. Brus et al. (eds.), *Modern Trends in Cartography*, Lecture Notes in
Geoinformation and Cartography, DOI 10.1007/978-3-319-07926-4_15

map that has ever been developed. Overlaying features on top of these maps provides a familiar and comfortable frame of reference for the map user. As a result, maps made in other ways, with other software, are viewed as unfamiliar and even foreign. This includes maps created with popular GIS software such as ESRI's ArcGIS.

Mapping APIs are closely integrated with map tiling. Used since the early days of the World Wide Web, image tiling speeds the delivery of graphics through the Internet. In comparison to text, images require more storage and therefore take longer to download. A solution is to divide the image into smaller segments, or tiles, and send each tile individually through the Internet. These smaller files often travel faster because each can take a different route to the destination computer. On the receiving end, the tiles are reassembled in their proper location on the web page. With a moderately fast Internet connection, all of this occurs so quickly that the user rarely notices that the image is actually composed of square pieces. With slower connections, the individual tiles are clearly evident—sometimes painfully so. In combination with AJAX, an alternative way of interacting with the server, the tiling of maps facilitates MSP maps by allowing maps to be easily panned and zoomed. In evaluating mapping APIs, we begin by reviewing some of the different APIs that are currently available.

Comparison of Application Programmer Interfaces for Mapping

Google Maps API

Introduced soon after Google Maps in 2005, the Google Maps API is by far the most commonly used. The API consists of a series of functions that control the appearance of the map, including its scale and location, and any added information in the form of points, lines or areas and associated descriptions. The use of Google Maps API is essentially free, provided the site does not charge for access. Google limits the number of maps that can be served: A site cannot generate more than 25,000 map loads a day for 90 consecutive days. A map load is one map displayed with the Google Maps API. Once loaded, the degree to which a user interacts with a map— panning or zooming—has no impact on the map load number. It would be extremely difficult for a non-institutional user of the Google Maps API to reach 25,000 daily map loads. Even if a site were to go "viral" with a topic that generates considerable interest, it would need to sustain 25,000 map loads per day for 90 consecutive days. Viral topics are fickle and have a much shorter lifespan.

The default Google marker is an upside-down raindrop symbol but a large number of alternative symbols are available. It is even possible to design symbols using a program like Adobe Photoshop™. The code in Fig. 1 places the basic Google marker at the center of the map. The event.addlistener option sets the zoom

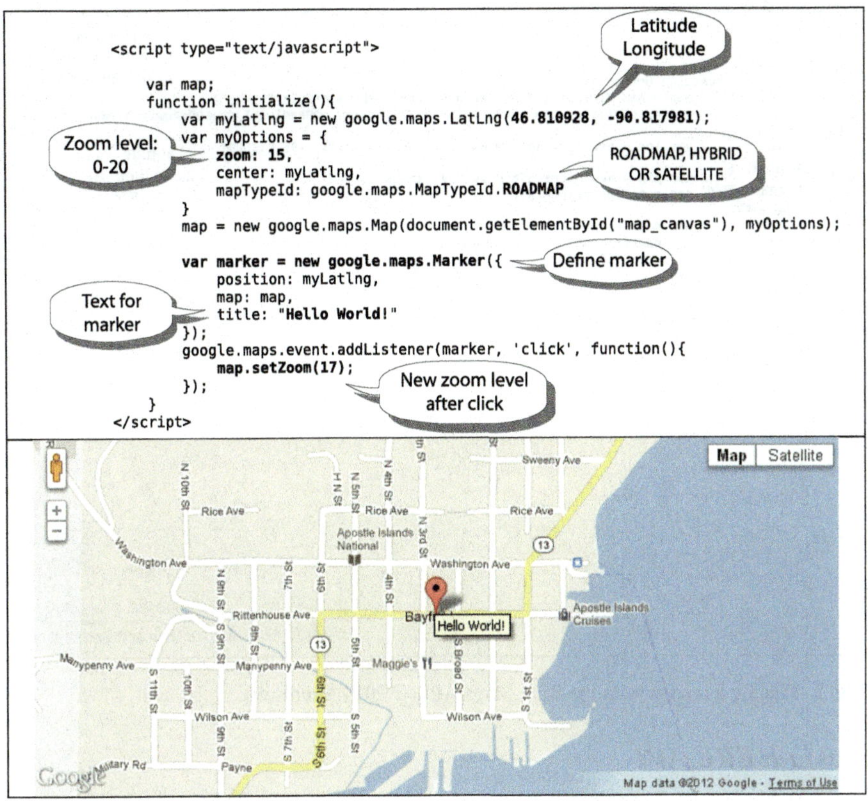

Fig. 1 A single marker with a mouseover function that displays "Hello World!" (© 2014 Google)

level to 17 when the marker is clicked. The initial zoom level is 15. The title text, "Hello World," is displayed when the mouse is hovered over the marker.

Bing Maps API

Microsoft's Bing Maps followed Google's lead with map tiles and developed its own API. A marker, called a PushPin, is displayed in Fig. 2. Like Google's initial API, each of these required the use of an electronic key, an alphanumeric string within the initial reference to the API. While the key is made freely available, it limits its use to the server that is specified when the key is requested. The primary benefit of the key is that it controls access to the API. A user who exceeds the limits on use, or uses the API for illegal purposes, can easily be denied access.

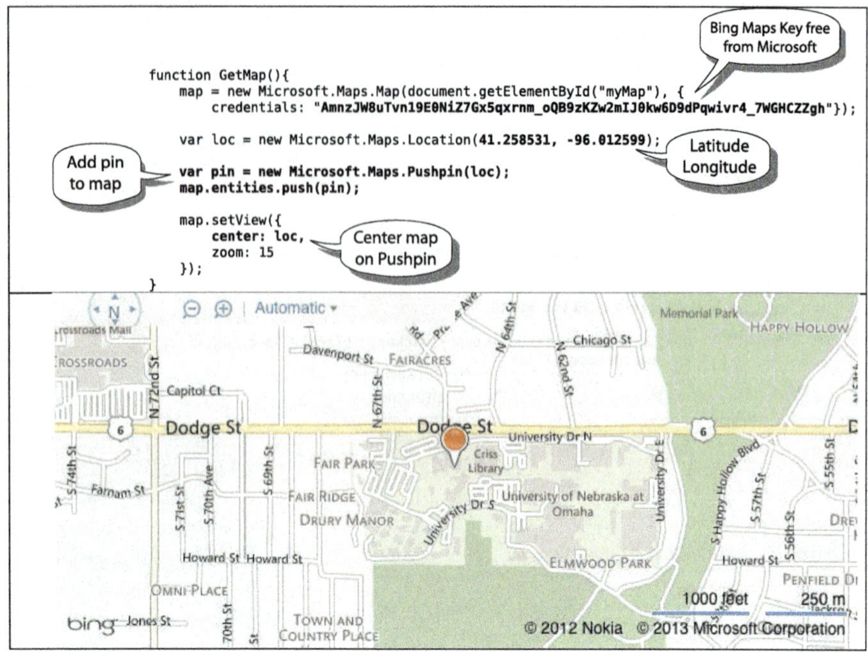

Fig. 2 Placing a marker with the Bing Maps API (© 2014 Microsoft)

Nokia HERE API

Mobile phone maker Nokia had concentrated on supplying a stand-alone mapping application for its phones, now a unit of Microsoft. Using a large map file that was stored in the memory of the phone—about 10 MB in size, the user was able to make a map without being connected to the Internet. With the purchase of the mapping giant NavTeq in 2007, the company turned its attention to the Internet with the introduction of OviMaps in 2007, renamed to Nokia Maps in 2011 and eventually to HERE in 2012—the latter also available as an "app" for mobile devices. Nokia assumed control of Yahoo! Maps in 2012. Yahoo had been a major map provider, with its own map tiles and API.

The Nokia API also requires the use of a key, as shown in Fig. 3 that adds a marker to the map. Nokia offers a number of options to change the default marker, including the display of text. Functions to control the appearance of the marker exceed those available through the Google Maps API.

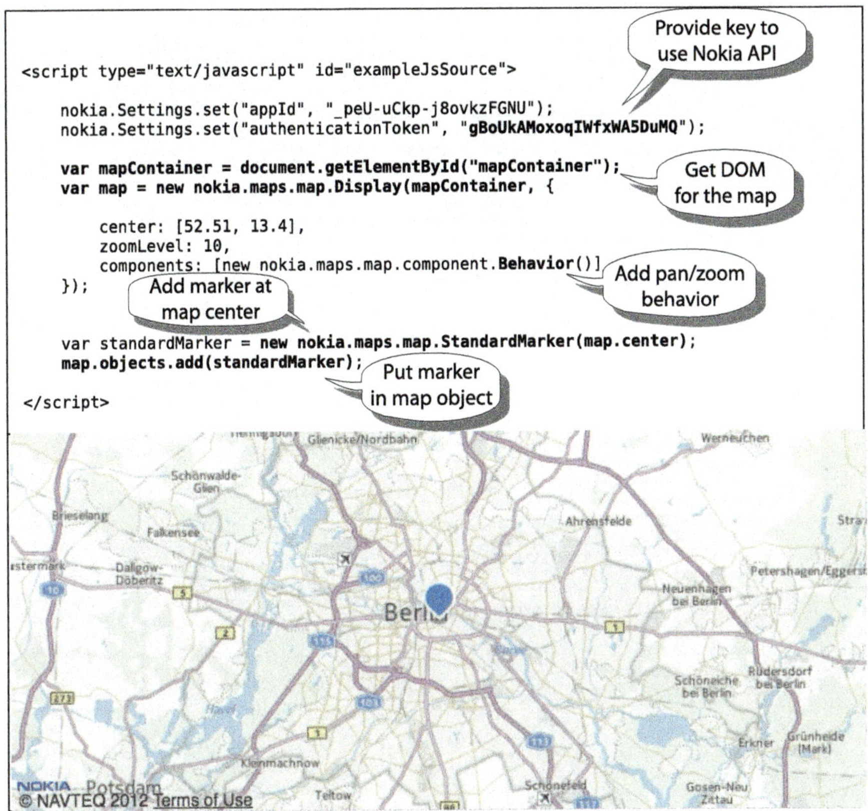


```
<script type="text/javascript" id="exampleJsSource">

    nokia.Settings.set("appId", "_peU-uCkp-j8ovkzFGNU");
    nokia.Settings.set("authenticationToken", "gBoUkAMoxoqIWfxWA5DuMQ");

    var mapContainer = document.getElementById("mapContainer");
    var map = new nokia.maps.map.Display(mapContainer, {

        center: [52.51, 13.4],
        zoomLevel: 10,
        components: [new nokia.maps.map.component.Behavior()]
    });

    var standardMarker = new nokia.maps.map.StandardMarker(map.center);
    map.objects.add(standardMarker);

</script>
```

Fig. 3 A Nokia map with marker (©2014 Nokia. Nokia map content is used with permission)

MapQuest API

MapQuest first introduced its mapping website in 1996 and enjoyed many years as the primary online mapping website. Caught off-guard by the introduction of Google Maps in 2005, it adapted slowly to the new tile-based, API-based mapping method. Even so, it maintained its number one map provider position until 2009. As with the Nokia API, the use of the MapQuest API also requires a numeric key. The example in Fig. 4 demonstrates how to add a marker to the map. Options for marker display are not well-developed.

OpenStreetMap API

OpenStreetMap is the major volunteered geographic information (VGI) website, a method of data acquisition also referred to as *crowdsourcing*. It has thousands of

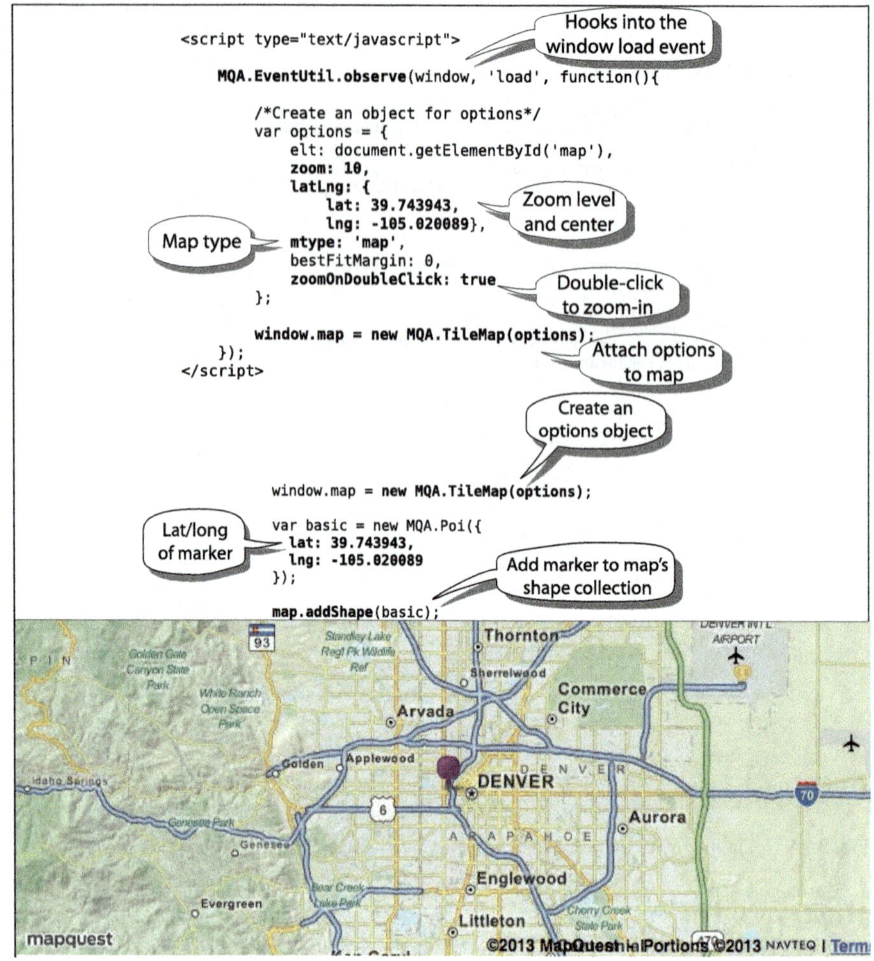

Fig. 4 Adding a marker with the MapQuest API (© 2014 MapQuest—Portions © 2014 NAVTEQ, Intermap)

contributors who have uploaded everything from GPS traces to new points of interest. One major advantage of OpenStreetMap is the ability to access the underlying vector data, although the rendered, tile-based map is used here. The example in Fig. 5 shows how a marker is added to a map with the OpenStreetMap API. Although perhaps more functional, the OpenStreetMap code is longer and more complicated than other mapping APIs.

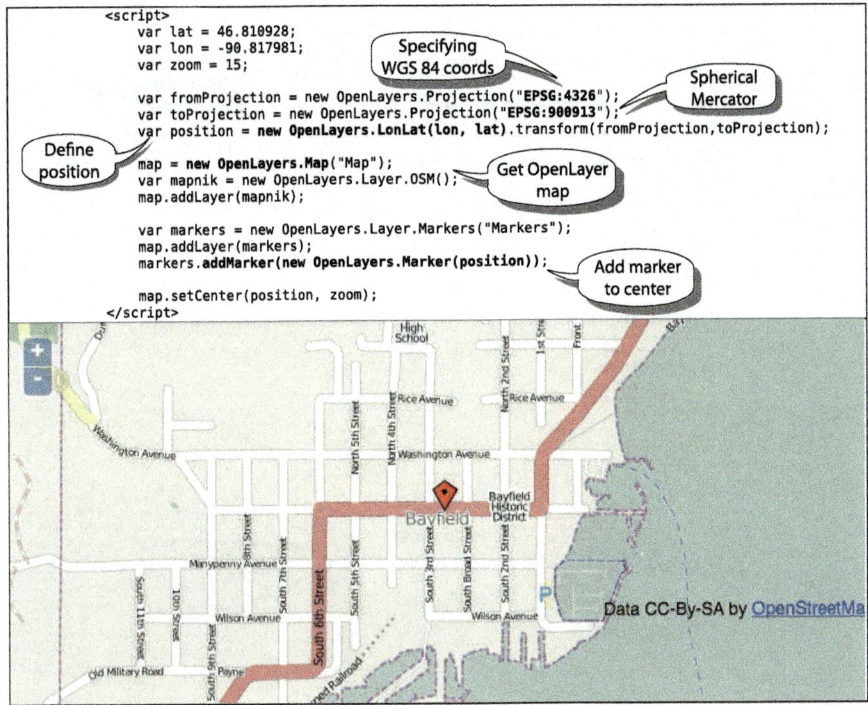

Fig. 5 Adding a single marker with the OpenStreetMap API (© 2014 OpenStreetMap contributors CC-BY-SA)

Leaflet API

The Leaflet API is specifically designed for mobile devices. Developed by Agafonkin (2013) and other contributors, it is just 28 kB of JavaScript code—much smaller than other APIs. The code is also more readable. Leaflet is closely integrated with CloudMade tiles that have been rendered from OpenStreetMap vector data although it can use other tile services as well. Figure 6 demonstrates the use of the Leaflet API and the mapping of multiple types of symbols.

Baidu Map API

The government of mainland China has exercised a level of control over the Internet that is unmatched by all other countries in the world. Access to Google is intermittent, making it essentially useless within China. Although Google Maps works better than the Google search engine and Google's gmail, the blocking of the Google servers—now located in Hong Kong—has fostered the development of

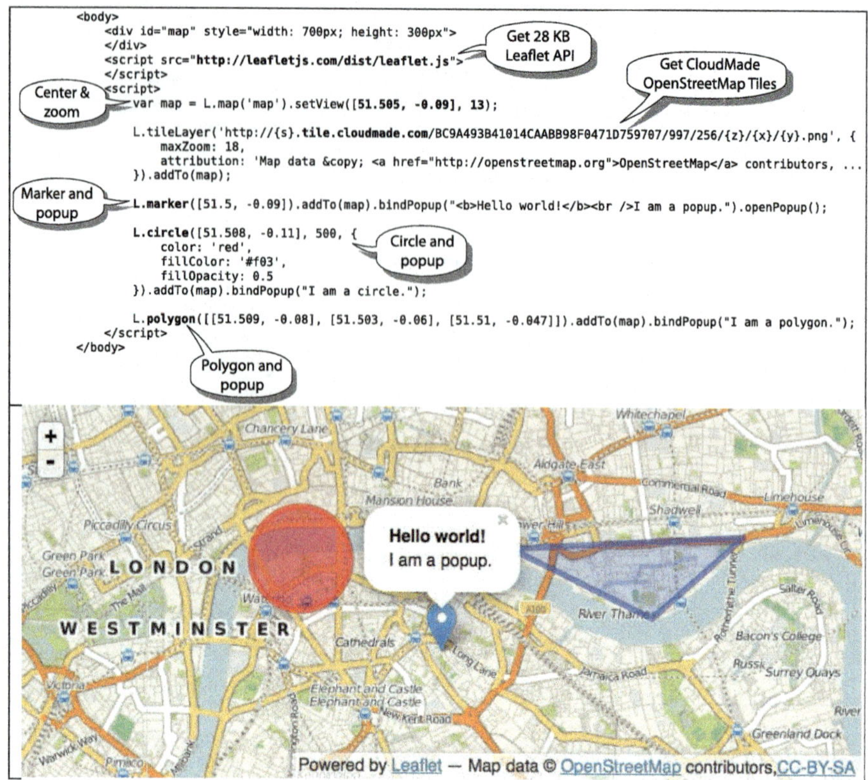

Fig. 6 A basic Leaflet API map with circle, marker, and triangle (© 2014 OpenStreetMap contributors CC-BY-SA)

online mapping entities within mainland China. The two major companies are Baidu and AutoNavi. Both companies provide detailed coverage for China and Taiwan but not for other parts of the world.

Baidu is the major search engine in China, analogous to Google for the rest of the world. With headquarters in Beijing, the company offers a Chinese language search engine for websites, audio files, and images, in addition to Baidu Map. Baidu became the first Chinese company to be listed on the NASDAQ-100 index. The Baidu mapping API is very similar to other APIs. Figure 7 shows how to add a marker to the Baidu map.

Mapstraction

Mapstraction is an open-source library that provides a common API for various online mapping services (Duvander 2010). The purpose is to make it possible to easily switch between mapping APIs without having to worry about the unique

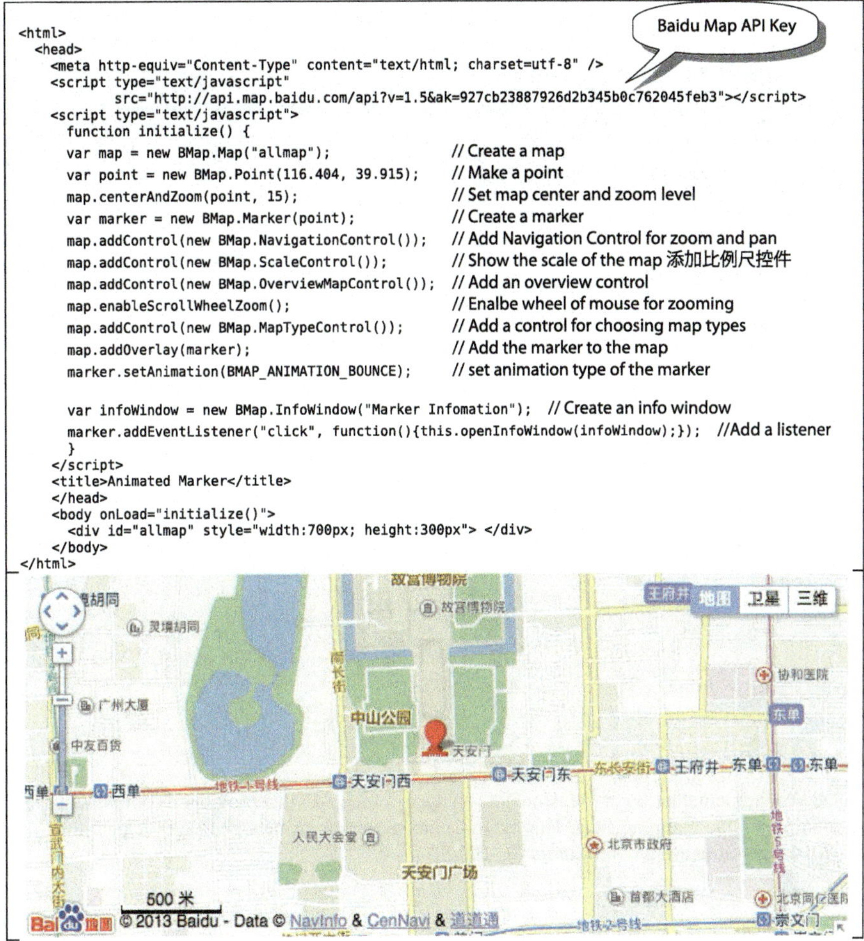


```html
<html>
  <head>
    <meta http-equiv="Content-Type" content="text/html; charset=utf-8" />
    <script type="text/javascript"
            src="http://api.map.baidu.com/api?v=1.5&ak=927cb23887926d2b345b0c762045feb3"></script>
    <script type="text/javascript">
      function initialize() {
      var map = new BMap.Map("allmap");                       // Create a map
      var point = new BMap.Point(116.404, 39.915);            // Make a point
      map.centerAndZoom(point, 15);                           // Set map center and zoom level
      var marker = new BMap.Marker(point);                    // Create a marker
      map.addControl(new BMap.NavigationControl());           // Add Navigation Control for zoom and pan
      map.addControl(new BMap.ScaleControl());                // Show the scale of the map 添加比例尺控件
      map.addControl(new BMap.OverviewMapControl());          // Add an overview control
      map.enableScrollWheelZoom();                            // Enalbe wheel of mouse for zooming
      map.addControl(new BMap.MapTypeControl());              // Add a control for choosing map types
      map.addOverlay(marker);                                 // Add the marker to the map
      marker.setAnimation(BMAP_ANIMATION_BOUNCE);             // set animation type of the marker

      var infoWindow = new BMap.InfoWindow("Marker Infomation");   // Create an info window
      marker.addEventListener("click", function(){this.openInfoWindow(infoWindow);});   //Add a listener
      }
    </script>
    <title>Animated Marker</title>
  </head>
  <body onLoad="initialize()">
    <div id="allmap" style="width:700px; height:300px"> </div>
  </body>
</html>
```

Fig. 7 Adding a marker to a Baidu map. Baidu is a major Chinese website and map provider (Map courtesy Baidu)

implementation of each. The example in Fig. 8 shows Mapstraction code that allows using either the OpenLayers or Google API. To access the Google Maps API, "openlayers" is simply changed to "googlev3" (see Fig. 9). Note that a link to the mapping API in the third line of the code is also changed.

```
<head>
  <title>Basic Mapstraction Map</title>
  <script src="http://openlayers.org/api/OpenLayers.js"></script>
  <script src="https://raw.github.com/mapstraction/mxn/master/source/mxn.js?(openlayers)"
        type="text/javascript"></script>
  <style type="text/css">
  #map {
        width: 700; height: 300px;
   }
  </style>

  <script type="text/javascript">
  var mapstraction;
  function create_map() {
      mapstraction = new mxn.Mapstraction('mymap', 'openlayers');
      mapstraction.setCenterAndZoom(
        new mxn.LatLonPoint(41.258531,-96.012599), 15);
   }
  </script>
</head>
```

Fig. 8 An implementation of Mapstraction, an open-source API that provides a common interface to multiple online mapping sites. This example creates a map using OpenLayers/OpenStreetMap (© 2014 OpenStreetMap contributors CC-BY-SA)

Fig. 9 An implementation of Mapstraction using Google Maps (© 2014 Google)

Evaluation of Mapping APIs

APIs have been evaluated in a number of different ways (Bloch 2006; Clarke 2004; de Souza et al. 2004; Farooq and Zirkler 2010; Gerken et al. 2011; Tulach 2008; Wong and Hong 2007). All of these studies have examined aspects related to API development and provide guidelines for API evaluation. The following describes alternative methods of evaluation for mapping APIs based on these and other studies.

Execution Speed In theory, the speed of execution of an API would measure its computational power, a measure of how well the code is written. APIs are mostly based on JavaScript, a programming language that is interpreted by the browser. The browser is compiling and executing the JavaScript code in a single step. Differences between browser's processing of JavaScript is thus much more significant than any differences in the quality of the API code.

A study undertaken by Weiser (2010) examined differences between mapping APIs and showed major differences in the execution of JavaScript code. Microsoft Internet Explorer (IE), the most-used browser, processed JavaScript code much

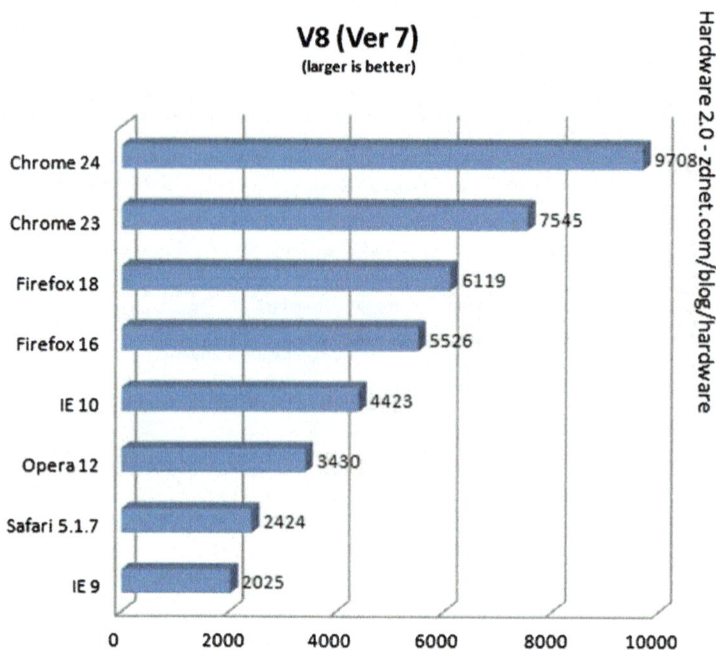

Fig. 10 JavaScript execution speed compared among the major browsers of the Version 8 JavaScript engine. The score across the x-axis is computed from the results of seven demanding tests. This tool is used by Google to optimize its Chrome browser (Kingsley-Hughes 2013)

more slowly than the other major browsers. JavaScript execution speeds are shown in Fig. 10 by browser and shows that IE is still the slowest in executing JavaScript. Speed-battle.com does an interactive test of the browser's performance, showing differences between browsers and computer operating systems.

Length-of-Code The "length-of-code" criterion evaluates how much JavaScript code is required to perform specific operations with an API. The underlying page-length-code premise is that any JavaScript code using an API should only be one page in length. Code that is longer than this indicates that the API is not properly written to support the needed operations, thus requiring extra JavaScript programming.

A general observation here is that the Leaflet API has much shorter code than any of the other mapping APIs. In contrast, the OpenStreetMap API has the longest code—while, perhaps, offering more functionality. The other mapping APIs fall somewhere between these two extremes.

Cartographic Functionality This criterion examines specific cartographic functions that are included in the API. All APIs include functions for adding points, lines, and areas to the map. Some, like the Google Maps API, have more advanced tools for making circles and other types of symbols. None have any of the advanced functions that one would expect for thematic cartography, such as range-grading of circles or perceptual scaling.

Cartographic Animation Although the multi-scale, panable MSP map is itself a form of interactive animation, the tiling of maps has made other forms of animation more difficult—such as moving a point across the map. Few mapping APIs have any significant functions to support animation and those that are included—such as bouncing markers—are very primitive. The lack of tools for animation is an indication of the paper-thinking that has crossed-over to these online maps. While panning and zooming go beyond the paper map interface, other aspects of the mapping environment are essentially the same as a paper map.

Map Cost The cost of online map delivery will continue to be a major concern. In 2011, Google settled on a free model of 25,000 map loads a day for 90-consecutive days. Although this number would be difficult for anyone to achieve, map providers began to look for alternatives. The solution for many was OpenStreetMap, although the delivery of these maps is usually slower (at least in North America). Leaflet was chosen as an API because it supported OpenStreetMap as well as other map providers. An additional wrinkle to the cost debate is CloudMade, a company that renders OpenStreetMap vector data into map tiles and closely integrated with the Leaflet API. While the first 500,000 map tiles are free, CloudMade charges $25 for downloading the next ½ million map tiles. It is difficult to compare the cost between CloudMade and Google Maps because of the different ways that cost is calculated but downloading 25,000 maps a month at a modest 42 tiles per map display would result in 1,050,000 total tiles. In this particular case, the use of Google Maps would be free while CloudMade would charge slightly over $25.

Of all of the many costs associated with the map in the cloud, the cost of updating the map is the most significant. The world depicted by large-scale online maps changes quickly. Even national, state, and local governments have difficulties maintaining their base maps. What is the best solution for the long-term viability of these maps? How can they best be updated?

For map updating, we are presented with two completely opposite solutions. Either, we rely on the OpenStreetMap model and crowdsourcing. Or, we accept the private sector approach that accurate maps are good for profitability, and that these companies will continue to provide updated maps for free because they can make money from them in other ways. Google has developed a significant revenue stream from its maps that has so far eluded other online map providers.

Long-Term Viability Of all of the ways of evaluating mapping APIs, the long-term viability of the API and the associated tiles is the most significant. Technology changes quickly and a particular API may no longer be available, or its continued development may cease. On the map side, the underlying vector base map or the rendered tiles may no longer be updated—or even be available. Updating maps and making them available through the Internet is expensive. It must be supported by a fairly significant revenue stream that is not directly associated with the map itself.

Any model of online map delivery that does not use such a model will inevitably fail because of the high costs of map updates and map delivery.

Summary

By providing the tools for creating user-defined maps, APIs represent the building-blocks for mapping in the cloud. The particular coding of each API varies and the open source Mapstraction API makes it possible to easily switch between many of them. The development of mapping APIs is in an early stage and it may be too early to try to define a standard set of calls as has been attempted with the Mapstraction project. For example, certain functions are implemented in some APIs that are not available in others. MapStraction can only support a common set of functions.

While there are slight variations in the coding, the major difference between the mapping APIs may be the rendering of the underlying base map, and the speed of map display. The delivery of the tiles to a large number of simultaneous users at an acceptable speed is probably the single most important factor in judging the any particularly API.

There are many ways to evaluate mapping APIs. While a thorough evaluation of functionality, execution speed, and length-of-code would be possible, the major evaluative criteria may be cost of use and long-term viability. The fastest and most functional API today is of little use if it is not still available a decade from now.

References

Agafonkin V (2013) Leaflet: an open-source javascript library for mobile-friendly interactive maps. http://leafletjs.com/

Bloch J (2006) How to design a good API and why it matters. In: Companion to the 21st ACM SIGPLAN symposium on object-oriented programming systems, languages, and applications, 22–26 October 2006, Portland, OR

Clarke S (2004) Measuring API usability. Dr. Dobbs Journal (May 2004), pp 6–9.

de Souza CRB, Redmiles D, Cheng L-T, Millen D, Patterson J (2004) Sometimes you need to see through walls: a field study of application programming interfaces. In: Proceedings of the 2004 ACM conference on computer supported cooperative work, 6–10 November 2004, Chicago, IL

Duvander A (2010) Map scripting 101: an example-driven guide to building interactive maps with Bing, Yahoo!, and Google Maps. No Starch Press, San Francisco, CA

Farooq U, Zirkler D (2010) API peer reviews: a method for evaluating usability of application programming interfaces. In: Proceedings of the 2010 ACM conference on computer supported cooperative work, 06–10 February 2010, Savannah, GA

Gerken J, Jetter H-C, Zöllner M, Mader M, Reiterer H (2011) The concept maps method as a tool to evaluate the usability of APIs. In: Proceedings of the SIGCHI conference on human factors in computing systems, 07–12 May 2011, Vancouver, BC. doi:10.1145/1978942.1979445

Google Maps JavaScript API Basics (2011) (search: *Google Maps Javascript API Basics*)

Kingsley-Hughes A (2013) The BIG browser benchmark (January 2013 edition). ZDNet., 14 January 2013

Tulach J (2008) Practical API design: confessions of a Java framework architect. APress, Berkeley, CA

Weiser P (2010) Performance analysis of vector overlays in online mapping APIs. Masters thesis, University of Nebraska at Omaha
Wong J, Hong JI (2007) Making mashups with marmite: towards end-user programming for the web. In: Proceedings of the SIGCHI conference on human factors in computing systems, April 28–May 03, 2007, San Jose, CA

Demography of Twitter Users in the City of London: An Exploratory Spatial Data Analysis Approach

Barbara Hofer, Thomas J. Lampoltshammer, and Mariana Belgiu

Introduction

Social media data like Twitter messages are increasingly used as source of information about people and their individual perspectives. There are numerous studies that investigate the content of posted messages to infer urban characteristics (Wakamiya et al. 2011), for policy making and scientific purposes (Craglia et al. 2012), or for advertising and social marketing (Pak and Paroubek 2010).

The question persists, if certain demographic groups dominate social media networks and if there is a digital divide expressing inequalities across the population. The digital divide has been studied in the context of Internet use and social network sites such as Twitter, Facebook or MySpace (Hwang and Park 2013). A recent study on Twitter news consumers found that almost half of the Twitter users are between 18 and 28 years young, have at least a bachelor's degree, and use Twitter on mobile devices (Mitchell and Page 2013). This image of the large group of Twitter users reminds of the general Internet users. Internet users are described as digital natives or "young people who grew up having constant Internet access and who use the Web constantly throughout their life" (Boyd and Crawford 2012).

Focusing on Twitter, this study analyses whether socio-demographic characteristics relate to numbers of geolocated tweets. Following the findings on Twitter

B. Hofer (✉) • M. Belgiu
Department of Geoinformatics – Z_GIS, University of Salzburg, Hellbrunnerstraße 34, 5020 Salzburg, Austria
e-mail: Barbara.Hofer@sbg.ac.at; Mariana.Belgiu2@sbg.ac.at

T.J. Lampoltshammer
Department of Geoinformatics – Z_GIS, University of Salzburg, Hellbrunnerstraße 34, 5020 Salzburg, Austria

School of Information Technology and Systems Management, Salzburg University of Applied Sciences, Urstein Süd 1, 5412 Puch/Salzburg, Austria
e-mail: Thomas.Lampoltshammer@fh-salzburg.ac.at

© Springer International Publishing Switzerland 2015
J. Brus et al. (eds.), *Modern Trends in Cartography*, Lecture Notes in
Geoinformation and Cartography, DOI 10.1007/978-3-319-07926-4_16

consumers by Mitchell and Page (2013), the variables of age and education are particularly considered. The study area is greater London, which is known to be a hotspot of Twitter users; a recent report showed that London is ranked third of the 20 cities world-wide with the largest number of posted tweets (Mediabistro 2012). The specific objective is to assess whether the distribution of tweets provides insights into the demographic group of Twitter users for greater London.

The employed methodology is based on techniques of exploratory spatial data analysis. The geolocated tweets of a period of 3 months (July to September 2013) are first separated into day-time and night-time tweets. This separation reveals differences in the patterns of tweet distribution across London's wards during day and night. The night-time tweets are then used for the subsequent steps of the analyses, because of the assumption that residents tweet from their home location rather at night than during the day.

A second step of pre-processing is the elimination of tweets from Twitter users with less than three tweets in the data covering 3 months. This elimination is thought to increase the chance of working with tweets from locals rather than tourists, although it is certainly an approximation to the issue only.

The analysis then focuses on identifying tweet hotspots. Tweet hotspots are locations with large numbers of tweets in comparison to population figures. The hotspots are then analysed regarding their socio-demographic similarities using the Exploratory Spatial Data Analysis tool GeoDA. Visual exploration supports the selection of candidate variables for an Ordinary Least Squares (OLS) regression analysis.

The visual exploration of London's wards reveals deviations in demographic characteristics for hotspots outside the city centre. These deviations can be confirmed in the statistical analysis, which does not indicate a strong relationship between tweet counts and socio-demographic characteristics of the population. Despite the fact that issues like tweeting tourists and lacking details in the demographic variables may influence these results, the analysis cannot confirm that tweet numbers can be explained by socio-demographic variables for London.

Twitter Data Analyses

Twitter messages are evaluated for various purposes based on their content and/or location. The objectives are either to understand the behaviour of Twitter users in detail or draw conclusions about reported phenomena. There are studies that examine the content of the Twitter messages to describe urban areas (Wakamiya et al. 2011), to assess the reaction of the community to physical events such as earthquakes (Crooks et al. 2013) or to perform sentiment analysis (Pak and Paroubek 2010; Kouloumpis et al. 2011). Another specific question frequently investigated is the influence of Twitter on society. For example, Huberman et al. (2008) examined social ties within the Twitter network. Their work showed that only a diminutive number of people among the friend and follower network is actively and frequently engaged in interactions. De Longueville et al. (2009) analysed the Twitter social network during moments of environmental crises. They identified a segmentation process of the user group into distinct types of communicators that handle and forward crisis information in different ways.

Several studies are focusing on understanding the demographic characteristics of the Twitter users at global level (Kulshrestha et al. 2012), at country level (Mislove et al. 2011) or at city level (Adnan and Longley 2013). Kulshrestha et al. (2012) evaluated the distribution of the Twitter users across the world and the Twitter adoption rate compared to the socio-economic status of the country's population such as the Human Development Index (HDI). The results of this study showed a highly unequal distribution of the Twitter users across countries and a high correlation between the HDI and Twitter adoption rate. Mislove et al. (2011) compared the number of Twitter users with the number of population in US counties and assessed the ethnicity and gender of the Twitter users. This study revealed that there exists a significantly larger number of male Twitter users in the US. Furthermore, sparsely populated areas are underrepresented in terms of tweeting activity.

Adnan and Longley (2013) performed a descriptive analysis of Twitter users in London, Paris and New York. In particular, they studied demographic aspects of Twitter users such as name, ethnicity and gender to get deeper insights into the ethnic diversity of the cities' population. The authors assessed also the hourly twitter activity and identified two tweeting activity peaks in London: between 10 a.m. and 11 a.m. and between 7 p.m. and 11 p.m. The geographic distribution of the day and night tweeting activity was not assessed. In a related article, Adnan et al. (2013) investigated the spatial distribution of tweets from different ethnic groups across London. They separated the messages into day-time and night-time tweets in order to gain insights in activity patterns across London.

This work aims at describing the socio-demographic characteristics of residents in locations with high tweet counts. The question is whether socio-demographic variables predict numbers of Twitter messages and whether the distribution of tweets provides insights into the demographic group of Twitter users. The specific contribution of this work is the exploration of representations and analyses for investigating who Twitter users in London are.

Data and Methods

The Twitter community produces tremendous amounts of data on a daily basis. To access these data for development and research purposes, Twitter offers a Streaming Application Programmable Interface (API).[1] Via this API, a subset of georeferenced tweets can be accessed at any point in time. To access the API in a

---

[1] Twitter Streaming API—https://dev.twitter.com/docs/streaming-apis (2013-12-05).

Fig. 1 Study area around the London city centre

convenient way, the authors employed the Twitter4J[2] Java programming library. The target place for this study is the city of London, UK. For a collecting area, the authors selected a 28 km radius around the city centre. The defined radius distance represents an approximation of London's bounding box. Some parts of the outskirts of greater London area were not completely covered in the collection procedure. Areas not covered with tweets were not included in the further analysis (Fig. 1).

The Twitter data for this work have been collected over the course of 3 months: July to September 2013. In total, about 250,000 Tweets were collected. These tweets are all geolocated, which means that each tweet is located in space with coordinates. Geolocated tweets make up a subset of around 1 % of the total number of tweets (Morstatter et al. 2013). The information per tweet contains: The date of the tweet as well as its associated time (HH:MM:SS), the Twitter user name, the unique internal Twitter user ID, the coordinates of the tweet, and finally the message of the tweet.

As a first step of pre-processing, the tweets were separated into day-time and night-time tweets. The range for day-time is from 07:00 until 18:59:59, while the range for the night-time Tweets is from 19:00 to 06:59:59. The size of the two subsets is almost exactly 50 % each.

The second step of pre-processing regards the requirement to separate tourists from locals as socio-demographic data are based on residents in an area. The approach chosen to address this issue is to work with tweets from users, who posted

---

[2] Twitter4J Library—http://twitter4j.org/en (2013-12-05).

at least three tweets in the covered period of 3 months. This subset of the dataset is supposed to increase the probability to work with tweets from residents, who spend more time in the area and have more chances to post tweets.

Besides geolocated tweets, socio-demographic and population density data are used for this study. The socio-demographic data for London's wards, which are subunits of boroughs, come from the Greater London Authority (GLA). They were downloaded from the GLA data store for the year 2011.[3]

The analysis of Twitter data is composed of several steps. First, the distribution of tweets during day-time and night-time were compared in order to identify wards with high tweet counts and geographic differences between the day and night patterns.

The subsequent step focused on identifying tweet hotspots. Tweet hotspots are locations with very large numbers of tweets in comparison to population figures. The tweets used are night-time tweets, which are more likely sent from locations were people are residents.

Having identified tweet hotspots, the analysis of socio-demographic variables can proceed. For this purpose, the Exploratory Spatial Data Analysis (ESDA) tool GeoDA[4] is used. The authors focused on a parallel coordinate plot in order to visually inspect similarities between variable distributions across tweet hotspots. This visual inspection of socio-demographic variables serves to identify candidate variables for an Ordinary Least Squares (OLS) regression analysis.

The choice of variables for the OLS regression model is based on the characteristics of Twitter users identified in a recent study (Mitchell and Page 2013) and the inspection of variables and their distributions in GeoDA. The OLS model is prepared for testing the statistical relationship between Twitter data and socio-demographic variables. The regression model provides a generalized assessment of the regression of the variables across the entire study area. Depending on the significance of the identified relationships, a spatially-explicit regression analysis can provide detailed insights. The OLS model therefore provides the basis for a Geographical Weighted Regression (GWR) analysis.

Results and Discussion

This section introduces the results obtained by the exploratory spatial data analysis and statistical analysis. Different forms of representations were used to explore and portray the findings.

By splitting the tweets into day-time and night-time sets and eliminating tweets of users sending less than three tweets, we obtained roughly 105.500 day-time tweets and 106.00 night-time tweets.

---

[3] Socio-demographic   data—http://data.london.gov.uk/datastore/package/ward-profiles-and-atlas (2013-12-16).

[4] GeoDa ESDA tool—http://geodacenter.asu.edu/projects/opengeoda.

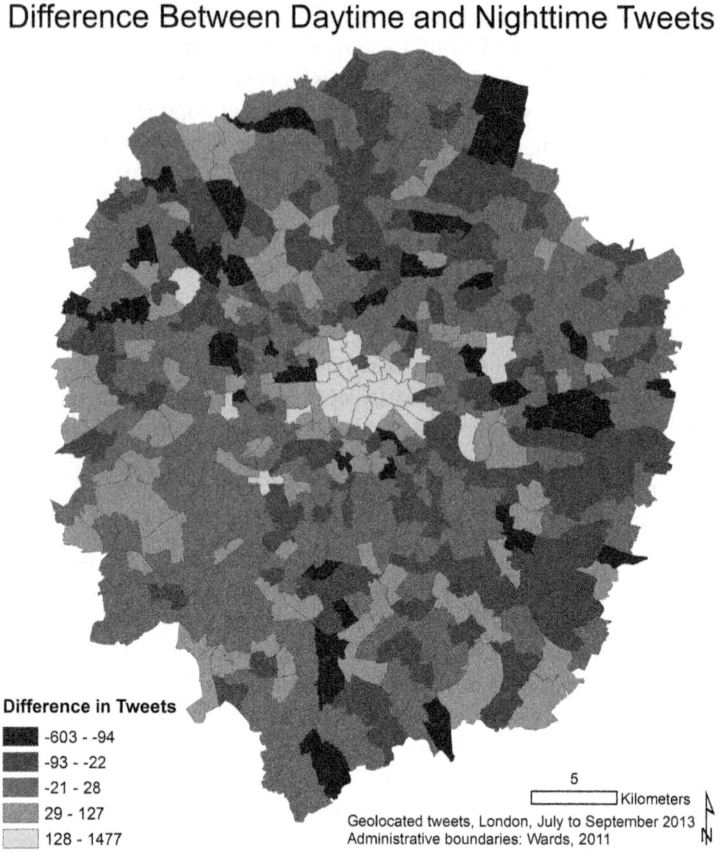

Fig. 2 Difference between day-time and night-tweets showing areas with greater activity during the day (*white*) and during the night (*dark grey*)

The difference between day-time and night-time tweet numbers is shown in Fig. 2. During day-time the Twitter activity in the city centre and nearby wards is higher than during night-time. Wards at some distance from the city centre show increased tweet numbers at night time. This pattern could confirm the assumption that Londoners work in the city centre and along the Thames, but live in the outskirts or bedroom towns.

Adnan and Longley (2013) stated that London's outskirts have low Twitter usage. Our study revealed that during the night, the tweeting activity intensifies in some of the wards away from the city centre. As depicted in Fig. 2, there are several *social hubs* (Bawa-Cavia 2011) that emerge especially in the North-Eastern, North-Western and Southern parts of London.

The next step in the analysis is the comparison of tweets to population figures. Taking night-time tweets as the basis, we show the tweets per capita in Fig. 3. This figure allows the identification of tweets hotspots, which are areas with high tweet

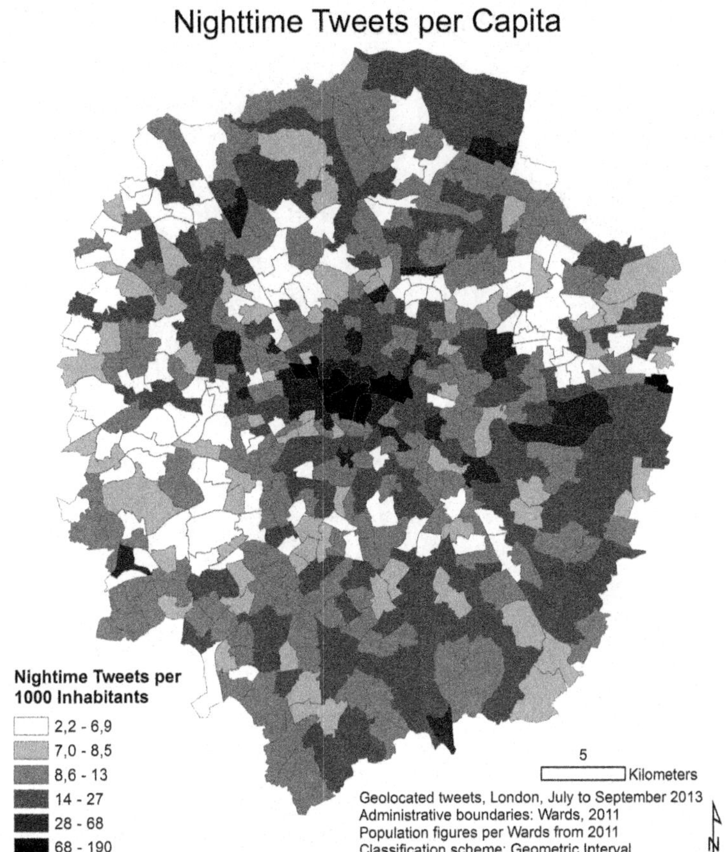

Fig. 3 Wards with at least 28 tweets per 1,000 inhabitants are considered tweet hotspots

counts per capita. Specifically, we consider wards with at least 28 tweets per 1,000 inhabitants as tweet hotspots.

Certainly, the identified hotspots are related to population density (Fig. 4). The assumption is that Twitter usage increases with the number of inhabitants. For example, Kulshrestha et al. (2012) found that as the population of a county increases, the Twitter Representation Rate calculated by dividing the Twitter users in a county by the number of people in that county increases as well. These findings are also backed by the work of Abrol and Khan (2010). Tweet hotspots are therefore often located in wards with lower population density values as in the case of the central wards. However, there are a few exceptions to this rule as the ward of Goresbrook in the very East of the study area (borough *Barking and Dagenham*) and the ward of Colindale (borough *Barnet*) in the North-West of the study area.

The tweets per capita map shows that the city centre and neighbouring wards have the highest counts of Twitter messages (Fig. 3). The same result was reported in the study conducted by Adnan and Longley (2013). The city centre and

Population Density

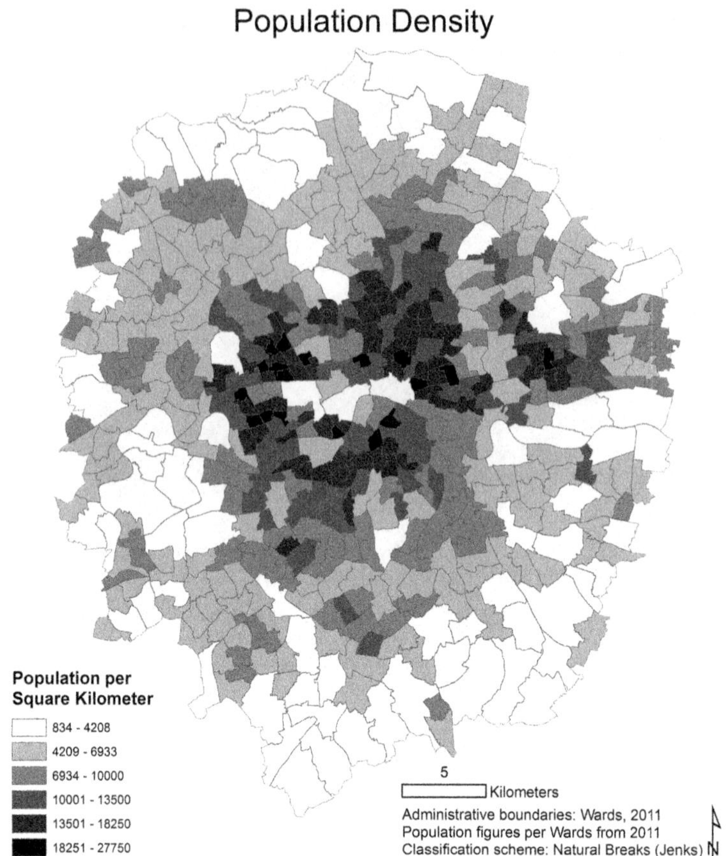

Fig. 4 Population density map for the study area

surrounding wards have a high level of commercial, touristic and work activities (Adnan and Longley 2013), which could explain the overrepresentation of tweet counts in that area. The explanation of hotspots in the East and North-East is less intuitive. Therefore, the subsequent step of the analysis investigates socio-demographic characteristics of hotspots. The objective is to understand whether population demographics explain the tweet hotspots.

A parallel coordinate plot of socio-demographic variables is shown in Fig. 5. Red and orange lines represent tweet hotspots. Visual exploration of the plot suggests that tweet hotspots tend to share similar characteristics especially regarding the percentage of inhabitants in working age (16 and 64 years), the qualification and access to public transportation. Tweet hotspots seem to be characterized by a large percentage of people in working age and a low percentage of people over 65 years. A substantial percentage of the population has a qualification of level 4, which corresponds to at least a Bachelor's degree. The fraction of inhabitants without qualification tends to be lower in tweet hotspots. However, there is

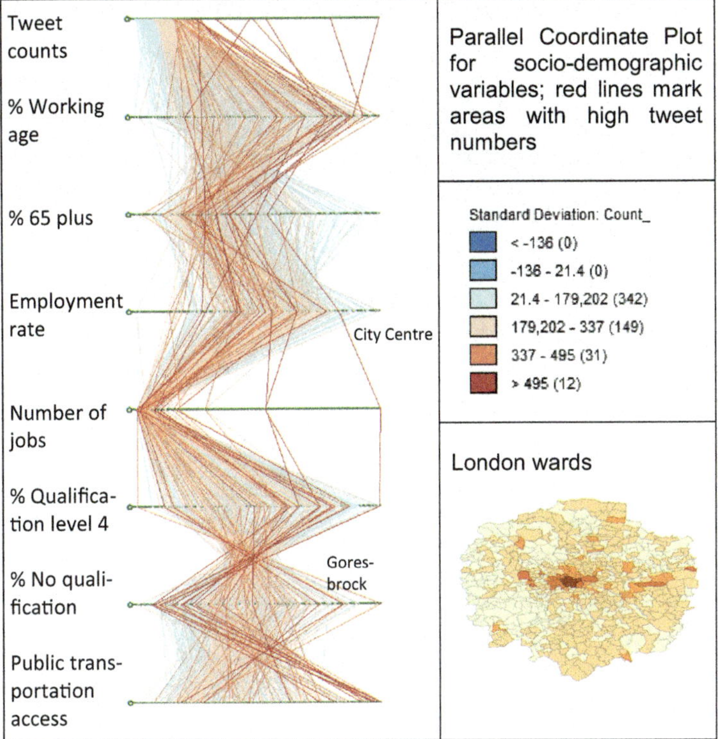

Fig. 5 Parallel coordinate plot of socio-demographic variables (*red* and *orange lines* refer to tweet hotspot locations)

considerable variability in single variables across tweet hotspots, which is indicated by the spread of the lines on the axes. In addition, single wards stand out as different. For example, the ward of Goresbrook, located in the borough *Barking and Dagenham* in the very East of the study area, reverses the relationship between inhabitants with no qualification and qualification of level 4. The number of jobs in the city of London is exceeding the job number of all wards in the study area, which links back to the high tweeting activity during the day.

Figure 6 shows bar charts with the variables of percent of inhabitants in working age, above 65 years, employment rate, percent with qualification of level 4 (at least Bachelor's degree) and no qualification. The display shows that Londoners in and around the city centre tend to have a better level of qualification than in the tweet hotspots in the East, North-East and South. Three hotspots at the outskirts of the study area show a higher rate of non-qualified inhabitants in comparison to inhabitants with level 4 qualifications. Similarities regarding the age of inhabitants and the employment rate seem to be given across hotspots. However, inhabitants above 65 years tend to be more strongly represented within the city centre. A few hotspots, especially in the East, show a lower number of people above 65 years.

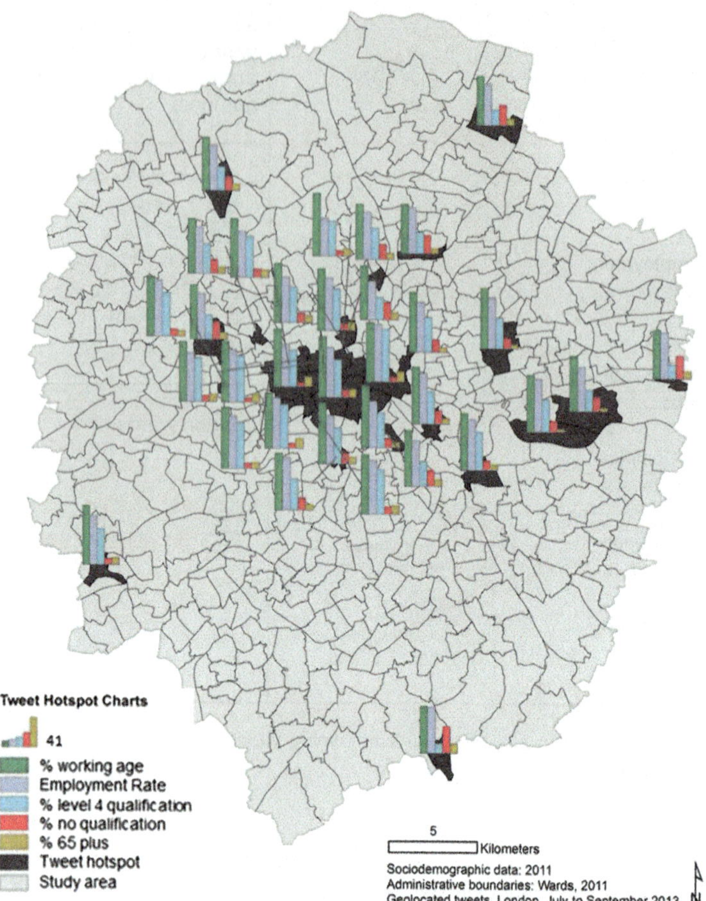

Fig. 6 Socio-demographic variables for tweet hotspots

The visual exploration of demographic variables serves the purpose of indicating candidate variables for an ordinary least squares (OLS) regression model with the purpose to explain tweet numbers through these variables. The OLS regression model of the variables show in Fig. 6 lead only to a low model fit, with R^2 being 21. That means that there is no strong statistical relationship between the selected socio-demographic variables and tweet counts. The result does not suggest that a spatially-explicit regression model is required as the variables have weak explanatory power on a global level. We conclude that socio-demographic variables do not directly explain tweet counts in London's wards.

The weak results of the statistical analysis could be influenced by the level of detail of the socio-demographic variables. The age groups are not sufficiently differentiated in these variables, as all inhabitants between 16 and 64 years are collected in the *working age* group. The study by Mitchell and Page (2013) identified the group of 18–28 year-olds as main consumers of Twitter news. A differentiation of inhabitants into more age groups may therefore lead to better results of the OLS regression analysis.

The use of Twitter data in this study is accompanied by certain caveats. First, the data used cover a period of 3 months only, which may not be entirely representative. Second, tweeting tourists may not have been perfectly excluded from the analysis with the chosen approach of working with tweets from users posting more than three tweets. Especially in the city centre, tourists could have a substantial influence on tweet counts. Third, we used the number of tweets in the analyses rather than the number of Twitter users. Twitter users are described as being mobile (Mitchell and Page 2013); working with Twitter users would therefore require identifying a potential home location through comprehensive filtering of the data. We assume, however, that the patterns of activity remain largely identical. Fourth, geolocated tweets are only a small subset of the total amount of tweets and may represent a particular user group, who is ready to share location information.

The study could be further improved by increasing its temporal resolution. The separation of tweets into day-time and night-time tweets is rather coarse and could be refined for supporting the findings.

Conclusions

The findings of this exploratory study showed that the use of Twitter varies substantially across greater London and that patterns differ for day-time and night-time. The exploratory analysis indicated similarities among identified tweet hotspots. For example, wards with a large percentage of inhabitants over the age of 65 are less likely to be tweet hotspots. However, the statistical analysis of tweet counts and socio-demographic variables could not confirm a definite result. Socio-demographic variables do therefore not directly explain tweeting activity.

A study of Twitter news consumers suggested the user group of being young, well-qualified and users of mobile devices (Mitchell and Page 2013). The characteristics of such a *typical* Twitter user could not be verified for London with the data used. This initial result may be repeated with data bearing more detail regarding age groups to verify the result of this initial study. However, the question persists, whether it is at all possible to relate socio-demographic data and Twitter usage. Twitter users are generally mobile and may tweet while being en route. A further filtering of tweets may be required to limit the analysis to locations that can be identified as residential areas or as home locations of Twitter users.

The largest tweeting activity takes place in the city centre of London. The city centre is certainly a place of a high level of activity (Adnan and Longley 2013). This observation suggests that the tweet counts show a map of activity

(continued)

centres across London. These activity centres should be analysed from different points of view than socio-demographic characteristics as well. For future work we suggest an analysis of the relationship between tweet counts and proxies for activity levels like tourist attractions, bars and stadiums.

Acknowledgments The authors would like to thank two anonymous reviewers for their helpful comments. Thomas Lampoltshammer is funded by the Austrian Science Fund (FWF) and the Salzburg University of Applied Sciences through the Doctoral College GIScience (DK W 1237-N23).

References

Abrol S, Khan L (2010) TWinner: understanding news queries with geo-content using Twitter. In: 6th workshop on geographic information retrieval (GIR '10), 8. ACM, New York

Adnan M, Longley P (2013) Analysis of Twitter usage in London, Paris, and New York City. In: 16th AGILE international conference on geographic information science, Leuven, May 14–17, 2013, pp 1–7

Adnan M, Lansley G, Longley PA (2013) Twitter geodemographic analysis of ethnicity and identity in Greater London. In: 12th international conference on geocomputation, LIESMARS. Wuhan University, China

Bawa-Cavia A (2011) Sensing the urban: using location-based social network data in urban analysis. In: the 1st workshop on pervasive urban applications PURBA '11, San Francisco

Boyd D, Crawford K (2012) Critical questions for big data: provocations for a cultural, technological, and scholarly phenomenon. Inf Commun Soc 15:662–679

Craglia M, Ostermann F, Spinsanti L (2012) Digital Earth from vision to practice: making sense of citizen-generated content. Int J Digital Earth 5:398–416

Crooks A, Croitoru A, Stefanidis A, Radzikowski J (2013) #Earthquake: Twitter as a distributed sensor system. Trans GIS 17:124–147

De Longueville B, Smith RS, Luraschi G (2009) OMG, from here, I can see the flames!: a use case of mining location based social networks to acquire spatio-temporal data on forest fires. In: Proceedings of the 2009 international workshop on location based social networks, pp 73–80

Huberman B, Romero D, Wu F (2008) Social networks that matter: Twitter under the microscope. First Monday [S.l.]. doi:10.5210/fm.v14i1.2317. Available at: http://firstmonday.org/ojs/index.php/fm/article/view/2317/2063. Accessed 13 Oct 2014

Hwang Y, Park N (2013) Digital divide in social networking sites. Int J Mobile Commun 11:446–464

Kouloumpis E, Wilson T, Moore J (2011) Twitter sentiment analysis: the good the bad and the OMG! In: ICWSM Fifth International AAAI Conference on Weblogs and Social Media, 17–21 July 2011, Barcelona, Spain

Kulshrestha J, Kooti F, Nikravesh A, Gummadi PK (2012) Geographic dissection of the Twitter network. In: ICWSM Sixth International AAAI Conference on Weblogs and Social Media, June 2012, Dublin, Ireland

Mediabistro (2012) Revealed: the top 20 countries and cities of Twitter [STATS]. http://www.mediabistro.com/alltwitter/twitter-top-countries_b26726. Last accessed on 13 Oct 2014

Mislove A, Lehmann S, Ahn Y-Y, Onnela J-P, Rosenquist JN (2011) Understanding the demographics of Twitter users. In: ICWSM Fifth International AAAI Conference on Weblogs and Social Media, 17–21 July 2011, Barcelona, Spain

Mitchell A, Page D (2013) Twitter news consumers: young, mobile and educated. Pew Research Center: http://www.journalism.org/files/2013/11/Twitter-IPO-release-with-cover-page-new2.pdf. Last accessed on 13 Oct 2014

Morstatter F, Pfeffer J, Liu H, Carley KM (2013) Is the sample good enough? Comparing data from Twitter's streaming API with Twitter's firehose. In: International AAAI conference on weblogs and social media ICWSM 17–21 July 2011, Barcelona, Spain, pp 400–408

Pak A, Paroubek P (2010) Twitter as a corpus for sentiment analysis and opinion mining. In: LREC Proceedings of the Seventh International Conference on Language Resources and Evaluation, Valletta, Malta

Wakamiya S, Lee R, Sumiya K (2011) Crowd-based urban characterization: extracting crowd behavioral patterns in urban areas from Twitter. In: Proceedings of the 3rd ACM SIGSPATIAL international workshop on location-based social networks, pp 77–84. ACM, Chicago

Visualization Problems in Worldwide Map Portals

Jana Stehlíková, Helena Řezníková, Hana Kočová, and Zdeněk Stachoň

Introduction

Worldwide map portals are very often used for identifying the geographical location of a target destination or searching for the best route to an unfamiliar place. There are a great number of web maps on the Internet which provide a map itself and many map tools, however much they differ in their level of user friendliness. Although web map portals have been analysed in numerous studies with respect to user interface and map visualization, no standard of web map design has been published so far.

This paper builds on and contributes to the work of Nivala et al. (2008), who examined four different mapping sites. Several usability tests were conducted on users, and usability problems were recognized and further analysed. However, these tests were carried out 5 years ago. Is it feasible to assume that all of these problems have already been solved?

This study provides additional insight into problems related to map visualization and designing the user interface of worldwide map portals. First, some problems of usability testing are identified. A usability test covering three worldwide map portals (Google Maps, Bing Maps and MapQuest) is then described. Finally, results of the testing and guidelines for designing highly usable map visualization are presented.

J. Stehlíková (✉) • H. Řezníková • H. Kočová • Z. Stachoň
Faculty of Science, Department of Geography, Masaryk University, Kotlářská 2, 611 37 Brno, Czech Republic
e-mail: jana.st@mail.muni.cz; 177665@mail.muni.cz; 170345@mail.muni.cz; 14463@mail.muni.cz

© Springer International Publishing Switzerland 2015
J. Brus et al. (eds.), *Modern Trends in Cartography*, Lecture Notes in Geoinformation and Cartography, DOI 10.1007/978-3-319-07926-4_17

Previous Studies

Progressive development of the Internet in the last two decades led to the need for testing of web site usability. Therefore, there have been many studies dealing with usability evaluation. Nielsen (1993), who defined the elements of usability as efficiency, learnability, memorability, error rate and satisfaction, is considered a father of usability testing. Subsequently, many usability evaluations of web sites, software or information systems have been made, e. g. van Waes (2000) and Battleson et al. (2001). Travis (1999) published a Userfocus web portal, which deals with various usability issues in general and provides guidelines for designing technologies and usability testing. A usability international standard ISO 9241-210 (2010) was also adopted.

An important part of general usability issues is the evaluation of map portals. There are many studies dealing with usability testing of web maps from different points of view. Schobesberger (2009) presented various types of usability evaluation of web mapping portals. Interview, user observation, user surveys, remote evaluation and eye-tracking were defined as the main techniques. You et al. (2007) studied issues related to zoom and pan functions. The author examined five web map portals by evaluating simulated user interface and designed an interface with ideal zoom and pan functions.

Hub et al. (2011) and Komarova et al. (2007) have presented heuristic usability evaluations of Geoweb, which consist of defining potential problems, i.e. heuristics, and examining user interface and judging its harmony through known heuristics. He et al. (2012) and Komarkova et al. (2010) used the think-aloud method for testing of national map portals and a user test, and found that users were influenced by Google Maps when working with the national map portals. Komarkova et al. (2009) carried out a written questionnaire and practical user tasks on GIS applications of Czech regional authorities. The most significant finding was that users prefer to have the option of switching on/off all layers on all tested map portals.

Skarlatidou and Haklay (2006) analysed the usability of some world-wide map portals with a user test and questionnaires. The time needed to perform the task was measured and the success rate was calculated. Skarlatidou and Haklay came to the conclusion that Google Maps is the most usable map portal and Map Quest is the least usable one. Correspondingly, Nivala et al. (2008) presented a usability evaluation of world-wide map portals similar to the analysis presented in this paper. Four world-wide mapping sites were evaluated employing user tests and expert evaluations. Altogether, 403 usability problems were identified. The problems were divided into four categories: user interface, map, search operations, and help and guidance. Based on the evaluation of web map portals, some design guideline suggestions were discussed and presented as a result of this study.

Usability studies focusing on web map portals also generate inspiration for further development or adoption of new tools and visualization methods, e.g., more user-centred options of cartographic visualization mentioned in Konečný et al. (2011), Stachoň et al. (2013), etc.

Another approach to examining usability is using eye-tracking to follow the user's eye. An example of an eye-tracking study focusing on the comparison of 2D and 3D cartographic visualization can be found in Popelka and Brychtová (2013).

Methods

It has been 5 years since Nivala et al. (2008) presented a usability evaluation of world-wide map portals. The present study deals with usability testing of the same map portals and tries to identify whether the problems described in 2008 have been solved or persist. Moreover, the potential deficiencies of up-to-date portals are indicated. Nivala examined Google Maps, MSN Maps & Directions, MapQuest, and Multimap. MSN Maps & Directions and Multimap have since been joined to Bing Maps, so the well-known map sites compared in this work are Google Maps (hereafter referred to as GM, http://maps.google.com/), MapQuest (MQ, http://www.mapquest.com/), and Bing Maps (BM, http://www.bing.com/maps/).

It was considered whether to include some alternative or local map portals in the study. The study was eventually limited to the above-mentioned map portals due to particular map portal functions, language dependencies that might limit future study, etc.

Visual Interpretation

The map portals were initially examined in terms of the problems analysed by Nivala et al. (2008), who defined four categories: user interface, map, search operations, and help and guidance. Since the map portals have seen certain improvements in the meantime, some of these problems have been solved while others remain unsolved.

The problems concerning user interface may be regarded as solved. Nevertheless, the main problem reported concerning user interface was the overloading of web map homepages with advertisements and additional information. Moreover, in some cases there was no link to the map on the homepage. Another problem was the fact that links opened in the same browser window as the map, and there was no quick way back to the homepage. The last major problem in this category was the form and placement of the function buttons.

Most of the defined problems were related to the map field. The principal problem was map visualization—maps looked like paper maps instead of web maps, and their visualization was messy and confusing. The other problems presented in 2008 were the absence of a print function or route direction tool and the fact that the scale bar showed only miles. Panning was sometimes found to be problematic and slow, or there was no possibility to add markers into the map. Most of these problems have since been resolved, but other problems in this category still persist: e.g., there is no legend; data is seen as inconsistent—some objects in a given category are shown on the map while others are not; there is no option to modify the map by checking boxes to show or hide different layers, and there is no option to highlight certain categories of objects, e.g., hotels or theatres. There are also problems with zooming and panning. For example, some symbols or text appear and disappear randomly with different scales, the step between map scales is too large, and the visualization between the scales is too varied. Some colours are considered unintuitive. Another problem is labelling—the placement of the text, its legibility, etc. The final group of problems in this category is the misinterpretation of some symbols.

The third category covered search operations. There were two search possibilities: a free search, which allows a user to use search criteria more liberally, and a search done using different boxes for a strictly defined input, e.g., a country or address. Some of these problems have been resolved, and others have not. The resolved problems include those related to default settings—if the user did not define the country of his location, the site would only give results from the USA. Another group of problems concerns route or address searching—there was no way to search for a route by clicking on the map itself, no possibility for multi-stop route searches and no good methods for route modification. A final problem was centring the map according to the result irrespective of its location. The unresolved problems include those concerning visualization of the search result—streets are indicated by a point instead of a line. There is no possibility of conducting several separate searches. Moreover, there is no option for enabling a direct return to the search result, and in some cases users do not know whether the search result is relevant or not.

The final category is help and guidance. A major problem was the absence of help, but this has already been settled. Some users identified problems with the comprehensibility of the help provided.

Following Nivala's study, the map portals were compared with each other by visual exploration. Elements of functionality were described for each portal, and its potential vulnerabilities were recognized.

Null-Hypotheses and Pre-test

Based on an analysis of the problems defined by Nivala et al. (2008) and an exploration of up-to-date map portals, null-hypotheses for further testing were formulated. The null-hypotheses were encouraged by a simple pre-test, which was performed on 20 geography students. First, weaknesses of each portal were determined and the students were subsequently asked a simple question. If they agreed about a potential problem, it was proposed that the vulnerability be evaluated in a user test. For example, the colouring of bodies of water on the MQ map was considered unsuitable, since the pools in the parks were difficult to distinguish.

The students were then asked how many pools they see in the park on GM, MQ, and BM.

The following null-hypotheses were formulated as a result of the pre-test and the exploration of the web map portals: GM will appear as the best map portal because users are used to working with it and the orientation is easy. The map visualization and symbology are also very familiar to the greatest number of users, and therefore they will consider its symbology to be the best. GM's labels will also be considered the best, since the labels on MQ are too small and difficult to read, and labels on BM are disproportionate and they overlay additional information. Moreover, users will prefer the possibility of turning layers on and off and regulating the amount of information contained in the map. On MQ, it will be difficult to distinguish bodies of water in the parks.

User Test

The user test in the MUTEP (MUltivariate TEsting Program) was drawn up according to the null hypotheses and the pre-test results as described above. MUTEP was developed by the Department of Geography at Masaryk University for the purposes of objective experimental exploration and evaluation of cartographical products. This application uses an innovative cognitive testing method that combines quantitative and qualitative information; the software enables us to test various inputs, from isolated cartographic symbols to a set of complex interactive maps. Moreover, it allows us to explore various types of operations or cognitive processes. MUTEP enables group administration, automatic evaluation of results and export to a form that is suitable for statistical analyses (Šašinka and Morong 2012).

MUTEP was installed on a server; preparation and administration of the test battery were therefore realized online. The test was conducted under homogeneous conditions via a web browser and took between 15 and 20 min. The test consisted of an introductory portion and two portions of testing tasks. The introductory portion was necessary in order to obtain the data for the subsequent statistical evaluation.

After filling in a personal data questionnaire, the users were asked to answer several questions about the analysed web map portals, e.g. which portals they use and for what purposes (see Fig. 1).

In the second portion of the test, users were asked to choose one of four buttons according to the assigned task. Only one option could be chosen. The first task in this portion tested the subjective preferences of the users: the preferred map on two scales (see Fig. 2) and the most suitable symbol for a cinema and a school. These categories were chosen because they were present in all three web map portals, yet they were visualized with different symbols. The same type of task was used for determining the number of bodies of water in a park on MQ to confirm or contradict the results of the pre-test (Fig. 3, task 1–3). The final task in this set of tasks tested whether the users misinterpret the violet polygons in a park on BM as bodies of

Fig. 1 Illustration from the user test—task 4 (How many pools are located in Güell Park?)

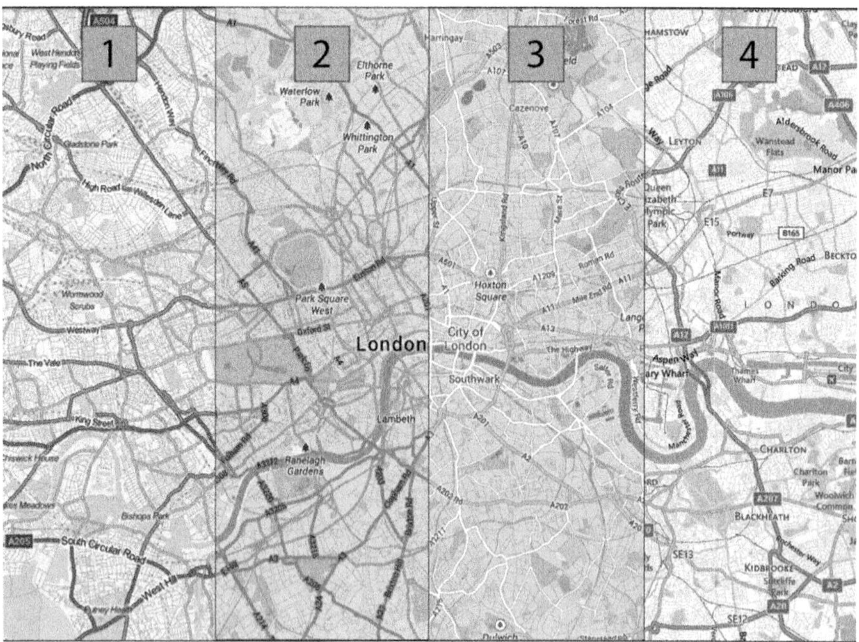

Fig. 2 Different visualizations used on (*1*) MQ, (*2*) the old version of GM (until the first half of 2013), (*3*) the new version of GM, (*4*) BM

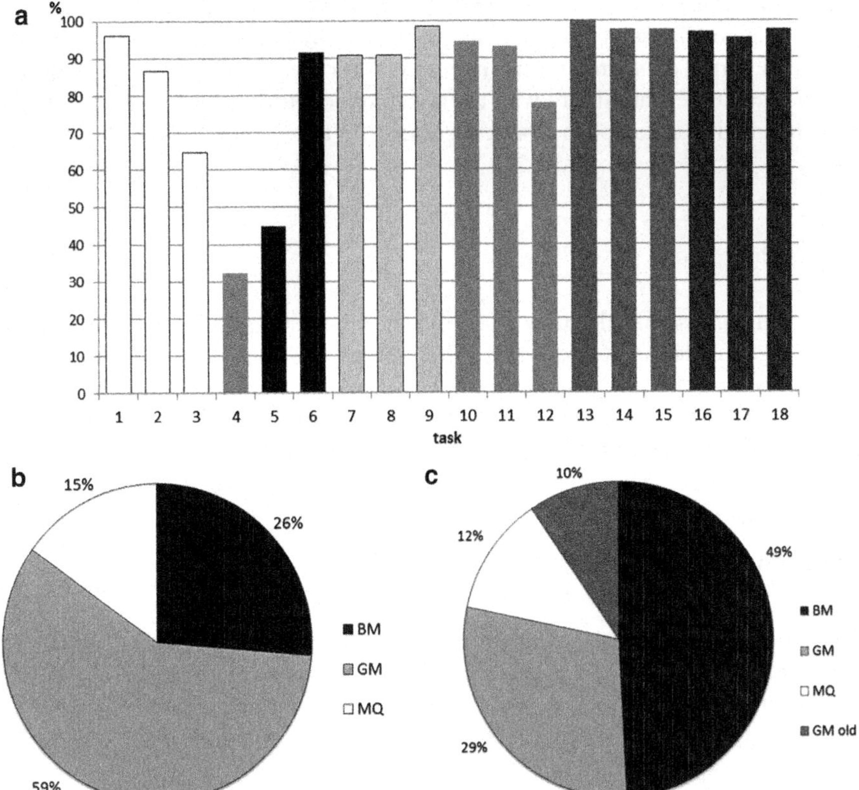

Fig. 3 (**a**) Correctness rate for the test tasks: *1–3* bodies of water on MQ, *4*–public buildings on BM, *5*, *6*–labels on BM, *7–9* post offices on BM, GM, MQ, *10–12* pharmacies on BM, GM, MQ, *13–15* parks on BM, GM, MQ, *16–18* intersection of river and railway on MQ, BM, GM; (**b**) Preferences of tested portals regarding visualization; (**c**) symbology preferences

water (they actually represent the interior structure of public buildings) (see Figs. 1 and 3, task 4). The legibility of map labels for large cities on BM was examined in the same way—the users chose the name of the largest city on the map by clicking on an appropriate button (Fig. 3, tasks 5 and 6).

The task in the third portion of the test involved marking an appropriate point by clicking on the map field. Only one point on the map could be marked. The purpose of this portion was to evaluate the distinctiveness of symbols and users' efficiency when orienting themselves on a map. The respondents were asked to find the nearest post office to a given street intersection (Fig. 3, tasks 7–9), the nearest pharmacy to a given hotel (Fig. 3, tasks 10–12) and parks (Fig. 3, tasks 13–15) on all three examined map portals. The final task was to mark the intersection of a watercourse and a railway (Fig. 3, tasks 16–18). This set of tasks used similar map segments to ensure the comparability of all three evaluated map portals. Selected types of tasks had to be answered repeatedly to eliminate the influence of the order of the map portals on the test. It usually took more time to perform a task using the first map portal, compared to the time needed for the second and third evaluated portal, due to user familiarity with the task. Correct performance of a task and the time required to accomplish it were measured for the selected tasks.

Results

The main purpose of this study was to identify current usability problems on the following map portals: Google Maps, Bing Maps, and Map Quest. In this section, we compare the potential usability problems arising from our research with the results of Nivala et al. (2008). The results are divided into two parts and correspond to the main research methods: qualitative and quantitative.

Participants in User Test

The user test described in the previous section was taken by 127 respondents (67/60 male/female). The data gathered from the personal data questionnaire shows that the majority of the participants were students of Geography at Masaryk University (average age 22.9 years). Most of the participants use a computer every day, so the user group was quite homogenous.

Qualitative Data/Research

Most of the problems identified relate to the maps themselves. The map is the most important part of a map portal, since it contains all the information that is provided. Moreover, the usability errors concerning the user interface, search operations and help and guidance seem to have been solved since Nivala et al. (2008) conducted her study. It is possible to divide the remaining problems into two groups: map visualization and map tools.

Map visualization is a rather subjective issue; nevertheless, differences can be recognized among the evaluated portals. Several problems described by Nivala et al. (2008) still persist and have already been reported in this paper in the section about methodology. Nevertheless, some other problems have been identified.

Figure 3 shows the differences in map visualization of the tested web map portals of a certain scale. Map visualization can influence users' ability to read information on a map. It is important to realize that a map is not only a presentation of spatial data, but it is especially a means of communication between the cartographer and the map user, as described by MacEachren (1979).

Qualitative information analysis focused on evaluating the most preferred visualization of the presented maps and on preferences for selected map symbols. Preferences are shown in Fig. 3b and reveal that most users prefer GM's map visualization, especially in its newest version. Tested symbology concerned map signs for a school and for a cinema. Figure 3c demonstrates that users prefer BM's symbology. There was no difference in preferences between male and female participants.

Users were also asked to reveal which map portals they use in everyday life. Options included the tested map portals as well as Mapy.cz, the favoured map portal in the Czech Republic. Mapy.cz was not a tested portal because we chose to concentrate on international map portals which are used world-wide. An interesting result is that almost all respondents use GM and Mapy.cz, but only 8.7 % of them work with BM and only 2.3 % use MQ. The popularity of GM may be explained by the amount of the data provided for the Czech Republic (or Europe), since the other map portals focus more on the USA. Another question was aimed at discovering the reasons respondents use map portals. The majority of respondents answered that they use map portals for searching addresses and routes, planning trips and looking for points of interest. An interesting result is that a large number of users (24.8 %) use topographic maps but not orthophotomaps.

Another subjective aspect related directly to Fig. 2 may be the map skeleton. The figures illustrate that the main orientation objects are rivers, parks and main roads on GM and BM, and the road network on MQ. The base map changes automatically when zooming on BM, which can cause changes in map orientation and thus confuse users.

The map content varies on each map and it is changing very quickly. During work on this study, the map content changed several times. For example, the point symbol categories on GM were reduced.

Quantitative Data/Research

Quantitative data provides more objective information, and so we focused on this data in the following parts of the test. First it was found out that the size of labels differs among the portals. It may be observed that the labels on MQ are rather small in comparison to labels on GM and BM. Moreover, labels on BM are disproportionate, so they overlap and obscure the content of the map itself. On GM, additional information about objects is provided using mouseover. The analysis of label size shows that labels in GM are the most usable ones. Another problem with BM is the layer order, which does not follow cartographic rules and makes the map less usable.

The chart in Fig. 3a shows the number of correct answers. Tasks focusing on map labels (tasks 5 and 6) proved that only 44.9 % of respondents were able to answer the question about city labelling correctly on BM (task 5), which confirmed the hypothesis that disproportionate labels may confuse the user. Task 6 has a high correctness rate, but the time required to perform the task was longer.

A potential problem could be found in the fact that map visualization varies geographically on GM. Buildings are presented on MQ and GM, while BM shows only a street network. GM provides shadowed 3D buildings, which cover a significant part of the map and might be considered disruptive. Moreover, on BM it is possible to display the interior structure of some public buildings (e.g. shopping malls, museums); buildings with an interior structure are violet, which is quite similar to the blue colour indicating bodies of water. This is confirmed by the fact that 67.8 % of respondents believed the violet colour denotes a body of water instead of a building, as shown in Fig. 3, task 4.

In tasks 1–3 in Fig. 3, respondents were asked to count bodies of water in parks on MQ. The bodies of water on MQ were ultimately recognized correctly by most users (over 80 % on average), but most said counting them was very difficult and the time needed to perform the task was too long.

Tasks 7–12 sought to analyse which type of searching logic on the map portals is more intuitive for users. MQ is the only portal with the possibility to switch layers on/off, and finding an object requires that the user first turn on a given layer. By comparison, on BM and GM all symbols are present on the map all the time. This implies different searching logics. Users were asked to find the post office nearest to a particular street intersection (tasks 7–9) and to find the pharmacy nearest to a particular hotel (tasks 10–12). The chart shows that the correctness rate was quite similar for all portals (around 90 %) with the exception of finding the pharmacy on MQ, which was 78 %.

Tasks 13–18 required users to find a particular location on all tested portals, such as a park or the intersection of a watercourse and a railway. As shown in Fig. 3, the results were similar for all portals. The correctness rate was around 95 %.

One interesting result was the fact that women required more time to perform the tasks than men did. The biggest difference was when locating the post office nearest to a given street intersection (Fig. 3, tasks 7–9), in which the average time for men was around 40 s and around 50 s for women.

Discussion

The test was designed to examine the above-mentioned persistent problems in map portal visualization. The secondary focus was on differences in map visualization among the portals tested. It can be said that there are some differences among the examined map portals, but they are not significant ones. GM seems to be the most usable map portal, since users needed the least time to perform the tasks when compared to BM and MQ. This fact can be explained by the popularity of GM in the Czech Republic, and respondents of a different nationality might produce different results. The results of a questionnaire established that the majority of respondents know and use only GM. By contrast, MQ may be considered the least intuitive map portal because the majority of respondents required the longest time to perform the tasks. MQ differs from the two other portals the most in the areas of map content and searching logic, and MQ is also the only map portal which offers the option of switching layers on/off. This may appear to be an advantage for users, since the map is not overloaded with map symbols. However, the results of the user test may prove that user orientation on a map is highly dependent on practice with the described scheme.

Further analysis focused on the differences between men and women in the user test. In general, women needed more time to perform most of the tasks. This finding needs to be further verified in extended studies. Another promising direction of future research lies in the area of cartographical visualization analyses of spatial data originating from spatial data infrastructures, such as INSPIRE (INfrastructure for Spatial InfoRmation in Europe) as described, for example, by Řezník (2013). At the same time, issues concerning the popularity and diversity of various map portals in different countries can be examined.

Conclusion

This paper presents a usability evaluation of the most popular world-wide map portals: Google Maps, Bing Maps, and Map Quest. The main aim of this study was to identify potential vulnerabilities in usability of these map portals and compare the results with a previous usability study by Nivala et al. (2008). Since it has been 5 years since Nivala conducted her tests, we asked which problems have already been solved, and we identified persistent problems and possible new weaknesses in map portals. The main methods used in this research were a combination of qualitative and quantitative tasks performed in an electronic user test conducted in a MUTEP environment.

We found that the greatest number of problems mentioned by Nivala et al. (2008) that still persist concern cartographic visualization. Problems regarding user interface, help and guidance and search operations have already been solved.

Table 1 summarizes the results obtained in the user test, establishing the main strengths and weaknesses in map visualization of each portal tested.

According to Table 1, BM has the most weaknesses of all tested portals. Nevertheless, most users liked the symbology used by BM. GM may be considered to be the best map portal, since most users preferred its general map visualization and its labels seemed most effective. It may be assumed that MQ is the least usable portal, since it uses searching logic that is different from that of the other two portals and users found it to be quite confusing. On the other hand, MQ is the only portal with the option of switching layers on and off.

Visual interpretation helped to identify null-hypotheses that could be proven by the user test. The user test showed that the majority of null-hypotheses were true, with the exception of the hypothesis that considered the GM symbology to be the best one.

(continued)

Table 1 Strengths and weaknesses of map visualization of GM, BM, and MQ

	Strength	Weaknesses
GM	Most preferred map	Visualization varies geographically
	Most usable labels	3D building layer cannot be switched off
	Most preferred symbology	Disproportionate labelling
BM		Wrong layer order
		Violet colour for buildings with interior structure
MQ	Option of switching layers off/on	Colour for bodies of water

The user test results demonstrated the popularity of GM in the Czech environment. On one hand, there are no significant differences between the examined map portals. On the other hand, differences between men and women were identified, which cartographers might respond to by e.g. developing the more user-centred options of cartographic visualization mentioned above. Gender divergences might be further verified by e.g. using eye-tracking tools.

There is great potential for further extension of this study, especially by involving alternative or local map portals or by examining gender differences. It would also be possible to cover various new map portal technologies, such as Google Street View, 45° images or mobile applications.

Acknowledgments This research has been supported by funding from a project of Masaryk University under the grant agreement No. MUNI/A/0902/2012, which is called 'Expression of Global Environmental Change in Component Earth's Spheres.'

References

Battleson B, Booth A, Weintrop J (2001) Usability testing of an academic library Web site: a case study. J Acad Librarianship 27(3):188–198. ISSN: 0099-1333

He X, Persson H, Östman A (2012) Geoportal usability evaluation. Int J Spatial Data Infrastructures Res 7:88–106. ISSN: 1725-0463

Hub M, Vísek O, Sedlák P (2011) Heuristic evaluation of Geoweb: case study. In: Proceedings of the European computing conference, Paris. ISBN: 978-960-474-297-4

ISO 9241-210 (2010) Ergonomics of human-system interaction. International Organisation for Standardisation, Geneva

Komarkova J, Jakoubek K, Hub M (2009) Usability evaluation of Web-based GIS – case study. In: Proceedings of IIWAS, Kuala Lumpur

Komarkova J, Jedlička M, Hub M (2010) Usability user testing of selected Web-based GIS applications. WSEAS Trans Comput 9(1). ISSN: 1109-2750

Komarova J, Vísek O, Novak M (2007) Heuristic evaluation of usability of GeoWeb Sites. Web and wireless geographical information systems. In: 7th international symposium, W2GIS 2007, Proceedings, Cardiff, 28–29 November 2007, pp 264–278. Springer, Berlin/Heidelberg. ISBN: 978-3-540-76925-5

Konečný M, Kubíček P, Šašinka Č, Stachoň Z (2011) Usability of selected base maps for crises management – users perspectives. Appl Geomatics 3(4):189–198

MacEachren AM (1979) Communication effectiveness of choropleth and isopleth maps: the influence of visual complexity. Unpublished Ph.D. dissertation, University of Kansas

Nielsen J (1993) Usability engineering. Academy Press Professional, Cambridge

Nivala A-M, Brewster SA, Sarjakoski LT (2008) Usability evaluation of web mapping sites. Cartogr J 45(2):130–140. ISSN: 1743-2774

Popelka S, Brychtová A (2013) Eye-tracking study on different perception of 2D and 3D terrain visualization. Cartogr J 50(3):240–246s. ISSN: 1743-2774 (Maney Publishing)

Řezník T (2013) Geographic information in the age of the INSPIRE directive: discovery, download and use for geographical research. Geografie 118(1):77–93. ISSN: 1212-0014

Šašinka Č, Morong K (2012) Původnívý zkumný nástroj pro oblast kartografie a psychologie – Multivariantní Testovací program (MuTeP). In: Halama P, Hanák R, Masaryk R (eds) Sociálne procesy a osobnosť 2012: Zborník príspevkov z 15. ročníka medzinárodnej konferencie. Bratislava: Ústav experimentálnej psychológie, 2012. s. 188–194, 6 s. ISBN: 978-80-88910-40-4

Schobesberger D (2009) Towards principles for usability evaluation in web mapping – usability research for cartographic information systems. In: Proceedings of 24th cartographic conference, Santiago de Chile

Skarlatidou A, Haklay M (2006) Public web mapping: preliminary usability evaluation. Presented at GIS Research UK 2005, Nottingham, 5–7 April

Stachoň Z, Šašinka Č, Štěrba Z, Zbořil J, Březinová Š, Švancara J (2013) Influence of graphic design of cartographic symbols on perception structure. KartographischeNachrichten, roč. 2013, č. 4, s. 216–220. ISSN 0022–9164

Travis D (1999) User focus (on-line: http://www.userfocus.co.uk). Accessed 5 Dec 2013

van Waes L (2000) Thinking aloud as a method for testing the usability of Websites: the influence of task variation on the evaluation of hypertext. IEEE Trans Prof Commun 43(3). ISSN: 0361-1434

You M, Chen C-W, Liu H, Lin H (2007) A usability evaluation of web map zoom and pan functions. Int J Des 1(1):15–25. ISSN: 1994-036

Geovisualising Unequal Spatial Distribution of Online Social Network Connections: A Hungarian Example

Ákos Jakobi and Balázs Lengyel

Introduction

Online social networks (OSN) are basically virtual creations, which stand for expressing real life friendships in forms of virtual space acquaintance. They serve also as major platforms of ICT-enabled communication, supporting place-independent social life; however, recent findings suggest that geographical location of users strongly affect network topology (Takhteyev et al. 2012). As a matter of fact OSNs may be related to locations, since the users themselves could be at large geographically located as well. By retrieving geolocated data from the network-database of an OSN it becomes possible both to visualise and to analyse geographical relations of participants.

Online social networks are large-scale networks from social network sites (SNS) in which users are the nodes and their connections with other users are the edges. SNSs are defined as web-based services that "enable users to articulate and make visible their social networks" (Boyd and Ellison 2007). The definition claims that SNSs are supplemental forms of communication between people who have known each other primarily in real life (Ellison et al. 2006, 2007). In other words, major SNSs are not used to meet new people, but rather to articulate relationships with people in their existing offline network. Furthermore, the tie-distribution of the largest network (Facebook) is very close to multi-scaling behaviour of real-life social networks (Ahn et al. 2007; Backstrom et al. 2012; Ugander et al. 2011). It

Á. Jakobi (✉)
Department of Regional Science, Eötvös Loránd University, Budapest, Hungary
e-mail: jakobi@caesar.elte.hu

B. Lengyel
OSON Research Lab, International Business School Budapest, Budapest, Hungary

Centre for Economic and Regional Studies, Hungarian Academy of Sciences, Budapest, Hungary
e-mail: blengyel@ibs-b.hu

© Springer International Publishing Switzerland 2015

J. Brus et al. (eds.), *Modern Trends in Cartography*, Lecture Notes in Geoinformation and Cartography, DOI 10.1007/978-3-319-07926-4_18

means, OSNs clearly differ from other web-based networks like Internet infrastructure. The latter are led by power-law tie-distribution: a small share of webpages accounts for an outstandingly high number of links (Barabási and Albert 1999).

In our understanding OSNs are "biased versions of real-life networks" (Backstrom et al. 2012; Ugander et al. 2011). Therefore, we claim that virtual space and physical world are strongly interrelated, since it is assumed that flesh and blood users document their offline friendships in the online environment. The linkage between virtual and physical spheres, namely geolocalisation of online content is typically based on the position of users that can either stem from voluntary geographic information that users attach to the content they upload (e.g. image posts) or can be based on IP addresses etc. The possible projection of virtual world on real geographies and the effect of location on shaping cyberplace poses several questions that are beyond the subject of this paper. For example, extending research is still missing on the correlation among online and offline social networks (Traud et al. 2008; Hogan 2009). We only use the statement that geographical location and physical distance are very important in OSNs.

In recent years there has been a growing public and scientific interest in analyzing OSNs; the mainstream of research covers a very wide area including learning- and communication processes (Greenhow 2011), online identity (Zhao et al. 2008), youth and digital media (Boyd 2008), online privacy (Acquisti and Gross 2009), network dynamics (Kumar et al. 2006), among others. Geography and geovisualisation has been also involved to the discussion, for example in the field of user-generated information mapping (Yardi and Boyd 2010; Graham and Zook 2011), infographic geovisualisation of network connections (Buttler 2010; Warden 2010) or in cartographic explorations of online and offline geographic relations (Gätz 2010). Additionally, interesting experiments could be mentioned in connection with user behaviour mapping (Fischer 2011) or geotagged information mapping (Graham and Gaffney 2012) as well.

Research works on basic OSN characteristics highlighted that geography matters in network formation. Liben-Nowell et al. (2005) stressed that only one-third of friendships realized on LiveJournal blogging SNS was independent of bounded geographical areas. Escher (2007) also found that majority of ego-networks are local. A mega-analysis of Facebook found that majority of connections are within country borders and the number of ties across countries accords with geographical distance (Ugander et al. 2011). Thus, the geographical location of a user strongly determines the geographical position of the friendship ties he/she documents on the website.

It came to the light also that distance seems to be of crucial importance in OSNs. For example, research on Facebook, LiveJournal and Twitter also found evidence on "small world phenomenon" (Backstrom et al. 2012; Liben-Nowell et al. 2005; Yardi and Boyd 2010). Users formulate strongly connected cliques with physically proximate other users whereas relatively few long distance ties make the whole network connected and establish short average paths between two random users. There might be differences among distance types as well; for example Takhteyev et al. (2012) found that the frequency of airlines between two cities has the strongest correlation with intercity Twitter ties while the network falls into sub-graphs characterized by regions, country borders and language usage.

Although there are already many outcomes and assumptions on the spatial structure of OSNs, several questions arise regarding their particularities especially on detailed level of information. This all naturally is originated from the novelty of the topic. Additionally, up-to-now only few cartographic representations were published on local network connections and topologies, therefore geovisualisation of the spatial distribution of online social network connections provides new insights on understanding cyberspace geographies.

Database Characteristics: The Case of iWiW

We have chosen iWiW (International Who Is Who), the largest Hungarian online social network in order to visualise and to analyse the relationship between virtual and real world geographies. The iWiW was launched in April 2002 and shortly became the most known SNS in Hungary and even the most visited national website in 2006. The number of users was limited in the first years but started to grow exponentially due to new functions introduced in 2005 (e.g. personal advertisements, picture upload, public lists of friends, town-classification, e-mail box etc). The system had 640,000 members with 35 million connections in April 2006, when Origo Ltd. (member of the Hungarian Telecom group) became the owner of the site. The number of registered users continued to rise afterwards; it counted for 1.5 million users in December 2006, more than 3.5 million users in October and exceeded 4 million in December of 2008, however, the number hardly increased since then. In 2013 the total number of registered users is around 4.7 million.

The competition among SNSs favours Facebook in Hungary as well. Though Hungarian Facebook users reached the level of 3 million in late 2011 only, Facebook's market share is growing dynamically and stagnating iWiW is forecasted to fail. For example, Facebook outnumbered iWiW in terms of daily visitors in October 2010. However, iWiW still has higher number of registered users, which provided a better approach in our case for demonstrating the role of geographical location and physical distance in network connections.

The following analysis is based on a data collection for January 2013. This database indicated the number of connections and the sum of users for an aggregated level of Hungarian settlements, as well as the number of intra- and intercity connections. This second one was later used to create settlement-to-settlement connectivity matrix.

The original individual level data were basically aspatial, which had to be therefore geographically located to perform cartographic and spatial analysis. We have processed user profile data for geolocalization. However, localization of users based on profile information is considered to be problematic in papers focusing on OSN user and social media content localization (Hecht et al. 2011). In iWiW, however, it is compulsory to choose location from a scroll-down menu when registering as user. This place of residence can be easily changed afterwards and certainly there is no eligibility check. Thus, one might consider our location data based on user profiles data is a biased and occasionally updated census-type measurement. After geolocating all individual user data and summing them up on settlement level, as a secondary step geocoding of settlements had to be done in order to be able to create network connectivity maps. This was made by locating all settlement data as points of the centroid of original settlement polygons. The final settlement centroid points were georeferenced by decimal degree coordinates in WGS-84 projection.

In each record of the geolocated connectivity database—as mentioned above—also city to city connections were stored. In other words the dataset contained information on the location of network nodes (settlements) and edges (number of connections between settlements). This was registered in the database in forms of settlement pairs, namely with paired coordinates of the "from node" and "to node" settlements (network connections were naturally undirected, hence from or to status does not count and treated as equal). The interconnectedness of nodes then made it possible to draw network edges as network topology. The database additionally contained information on the volume of connections between two settlements (the number of users who have friendships between the two settlements) which data were applied as weighting attributes of city to city edges.

The nature of OSNs basically assumes that people have online friends often from the same city (the "from node" and "to node" is identical). In a network database intra-city connections appear as loops that have to be treated differently from city to city edges. In our case, therefore, loops were many times left out from connectivity analyses and mapping.

In our dataset out of the total number of 3,135 Hungarian villages and cities there were 2,562 settlements, which had active user data. The remaining 573 settlements did not have iWiW users; majority of these latter locations are very small villages. Since there were people who did not signify own location precisely, or choose place of residence outside of Hungary, the number of users in our dataset was somewhat smaller than the total number of iWiW users. In the analysed database altogether 4,058,505 users were scattered along 2,562 Hungarian settlements. The users have established 785,841,313 friendship ties in the website, out of which 369,789,373 ties remain within settlement borders (intra-city loops) and 415,653,749 ties are established between users from two distinct settlements. The network database covered 1,369,978 settlement-settlement pairs (Table 1).

Although interesting and useful information are possible to be derived from raw numbers of intercity ties, in order to deal with size differences of settlements also log-likelihood ratios were calculated from the raw connectivity data. This measure is a basic statistical concept; in our case this is the logarithm of the ratio of observed and randomly expected settlement-to-settlement tie weights (Eq. 1):

Table 1 Main database characteristics

	Users	Settlements
Number of nodes	4,058,505	2,562
Number of ties	785,841,313	1,372,540
Number of intracity ties (loops)	369,789,373	2,562
Number of intercity ties	415,653,749	1,369,978

$$\text{LLR}_{ij} = \text{Log}\left(\frac{w_{ij}}{\frac{w_i * w_j}{\sum_{i=1,j=1}^{n} w_{ij}}} \right) \tag{1}$$

in which LLR_{ij} refers to the log-likelihood ratio between settlement i and j, w_i and w_j are the sum of tie weights (total number of connections) of settlement i and j and w_{ij} is the raw weight of tie between settlement i and j. The higher positive LLR refers to strong settlement-to-settlement ties, while negative LLR represents weak intercity connections. In the following both raw weight ties and log-likelihood ratios are applied to explore absolute and relative connectivity weight characteristics.

Since raw weight ties are strongly biased by settlement size (cities with larger population numbers have higher absolute number of connections) the histogram of raw weight tie values show power-law distribution; the majority of settlements have relatively small number of connections, while some have very large connection numbers. The log-likelihood ratios of the overall 1,369,978 settlement connections had on the other hand almost normal distribution with a skewness test 1.204 and kurtosis result 2.244. LLR values range from -5.08 to 8.61 (Fig. 1).

The peak of the histogram is around -1.5 which means that the majority of intercity connections are weak and strong connections are rare. This is also because expected frequencies were calculated from the total number of connections of cities, which also includes loops (intra-city ties) that are always higher than the randomly expected. This pushes the distribution to the negative range.

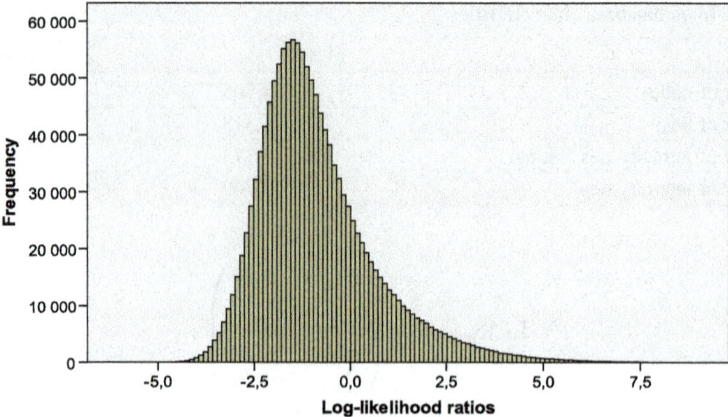

Fig. 1 Histogram of log-normalized settlement-to-settlement weights (LLR distribution)

Geovisualisation of Connections and the Analysis of Spatial Connectivity

A map many times tells more about spatial relations of data than the dataset itself. Therefore we also expect from geovisual outcomes that geographical motives of online social network connections could be understood deeper. There are many possibilities to visually represent network topology, out of them some software techniques are only capable to draw the topological structure of network (Gephi, Netdraw), while it is possible with others to visualise geolocated flow-type information (Flowmapper) or to perform geolocated data analysis and mapping simultaneously (ArcGIS). Since our needs coincide with both geovisualisation and analysis of data we have chosen ArcGIS 10.0 to perform tasks.

Since our dataset contained information only for point-type data (coordinates of point pairs) the line-type edges of network connection had to be calculated and generated in order to visualise spatial connectivity. A proper solution for that was to apply ArcGIS Data Management Tools, where XY To Line tool was useful to convert point pairs to lines.

As a result of transformation more than 1.3 million lines were created. To visualize network connections, however, it is not advised to draw all 1.3 million settlement-to-settlement lines simultaneously, since the result map would become confusing and non-interpretable. Therefore, we have chosen only the most important ties to be drawn on the first map. One solution for that was to select ties which were particularly often codified by users, namely to visualize settlement-to-settlement lines only that covered large absolute number of connections. According to that Fig. 2 shows network ties, which represented at least 15,000 connections. The result map contains altogether 1996 settlement-to-settlement lines, while line width signifies categories according to the absolute number of connections.

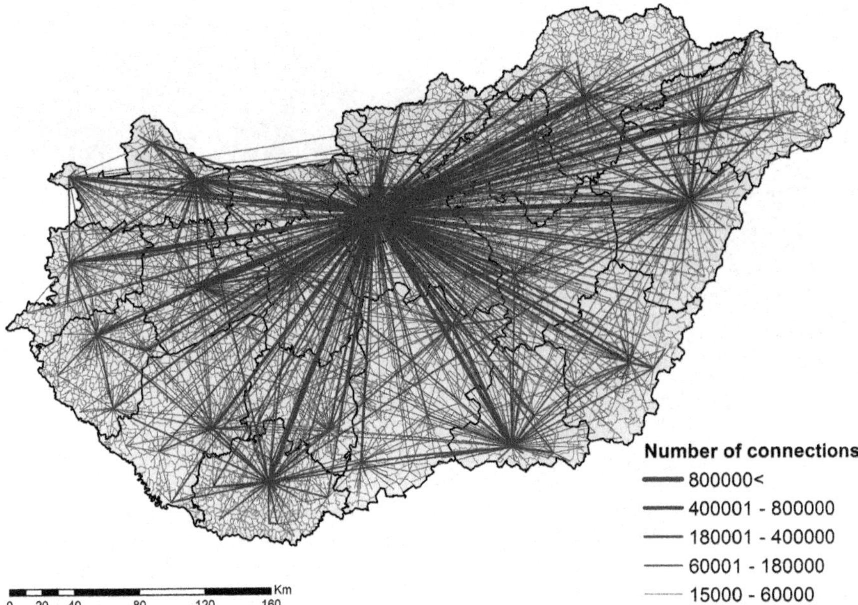

Fig. 2 The strongest settlement-to-settlement ties (over 15,000 connections)

When looking at maps of absolute connection numbers it seems obvious that the largest numbers appear between the most populated cities. The overall record was measured between Budapest and Debrecen (the first and second largest cities in Hungary) with more than 1.8 million individual intercity connections. Other major connections are typically located between regional or subregional centres, which perform as hubs in the network. Certainly the most important hub is Budapest with naturally the highest total number of connections.

One may argue, however, that maps of raw connection numbers represent only the ties between populated places and not the strength of a connection between any two settlements. On the contrary, maps of edges with log-normalized weights could set off important connections regardless of settlement size. Figure 3 shows the strongest ties between settlements according to the log-likelihood ratios of connections, however, only those over the ratio of 4.0 are visible. This limit was chosen since it really covers the strongest links of the network and also because the overall number of connections is still manageable at this point in order to be properly visualized. The map shows altogether 34,968 connections, hence reflects the spatial pattern of the most important ties from a local perspective.

Figure 3 can be considered also as a combination of different level log-likelihood ratio weight maps. If we separate the LLR maps as individual layers (Fig. 4) the main characteristics of iWiW network topology would outcrop. It seems that the higher the LLR weight between two settlements the shorter the connecting line. In other words it is definitely observable on the map of connections with LLR weights over 7 that extreme strong network connections are between settlements

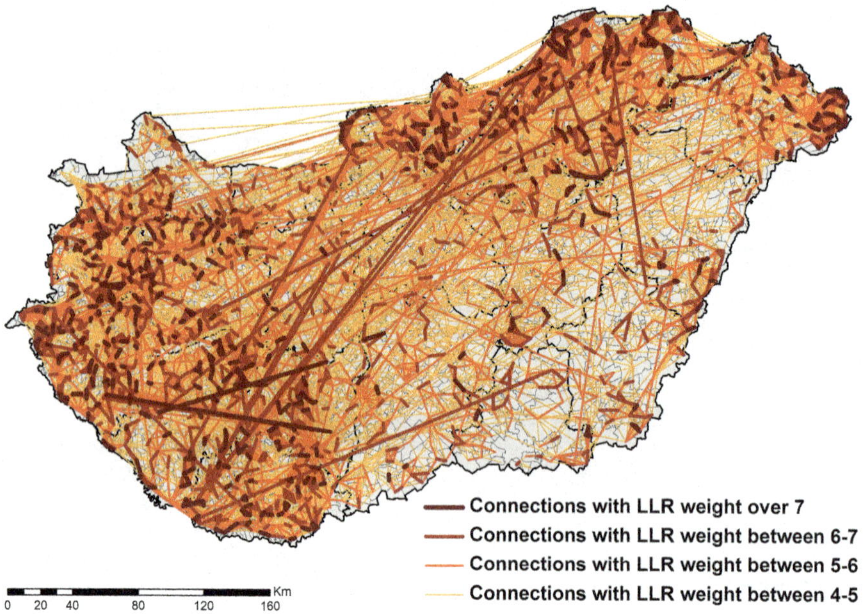

Fig. 3 The strongest weighted settlement-to-settlement connections (the highest LLR weights)

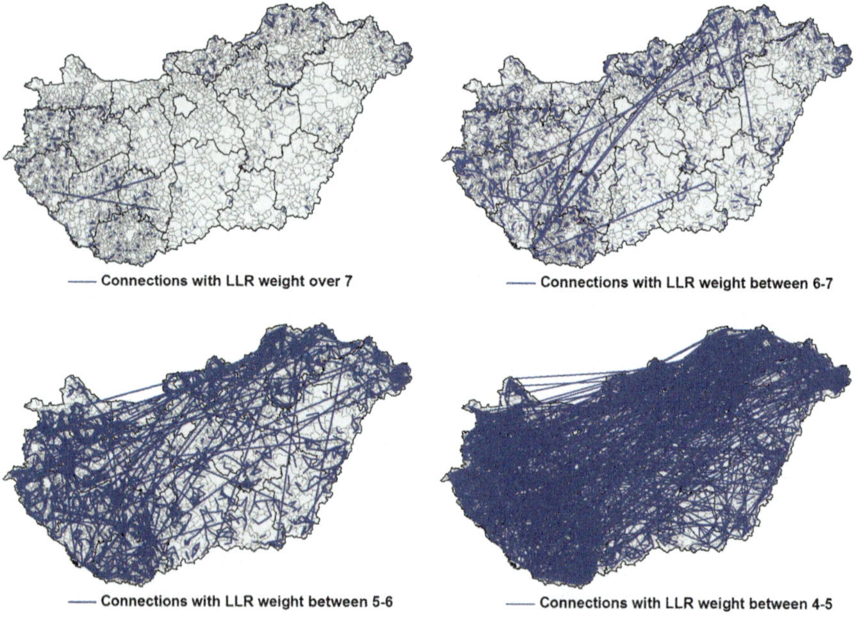

Fig. 4 Maps of settlement-to-settlement connections with LLR weights over 7, between 6–7, between 5–6 and between 4–5

that are geographically close to each other (there are only few exceptions). The average length of connections somewhat increases on the map of connections with LLR weights between 6–7 or 5–6, while it seems much larger on the map of connections with LLR weights between 4–5. The map with LLR weights over 7 looks much clearer due to short distance of relations than those having smaller LLR weights but longer lines. We should note, however, that the number of ties is also increasing when stepping on lower LLR weight levels: there are only 2,281 connections, which have LLR weights over 7, 3,951 connections have LLR weights between 6–7, 8,917 connections have LLR weights between 5–6 and 19,819 connections have LLR weights between 4–5.

It is also observable (both on Figs. 3 and 4) that the highest LLR weights are primarily perceptible in western and south-western parts of the country (in West-Transdanubia and South-Transdanubia regions), as well as among northern settlements (in the region of North-Hungary). These regions are basically typified by small sized settlements, and it is also observable that the largest LLR weights are first of all connected to small settlements (even to very tiny ones). It additionally reflects that small settlements, or more precisely settlements with the highest LLR weights, have only connections with very few other settlements, therefore the chance that two settlements are (relatively) very strongly connected is rather high. Two small settlements could be very tightly related to each other in the log-normalized network space of iWiW.

On the other hand there are of course weak connections in the network as well, which stand for ties between settlements with the lowest number of friendships. By drawing the map of the weakest settlement-to-settlement links with the lowest local log-likelihood ratio weights, very distant relations appear (Fig. 5). The map of extreme weak connections reflect that basically settlements, which are located very far form each other, are having very few number of acquaintance. In our dataset there were 1,529 connections under the LLR weight of −4.0, of them 3 were even under −5. The result map shows this time hardly visible spatial pattern of connections, or rather a haystack of lines.

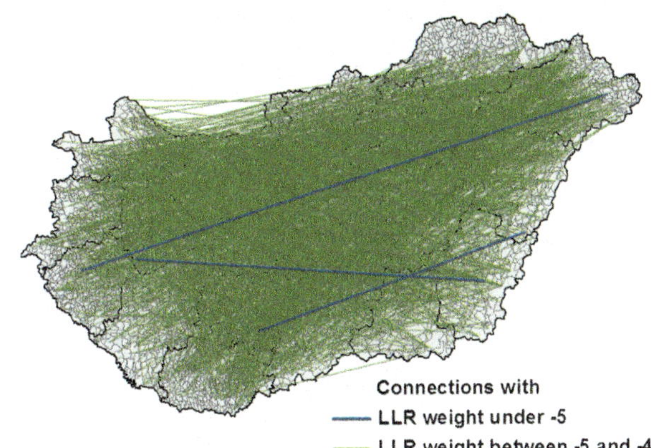

Fig. 5 The weakest settlement-to-settlement connections (with the lowest LLR weights)

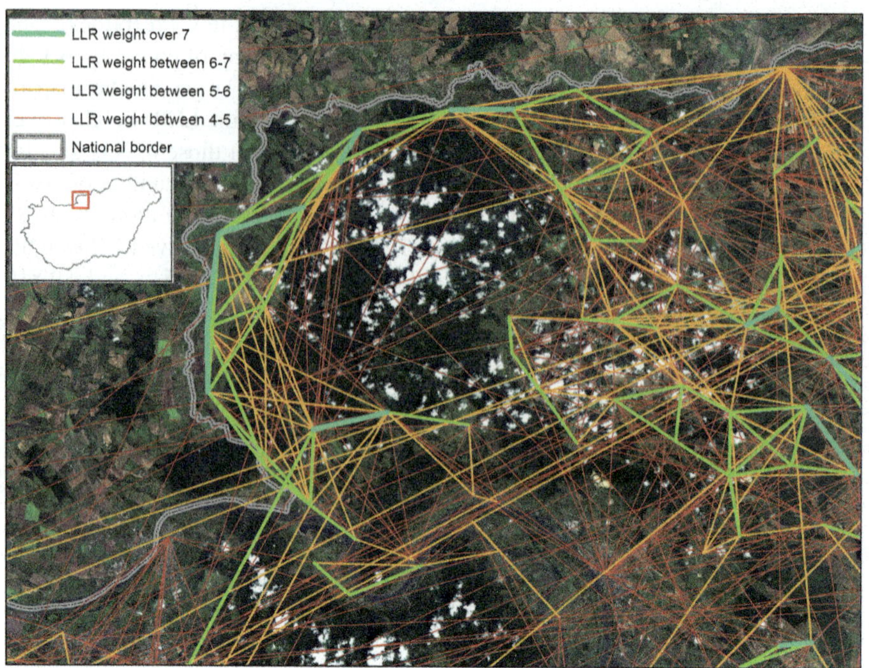

Fig. 6 Special characteristics of network topology around Börzsöny mountainous area

Furthermore, when going into details of network topology, quite interesting geographical motives of network organization are possible to be discovered. Figure 6 for instance mirrors some special characteristics of network topology by examples of connections around Börzsöny mountainous area. The map confirms that virtual space connections largely follow connections of offline geographies, since settlements around the mountain have observably stronger network ties with close neighbours of the same mountain side than with those on the other side. Lines with larger LLR weights are typically located on the same side of the mountain, while there are only rare and weak ones crossing over it.

Derived Maps and Geospatial Aggregation of Data

Beside visualisation of network topology our dataset offered the chance also to draw derived maps of connectivity. The following thematic maps accordingly were created by aggregating data on the level of settlements, hence reflecting further main characteristics of online social network performance.

Figure 7 shows settlement level patterns of four different and important OSN indicators. The first map (top left) shows the average LLR weight values by settlements, where darker areas with higher values refer to settlements that have

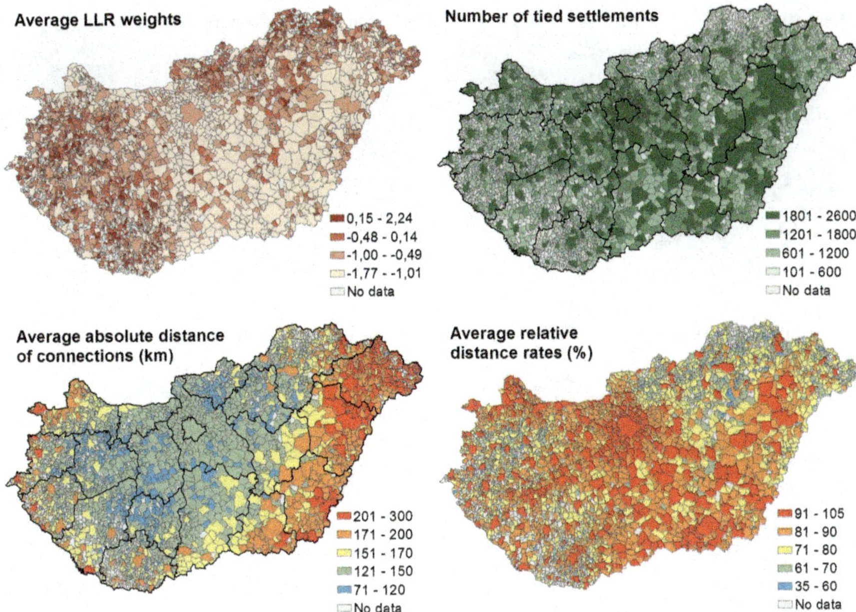

Fig. 7 Settlement level patterns of main OSN performance indicators

usually strong connections with others. The second thematic map (top right) reflects the aggregated number of settlements, with that a certain settlement is in connection. The more settlements a certain location is connected with, the larger number of tied settlements it has and the darker it appears on the map. It is interesting to see by the comparison of maps that higher values appear typically at western and northern areas on the first map (usually in case of small settlements), while at central and south-eastern settlements on the second map (typically at larger cities). The third map (bottom left) shows the average distance of connections, while the last one (bottom right) indicates the average relative distance rates. The latter map was created since average absolute distance of connections largely depends on central or peripheral geoposition of a settlement, while relative distance rates denote non-biased network proximity. Consequently the observed and expected distance averages were compared as follows (Eq. 2):

$$\mathrm{ARD}_i = \left(\frac{\sum_{i=k}^{k} d_{ij}}{k} \middle/ \frac{\sum_{i=1}^{n} d_{ij}}{n} \right) * 100 \tag{2}$$

in which ARD_i refers to the average relative distance rate of settlement i, d_{ij} is the distance between settlement i and j, k is the observed number of connected settlements and n is the total (or expected) number of settlements.

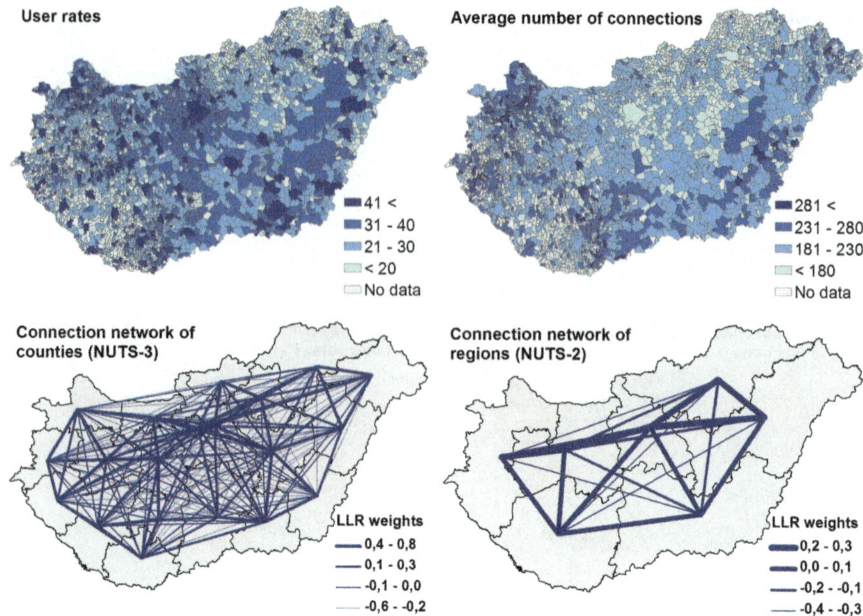

Fig. 8 Settlement level user rates and connectivity values, as well as NUTS-3 and NUT-2 level aggregations of raw connection data

In other words average relative distance rates are calculated as a percentage of average observed distances compared to average expected distances, or could be treated as normalised versions of absolute distance values. While the average absolute distance map shows rather concentric spatial pattern of values (as expected) the other map performed differently, indicating how far a connection network is reaching from a certain location. It is also observable that larger settlements have more distant connections on average, while smaller ones are usually tied with only closer partners.

Thematic map of user rates (the rate of iWiW users among total population) reflects also that intensity of OSN activity is pretty much depending on settlement size (top left map on Fig. 8). It seems that larger cities have not just higher total number of users, but higher user rates as well. On the other hand, if we look on the map of the average number of connections, namely the average number of iWiW friendship ties among co-located iWiW users (users with the same settlement designation) it can be observed that peripheral areas in the south and west, and sometimes also smaller settlements have higher connectivity values (Lengyel and Jakobi 2014). Differences and spatial patterns are, however, not too significant.

By taking advantage of ArcGIS analytical techniques geospatial aggregations of data were performed on the level of NUTS-3 and NUTS-2 regions as well. As a result, generalized network maps (bottom left and right maps on Fig. 8) were created indicating macro-regional network characteristics. Although the maps show all ties among nodes, strong or weak connections can be easily distinguished.

Conclusions

Online social networks have been opened new opportunities for empirical research and of high probability will count for a growing share of scientific interest aiming closer understanding of online communication and human development. In this paper we presented an initial attempt in establishing a geography-related research line in this promising field of interdisciplinary focus. With a special emphasis on geovisualisation of unequal spatial distribution of network connections we tried to look behind the data and to discover geospatial characteristics of OSN performance.

We demonstrated here that OSNs are definitely place-dependent, because many aspects of network connectivity happened to show geographical relatedness as well. By mapping the strongest and most important settlement-to-settlement connections it turned out that distance does significantly matter in network formation. Although OSNs are virtual space creations, it seems that offline factors of settlement size and location still have decisive effects in network connections.

Acknowledgments Research work of Ákos Jakobi was supported by the János Bolyai Scholarship of the Hungarian Academy of Sciences. Balázs Lengyel acknowledges financial support of IBS Research Grant.

References

Acquisti A, Gross R (2009) Predicting social security numbers from public data. Proc Natl Acad Sci U S A 106(27):10975–10980

Ahn YY, Han S, Kwak H, Eom YH, Moon S, Jeong H (2007) Analysis of topological characteristics of huge online social networking services. In: Proceedings of the 16th international conference on World Wide Web. http://citeseerx.ist.psu.edu/viewdoc/download?doi=10.1.1.87.2864&rep=rep1&type=pdf. Accessed 1 Aug 2013

Backstrom L, Boldi P, Rosa M, Ugander J, Vigna S (2012) Four degrees of separation. arXiv:1111.4570. http://arxiv.org/pdf/1111.4570v3.pdf. Accessed 14 Oct 2014

Barabási AL, Albert R (1999) Emergence of scaling in random networks. Science 286:509–512

Boyd D (2008) Why youth ♥ social network sites: the role of networked publics in teenage social life. In: Buckingham D (ed) Youth, identity, and digital media. MIT, Cambridge, pp 119–142

Boyd D, Ellison NB (2007) Social network sites: definition, history, and scholarship. J Comput Mediat Commun 13:210–230

Buttler P (2010) Visualizing Facebook friends: eye candy in R. http://paulbutler.org/archives/visualizing-facebook-friends/. Accessed 13 Dec 2013

Ellison N, Steinfield C, Lampe C (2006) Spatially bounded online social networks and social capital: the role of Facebook. Paper presented at the annual conference of the International Communication Association, 19–23 June. http://www.icahdq.org/conf/2006confprogram.asp

Ellison N, Steinfield C, Lampe C (2007) The benefits of Facebook friends: social capital and college students' use of online social network sites. J Comput Mediat Commun 12:1143–1168

Escher T (2007) The geography of online social networks. Presentation at Oxford Internet Institute September 5. http://people.oii.ox.ac.uk/escher/wp-content/uploads/2007/09/Escher_York_pre sentation.pdf. Accessed 6 Jul 2013

Fischer E (2011) The geotaggers' world atlas. http://www.flickr.com/photos/walkingsf/4621754005/in/set-72157623971287575. Accessed 6 Jul 2013

Gätz T (2010) Facebook vs. the rest of the world. http://www.flickr.com/photos/thorstengaetz/5261467662/. Accessed 13 Dec 2013

Graham M, Gaffney D (2012) Where do tweets come from? http://www.zerogeography.net/2012/04/where-do-tweets-come-from.html. Accessed 13 Dec 2013

Graham M, Zook M (2011) Visualizing global cyberscapes: mapping user generated placemarks. J Urban Technol 18(1):115–132

Greenhow C (2011) Learning and social media: what are the interesting questions for research? Int J Cyber Behav Psychol Learn 1:36–50

Hecht B, Hong L, Suh B, Chi EH (2011) Tweets from Justin Bieber's heart: the dynamics of the "location" field in user profiles. In: Proceedings of the ACM conference on human factors in computing systems, pp 237–246. ACM Press, New York

Hogan B (2009) A comparison of on- and offline networks through the Facebook API. http://papers.ssrn.com/sol3/papers.cfm?abstract_id=1331029. Accessed 6 Jul 2013

Kumar R, Novak J, Tomkins A (2006) Structure and evolution of online social networks. In: Proceedings of 12th international conference on knowledge discovery in data mining, pp 611–617. ACM Press, New York

Lengyel B, Jakobi Á (2014) Online social networks and location: the dual effect of distance on spread and involvement. Tijdschrift voor Economische en Sociale Geografie (in press)

Liben-Nowell D, Novak J, Kumar R, Raghavan P, Tomkins A (2005) Geographic routing in social networks. Proc Natl Acad Sci U S A 102:11623–11628

Takhteyev Y, Gruzd A, Wellman B (2012) Geography of Twitter networks. Soc Netw 34:73–81

Traud A, Kelsic E, Mucha P, Porter M (2008) Community structure in online collegiate social networks. http://cds.cern.ch/record/1124061. Accessed 10 Jul 2013

Ugander J, Karrer B, Backstrom L, Marlow C (2011) The anatomy of the Facebook social graph. http://arxiv.org/abs/1111.4503. Accessed 6 Jul 2013

Warden P (2010) How to split up the US. http://petewarden.com/2010/02/06/how-to-split-up-the-us/. Accessed 13 Dec 2013

Yardi S, Boyd D (2010) Tweeting from the town square: measuring geographic local networks. Paper presented at AAAI of the international conference on Weblogs and Social Media, 23–26 May. http://yardi.people.si.umich.edu/pubs/Yardi_TownSquare10.pdf. Accessed 11 Jul 2013

Zhao S, Grasmuck S, Martin J (2008) Identity construction on Facebook: digital empowerment in anchored relationships. Comput Hum Behav 24:1816–1836

The Competitive Analysis Method for Evaluating Water Level Visualization Tools

Robert E. Roth, Chloë Quinn, and David Hart

Introduction

A *competitive analysis study* is a usability engineering method administered to critically compare a suite of similar applications according to their relative merits (Nielsen 1992). While most usability engineering methods solicit feedback directly from targeted end users, a competitive analysis study is a *theory-based* method in which the design/development team leverages established theoretical principles to evaluate the collected suite of applications (Roth 2011). In other words, a competitive analysis study is a content analysis of secondary sources—common to archival research in social science—conducted for the purpose of usability engineering. While not a replacement for user-based evaluation, a competitive analysis study may be beneficial in a variety of mapping and visualization contexts, such as when the design/development team knows little about the application domain, when a user-based needs assessment study cannot be completed due to limited project resources or limited access to targeted end users, when there are a large number of existing applications that implement similar functionality, and when there is a previous version of the visualization tool already in use. Thus, a competitive analysis study primarily is appropriate during the early, formative stages of design and development (see Robinson et al. 2005, for a discussion of formative versus summative usability assessment).

In this paper, we demonstrate the potential of the competitive analysis method for cartography through the case study of *water level visualization*, or map-based visualization tools depicting the exposure or flooding of land as a result of

R.E. Roth (✉) • C. Quinn
Department of Geography, University of Wisconsin–Madison, Madison, WI, USA
e-mail: reroth@wisc.edu

D. Hart
Sea Grant Institute, University of Wisconsin–Madison, Madison, WI, USA

© Springer International Publishing Switzerland 2015

J. Brus et al. (eds.), *Modern Trends in Cartography*, Lecture Notes in Geoinformation and Cartography, DOI 10.1007/978-3-319-07926-4_19

historical/current storm events or future climate change predictions (Kostelnick et al. 2009). The competitive analysis was completed during the formative stages of design and development of the U.S. National Oceanic & Atmospheric Administration (NOAA) *Lake Level Viewer*, a map-based visualization tool supporting adaptive coastal management of hazards related to future water level change across the Great Lakes (USA). The NOAA Lake Level Viewer is a sibling visualization tool to the NOAA *Sea Level Rise Viewer* (Fig. 1a), which supports adaptive management of coastal hazards along the Atlantic, Pacific, and Gulf of Mexico shorelines in the U.S. While climate modeling suggests a likely increase in global sea levels over the next century (IPCC 2007), regional climate modeling in the Great Lakes suggests a possible decrease in lake levels, but an increase in the annual variation of water levels (Angel and Kunkel 2010; Hayhoe et al. 2010). The water levels across the Great Lakes already set or approached record lows in 2012–2013. Thus, a fundamental redesign of the Lake Level Viewer was necessary to support the very different adaptive management context in the Great Lakes.

The purpose of the competitive analysis was threefold. First, the competitive analysis captured existing best practices in water level visualization, allowing for identification of common design and development solutions to the end of determining key user needs that must be supported in the Lake Level Viewer. Second, the competitive analysis suggested possible opportunities for the Lake Level Viewer, pointing out currently unmet user needs that may be supported by the tool. Finally, because the tools are compared according to theoretical principles in cartography, the competitive analysis revealed important gaps between theory and practice, helping to problematize suboptimal solutions and to stimulate discussion about functional and technological innovation.

The paper is structured in three additional sections. Our method design is described in the following section. A total of twenty-five (n = 25) water level visualization tools were compared across two broad themes in cartography: representation and interaction. We provide the results of the competitive analysis in the fourth section, with discussion split between insights related to representation design versus interaction design. In the final section, we provide a summary of design recommendations derived from the competitive analysis and report on future work to bring the Lake Level Viewer online.

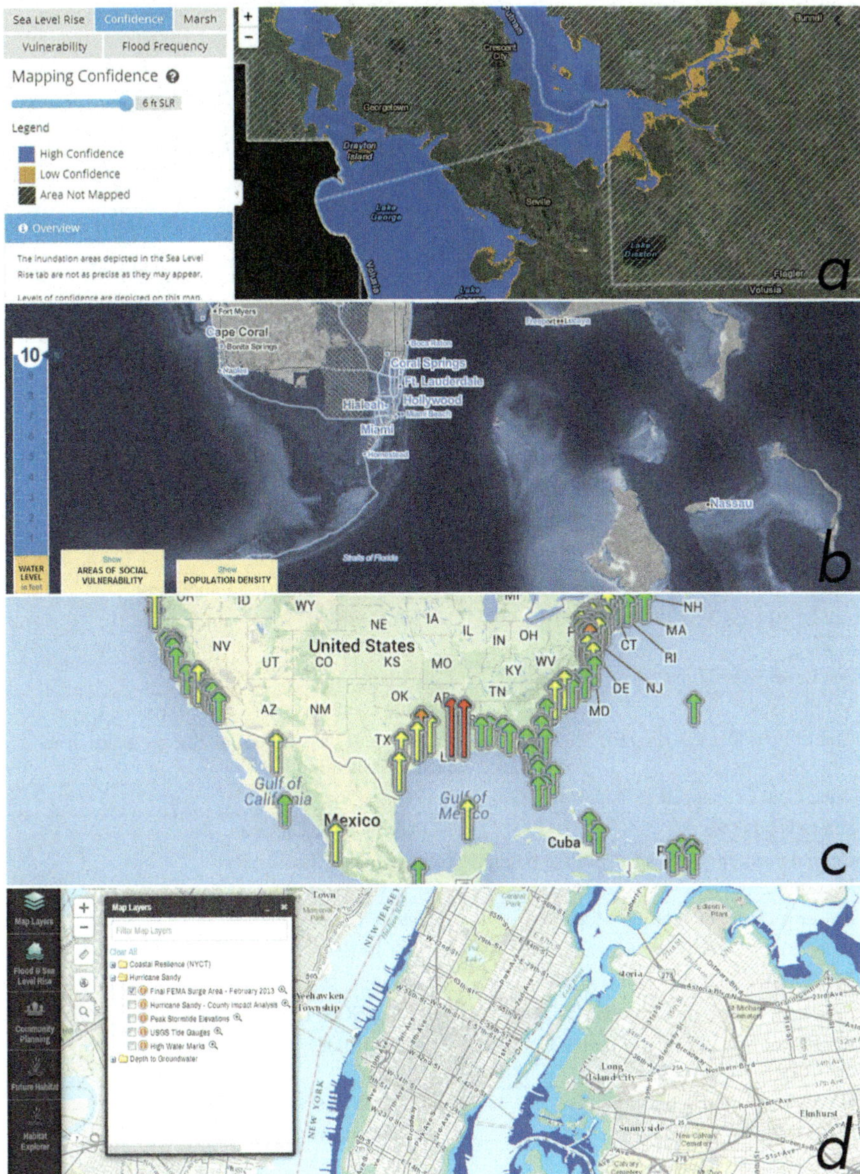

Fig. 1 (a) The NOAA Sea Level Rise and Coastal Flooding Impacts Viewer (http://www.csc.noaa.gov/slr/viewer/); (b) Surging Seas (http://sealevel.climatecentral.org/); (c) Sea Level Trends (http://tidesandcurrents.noaa.gov/sltrends/); and (d) Coastal Resilience Future Scenarios Map, New York & Connecticut (http://maps.coastalresilience.org/nyct/)

Table 1 The twenty-five water level visualization tools included in the competitive analysis

Name	Agency
Sea Level Rise & Coastal Flooding Impacts Viewer (v1)	NOAA Coastal Services Center
Sea Level Rise & Coastal Flooding Impacts Viewer (v2)	NOAA Coastal Services Center
New Jersey Flood Mapper	JCNERR, CRSSA, NOAA CSC, Sustainable NJ, Rutgers University
Green Bay LakeViz	UW-Madison, Department of Geography
Lakes Entrance Visualization	Monash University
Explore SahulTime	Monash University
Coastal Resilience Future Scenarios Map, NY/CT (v1)	The Nature Conservancy; Coastal Resilience
Coastal Resilience Future Scenarios Map, NY/CT (v2)	The Nature Conservancy; Coastal Resilience
Coastal Resilience Future Scenarios Map, Gulf Coast (v1)	The Nature Conservancy; Coastal Resilience
Coastal Resilience Future Scenarios Map, Gulf Coast (v2)	The Nature Conservancy; Coastal Resilience
Interactive Sea Level Rise Web Map	SBEP & MOTE Marine Laboratory
Sea Level Rise Visualization for AL, MS, & FL	NOAA—MS-AL Sea Grant, USGS
Sea Level Rise Threatened Areas Map	Cal-Adapt
Surging Seas	Climate Central
Connecticut Coastal Hazards Viewer	Connecticut Department of Energy & Environmental Protection
What Could Disappear	NY Times
Sea Level Trends	NOAA Tides & Currents
Coastal Flooding & Sea Level Rise Impact Viewer	George Mason University
Sea Level Rise Inundation Maps	Delaware DNR
Flood Map Water Level Elevation Map	Sameer Burle
Flood Maps	firetree.net
SLR Impacts for Wilmington, Delaware	NOAA, USGS, Delaware DNR
Impacts of Sea Level Rise on the California Coast	Pacific Institute
USGS Flood Inundation Mapper	USGS, NWS, USACE, FEMA
SLAMM View	U.S. Fish and Wildlife Service

Method Description

The competitive analysis study was completed on a sample of twenty-five (n = 25) water level visualizations tools available online, including two versions of the NOAA Sea Level Rise Viewer. The sample of visualization tools was gathered through recommendations from the University of Wisconsin Sea Grant Institute and feedback from NOAA partners. The only requirement for inclusion in the sample was that the visualization must be map-based, with the water level represented cartographically. Six (24 %) tools were developed by U.S. federal agencies, three (12 %) by U.S. state agencies, four (16 %) by university research centers, six (24 %) by non-profit agencies, and three (12 %) by private industry or independent consultants; the remaining three (12 %) tools were developed through partnerships of federal, state, municipal, and/or university stakeholders. Two tools (8 %) have a geographic coverage of the entire globe, seven (28 %) by a single country (six within U.S., one within Australia), nine (36 %) by one or several states (within this category, four displayed states in the Atlantic Northeast, three in the Gulf of Mexico, and two along the Pacific coast), and six (24 %) by a single municipality (five within the U.S., one in Australia). Table 1 lists each of the evaluated applications.

Several patterns exist within the sample regarding the purpose of the visualization tool, or the user goal that the visualization tool was intended to support. Following the MacEachren (1994) Cartography[3] framework, thirteen (52 %) of the tools primarily support the goal of presentation, constraining the user interface to ensure that communication of the waterline or flood extent is clear. In contrast, the remaining ten tools (40 %) primarily support the goal of exploration, enabling the user to formulate 'what if?' questions by interactively building user-defined scenarios; four (12 %) of these ten tools also provide basic support for analysis, allowing for the computation of user-defined statistics. The vast majority of tools (24/25; 96 %) emphasize prediction, depicting the future threat of flooding and/or storm surges. A small subset of tools (3/25; 12 %) present historical information, allowing users to visualize the future waterline or flood extent in the context of past events. A majority of tools (13/25; 52 %) depict the potential damage to physical and social infrastructure, indicating a need to support planning and preparedness for at-risk areas. A large minority of tools (11/25; 44 %) explicitly support adaptive management in response to climate change, symbolizing areas that will be impacted by rising sea levels according to different climate change predictions.

The sample of visualization tools were compared across a fundamental distinction within cartography: (1) *representation*, or the graphic encoding of information in the map display, and (2) *interaction*, or the means by which the user is able to manipulate the map display (Roth 2013b). Within the cartographic representation theme, coding emphasized three topics: (1a) variation in the way in which the waterline or flood extent is symbolized, (1b) inclusion of uncertainty information about the waterline/flood extent and variation in the way this uncertainty information is symbolized, and (1c) variation in the basemap or overlay context information provided to enrich the interpretation of the waterline or flood extent. Within the cartographic interaction theme, coding emphasized two topics: (2a) variation across supported interaction operators (i.e., the basic system functionality) and (2b) variation in the web mapping technology used to implement the visualization, and the opportunities and constraints therein. The analysis was completed in September and October of 2013.

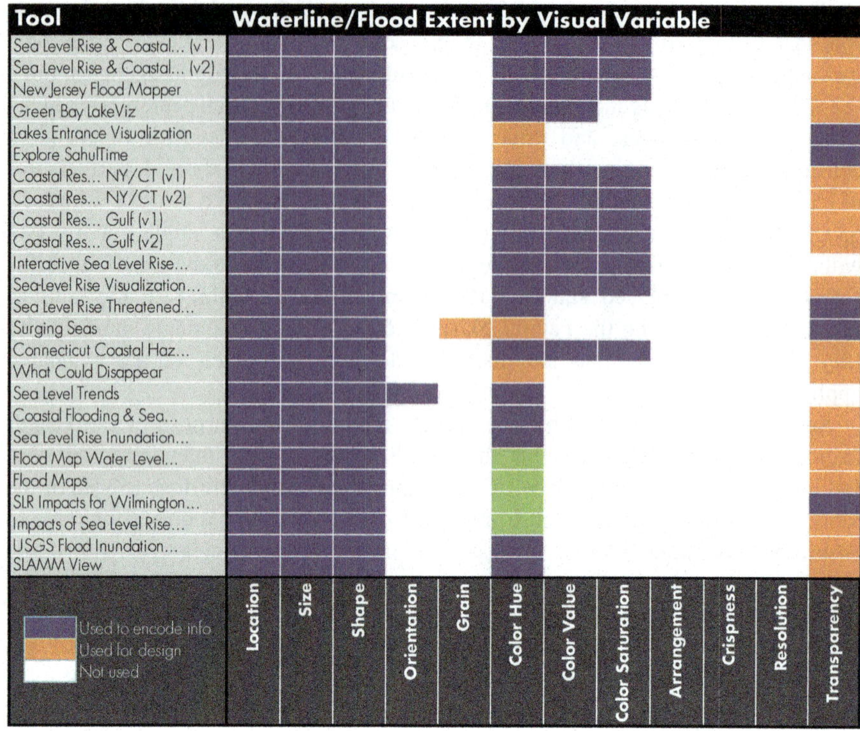

Fig. 2 Waterline/Flood Extent by Visual Variable

Results

Representation of the Waterline/Flood Extent

Given the emphasis of the competitive analysis on water level visualization tools, we first analyzed the way in which the waterline or flood extent is represented across the tools. Graphic representations signify information by leveraging one or several *visual variables*, or basic buildings blocks of the visual scene (Bertin 1967–1983). Commonly employed visual variables for vector-based signification include: location, size, shape, orientation, grain, color hue, color value, color saturation, arrangement, crispness, resolution, and transparency. Figure 2 provides an overview of the waterline/flood extent representation by visual variable.

The competitive analysis revealed that the visual variable location (25/25; 100 %) is used across all tools to represent the location of the waterline and the visual variables size (25/25; 100 %) and shape (24/25; 96 %) are used to represent the flood extent in all or most tools. The ubiquitous use of size and shape to represent the flood extent suggests that existing tools highlight not the predicted position of the waterline, but the areas that will be flooded as a result of the shifting waterline. Because a decrease in future water levels for the Great Lakes is possible, the conceptualization of the Lake Level Viewer as a 'flood' visualization is inappropriate. Therefore, different symbolization is needed for newly exposed land versus newly inundated land.

Many of the visualization tools encode the flood depth (a numerical variable) using an additional visual variable. Six (24 %) tools use color hue to represent water depth, two (8 %) use transparency, and one (4 %) tool uses a combination of color value and color saturation. Drawing from semiotics, the use of value + saturation and transparency are predicted to be effective solutions for representing a numerical variable, while the use of color hue is not (MacEachren 1995).

Finally, two unique representation solutions are worth noting. First, the Surging Seas visualization loads basemap layers of different detail for areas within the flood extent (satellite imagery) versus beyond the flood extent (a generalized vector map) (Fig. 1b). This solution allows for impacted areas to be viewed in more detail without a flood symbol obfuscating the area of interest, a limitation of other tools. Unfortunately, this solution may not be as useful for representing a declining water level in the Great Lakes, as imagery is not available for areas currently inundated. Second, the Sea Level Trends tool makes use of the visual variable orientation to represent water level change at a small cartographic scale, with the amount of change represented redundantly using color hue and size (Fig. 1c). The Lake Level Viewer may benefit from such an 'overview' (Shneiderman 1996), as land flooding or exposure on the Great Lakes is confined to a relatively small area along the coast that is viewable at large cartographic scales only.

Representation of the Certainty of the Waterline/Flood Extent Prediction

The term *uncertainty* describes any cause for a mismatch between reality and the user's understanding of reality (Roth 2009) and may be considered as a series of filters within the reality → variable-definition → data-collection → information-assembly → knowledge-construction pipeline (Longley et al. 2005). Effective uncertainty representation is essential to the design of visualizations that support decision making (Agumya and Hunter 2002). In GIScience, information uncertainty is considered multifaceted, exhibiting at least three components: (1) *accuracy/error*, or the correctness of a measurement or estimate, (2) *precision/resolution*, or the exactness of a measurement or estimate, and (3) *trustworthiness*, or the confidence that the user has in the information (MacEachren et al. 2012). Trustworthiness typically is conceptualized as a 'catch-all' category that includes aspects of the currency, completeness, internal consistency, credibility, subjectivity, interrelatedness, and lineage of the represented information (MacEachren et al. 2005).

                                                     R.E. Roth et al.

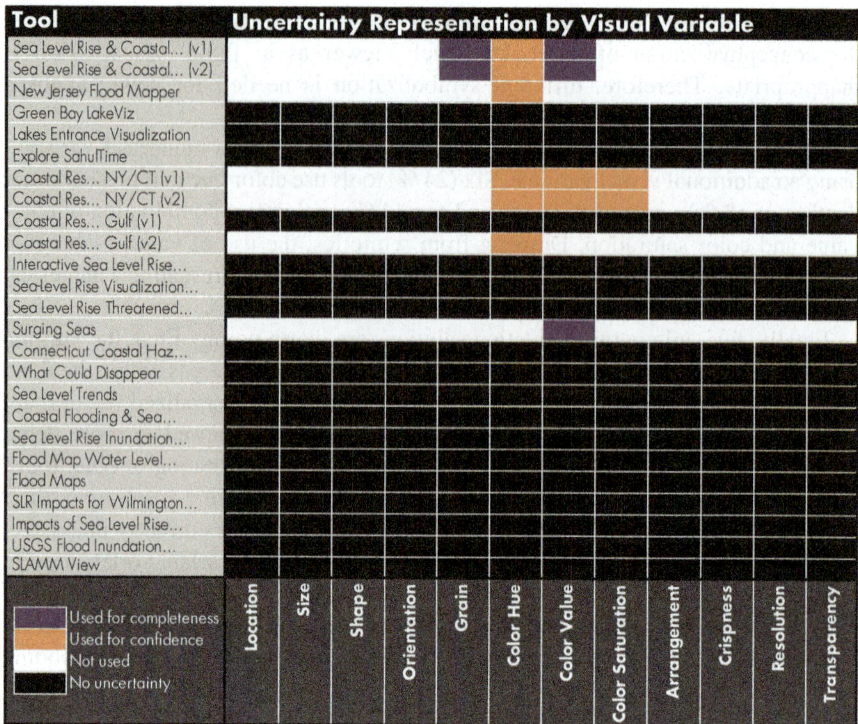

Fig. 3 Uncertainty Representation by Visual Variable

Figure 3 provides an overview of the uncertainty representation by visual variable. Seven (28 %) of the twenty-five visualization tools represent some form of information uncertainty, with three (12 %) tools representing completeness, or the extent to which information is comprehensive for the area, and six (24 %) tools representing confidence, using the term in a way that is similar to trustworthiness. Completeness is represented using the visual variables grain (2/25; 8 %) or color value (3/25; 12 %), while confidence is represented using the visual variables color hue (6/25; 24 %) or a combination of color value and color saturation (2/25; 8 %).

The NOAA Sea Level Viewer is the only visualization tool in the sample to represent two kinds of information uncertainty (Fig. 1a). Completeness is represented using the visual variables grain ('Area Not Mapped') and color value (counties not touching the coast), while confidence is represented using the visual variable color hue. The concept of confidence is defined in the Sea Level Rise Viewer as the "Level of certainty that a mapped [sea level rise] scenario is correct, taking into account topographic and tidal surface errors". This confidence metric will require revision for the Lake Level Viewer, as the seasonal processes of ice cover and snow melt on the Great Lakes influence water levels as much as tidal processes and storm events. Minimally, use of a blue color hue to represent 'confidence' requires reconsideration in order to encode uncertainty in both exposed and inundated land.

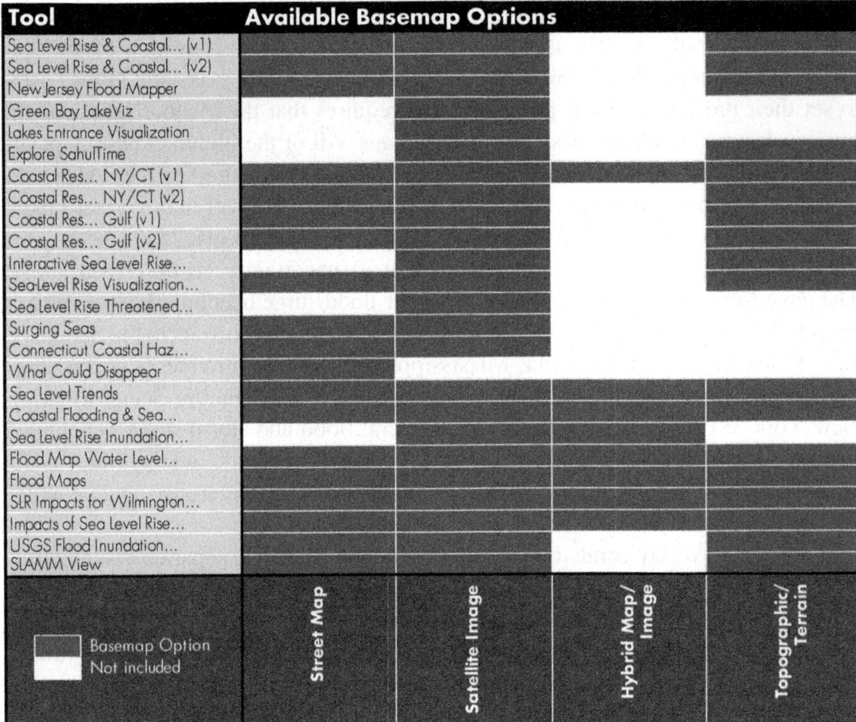

Fig. 4 Available Basemap Options

Basemap and Overlay Context Information

Provision of context information is essential for successful interpretation of the waterline or flood extent. Traditionally, the relative importance of different map features was communicated to the map user through a carefully designed *visual hierarchy* (Slocum et al. 2009); in the case of a water level visualization, the position of the waterline should rise to the forefront to ensure immediate visual inspection, with other context information receding into the background. The possibility of interactivity, along with historical constraints in web mapping technology, have transformed this traditional paradigm, with context information typically organized into *basemap* versus *overlay* layers for maps published online. Typically, the basemap layers are rasterized and served as a set of tiles to support instantaneous panning and zooming, while the overlays are drawn as vectors to enable retrieval of additional details about the features.

Figure 4 provides an overview of commonly available basemap options. The competitive analysis identified four basemap layers commonly available in water level visualization tools: satellite or aerial images (24/25; 96 %), street maps (19/25; 76 %), a map-image hybrid that includes labels and some vector features (8/25; 32 %), and a topographic map or other terrain representation (18/25; 72 %).

Twenty (80 %) of the tools provide more than one basemap option, while seven (28 %) of the tools provide all four basemap options. While provision of multiple different basemap options supports a wider array of map use tasks and allows users to set their preference, such provision also requires that the symbolization of the overlay features works across various basemaps. All of the included basemaps are much more detailed on the land side of the coastline compared to the water side, again a concern given the possibility of a declining water level in the Great Lakes.

Most of the tools provide additional overlay context layers (18/25; 72 %) beyond the waterline/flood extent or an indication of its uncertainty, as described above. The most frequent overlay layers provided are flood/surge benchmarks specific to a notable flood scenario or historical event (11/25; 44 %). For instance, the Sea Level Rise Visualization for Alabama, Mississippi, and Florida provides a storm surge overlay for Hurricane Katrina and the Coastal Resilience Future Scenarios Map, New York & Connecticut (v2) provides several flood and storm surge benchmark overlays for Super Storm Sandy (Fig. 1d). Such benchmark overlays are useful because they provide a meaningful and memorable point of reference against which to compare future sea level rise scenarios (Harrower 2002).

Additional overlay context layers provided by at least two separate visualization tools include: marsh/wetlands (9/25; 36 %), critical facilities or infrastructure (8/25; 32 %), socioeconomic vulnerability (7/25; 28 %), land use or land management (4/25; 16 %), populated areas (5/25; 20 %), photos of historic or simulated flooding (4/25; 16 %), parks or protected natural areas (3/25; 12 %), and erosion susceptibility (2/25; 8 %). The overlay options are perhaps the best way to infer the intended use case scenarios of the visualization tool using the competitive analysis method. There appears to be a split in emphasis between visualization tools supporting adaptive management of the human or built environment—with layers including critical facilities/infrastructure, socioeconomic vulnerability, and populated areas—and visualization tools supporting adaptive management of the physical or natural environment—with layers including marsh/wetlands, land use or land management, parks or protected areas, and erosion susceptibility. An integrated approach providing overlay layers about both the human and physical environment, and the interaction therein, supports a more robust geographic dialogue about the impact of changing water levels across the Great Lakes.

Supported Interaction Operators

Interaction operators describe the generic kinds of interactive functionality implemented in the visualization (Roth 2012, 2013a). Interaction operators can be delineated into *work operators*, or operators that are performed explicitly to accomplish the user's goal or objective, versus *enabling operators*, or operators that are performed to prepare for, or clean up from, a work session (Whitefield et al. 1993).

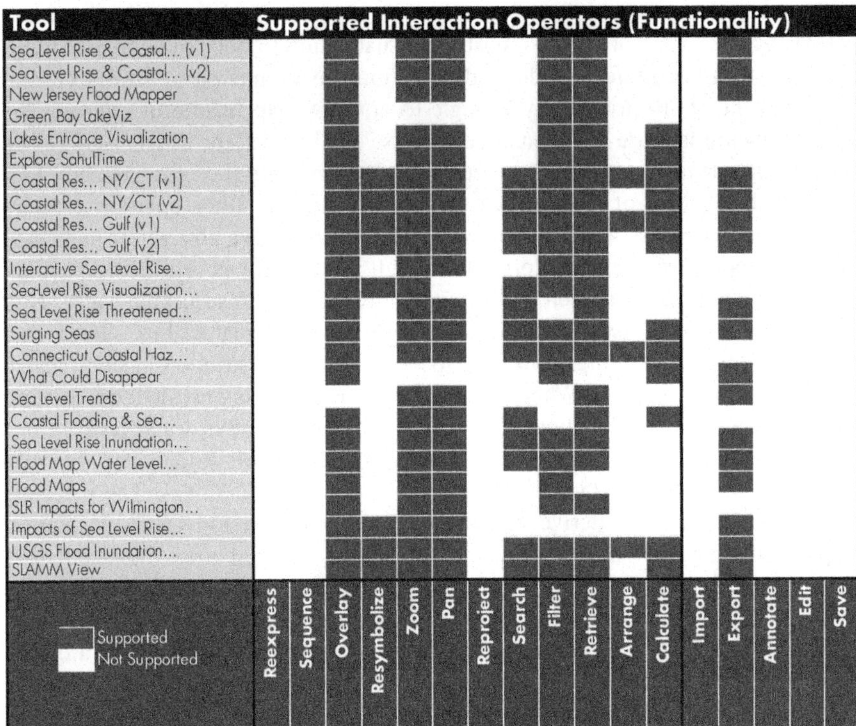

Fig. 5 Supported Interaction Operators

Figure 5 provides an overview of the supported interaction operators (i.e., functionality) across the sample of visualization tools. The most commonly implemented operator is *overlay* (24/25; 96 %), which allows users to toggle the visibility of the overlay context layers shown in the display. While only nineteen (76 %) of the evaluated visualization tools include additional overlay context layers, the other five (20 %) tools implement overlay functionality for toggling of the waterline/flood extent itself. Such a use of overlay for the waterline/flood extent overcomes the aforementioned problem of obfuscating an area of interest with a polygonal symbol, albeit the basemap and flood extent still cannot be viewed in concert. Twenty-three (92 %) of the maps support *zoom*, or the ability to change the map scale, and twenty-two (88 %) of the maps support *pan*, or the ability to change the map centering, typically after zooming into the map. Also tied for the third most common operator is *retrieve* (21/25; 84 %), or the ability to request specific details about a map feature in the visualization. The provision of overlay, pan, zoom, and retrieve to manipulate a multiscale basemap is increasingly common today due to the ease in implementing these features with contemporary web mapping technologies (Roth et al. 2014); a web map with this basic functionality often is described informally as a 'slippy' map. It therefore is not unexpected that the large majority of evaluated visualizations support these four operators.

Somewhat surprising is the frequency that the *filter* operator is implemented (21/25; 84 %), or the ability to adjust the visualization to only show map features that match one or more user-defined conditions, as compared to *search* (13/25; 52 %), or the ability to identify a single location or map feature of interest. The search operator is more common in general use applications for which users have a single, concrete task, and therefore need a single entry point (i.e., a 'search box') for locating the feature of interest; on the other hand, the filter operator is more common in expert use applications for which the users have abstract or undefined tasks and require iterative exploration through small changes to filtering parameters. Many of the visualization tools use the filter operator to adjust the water level. The one evaluated visualization tool specific to the Great Lakes, Green Bay LakeViz, supports filtering from -12 to $+9$ ft based on variation in flood gauge data from 1996-to-present. When search is implemented, it is provided to reposition the map to a particular location, not a particular map feature or water level.

A small majority or large minority of visualization tools implement the work operators *calculate* (12/25; 44 %) and *resymbolize* (8/25; 32 %). The calculate operator allows users to derive custom information about map features of interest. Implementations include the dynamic calculation of total area impacted by a hypothetical flood (5/25; 20 %), a spatial measurement tool (4/25; 16 %), and the dynamic calculation of unique land use types impacted by a hypothetical flood (3/25; 12 %). The resymbolize operator allows the user to set or change a design parameter of the map representation without changing the features displayed on the map (as with the filter operator). The resymbolize operator exclusively is provided to adjust the transparency of overlay context layers. Finally, the arrange operator is implemented in four (16 %) of the visualization tools, allowing the user to adjust the position of map elements and interface functionality to avoid overlap with the map.

Importantly, the Fig. 5 analysis reveals several opportunities for the Lake Level Viewer that could set it apart from other water level visualization tools. First, none of the visualization tools implement the *reexpress* operator, which produces a new visualization of the same information, effectively 'showing it another way'. Viewing the inundated or exposed land from a profile view along a user-defined transect, for instance, is one way in which the visualization can be reexpressed to generate new insight. Second, only the SLAMM View tool implements the *sequence* operator, allowing for creation and comparison of side-by-side small multiples of different future scenarios (Tufte 1983). However, no tool implements the sequence operator to control animations of the waterline or flood extent. Finally, while the *export* enabling operator is commonly supported (17/25; 68 %) as a way to share the link of the current map view, the implementation of additional enabling operators may improve analytical work across use sessions. In particular, the *annotate* operator could support collaborative decision making, allowing users to externalize their thoughts into the map display for sharing with their project team.

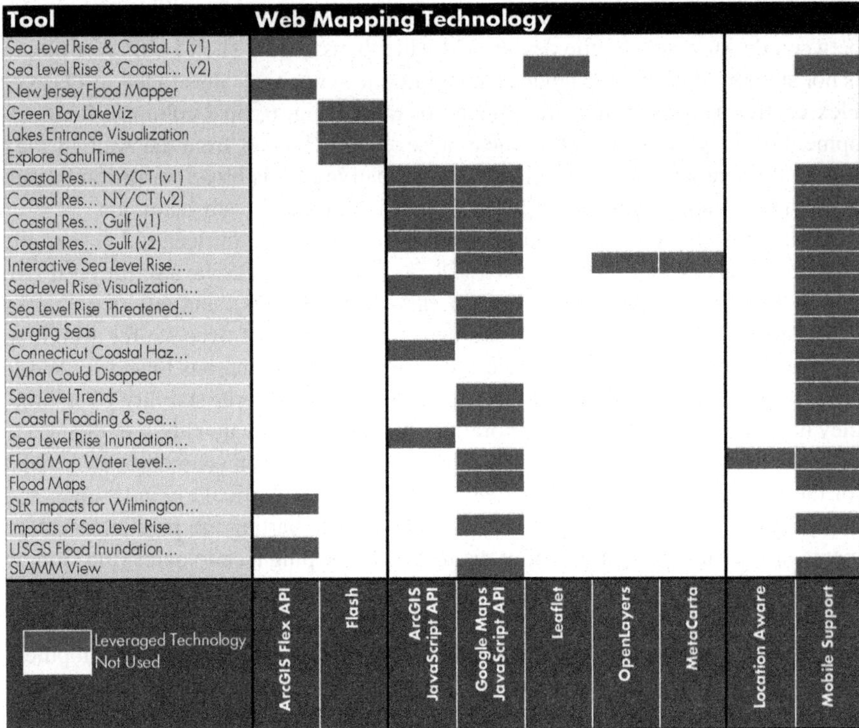

Fig. 6 Leveraged Web Mapping Technologies

Web Mapping Technologies

Following analysis of the interaction operator functionality provided in the visualization tools, we then inspected the underlying technology used to implement this functionality. *Web mapping technology* describes the amalgam of frameworks, libraries, APIs, and web services that enable the creation and dissemination of web maps (Peterson 2003). Our evaluation was limited primarily to the client-side or front-end implementation of the tool, given the focus on visualization and not processing.

Figure 6 provides an overview of the client-side web mapping technologies leveraged by the visualization tools included in the sample. Of the twenty-five visualization tools, eighteen (72 %) rely upon modern web standards (e.g., the browser-native definitions of HTML, CSS, and JS) while seven (28 %) rely upon a proprietary plugin to run a binary executable. For nearly a decade, a large number of web maps leveraged the FlashPlayer plugin by developing in the Adobe Flash or Flex authoring environments. Use of FlashPlayer resulted in a relatively small file size, a benefit for vector-based mapping, and improved cross-browser/cross-platform dependency. The use of FlashPlayer has waned in recent years, however, due to the pervasiveness of AJAX and the increased emphasis on responsive design between desktop and mobile devices (Muehlenhaus 2013). The FlashPlayer plugin is not supported by mobile devices, meaning the seven tools developed in Flash or Flex cannot be loaded on a smartphone or tablet (Fig. 6, final column). Redevelopment of the second version of the Sea Level Rise Viewer from the ArcGIS Flex API to the Leaflet open source library is indicative of this broad transition in web mapping technologies from proprietary plugins to modern web standards.

Of the eighteen tools leveraging modern web standards, thirteen (52 %) use the Google Maps JavaScript API, seven (28 %) use the ArcGIS JavaScript API, two (n = 8 %) use open source solutions (Leaflet, OpenLayers), and one (n = 1) uses MetaCarta. There is an emerging and active community of open source web map developers contributing their source code to the public commons for reuse. While open source solutions historically have suffered from poorer stability over time, they have the advantages of incorporating innovations more quickly into their code base and are free or near free to use. The choice of the open source library Leaflet for the second version of the Sea Level Rise Viewer is particularly intriguing, and likely fruitful. A recent study by Roth et al. (2014) charting the parallel developments of the same web map in four distinct web mapping technologies (the Google Maps JavaScript API, D3, Leaflet, and OpenLayers) found that Leaflet was able to produce a web map of comparable functionality to the web map leveraging the Google Maps JavaScript API, but resulted in a much more satisfying development experience given the openness and extensibility of the code repository. The ArcGIS JavaScript API remains a viable option, particularly when the GIS functionality provided by the ArcGIS suite is needed.

Finally, only the Flood Map tool is explicitly *location-aware*, drawing on the user's IP location to recenter the map to his or her current position. Overall, this may be a missed opportunity, as users are increasingly encountering web maps that are updated to their specific use context (e.g., their geographic location, their past interactions, etc.). However, there may be privacy or accountability concerns explaining the lack of location-aware technologies in water level visualization.

Conclusion and Outlook

This paper provides a functional and technological comparison of map-based water level visualization tools to inform the design of the NOAA Lake Level Viewer. A competitive analysis of twenty-five (n = 25) visualization tools was conducted according to criteria related to the representation or interaction design of the evaluated tools: (1a) variation in the waterline or flood extent symbolization, (1b) variation in included uncertainty information and uncertainty symbolization, (1c) variation in the provided basemaps and overlay layers, (2a) variation in the supported interaction operators, and (2b) variation in the underlying web mapping technology.

(continued)

Overall, we deem the competitive analysis as successful in meeting the project goals. First, we were able to identify and assess current practices in water level visualization, such as the use of a blue gradient in flood-centric representations, the inclusion of a common set of basemap and context options, the widespread support of the pan, zoom, overlay, retrieve, and filter interaction operators, and the general move away from web mapping technologies using proprietary plugins to those leveraging modern web standards. Second, we were able to identify unique solutions that potentially represent unmet user needs, including representation of exposed as well as flooded land, design of a flood representation that does not obfuscate the area of interest, provision of an informative overview at small cartographic scales, representation of multiple kinds of uncertainties, inclusion of meaningful and memorable benchmarks, and support of the reexpress, sequence, and annotate interaction operators. Finally, the competitive analysis helped to identify important gaps between theory and practice, namely the relative lack of uncertainty communication across the reviewed tools, the overall separation of tools designed to manage the build or human landscape from those designed to manage the natural or physical landscape, and the surprising implementation of filter rather than search for a public-facing visualization tool. Notably, the higher-level distinction between representation and interaction proved to be a useful way for coding the similarities and differences across the evaluated set of water level visualization tools; we anticipate that this distinction will be remain useful when completing a competitive analysis of visualization tools purposed for a different domain.

The competitive analysis represents the first stage in a broader user-centered design and development process for the NOAA Lake Level Viewer. Insights generated through the competitive analysis currently are being combined with stakeholder feedback received through a set of needs assessment interviews to generate a first draft of a requirements document. Two additional stages of user feedback are planned in the future: a cognitive walkthrough study on wireframe designs of the Lake Level Viewer and an interaction study on an alpha version of the tool. The Lake Level Viewer is expected to be published online at the end of 2014.

Acknowledgments The research was supported by the U.S. National Oceanic & Atmospheric Administration through Award #167152.

References

Agumya A, Hunter GJ (2002) Responding to the consequences of uncertainty in geographical data. Int J Geogr Inf Sci 16(5):405–417

Angel J, Kunkel K (2010) The response of Great Lakes water levels to future climate scenarios with an emphasis on Lake Michigan–Huron. J Great Lakes Res 36(Suppl 2):51–58

Bertin J (1967–1983) Semiology of graphics: diagrams, networks, maps. University of Wisconsin Press, Madison

Harrower MA (2002) Visual benchmarks: representing geographic change with map animation. Penn State, University Park

Hayhoe K, VanDorn J, Croley T, Schlegal N, Wuebbles D (2010) Regional climate change projections for Chicago and the US Great Lakes. J Great Lakes Res 36:7–21

IPCC (International Panel on Climate Change) (2007) IPCC Fourth Assessment Report: Climate Change 2007. http://www.ipcc.ch/publications_and_data/publications_and_data_reports.shtml#1

Kostelnick JC, McDermott D, Rowley RJ (2009) Cartographic methods for visualizing sea level rise. In: International cartographic conference, Santiago

Longley PA, Goodchild MF, Maguire DJ, Rhind DW (2005) Geographic information systems and science. Wiley, West Sussex

MacEachren AM (1994) Visualization in modern cartography: setting the agenda. In: MacEachren AM, Taylor DRF (eds) Visualization in modern cartography. Pergamon, Oxford, pp 1–12

MacEachren AM (1995) How maps work. The Guilford Press, New York

MacEachren AM, Robinson A, Hopper S, Gardner S, Murray R, Gahegan M, Hetzler E (2005) Visualizing geospatial information uncertainty: what we know and what we need to know. Cartogr Geogr Inf Sci 32(3):139–160

MacEachren AM, Roth RE, O'Brien J, Li B, Swingley D, Gahegan M (2012) Visual semiotics & uncertainty visualization: an empirical study. IEEE Trans Vis Comput Graph 18(12): 2496–2505

Muehlenhaus I (2013) Web cartography: map design for interactive and mobile devices. CRC, Boca Raton

Nielsen J (1992) The usability engineering life cycle. Computer 25(3):12–22

Peterson MP (ed) (2003) Maps and the Internet. Elsevier, Amsterdam

Robinson AC, Chen J, Lengerich EJ, Meyer HG, MacEachren AM (2005) Combining usability techniques to design geovisualization tools for epidemiology. Cartogr Geogr Inf Sci 32(4): 243–255

Roth RE (2009) A qualitative approach to understanding the role of geographic information uncertainty during decision making. Cartogr Geogr Inf Sci 36(4):315–330

Roth RE (2011) A comparison of methods for evaluating cartographic interfaces. In: International cartographic conference, Paris, 5 July 2011

Roth RE (2012) Cartographic interaction primitives: framework and synthesis. Cartogr J 49(4): 376–395

Roth RE (2013a) An empirically-derived taxonomy of interaction primitives for interactive cartography and geovisualization. Trans Vis Comput Graph 19(12):2356–2365

Roth RE (2013b) Interactive maps: what we know and what we need to know. J Spatial Inf Sci 6(59–115)

Roth RE, RG Donohue, CM Sack, T Wallace, T Buckingham (2014) Keeping pace with evolving open source web mapping technologies. Cartogr Perspect (in press)

Shneiderman B (1996) The eyes have it: a task by data type taxonomy for information visualization. In: IEEE conference on visual languages, Boulder. IEEE Computer Society Press, pp 336–343

Slocum TA, McMaster RB, Kessler FC, Howard HH (2009) Thematic cartography and geographic visualization, 3rd edn. Pearson Prentice Hall, Upper Saddle River

Tufte E (1983) The visual display of quantitative information, 2nd edn. Graphics Press LLC, Cheshire

Whitefield A, Escgate A, Denley I, Byerley P (1993) On distinguishing work tasks and enabling tasks. Interact Comput 5(3):333–347

Part III
Advanced Methods in Map Use

Models and Methods to Represent and Explore Phenomena on GIS

Anne Ruas

Mapping Phenomena at the District Level: Needs and Diagnostics

The increasing cost of energy cumulated with the attractiveness of the cities incites scientists and politicians to imagine smart cities for the future. These smart cities would concentrate the population and manage, as well as possible, all the drawback of this concentration such as the noise, the urban heat island effect or the various chemical pollutions. In the following of the paper we name these effects the phenomena. They are continuous, they fill the space and their quantity varies according to space and time.

To compute a phenomenon we need at least some sensors to measure real data and models to extrapolate measures spatially and temporally. Today many types of sensors exist and research and technology is focusing on reducing their price and their size and to encapsulate sensors in boxes that are able to trigger the measure and to send the result to a server according to appropriate protocols. These boxes contain a set of sensors like the IGN France GeoCube that can measure, store and send to a sever the location, the accurate time (GPS module), the wind and any other measured quantities. All of these kinds of components are fast improving.

To extrapolate measures, two classical types of method exist. Either the sensor network is dense enough so values can be interpolated and extrapolated with methods such as the Kriging (using real and virtual stations as in Ung et al. 2001), or the phenomenon is built thanks to few measures, a specific model and wind information for propagation. Models are interesting because they allow forecasting a phenomenon (see for example the importance of meteorology forecasting for everyday life). This last solution requires also fewer sensors, which is

A. Ruas (✉)
LISIS Laboratory, IFSTTAR, 14-20 Boulevard Isaac Newton, 77447 Marne la Vallée Cedex 2, France
e-mail: anne.ruas@ifsttar.fr

© Springer International Publishing Switzerland 2015

J. Brus et al. (eds.), *Modern Trends in Cartography*, Lecture Notes in Geoinformation and Cartography, DOI 10.1007/978-3-319-07926-4_20

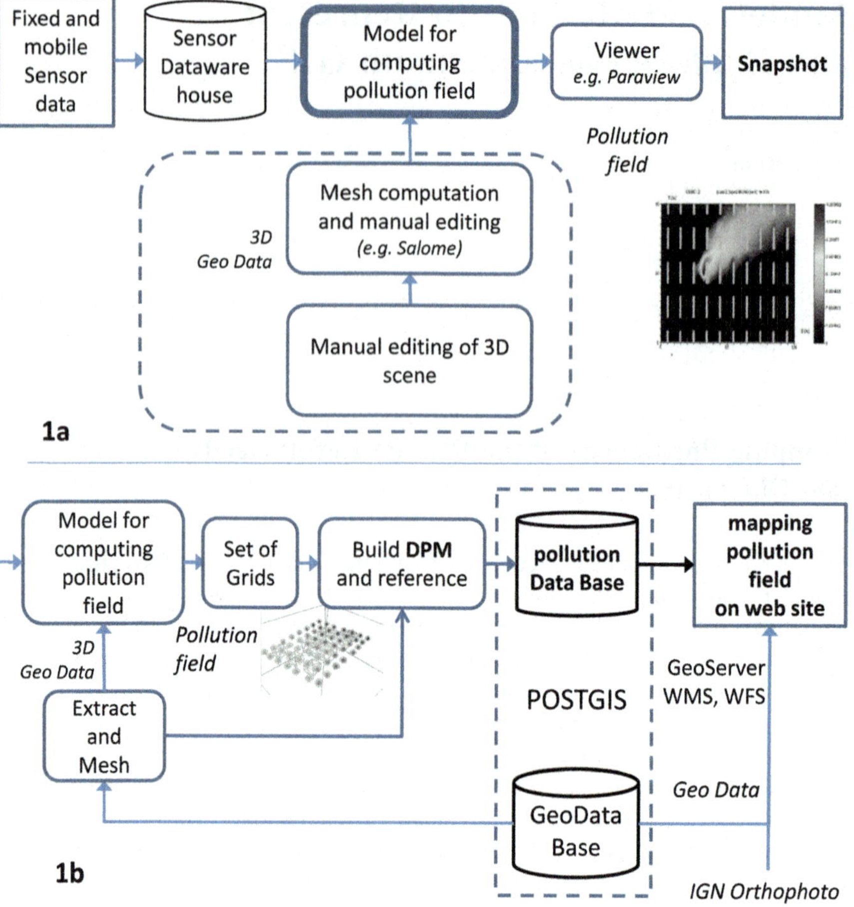

Fig. 1 (**a**) The classical data flow for expert analysing pollution propagation. (**b**) A new data flow that allows to map the results in 2D on the web

today somehow a more realistic solution according to the difficulty we have to obtain authorization to set sensors in a city. An important point is that these models, using fluid dynamic, require a representation of the space in 3D (see Fig. 1a, b).

Many models have been developed by experts to compute specific phenomena from a set of measures. In France for example the model Saturne (see http://cerea.enpc.fr), allows to study either the effect of a phenomenon on the environment (such as the pollution on a district) or the effect of the environment on a phenomenon (such as the impact of buildings on the local rising of temperature). Saturne software is based on fluid dynamic (CFD) model, developed by EDF, and contains an atmospheric module, developed by the CEREA laboratory, which allows the computation of different pollution propagation (Winiarek et al. 2013). This module has been used in the *Immanent* project to compute the propagation of benzene around a gasoline station from fictive measures (Cheaib et al. 2013).

Having sensors, protocols and models, we can imagine an easy access to the representation of phenomena at different levels of detail including the district level, at any time. But if many web sites are mapping phenomena such as Air Quality, most of them are spatially not very accurate (e.g. ESRI 2007). In our research on the representation of phenomena such as Air Quality or urban heat island effect, we wish to work at the district level: we wish to represent phenomena in the street, surrounding the buildings where people are working or living. In order to map real time data, we need to set architecture to connect all the components (second section) and we need dedicated models and methods to map the phenomena in a meaningful way (third and fourth sections).

Improving the Data Flow to Take Advantage of the Existing Models

Researchers who conceive models are mainly focusing on the quality of the estimation and on the efficiency of the computation, not on their interfaces. As a consequence, these models, even though they use geographical information, do not read nor write OGC standard or even commercial standard such as shape. Figure 1a illustrates difficulties we had to face in the Immanent project: the system could not read 3D data in GIS format (we had to digitized existing data through a CAD software!), and by default, the output is a snap shot capture on a physical viewer software such as Paraview. These types of viewer are perfect to classify and map the values is different ways. They are excellent tools for the expert but they are not adapted for decision making. First of all, there is no geographical coordinate system (the 3D field is in the image local coordinate system), and the symbolisation is very far from being rich as a GIS can be. The graphical representation is very poor as illustrated in Fig. 1a, limited to the representation of the values of the phenomenon but neither on its effect nor on its geographical context which is very simplified. We could say that this kind of representation is not contextualised, there is nearly no semantic. Nothing but the phenomenon is detailed, although the impact of the phenomenon depends on its context.

In order to improve the representation as well as the access to the information, it was necessary to modify the data flow. We introduced the concept of *pollution data base* in order to store the results, we modified the output according to GIS format and we developed web services to allow an easy access to the data (Fig. 2).

The first task consisted in being able to project the output of the model (the *pollution field*) on top of geographical data. A simple solution was developed to write a set of files that looked like DTMs. Even if the phenomenon is 3D, we extracted a grid defined by a *grid threshold* and an altitude ($Z0$). The system

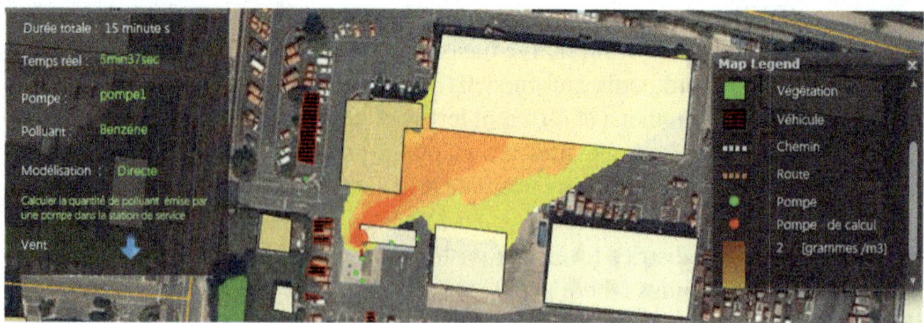

Fig. 2 Web services to view Benzene propagation around a gasoline station from fictive measures

generated several files as the phenomenon varied in time. There was one file for each computed time. Each file was then modified to integrate three constraints:

- the beginning of the file was modified to follow exactly a DTM format,
- a translation was integrated to respect a geographical reference system (Lambert 93),
- Values were analysed and normalised.

We name this result a DPM: a *digital phenomenon model*. A phenomenon which lasts a certain duration is symbolised by a set of DPMs.

These DPMs, qualified by specific information (see third section), are stored in a data base, named the *pollution data base*. Figure 1b, compared with Fig. 1a, illustrates the changes we had to make to easily map the pollution on top of geographical information extracted from the IGN-France BDTopo ©.

A classical architecture based on PostGis, GeoServer, WMS and WFS was set to view the DPMs on top of Geographical data from IGN-France (BDtopo and Orthophotographies) (Boukhechba 2013). Figure 2 illustrates the result which can also be accessed on http://representation-phenomenes.ifsttar.fr;

In the following we present data models to map and explore the data.

Data Models to Map Pollution Data in 2D

Figure 1b illustrates the data flow to map the pollution field by means of DPMs. It requires a data model illustrated in Fig. 3a. We also present a second data model (Fig. 3b) which allows more possibilities.

The core of the model is the *Pollution Episode*. The pollution episode is the description of the phenomenon at a specific date and during certain duration. It is computed by means of measures of pollution and wind and a model, parameterised in a specific way. The maximum and the minimum episode values are useful for the symbolisation (*categorisation method*).

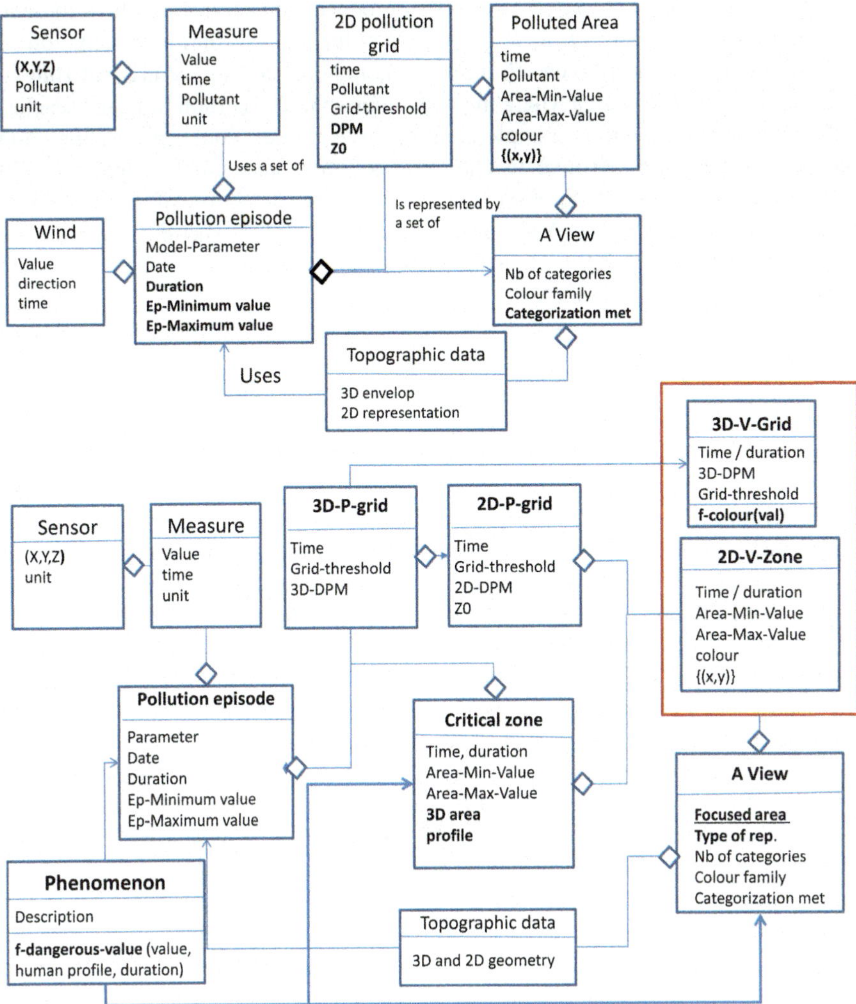

Fig. 3 (**a**) (*Top*) Data model for web services in 2D. (**b**) (*Bottom*) From pollution field to 2D and 3D visualisation

The *2D Pollution grid* is the DPM with information related to the grid-threshold, the altitude (Z0) and the time. A pollution episode is described by a set of 2D Pollution grids.

The *view* describes the graphical rules to map the DPMs: the categorization method to classify the values, the number of colours and the colour family (in Fig. 2, the family colour used is the *Yellow–Orange–Red*);

The *Polluted Areas* are computed from a *2D pollution grid* with graphic rules described in a *view*. An area is described by its geometry ({(x, y)}), its colour, a range of values, a time and the pollutant.

  A. Ruas

This data model (Fig. 3a) is efficient to produce web services such as the one presented in Fig. 2. If we want to go further in the analysis and representation of pollution fields, we need to enrich it and to include the concept of **critical zone.**

A *critical zone* is an area where the phenomenon lasts a certain time and reaches values which are dangerous or unpleasant for some profiles. For example babies and old people are more sensitive to chemical pollution than other people; but old people are less sensitive to noise than babies. The detection of critical zones requires the analysis of 3D Pollution field series. Compare with Fig. 3a, we need thus to store the 3D pollution field in a Pollution Data Base to make appropriate analysis, and not only 2D-DPMs.

Some cut, aggregation, or categorisation are applied on the pollution DB to detect critical zones or to map specific *focused areas*. These methods are classical in spatial analysis. They can be implemented thanks to existing methods on R for example. We did not represent these methods but they would be used for example on a set of 3D-P-Grids to detect Critical zones. The *focused-area* is defined as an attribute of a *view*.

As a consequence, the data model should be enriched (see Fig. 3b). It contains *3D pollution grids* (3D-P-grid, also named the 3D pollution field), *critical zone*, and a class describing the *phenomenon*, as well as objects that allow a better representation (thanks to objects in the classes *View*, *3D-V-Grid* and *2D-V-Zone*).

The *phenomenon* (like the noise or the chemical pollution) holds information which describes what a dangerous (or unpleasant) value is according to the value of the phenomenon and at least the human profile. If dangerous values are described on a scale going from 0 to 6, each value is (at least) a function depending on the value(s) of the phenomenon (e.g. the DB and frequency for noise; the number of gr/m^3 for Benzene), the *human profile*, the *duration* and eventually other criteria such as the hour in a day (for noise). These functions are essentials to go from quantitative to qualitative values, as only the expert can interpret quantitative values. We suppose that one function (named *f-dangerous-value* in Fig. 3b) is defined for each phenomenon. It allows computing critical zones. Thus *critical zones* are computed by analysing the 3D-P-Grid through the filter of the information given by f-dangerous-value.

A *Critical zone* object is a 3D area. It can be mapped in 2D by means of a *2D-V-Zone* object (see below), but it holds specific information describing the profile for which the area is critical and time information (when the zone is critical).

At least an object *View* is described by a focused-area and as defined previously information to categorise values (a method and a number of categories) as well as a *family-colour*. A *view* is composed by a set of *2D-V-Zone* objects or *3D-V-Grids* objects. The type of representation describes if the representation of pollution is done by means of zones (objects from *2D-V-Zone*) or grids (objects from *3D-V-Grid*). We add the attribute '*focused area*' for efficiency purposes, as it is some time more efficient to map the pollution on a focused area only and not on the entire geographical area. Focused-area is used mainly for 3D-V-Grid objects (see below).

A *2D-V-Zone* object is a Visual object (V for visual) that is flat but could be horizontal, vertical or whatever orientation. In Fig. 2 the *2D-V-Zone* objects are horizontal and we describe in Fig. 3b the geometry by a simple plan $\{(x, y), z0\}$. An area represents values included in a range of pollution between an *area-min-value* and *area-max-value*. The area is mapped in a specific *colour that* belongs to the *colour-family* of the view.

A *3D-V-Grid* object is a Visual object (V for Visual) that is a 3D grid. It is defined in fourth section.

Statistics and 3D Visual Grids to Map Phenomena Such As Pollution

Several studies are proposing solutions to map pollution thanks to GIS (ESRI 2007) or (Al Koas 2010). One of the key points is to be able to analyze the data in order to detect critical zones or specific events by means of appropriate methods. Rude and Beard (2012) explore the detection of high-level events from spatially distributed time series. Their goal is to extract interesting occurrences from time series as aggregate patterns of primitive patterns. Peuquet (2012) proposes a statistical temporal pattern discovery technique called T-pattern analysis via a prototype called Stempo. So, we need methods to make spatio-temporal analysis on 3D grid data. The first conclusion is then that it is not possible to master the representation of phenomena without solid spatio-temporal analytical tools (such as R).

In terms of graphical representation, in third section, we project the 3D information in 2D. 2D is convenient as most of today tools (including web services) are dealing with 2D data. However phenomena varies in Z and this information is important as space is not lived in the same way for all altitude. Pollution at 50 cm height has obviously not the same impact as at 1.50, 4 or 20 m.

In order to perceive this variation of values according to z, a possibility is at least to view the topographic data in 3D as illustrated in Fig. 4. In this Fig. 4 both left images are not GIS but image viewers. Pollution is perceived as a plan grid but at least a 3D view allows seeing at which altitude the pollution is computed and mapped.

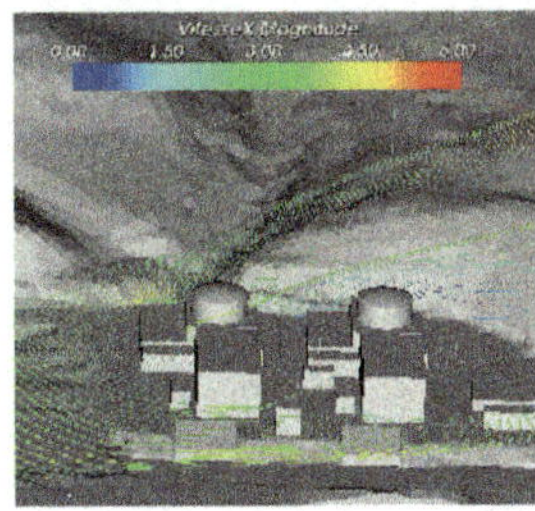

Dispersion around plant
© Cerea (Saturne & Paraview)

Pollution representation on EveCity
© CSTB

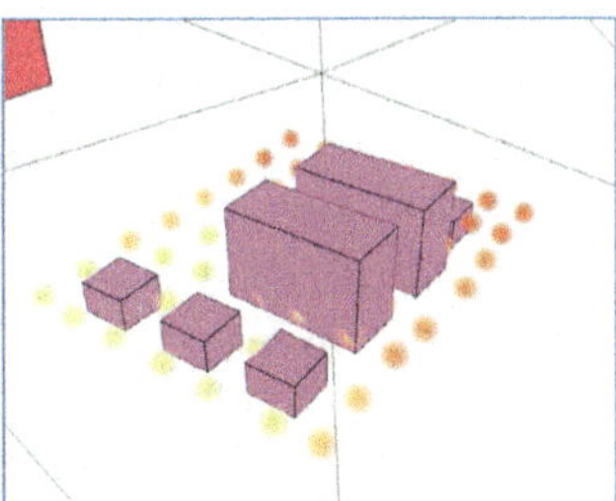

Experiments on GeOxygène 3D
© IFSTTAR

Fig. 4 3D representations of pollution on different tools

The right image is a first attempt to introduce a grid structure in the classical 3D GIS GeOxygène. GeOxygène3D was initially conceived to map solid objects in 3D (such as road and building) but not field information. If it can map a DTM and match 3D objects as buildings on this DTM, as many other 3D GIS like TatukGIS does, it is a priori not adapted to map the 'empty space'. However this empty space is the space of phenomena. So on GeOxygene3D we enriched our software with appropriate data structures to map empty space on demand. It is not so difficult: it requires an easy and efficient construction of 3D grids, on focused areas, with a chosen grid threshold. Then we can consider that each node of this grid is an object that has its own symbolization according to the value of the phenomenon. First results of our developments are illustrated in Fig. 4 (right).

In order to master this type of representation we add the class *3D-V-Grid* to our model. A *3D-V-Grid* object is a Visual object (V for Visual) that is a grid defined in the *focused-area* and that have a certain *grid-threshold*. A *3D-V-Grid* object is computed from one of several *3D-P-Grid* objects. If it is several objects, it means that the *3D-V-Grid* object has a time and duration; else it has just a time. A 3D-V-Grid object is composed of nodes located on the grid structure. Each node is symbolised according to its value by means of the method *f-colour* (val). The value of each node of computed by means of specific aggregation and interpolation methods applied on *3D-P-Grid* objects.

A current project is to re-implement this model and methods on a new 3D open source GIS software developed by Oslandia Company on top PostGis3D.

> **Conclusion**
>
> In this paper we present a diagnostics of difficulties we had to face to map pollution data. Many solutions exist outside our community mixing a set of tools and ending with a poor symbolization on image viewers (Fig. 4, left). However it is important to map phenomena in a rich way for different reasons. The geographical context should be well represented because it contextualizes each phenomenon. This contextualization gives meaning and allows better understanding. Then the phenomena itself should be represented in a qualitative way to integrate human profile as well as the impact of the duration of the phenomena on human. Last but not least, a 3D representation would help understanding the variation of the impact according to the altitude. This is not nothing. A phenomenon and its impacts vary in altitude. This variation should be communicated.
>
> Many projects related to life quality in cities are going on. Valente et al. (2012) show a computation of particle matter exposure around and in a school. This research illustrates the use of geographical data to compute the dispersion of pollutants and the exposure of children to these pollutants in specific and exposed places. *Into the air* project (http://intheair.es) aims to map air pollutants over the Spanish city.

(continued)

So either the GIS community only concentrates on the mapping of 'solid' objects, and let other experts map phenomena on image viewers, or we (cartographers) need to enrich our software with appropriate data structures to map empty space on demand. Will cartographers miss the train of smart cities? I hope no, but it requires hurrying up in the enrichment of our GIS software.

References

Al Koas K (2010) GIS-based mapping and statistical analysis of air pollution and mortality in Brisbane. QUT, Australia

Boukhechba M (2013) Mise en œuvre d'une architecture logicielle à base de Services web pour la visualisation de données de pollution dans une scène géographique 2D en ligne. Rapport de stage Master Géomatique 2013 – IFSTTAR

Cheaib N, Ruas A, Gaborit N (2013) From sensor data to the perception of phenomena: software architecture for online access and offline analysis. In: Proceeding conference ICC, Dresden, Paper no. 166 (10B.2). http://acica.org. ISBN: 978-1-907075-06-3

ESRI (2007) GIS for air quality

Peuquet DJ (2012) A method for discovery and analysis of temporal patterns in complex space-time event data. In the extended abstracts of the 7th international GIScience 2012 conference, Springer

Rude A, Beard K (2012) High-level event detection in spatially distributed time series. In: Proceedings of the 7th international GIScience 2012 conference, Springer, pp 160–172

Ung AC, Weber C et al (2001) Air pollution mapping over a city – virtual stations and morphological indicators. In: Proceedings of 10th international symposium "transport and air pollution", 17–19 September 2001, Boulder

Valente J, Amorim JH, Cascao P, Rodrigues V, Borrego C (2012) Children exposure to PM levels in a typical school morning. In usage, usability and utility of 3D city models conference, Nantes

Winiarek V, Bocquet M, Duhanyan N, Roustan Y, Saunier O, Mathieu A (2013) Estimation of the caesium-137 source term from the Fukushima Daiichi nuclear power plant using a consistent joint assimilation of air concentration and deposition observations. Atmospheric Environment 82, 268–279 (2014)

Assessing Cartographic Products for Visual Usability

William Cartwright

Introduction

A project was undertaken to review a representative sample of the mapping products provided via the Department of Sustainability and Environment [recently re-named the Department of Environment and Primary Industries (DEPI)], Victoria Australia. It covered the maps provided via the Department's Web sites with regard to 'good practice' related to equity of accessibility. The project covered:

- Existing map design specifications for Web-delivered products;
- Overview of the suitability for using Web-delivered mapping resources, such as *Bing*® or *Google Maps*®, as 'foundations' for generating mash-up mapping applications (w.r.t. equity and accessibility); and
- General recommendations of further activities/actions needed to ensure usable maps

The Department provides access to its Web mapping products via a Web portal. This enables potential users to gain immediate access to appropriate mapping products. Users require access to the Internet and the ability to use a Web browser and interact with the mapping products presented. Also, where printable versions of the maps are made available, users need access to a colour printer to be able to print and then use printed versions of the map.

Evaluations were conducted to assess the general designs and map delivery. The maps were assessed according to an evaluation proforma developed expressly for this purpose. As well as general design elements, each map was evaluated for their usability by the colour blind.

W. Cartwright (✉)
School of Mathematical and Geospatial Sciences, RMIT University, Melbourne, VIC, Australia
e-mail: william.cartwright@rmit.edu.au

© Springer International Publishing Switzerland 2015

J. Brus et al. (eds.), *Modern Trends in Cartography*, Lecture Notes in Geoinformation and Cartography, DOI 10.1007/978-3-319-07926-4_21

This paper reports on the outcomes of the considerations of the usability of the Web-delivered maps for colour-blind users.

Considering the Colour-Blind

Humans perceive colour using photosensitive cones. The grey component of colours is perceived through the bleaching of photosensitive rods. Users lacking certain colour-receptive cones or who have damaged cones will not be able to see some of the red, green and blue components of the visible colour spectrum. 0.01 % of all Caucasian females and 2 % of all Caucasian males have deuteranopia/ deuteranomaly, where the retina lacks red-sensitive cones (Light and Bartlein 2004). The more common forms of colour impaired vision are deuteranopia/deu-teranomaly/Dalonism (green-weakness) (0.25 % of Caucasian females and 6 % of Caucasian males). These users are less sensitive to medium wavelengths—greens. They have a decreased ability to discriminate the green component of colours (Stephenson 2005). Protanopia is red-blind. Tritanopia is the rarest form of colour-blindness and is the lack of functional cones altogether.

The greatest confusion area is red-green confusion. Of the three colour-blindness examples illustrated, only Tritanopiates would be able to discriminate between the different zones of grasslands. Tritanopia is the rarest form of colour-blindness. Therefore colour design strategies are needed to ensure that different colours can be discriminated.

Design Strategies

One simple strategy is to ensure that all information conveyed with colour is also available as one colour tone. With paper mapping a number of innovative works have been undertaken to produce products that work for all users, irrespective of colour deficiencies. This has usually been achieved by producing a monochrome map. One example is the 'Color Blind Subway Map' for Manhattan. The current New York Subway map, designed by Michael Hertz, was introduced in 1979 (Hogarty 2007). It replaced the earlier (lauded) version that had been designed in the 1970s by acclaimed graphic designer Massimo Vignelli (Rawsthorn 2012).

A prototype Subway map for the colour-blind was produced by Brooklyn based designers Triboro Design (www.triborodesign.com/). It was produced in just one colour: florescent red. The 45 × 58 in. poster is the same size as the MTA Subway maps that are placed on the walls at New York Subway stations. This map is shown in Fig. 1.

Jenny and Kelso (2007) suggested some basic methods to ensure that colour-blind users are able to distinguish between different mapped elements, viz:

Fig. 1 Triboro design's
New York subway map.
Source: Heller. http://www.
printmag.com/Article/
Myopic-Subway-Map/

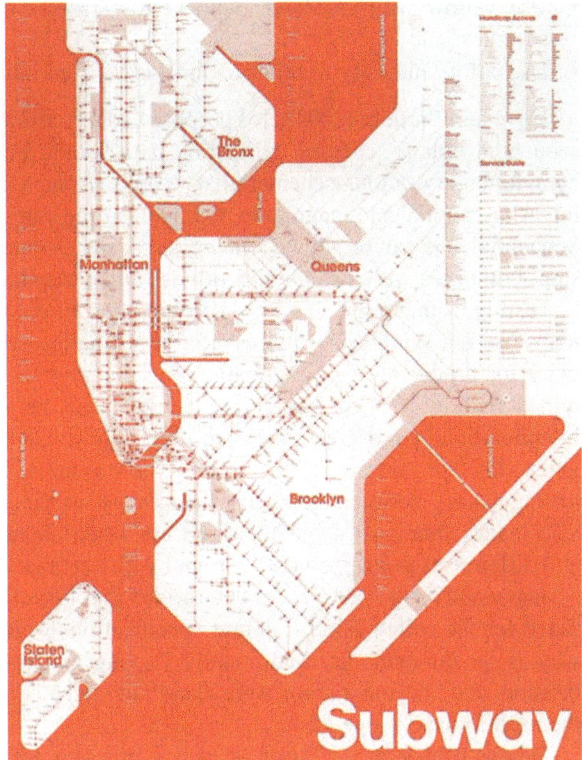

- choosing unambiguous colour combinations;
- using alternative visual variables; and
- directly annotating features.

Arditi (2010) also proposed three basic rules for better choosing colour schemes for colour-blind users:

- Exaggerate lightness differences between foreground and background colours, and avoid using colours of similar lightness adjacent to one another, even if they differ in saturation or hue;
- Choose dark colours (with hues from the bottom half of the hue circle shown below) against light colours from the top half of the circle. Avoid contrasting light colours from the bottom half against dark colours from the top half.
- Avoid contrasting hues from adjacent parts of the hue circle, especially if the colours do not contrast sharply in lightness.

Light and Bartlein (2004) offered three suggestions that might make colour graphics and maps more accessible to all. These suggestions were:

- Avoid the use of spectral schemes to represent sequential data, because the spectral order of visible light caries no inherent magnitude message;

- Use yellow with care and avoid yellow-green colours altogether in spectral schemes; and
- Use colour intensity to reinforce hue as a visual indicator of magnitude.

Light and Bartlein (2004) also proposed solutions for representing diverging and sequential colour schemes. Diverging data should be represented by using two complimentary colour schemes that diverge from a common hue. Here, colour intensity can indicate magnitude and hue can indicate sign (increase or decrease). Sequential data can be shown with colour schemes that use a sequence of lightness steps combined with a single hue or with a hue transition.

Distinguishing point classes is a problem for the colour-blind. As many thematic maps use dot symbols for representing information strategies need to be put into place for producing legible graphics. Research by Jenny and Kelso (2007) determined that by varying saturation, contrast can be increased, which improves interpretation by red-green impaired users. They also found that shifting hue from green to blue improves legibility. The optimum design was achieved by distinguishing between geometric shapes by varying hue and saturation. They also made comments about their success when colour was discarded altogether and differences in mapped data shown by different shaped symbols.

Jenny and Kelso (2007) also determined that line classes can be better differentiated for the colour-blind if colour specifications were also amended in a similar way as for dot symbols. Additionally varying line width also improved comprehension, and line annotations were also found to be beneficial, as well as removing the need to have a legend.

Distinguishing area classes is also a problem for colour-blind map readers. Different hues can be employed, as long as each hue has a unique saturation (the density of colour applied) and value (intensity). The addition of hachuring can also assist to improve legibility.

In some instances colour visualizations are designed so that one colour merges into another. Here, to improve legibility Brewer (1997) suggests the following design strategies:

- Vary lightness on the red-orange-yellow end of the rainbow;
- Omit yellow-green to avoid confusion with orange; and
- For bipolar data, omit green and use a scheme with red, orange, yellow, light blue and dark blue; and align the yellow-blue transition-diverging data range.

The needs of the visually impaired and blind for Web access and interface design are being addressed by the World Wide Web Consortium's accessibility initiative (W3C 2000) that includes 'translations' from graphics into audio for the blind.

A number of mapping organisations have addressed colour-blindness and map design specifications. For example, the United Kingdom's Ordnance Survey re-designed their topographic maps to ensure that they are readable by the colour-blind (The Map Room 2009; Ordnance Survey 2009). Digital mapping from the Ordnance Survey can now be customized to create 'colour-blind-friendly styles'.

This feature is delivered as part of the Ordnance Survey product 'OS VectorMap Local' (Ordnance Survey 2009).

Guidelines and Tools

General Web accessibility guidelines are published under Web Content Accessibility Guidelines (WCAG). Also, Checkpoint 2.2 of the guidelines requires that foreground/background colour schemes must have sufficient contrast when viewed by someone with colour vision defects or viewed on a black and white screen.

A number of tools exist to simulate how colour-blind users would see maps and other tools that assist in choosing appropriate thematic mapping colour schemes to ensure inclusiveness.

- *Color Oracle* is a free Java tool (Windows, Linux and Mac) developed by Dr Bernhard Jenny, of ETH Zurich, Switzerland. It can be used for evaluating the effects of several kinds of colour blindness. http://colororacle.cartography.ch/ (Authors note: the author was the second supervisor of Dr Jenny's Ph.D.).
- *Color Brewer* (www.colorbrewer.org) generates GIS colour schemes that are distinguishable by users who are colour-blind (Nugent 2011). The site provides schemes that are 'friendly' to colour-blind users.
- *Colour Vision* (www.iamcal.com/toys/colors/) provides a tool for simulating colour deficiencies.
- *Colorblind Web Page Filter* (http://colorfilter.wickline.org/) is a tool for checking Web pages for colour deficient or colour-blind users. A Web site URL is input into the package and the simulation is automatically generated online.
- *Colour-blindness Simulator* (http://www.etre.com/tools/colourblindsimulator/) uses an uploaded JPEG image ($<1{,}000 \times 1{,}000$ pixels and <100 kb) to generate a new image, simulating how a colour-blind person would see the image.
- *Colour Check* (http://www.etre.com/tools/colourcheck) checks colours for conformance to the WCAG (Web Content Accessibility Guidelines). It determines the colour difference and contrast between any two colours chosen for background/foreground applications.
- *Colourmaps* (www.tsi.enst.fr/~brettel/CRA24/fig3java.html) for checking the legibility of displays by dichromats,
- *Safe Web colors for colour-deficient vision* (http://www.btpic.com/age_disabiity/technology/RandD/colours/index.htm)—translations, definitions, etc.
- *What do colour-blind people see?* (http://www.tsi.emst.fr/~brettel/colourblindness.html) is an interactive java applet that demonstrates colourblindness for protanopia and deuteranopia.
- *Vischeck* (www.vischeck.com) is a resource for information about colourblindness. Map designers are able to upload an image file and Vischeck simulates how the map would appear to a colour-blind person. The real limitation of

Vischeck for map designers is the relatively small file sizes that can be accommodated.

Color Oracle was selected to generate colour blindness simulations.

Map Evaluations

Map evaluations were undertaken using an evaluation proforma. Each map was evaluated for their usability by the colour blind. Sample maps were generated to simulate how colour-blind viewers would see the colours used in the maps. These were generated using *Color Oracle* to simulate:

- Deuteranopia;
- Protanopia; and
- Tritanopia

This is done by firstly providing samples of each map: in its original form, as a greyscale, as three colour-blindness simulations—deuteranope, protanope and tritanope.

The maps evaluated were:

- Fire Recovery—Road, Track, and Recreation Site Status;
- Biodiversity Interactive Map;
- Fireplan;
- Forest Explorer Online;
- Interactive map;
- Office of Water, In Your Region;
- One Source One Message (OSOM) bushfire information;
- GPSnet Continuously Operating Reference Stations (CORS);
- Swoop magpie map;
- Map of the new scientific trial of Alpine grazing;
- Maps from the February 2009 bushfires, including the Kilmore East—Murrindindi complex;
- Planned burning operations plans;
- Future Coasts Digital Elevation Models;
- Hunting maps;
- Proposed Western Grasslands Reserves PDF;
- Topographic T7925-2-2-N;
- Topographic T8223-2-4-N; and
- Planned Burns Today, map and status of current burn locations.

All maps listed were evaluated, however, just one sample evaluation map is shown to provide a sample. Table 1 provides a summary the evaluation for this map (Fig. 2 and Table 2).

Table 1 Evaluation for the planned burning operations plans

Item	Sub-item	Acceptable	Border line – needs review	Unacceptable	Comments
Colour					
	Colour-blind alternative—greyscale version			✗	**Needs greyscale alternative option added to Web site**
	Different saturation for each hue enhance discriminabilty by colour-blind			✗	
	Colour-blind alternative—generate alternative colour scheme for colour-blind users			✗	**Need to provide alternative maps for colour-blind users or re-design colour scheme to ensure that colour-blind users can use the 'standard' map provided.**
	Unambiguous colour combinations			✗	
	Direct feature annotation			✗	
	Avoid use of spectral schemes to represent sequential data				**N/A**
Symbols					
	Shape variation to enhance discriminability by colour-blind				**N/A**
Lines					
	Varying line width to enhance discriminability by colour-blind		✗		
	Line annotations to enhance discriminability by colour-blind				**N/A Areas only**

A summary of all colour blindness findings is provided in Table 2

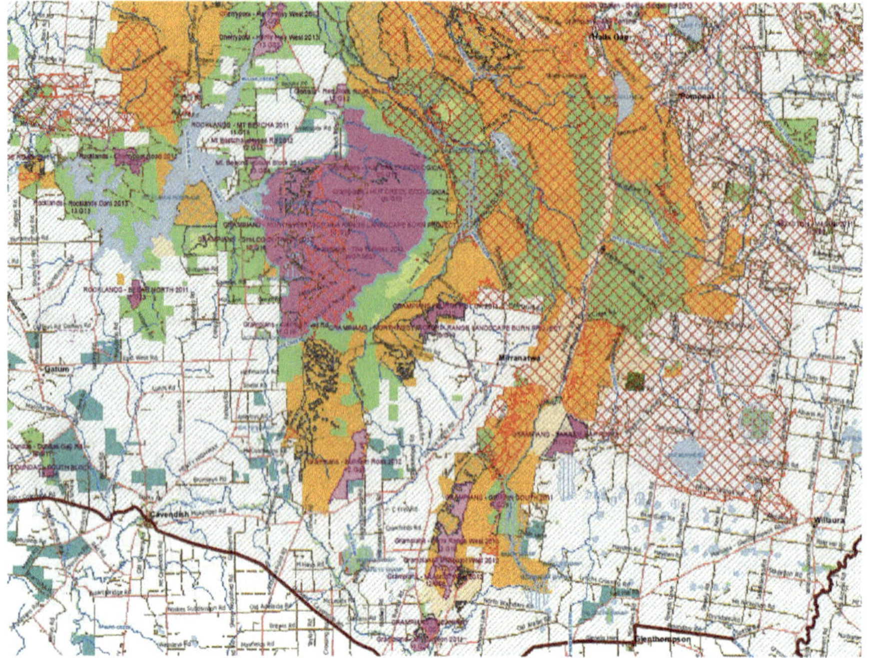

Fig. 2 Planned burning operations plans

General Findings

Some general findings were:

- Basically, none of the maps appeared to have been designed considering colour-blindness at all;
- The delivery of standard topographic maps requires conforming to standard symbology and colour specifications. There is little room to move and little latitude is available to change the design of these maps from formal specifications. However, in some cases, maps were produced that were nor legible to the colour-blind or users with other visual impairments that might be found in an aging population. As well, the greyscale version of these maps do not have differing grey values in the original colours to enable differentiation when displayed as shades of grey.
- All maps need greyscale alternative options on the Web site. This is necessary to ensure that the map can be read by the colour-blind and for users with access only to a black and white printer are able to generate a usable paper product. No greyscale options were made available.
- An ideal design is one where the grey component of each colour has been chosen to ensure that the map can work as a greyscale version without the need to

Table 2 Summary of colour blindness evaluations

Item	Sub-item / MAP	Evaluation																	
		1	2	3	4	5	6	7	8	9	10	11	12	13	14	15	16	17	18
Colour	Colour-blind alternative–greyscale version	✗	✗	✗	✗	✗	✗	✗	✗	✗	✗	✗	✗	✗	✗	✗	✗	✗	✗
	Different hue saturation	~	~	✗	✗	✗	✗	✗	✗	✗	✗	~	✗✗	~	✗	✗	✗	✗	✗
	Colour-blind alternative–generate alternative colour scheme for colour-blind users	✗	✗	✗	✗	✗	✗	✗	✗	✗	✗	✗	✗	✗	✗	✗	✗	✗	✗
	Unambiguous colour combinations	~	~	~	~	✗	✗	~	✗	✗	~	~	✗	~	✗	✗	✗	✗	~
	Direct feature annotation	✗	✗	n/a	✗	✗	n/a	n/a	✔	✗	✗	✗	✗	n/a	n/a	n/a	n/a	n/a	✗
	Avoid use of spectral schemes to represent sequential data	n/a	n/a	n/a	n/a	n/a	n/a	n/a	n/a	n/a	n/a	n/a	n/a	n/a	n/a	n/a	n/a	n/a	n/a
Symbols	Shape variation to enhance discriminability by colour-blind	✔	n/a	~	✗	✗	✔	✔	~	~	✗	n/a	n/a	n/a	n/a	✗	✗	n/a	✗
Lines	Varying line width to enhance discriminability by colour-blind	~	✔	~	~	~	✗	~	~	~	✗	~	~	~	✗	~	~	~	✗
	Line annotations to enhance discriminability by colour-blind	✗	n/a	✗	✗	✗	✗	✔	~	~	~	✗	n/a	n/a	✗	n/a	✗	✗	✗

✔ Acceptable
~ Needs review
✗ Unacceptable

generate a second unique greyscale map. In many maps evaluated they contained some colours that were too close in value to work effectively as a greyscale map.

- There exists the need to provide alternative maps for colour-blind users or re-design colour schemes to ensure that colour-blind users can use the 'standard' map provided.
- Colour schemes on all maps do not work at all for colour-blind users. An option to generate a unique map for viewing by deuteranopes, protanopes and tritanopes would contribute to the provision of 'maps for all'.
- The sheer number of colours on some maps makes it almost impossible to provide a standard map that works for all users, irrespective of their vision impairment.

Conclusion

The main recommendations that can be made are twofold, namely:

1. Map specifications must be reviewed and all colour schema re-specified to ensure that all colours are reproducible in monochrome and
2. For all maps in the public domain, make available the option for users to request another map format. This could include large print, a special colour scheme or monochrome maps for the colour-blind and outputs with fewer colour-representations of unique features. (This last point may necessitate 'breaking' some maps into a number of products, whereby original maps containing numerous colours, which would be impossible to reproduce effectively for colour impaired users, would be provided as a map set. The map set would contain all of the information from the original product, but deliver this information across a number of maps.)

References

Arditi A (2010) Effective color contrast. Lighthouse International. http://www.lighthouse.org/accessibility/design/accessible-print-design/...

Brewer CA (1997) Spectral schemes: controversial color use on maps. Cartogr Geogr Inf Syst 24 (4):203–220

Hogarty D (2007) Michael Hertz, designer of the NYC subway map. Gothamist, August 2007. http://gothamist.com/2007/08/03/michael_hertz_d.php. Accessed Dec 2013

Jenny B, Kelso NV (2007) Color design for the color vision impaired. Cartogr Perspect (58):61–67. http://jenny.cartography.ch/pdf/2007_JennyKelso_ColorDesign_lores.pdf

Light A, Bartlein PJ (2004) The end of the rainbow? Color schemes for improved data graphics. EOS 85(40):385 and 391

Nugent J (2011) Designing maps for the color blind. http://vcgiblog.wordpress.com/2010/11/02/designing-maps-for-the-color-blind/

Ordnance Survey (2009) Maps for the colour-blind are a real eye opener. http://www.ordnancesurvey.co.uk/oswebsite/media/news/2009/october/colourblind.html

Rawsthorn A (2012) The subway map that rattled New Yorkers. The New York Times, 5 August 2012. http://www.nytimes.com/2012/08/06/arts/design/the-subway-map-that-rattled-new-yorkers.html?_r=0. Accessed 4 Dec 2013

Stephenson DB (2005) Comment on 'Colour Schemes for Improved data Graphics', by A. Light and P. J. Bartelin. EOS 86(20)
The Map Room (2009) Ordnance survey announces colour-blind mapping. http://www.maproomblog.com/2009/10/ordnance_survey_announce...
World Wide Web Consortium (W3C) (2011) Web content accessibility guidelines. http://www.w3.org/WAI/GL/

Perception and Recall of Landmarks for Personal Navigation in Nature at Night Versus Day

Pyry Kettunen, Katja Putto, Valérie Gyselinck, Christina M. Krause, and L. Tiina Sarjakoski

Introduction

People usually find wayfinding at night in nature challenging due to limited lighting and difficulty of identifying landmarks. Supporting wayfinding in such demanding conditions with landmark-based route directions, maps or navigators would be very useful for many wayfinders (see Rehrl et al. 2010). The present study empirically identifies landmarks that people easily perceive in typical night conditions in nature and that could thus be prioritised for giving real-time route directions in these conditions. We also study the recall of landmarks and compare the results of perception and recall in the day and night conditions in order to identify the need of adaptation in route directions between the times of day.

We begin with the background of our research and review previous research related to limited lighting conditions and landmarks (section "Introduction"). Next, we describe the experiments performed in order to study the landmarks in nature during the day and at night, and present the conducted analysis (section "Methods"). We briefly discuss the results (section "Results") and then move to general discussion (section "Discussion") where we present a synthesis of the

P. Kettunen (✉) • L.T. Sarjakoski
Department of Geoinformatics and Cartography, Finnish Geodetic Institute, PO Box 15, 02431 Masala, Finland
e-mail: Pyry.Kettunen@fgi.fi; Tiina.Sarjakoski@fgi.fi

K. Putto • C.M. Krause
Cognitive Science, Institute of Behavioural Sciences, University of Helsinki, PO Box 9, 00014 Helsinki, Finland
e-mail: irvankoski@gmail.com; christina.krause@helsinki.fi

V. Gyselinck
Laboratoire Mémoire et Cognition EAU 2, Institut de Psychologie, Université Paris Descartes, 71 Avenue E. Vaillant, 92100 Boulogne-Billancourt, France
e-mail: valerie.gyselinck@parisdescartes.fr

© Springer International Publishing Switzerland 2015
J. Brus et al. (eds.), *Modern Trends in Cartography*, Lecture Notes in Geoinformation and Cartography, DOI 10.1007/978-3-319-07926-4_22

knowledge gained and address the limitations of the study. Finally, we state our conclusions and suggest future directions of research (section "Conclusions and Future Work").

Background and Motivation

People inherently employ physical features or objects in the environment in order to structure in their minds the routes that they move on. These objects or *landmarks* constitute a fundamental basis for *cognitive maps* encoded and processed in memory (Tolman 1948). Landmarks are used to describe the environment or routes to others, to analyse the properties of the environment, to plan routes and to navigate along routes in the environment (Presson and Montello 1988). People form landmark ontologies that are used for thinking spatially and for creating external spatial representations, such as maps and navigation applications (Smith and Mark 2001). The landmark ontologies in the spatial thinking vary according to the application domain, the aim of the task and the conditions of use (Winter et al. 2005; Snowdon and Kray 2009; Kettunen et al. 2013). In order to understand and support such multifaceted spatial perception and memory for navigation, the landmarks must be studied in different kinds of scenarios.

In spatial cognition research, the term "landmark" has many meanings. Lynch (1960) found landmarks to be one of the basic elements that people utilise for spatially perceiving a city environment, defining them as particular external reference points for wayfinders. Since then, the term landmark has been commonly used to refer to a particularly prominent feature in the environment. However, while studying the characteristics of landmarks, researchers have often adopted the broader meaning of a landmark as any feature in the environment to which spatial thinking refers (e.g., Presson and Montello 1988; Denis 1997; Brosset et al. 2008; Caduff and Timpf 2008; Rehrl et al. 2009). We employ this broader meaning in the present study because our motivation of wayfinding support requires the consideration of not only the most prominent global landmarks but also local landmarks that are often less prominent in nature.

Most of the empirical landmark research conducted so far has been restricted to urban environments and daytime conditions (e.g., Denis 1997; Rehrl et al. 2009). Nature sets particular challenges for wayfinding due to the difficulties in estimating travelled distances because of landmarks that are easily confused due to their resemblance (e.g., Cohen et al. 1978; Okabe et al. 1986). The changing vegetation and conditions also challenge a wayfinder, as shown in studies on novice nature hikers (Kaplan 1976), orienteers (Omodei and McLennan 1994), rescue cases (Heth and Cornell 1998) and experienced, but lost wayfinders (Whitaker and Cuqlock-Knopp 1992; Hill 2013). The night time makes the wayfinding challenges even more apparent and sets further difficulties mostly related to visibility in terrain (Kumagai and Tack 2005). In the present article, we investigate the weakly studied role of lighting in the perception and recall of landmarks in nature.

To understand and technically support wayfinding during all times of day is important for round-the-clock activities that require active navigation in such domains as the rescue services, police and army. Scientific research on the perception and navigation between lighting conditions has been rare but everyday experience shows that changes are drastical. People see different kinds of landmarks and apply different wayfinding strategies between day and night (Winter et al. 2005; Kumagai and Tack 2005). Geospatial applications already exist in which map colours adapt to night lighting. However, maps and other geospatial applications that provide landmark ontologies do not change according to the change in the lighting conditions. The motivation of our study is to address these changes in the ontologies used during the day and at night. This can later help to develop adaptation in geospatial applications, for example a terrain navigator that could more effectively support people's wayfinding in the varying lighting conditions by providing the user with easily perceptible landmarks related to navigation decisions.

Previous Studies on Landmarks and Wayfinding at Night

There have been only a few studies that have investigated landmarks in real environments under varying lighting conditions, particularly at night. The investigations that have been carried out in nature are even fewer. We thus base our present literature review mainly on landmark studies that consider experiments in virtual environments and daytime conditions.

Kumagai and Tack (2005) conducted a wayfinding experiment at night in nature with soldiers using night vision goggles. Based on performance time, traversed distance and direction estimations, wayfinding proved to be significantly more challenging at night compared to the day. The experiment did not guide participants to rely on landmarks for navigating but instead the participants were asked to detect enemy targets in the woods, a task which resembles a visual landmark search. The detection of targets was significantly weaker under night conditions since night vision goggles provided only a low contrast view at close distances.

Gauthier et al. (2008) also focused on night vision goggles and conducted their landmark search experiment indoors in a small artificial maze where they could set the lighting level similar to half moonlight for the goggle group. The control group participated without night vision goggles in full lighting and performed significantly better in the search, direction estimation and map drawing task. This indicated that night vision goggles affected negatively both wayfinding performance and the acquisition of spatial knowledge. The decrease in spatio-cognitive performance while using night vision goggles suggests that similar restricted vistas, such as while using a headlamp at night, are also likely to bring lower performance.

Winter et al. (2005) showed their participants panoramic images of city intersections photographed during the day or at night and asked them to score the prominence of facades in the images. The scores resulted in significant differences between day and night groups, which indicated that participants would use the facades as landmarks differently for day than for night conditions. Winter et al. (2005) also found that the participants ranked the criteria for the prominence of facades differently between day and night conditions.

We know from experience that day and night appear differently when navigating in relation to local and global landmarks. Darkness causes distant global landmarks largely to disappear while it emphasises local landmarks at close distances as well as all illuminated features. Presumably, this has an important effect on wayfinding at night, because local and global landmarks play divergent roles when navigating on the route. Local landmarks support route actions in vista space, whereas global landmarks support the conceptualisation of environmental space and construction of a cognitive map (Steck and Mallot 2000). Steck and Mallot (2000) created day- (local and global landmarks), night- (local landmarks) and dawn-like (global landmarks) conditions in a virtual street environment in which they investigated human navigation strategies. The participants made turning decisions only slightly worse in the night-like and dawn-like conditions, when only one type of landmark was present, compared to day-like conditions. We hypothesise that in reality, both day and night environments provide people with both local and global landmarks, but these may be different for the two lighting conditions. A similar change of navigation strategies between environments may be necessary, as observed in the experiment of Steck and Mallot (2000): those participants who relied on only one landmark type were readily able to start using the other landmark type if the preferred type was not available.

The lack of ambient light and long-distance visibility makes night navigators highly dependent on simultaneously available spatial information in personal memory or technical navigation equipment, such as maps or navigators. Waller et al. (1998) showed that blindfolded participants traversed an indoor maze significantly faster if they had a priori spatial knowledge from a map or virtual environment than those without prior knowledge. In addition, repetition did not make the participants without prior spatial knowledge advance to the level of the participants who had studied the maze initially. This implies that blindfolded navigation only gave access to information on the closest spatial features and not even to the extent of vista space. Presumably, the similar difficulty of constructing complete spatial knowledge for unfamiliar environments also applies to limited light conditions even when the perceptually accessible environment is larger.

Low-lighted night environments set challenges not only for wayfinding but also for directing locomotion. Adams and Beaton (2000) showed that people become significantly slower in approaching stairs and sharp turns in an urban environment at night and at twilight than during the day. Nature as a locomotion environment is full of obstacles of diverse sizes, which presumably slows down locomotion in a similar manner at night. The insufficient support for visual perception presumably also causes different landmarks to be observed at night than in day since the perceptual salience (Caduff and Timpf 2008) of the landmarks changes. Cognitive and contextual salience (Caduff and Timpf 2008) certainly also play their roles at night, but these may not change much compared to the daytime.

Some studies have addressed useful types of landmarks in nature during the day. Whitaker and Cuqlock-Knopp (1992) interviewed orienteers and military scouts for their particular memories of navigation experiences in their personal history and analysed the named landmarks. Man-made cues were mentioned the most frequently due to their particularity in the environment, then elevations as marked by contours and next, water and vegetation landmarks. Brosset et al. (2008) found orienteers in nature refer more often to linear features than in urban settings. Rehrl and Leitinger (2008) observed that landform-related landmarks dominate the navigation expressions used when ski touring. Snowdon and Kray's (2009) questionnaire revealed that people consider peaks and water courses as the most typical landmarks in nature, with woods, rocks and lakes being less important.

Montello et al. (1994) as well as Pick et al. (1995) observed the reading of elevation contours by experienced map users in hilly terrains and found that the users often relied on hills and large valleys (Pick et al. 1995) as well as in flat areas (Montello et al. 1994) that were easily distinguishable on the map. Montello et al. (1994) also investigated features recalled from landscape photographs and reported terrain and vegetation features as clearly the most referred to. In contrast, atmospheric, geological or other features were rare in the collected sketch maps and protocols.

Sarjakoski et al. (2012) and Kettunen et al. (2013) studied the differences in human landmark use in nature between seasons. The participants walked through a route in a national park while thinking aloud about the prominent features around. They perceived structure and passage landmarks most readily, followed by trees, waters, land cover, rocks, signs and landforms (Sarjakoski et al. 2012). Overall, it must be noted that the landmarks highlighted in all the cited studies above reflect to some extent the types of terrain in the experiments.

The aim of this study is to achieve new knowledge of the role of lighting in the perception and recall of landmarks in nature. The final goal of our research is to gather information about the need of adaptation for the employed landmark sets in navigation applications according to the seasonal and time of day conditions. We hypothesise that ontologies used for describing landmarks in nature differ from those in urban environments and that the importance of different landmark categories varies with the time of day conditions.

Methods

We studied the effect of night on the perception and recall of landmarks in nature trail experiments in which we brought groups of participants to hike in a forested national park area both in day and night. In this section, we describe the set-up and analysis of the experiments.

Experimental Set-Up

Participants

Our experiment included 22 participants (10 men and 12 women) who were evenly distributed into the day (5 men and 6 women) and night (5 men and 6 women) condition groups. The participants were 19–68 years old (median 42 years). None of them reported that they had walked the nature trail used in the experiment before and seven reported to have previously visited the area. During the experiments, we noted no prior spatial knowledge, such as recognition of places or scenery, that would have caused bias in the results. The participants were rewarded for their participation with travel costs and two recreation tickets.

In the background questionnaire, we asked if the participants were bilingual (McLeay 2003) and how often they used to visit nature and utilise common types of maps (range of 0–4: never, less often, monthly, weekly, daily). Only one of the participants was bilingual, so no bias by bilingualism is assumed to occur in the results. The participants averaged monthly nature visits (mean 2.2) and used maps on a weekly to monthly basis (mean of maximums of map type use values by a participant, 2.7). They had used city maps most (mean 2.6), followed by road maps (1.8), terrain maps (1.4) and, much less frequently, orienteering maps (0.7). Statistical tests did not highlight differences between the day and night groups in these measures ($W < 79.5$, $p > 0.19$ in the two-tailed Wilcoxon rank sum test).

The participants filled in the Santa Barbara Sense-of-Direction Scale (SBSOD) questionnaire, which is a self-report measure designed for assessing spatial abilities in the environmental scale (Hegarty et al. 2002). The SBSOD form was translated as part of the study. Based on this measure, the day (median score 70, mean 66.18) and night groups (median score 60, mean 67.73) had no difference in the spatio-cognitive abilities ($W = 56$, $p = 0.79$ in the two-tailed Wilcoxon rank sum test).

Environment

The route of the experiment followed a marked nature trail on footpaths and outdoor tracks that go around a low brook valley in woods. In this article, we call the environment "nature" because it is dominated by wild natural growth, such as spruce and birch trees, with only some roads and constructions along the route. There was a significant difference between the two routes. The first route followed a lakeshore, crossed a road and a river along a dam and contained no steep slopes. The second route followed an outdoor track and ran over a forested hill with considerably steep slopes and cliffs. The terrain conditions were considerably wet during the whole experiment because the preceding summer had been very rainy, and the footpaths were muddy and slippery.

Procedure

Test sessions began with participants filling in the consent form, background questionnaire and the SBSOD form. Next, the experimenter instructed the participants on the thinking-aloud method (Ericsson and Simon 1998) and gave them the first of the two route traversal assignments on a written form: "Walk a route with the experimenter and memorise the route so that you are able to *walk through it again without guidance*. The experimenter walks after you and guides when necessary." The experimenter asked the participants to think aloud their observations along the route while performing the task. Before the actual task, the participants practised thinking aloud while walking to the beginning of the route (150 m, 2–3 min). The first route traversal task ran on a 650 m long route and lasted for 11–19 min, after which the experimenter interrupted the task and gave another written assignment, modified from the previous one: ". . . memorise the route so that you are able to *describe it to another person who is to walk through the same route. . .*" The second part of the route was also 650 m long and took 8–16 min. After the second part, the experimenter asked participants to tell whether they found any difference between the two tasks during the traversal. We gave the two different task assignments for the thinking-aloud tasks in order to investigate if memorising for oneself or to another person would change the manner of thinking aloud. We made all the participants complete the tasks in the same order so that they focused similarly on the same parts of the routes and the contents of the collected data were comparable. Finally, the participants walked back to the starting point of the session guided by the experimenter, still memorising but without thinking aloud (150 m, 3–5 min).

At night, the participants wore a 900-lumen LED headlamp that provides a bright and targeted view up to several dozens of metres in an open area. This kind of lighting condition is typical to night-time activities in nature, in which similar light sources are typically used. In both day and night, the participants carried an audio recorder for saving the thinking-aloud recordings. The experimenters recorded video of the participants while walking after them.

After the route walkthrough, the participants had a break for 15 min in order to ensure that the short-term memory would not affect the recall tasks. The next task was to draw a sketch map on a blank paper according to a written assignment: "Draw the route you walked and explain your markings thinking aloud". We set no time restrictions for drawing, and it took 2–22 min for the participants to complete. We recorded the drawing both in audio and video.

Analysis

By landmarks, we mean all permanent and distinguishable features in the environment that participants noted during the tasks (as in, e.g., Denis 1997; Rehrl et al. 2009). We counted clearly determinate mentions of vegetation as landmarks, such as "spruce trees" or "snag", but ignored indeterminately introduced mentions of unbounded vegetation such as "grove" or "moss". We also ignored temporary features, which would be unreliable to use in route guidance at a later date due to common but occasional natural changes. For example, with "mud", participants described the underfoot condition of the path due to an exceptionally rainy summer.

Thinking Aloud During Route Traversal

In order to analyse the thinking aloud during the route traversal, we transcribed the recordings and applied a previously developed natural language processing (NLP) analysis on the transcripts (Sarjakoski et al. 2013). The NLP analysis was carried out by a team of two researchers, who made joint decisions on the selection and classification of landmarks, proceeding as follows:

1. Transformation of the inflected words into the basic form (Helsinki Finite State Transducer: Lindén et al. 2009);
2. Collection of the landmark words from the list of all words (Python scripts, Natural Language Toolkit NLTK: Bird et al. 2009);
3. Checking that the landmark words were really used for denoting landmarks in the transcripts (string searches in the transcript files);
4. Gathering of the landmark word synonyms together into landmark concepts,
5. Identifying bigrams that the participants used as landmark concepts, such as "fallen tree" (the two words preceding and the two words following the landmark words in the transcripts; Python scripts, NLTK);
6. Grouping of the landmark concepts under the previously defined landmark groups that fitted well in the present case: "Structures", "Passages", "Trees and parts of trees", "Waters", "Land cover", "Rocks", "Signs" and "Landforms" (Kettunen et al. 2013); and
7. Counts of the landmark concepts in the transcripts (Python scripts, NLTK).

We based the comparison of the two route-perception tasks (memorising for oneself and for another person) on a qualitative comparison of the answers that the participants gave when we asked about the differences between the tasks. The question was added in the experiment just after the first sessions, which is why there are no answers from three participants in the night group who were not asked the question.

Sketch Map Drawing

We analysed the sketch maps similarly to the analysis of our previous experiment (Kettunen et al. 2013): drawn and written features were classified according to the categorisation of the thinking-aloud part of the experiment. Two individual researchers completed the classification, using both the finished sketches and a

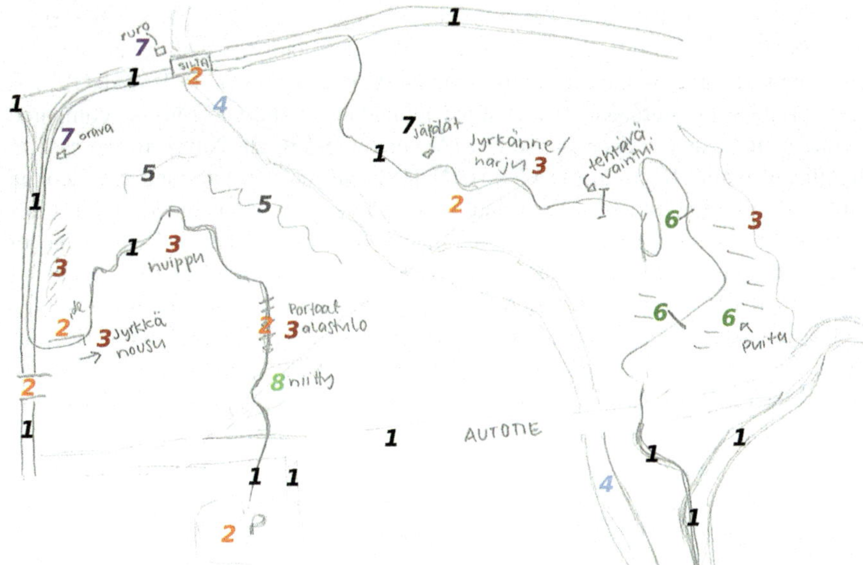

Fig. 1 An example of counting landmarks on a sketch map: (*1*) passage, (*2*) structure, (*3*) landform, (*4*) water landmark, (*5*) rock landmark, (*6*) tree or a part of tree, (*7*) sign and (*8*) land cover

video recording of the drawing participants. In case of the few differences that occurred in the classifications, the classifications were synthesised through the group work. Every marked feature was regarded as a landmark and separate sections of continuous landmarks were treated as individual landmarks in the "Passages", "Waters", "Land cover" and "Landforms" landmark groups (Fig. 1). We did not count landmarks that were mentioned during the thinking aloud while drawing if they were not also actually drawn on the sketch map.

Statistical Calculations

In order to identify landmark concepts that the participants used in significantly different frequencies between conditions, we ran two-tailed Wilcoxon rank-sum tests for each landmark using the SciPy Python package (Jones et al. 2001). In all the statistical calculations, we used significance level $\alpha = 0.05$.

In order to detect differences in the use frequencies of landmark groups between different conditions, we ran permutational multivariate ANOVA (PERMANOVA; Anderson 2001) using the R software (R Core Team 2013) to test the main effects of time of day and task as well as their interaction effect (50,000 permutations). We chose the non-parametric PERMANOVA because our samples are small and the normality of landmark frequency distributions is doubtful: Shapiro-Wilk multivariate normality test showed non-normality for the thinking aloud data from night

($W = 0.68$, $p = 0.001$) and for the sketch map data from day ($W = 0.54$, $p = 0.00003$). The high number of variables in relation to observations also prevented the use of the parametric MANOVA. The use of PERMANOVA is appropriate as the statistical power of the applied test is similar or higher compared to that of the exact version of the test (Anderson and Braak 2003). In the case of significant main effects in the PERMANOVA, we ran non-parametric Wilcoxon rank-sum tests for comparing day and night and the Wilcoxon signed rank test for comparing thinking aloud and sketch maps. We used relative frequencies of landmark groups for calculations (the count of a group divided by the total count of landmarks by a participant), in order to prevent verbosity affecting the results as well as to enable comparability between the two tasks. We tested for main effects of the background variables of the participants (section "Participants") one by one against the measured relative landmark group frequencies using the PERMANOVA (50,000 permutations within the tasks).

Results

This section presents the results of the experiments together with direct notes of their causes, whereas the synthesising overall discussion is provided in section "Discussion".

Thinking Aloud During Route Traversal

The participants used on average 2 min 39 s more time (median difference) for walking through the two routes in the night condition than during the day ($W = 32$, $p = 0.02$). The slower locomotion previously found in an urban night environment (Adams and Beaton 2000) also seems to occur in nature because people are more attentive to hazards on the route.

The participants spoke on average more words at night (all words counted; median 1,120; 1st and 3rd quartiles 908 and 1,532) than during the day (921; 676, 1,320). They mentioned 55 different landmark concepts during the day and 56 concepts at night (overall 62). The participants used these landmark concepts on average 105 times during the day (80, 128) and 123 times at night (100.5, 164.5).

Six landmark concepts were used by every participant in the thinking aloud during route traversal: "road", "fallen tree", "hill", "footpath", "bridge" and "boat shore". These were located directly on the route and were clearly visible under both lighting conditions. In the day condition, everyone also used "stairs", which were not necessarily visible at night as they were wooden and worn out. At night, everyone also used "route marking", "river", "outdoor track", "information sign", "spruce trees", "underpass" and "streetlamp". These were mostly features close to the route that were well lit in the headlamp spot.

Table 1 The 15 landmarks that were most used in the day and night conditions

Number of participants[a]	Relative frequency[b] (%)	Day landmark	Rank	Night landmark	Relative frequency[b] (%)	Number of participants[a]
11	13.3	Road	1	Route mark	13.7	11
9	9.1	Route mark	2	Road	11.1	11
9	7.1	River	3	River	5.1	11
11	6.2	Fallen tree	4	Signboard	4.6	10
8	4.8	Info board	5	Fallen tree	4.3	11
11	4.3	Hill	6	Hill	3.9	11
11	3.6	Footpath	7	Outdoor track	3.9	11
10	3.4	Outdoor track	8	Info board	3.4	11
7	3.2	Signboard	9	Spruce trees	3.3	11
8	2.8	Hillside	10	Bare rock area	3.2	9
11	2.7	Boat shore	11	Hillside	3.2	9
10	2.6	Underpass	12	Underpass	2.8	11
10	2.6	Spruce trees	13	Footpath	2.7	11
11	2.6	Bridge	14	Boulder	2.7	9
9	2.6	Water	15	Bridge	2.5	11

[a]The number of participants who mentioned the landmark (out of 11 per condition)
[b]The relative frequency of a landmark compared to the total number of landmarks in the respective time of day condition

The participants used some landmarks only during one of the conditions. During the day, these were "water slide", "graffiti", "leaning tree", "traffic island", "courtyard" and "slope ramp" (3 out of 6 in the "Structures" group). Most of these landmarks were distant or wide and thus impossible to see in the darkness even with the headlamp. At night, "pine", "conifer trees", "tall grass", "bushes", "goat willow", "flat" and "boulder field" (3 out of 7 in the "Land cover" group and 2 in the "Trees or parts of trees"). It seems that vegetation landmarks hit by the spotlight were often mentioned in the night condition, whereas they did not stand out during the day.

The most used landmarks were mainly the same during the day and at night (Table 1). Among the 15 most used landmark concepts, 13 were exactly the same in both conditions (exceptions being "boat shore" and "water" during the day and "bare rock area" and "boulder" at night).

Four individual landmarks showed statistical significance when testing the differences in use frequencies between day and night among the participants: they used "footpath" (median difference 0.9 pps, $p = 0.05$; no difference in number of users) more during the day and "boulder" (1.9 pps, $p = 0.01$; 6 users more), "standing rootstock" (1.2 pps, $p = 0.005$; 6 users more) and "streetlamp" (0.9 pps, $p = 0.008$; 2 users more) more at night. "Footpath" gained more attention during the day, probably due to its greater presence in the field of view, which was restricted at night. "Streetlamp" was noted more frequently at night, often distantly, due to its emission of light. "Boulder" and "standing rootstock" were probably more salient when seen in the spotlight than in daylight.

"Passages" and "Signs" were the two most used landmark groups for both day and night conditions, but in different order (Fig. 2). "Structures" was the third largest group in both conditions, while "Rocks" was the least used in both cases. The ranks of the landmark groups reflect the route and its environment: "Passages" was frequently used due to the route running on passages, and "Signs" were present all along the route to guide walkers. The higher use frequency of "Signs" compared to "Passages" at night was probably caused by the higher visual salience of signs in the spotlight of the headlamp.

The two-way permutational multivariate ANOVA resulted in significant main effects for both the time of day ($F = 2.92, p = 0.03$) and task ($F = 26.99, p < 0.001$). No significant interaction effect occurred ($F = 0.00, p = 0.99$). Consequently, we continued statistical analyses using univariate Wilcoxon tests, in which two landmark groups among eight scored statistical significance for the differences of

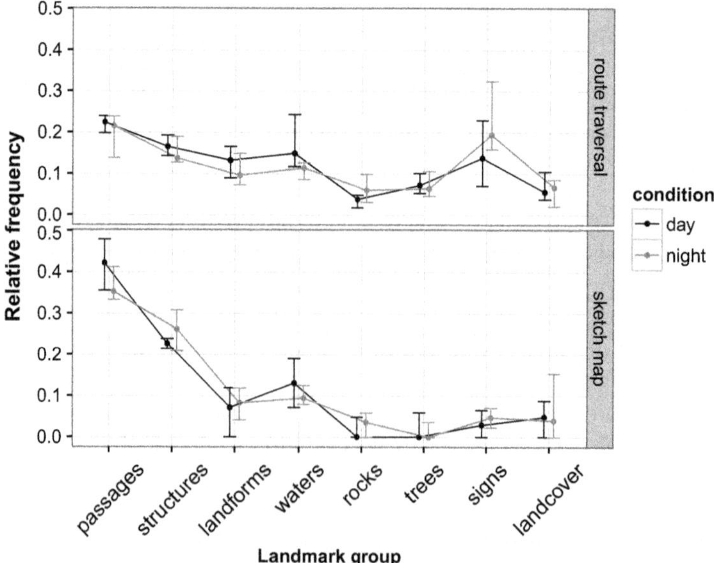

Fig. 2 Comparison of the relative frequencies of the landmark groups between day and night. The *error bars* depict a 95 % confidence interval for percentile bootstrap medians

relative frequencies between day and night. The participants used the "Waters" landmark group 3.4 pps (median difference) more during the day ($W = 91, p = 0.05$) and "Rocks" 2.3 pps more at night ($W = 24.5, p = 0.02$).

The one-way PERMANOVAs for demographic variables showed significant differences for age ($F = 2.43, p = 0.002$) and experience with orienteering maps ($F = 1.40, p = 0.05$). No significant differences were present for gender or on the SBSOD scale and not for the experience of nature, area, or with other map types. Pair-wise regression analyses showed a slightly positive correlation between age and the frequency of water landmarks ($\rho = 0.35$) and a slightly negative correlation with the frequency of sign landmarks ($\rho = -0.34$). Experience with orienteering maps correlated negatively with the frequency of water landmarks ($\rho = -0.40$).

Pair-wise regression analyses between the demographic variables showed that age correlated negatively with experience with orienteering ($\rho = -0.47$) and topographic maps ($\rho = -0.48$) and positively with the frequency of going out in nature ($\rho = 0.49$). These findings suggest that although the younger participants went out in nature less frequently, they used orienteering and topographic maps more often compared to the older participants, resulting in slight differences in the perception of water and sign landmarks between age groups.

With regards to the question about the differences of the task assignments (memorising for oneself and for another person), the day participants were divided (yes 6, no 4, 1 answer lacking) whereas the night participants answered mostly positively (yes 8, no 1, 2 answers lacking). The answers suggest that the dark night environment made the participants more attentive when perceiving the route for guiding another person, and they also noted the change themselves. The darkness possibly facilitated perceiving the route in the environment from someone else's perspective or otherwise made cognitive processing unconstrained and diverse, similarly to the studies of Steidle et al. (2011).

Sketch Map Drawing

The participants drew on average 21 landmark features on the sketch maps during the day (median; 1st and 3rd quartiles 17 and 32) and 24 features at night (22, 27). The rank of the landmark groups was similar in both conditions: "Passages" was the largest group, followed by "Structures" and "Waters" (Fig. 2). "Trees and parts of trees" was the least used landmark group in the sketch maps. Video recordings revealed that the participants often framed the sketch map with route and water landmarks and then added structure and other landmarks.

For sketch maps, there were no statistically significant differences between day and night in the frequencies of any of the landmark groups.

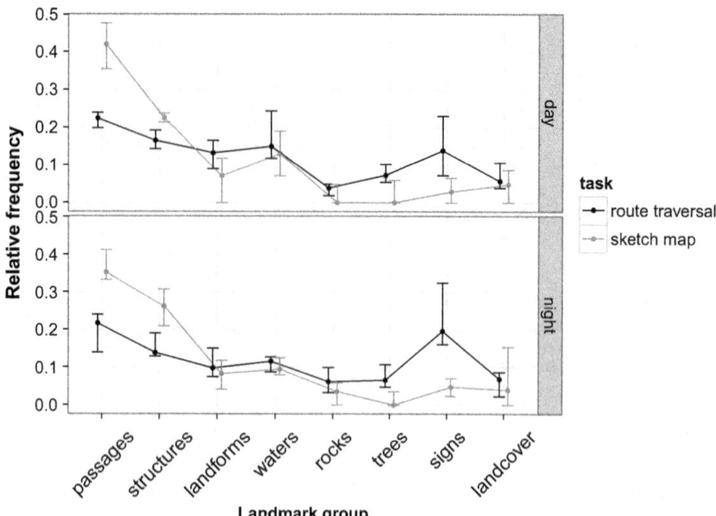

Fig. 3 Comparison of the relative frequencies of the landmark groups between the thinking aloud during route traversal and sketch maps. The *error bars* depict a 95 % confidence interval for percentile bootstrap medians

Comparison Between Tasks

The participants used the landmark groups with differing relative frequencies while thinking aloud during the route traversal and in the sketch maps (Fig. 3). In the day condition, we recorded statistically significant differences for the "Passages" (median difference 19.7 pps; $V = 0$, $p = 0.001$) and "Structures" (6.1 pps; $V = 10$, $p = 0.04$) landmark groups, which were used more in the sketch maps and "Signs" (10.8 pps; $V = 45$, $p = 0.01$), "Trees and parts of trees" (7.2 pps; $V = 65$, $p = 0.002$) and "Landforms" (6.0 pps; $V = 57$, $p = 0.03$), which were used more while thinking aloud during the route traversal. In the night condition, the participants used "Passages" (13.7 pps; $V = 0$, $p = 0.001$) and "Structures" (12.4 pps; $V = 4$, $p = 0.007$) more in the sketch maps at the statistically significant level, and "Signs" (14.7 pps; $V = 66$, $p = 0.001$), "Trees and parts of trees" (6.5 pps; $V = 65$, $p = 0.002$) and "Rocks" (2.5 pps; $V = 52$, $p = 0.01$) more in the thinking aloud during route traversal.

Discussion

The presented study investigated differences in the perception and recall of landmarks along nature routes between day and night. Significant differences were found in the perception of individual landmark types and landmark groups, whereas no differences were present in the recall as measured by sketch maps.

The limited lighting condition at night was characterised by the brightly lit spotlight of the headlamp and weak ambient light originating from the diffusion of lights in the surrounding urban region. The contrast between the spotlight and ambient light was high, causing participants to mostly see features under the spotlight. The results of the study reflected this restricted vista through an increased perception of close features with notable visual salience under the spotlight. On one hand, these features were relatively small and clearly bounded, such as "boulder" and "standing rootstock", which were lighted one by one at night whereas perceived more as groups in daytime. On the other hand, the spotlight highlighted vegetation features at close distance that the spotlight did not penetrate, such as "spruce trees" and "tall grass". These types of features did not necessarily contrast in the more extended ambient light vista by daytime. The significantly higher use frequency of the "Rocks" landmark group in the night than in the condition can thus be understood to be caused by the restricted spotlight vista at night. The same applies vice versa for the "Waters"; the dark colour of the water surface due to the lack of the reflected ambient light made "Waters" invisible at night compared to during the day. "Waters" were also distant from the route. Acoustic salience was only little involved in the study, mainly related to the noises of vehicles.

Although the vistas at night were restricted to the spotlight of the headlamp, the perception of distant and global landmarks was not completely missing, which agrees with our hypothesis. We observed the use of distant lights as orientation landmarks during the route traversal, most importantly lines of streetlamps that efficiently provided the participants with the directions of distant roads and outdoor tracks. The observation was confirmed statistically with the landmark concept "streetlamp" being used significantly more frequently by the night participants compared to the day participants. Even more convincingly, each night participant mentioned "streetlamp" during the route traversal.

In daylight, the participants used another set of distant and global landmarks. No individual landmark or landmark group was highlighted quantitatively, but the experimenters' observations confirmed that when people can see far away, they take the distant features in use as landmarks. The day participants used distant landmarks, such as "water slide" and "traffic island", in the "Structures" landmark group during the route traversal, which did not occur in the night data at all. The high salience of structures as landmarks in nature was confirmed in the experiment (previously found in Sarjakoski et al. 2012). The perception of some spatially extensive features was also notable during the day, most clearly water landmarks that were significantly more used compared to night. In addition, some spatially extensive surface-related landmarks, such as "courtyard" and "slope ramp", were only mentioned in the day condition when they were visible over a wide area.

Surprisingly, the differences in the perception of landmarks between day and night did not transfer to differences in the recall. With regards to the use amounts of landmark groups, sketch maps were drawn similarly in both conditions. A similar lack of difference in sketch maps between conditions occurred also in our previous study considering different seasons (Kettunen et al. 2013). People seem to recall and choose similar kinds of landmarks to draw in sketch maps of routes, independent of the conditions. Naturally, route-like sketch maps come to contain the route as the frame of the map, but there seems to be more in the observed similarity. A likely explanation for this phenomenon is that people apply common ontologies of landmarks while drawing sketch maps. This impression is supported by a qualitative observation from the present study that during the map drawing, several participants thought aloud about many landmark concepts belonging to the "Land cover" and "Landforms" landmark groups but still did not draw these features. Commonly used maps probably play an important role in people's conceptions of landmark sets that should be used when drawing sketch maps. In the present study, the applied landmark ontologies may have been importantly based on the landmarks that are typically presented on topographic maps.

In addition to the preconceptions that the participants seemed to have on what to draw on sketch maps, the differences in the frequencies of landmarks between the thinking aloud during route traversal and sketch maps were partly related to the characteristics of the two media. Most of the participants drew sketch maps based on the course of the route, which caused a surplus of passage landmarks compared to the thinking aloud during route traversal, even if the passage under foot was mentioned often also during the route traversal. On the other hand, salient features belonging to the "Trees and parts of trees" and "Rocks" landmark groups were repeatedly mentioned during the route traversal, which played a role for the significance.

The rest of the significant differences between the tasks most probably account for salience aspects. Structures were probably more noted in the sketch maps due to their visual salience (Caduff and Timpf 2008), for which reason they were readily recalled while drawing. The visual salience potentially explains at least part of the significantly lower degree of use for trees and parts of trees in the sketch maps, also known from our previous study (Kettunen et al. 2013): such common nature features were not effectively recalled even if many distinguishable instances were mentioned during the route traversal. In the case of signs, which were significantly less used in the sketch maps than had been expected, the cognitive salience (Caduff and Timpf 2008) may have played a role: people may have mentally merged the constantly observed sign landmarks as self-evident parts of the route that they drew on the maps as passage landmarks.

The discrepancy between the results obtained using the thinking-aloud protocol and those obtained using sketch maps is a central finding of the study and apparently caused by the nature of the tasks. Thinking aloud is an online task that reflected what the participants perceived and found important to mention. In contrast, the sketch maps reflected memory performance and they emphasise the most memorisable features on which the participants used to construct their spatial representation of the environment and that they would use later to find their way or to describe the route. The result that the memory performance does not differ between night and day conditions, although the online perception measure does, suggests that the absolute importance of different landmark groups is similar for the spatial representations of individuals even though the perceptive input differs significantly.

The present study applied a recently introduced method for analysing thinking aloud protocols using natural language processing (Sarjakoski et al. 2013). Its results reflect the distribution of selected landmark terms in the collected thinking-aloud protocols. The thinking-aloud method has been extensively used in probing human thinking processes (Ericsson and Simon 1980), but doubts have been presented on its validity in concurrent use (e.g., Smagorinsky 1998). Thinking aloud concurrently easily affects task performance if a participant strays to free-flow speech and forgets the task at hand. However, properly instructed, trained and undisrupted, thinking aloud has been shown to provide reliable data on the thoughts of test participants (Ericsson and Simon 1998; Boren and Ramey 2000). We took these prerequisites into account when directing the thinking-aloud tasks and consider the collected data to be reliable with regards to people's perception of landmarks along the route. Another shortcoming of the concurrent thinking-aloud method is the probable incompleteness of the data: participants may be unable to verbalise all of their thoughts and actions related to the task, particularly in non-verbal practices, such as spatial thinking (e.g., Whitaker and Cuqlock-Knopp 1992). In the present study, we instructed participants to memorise the route and to think-aloud about their perception of it, which directed the participants' concentration to the surrounding features along the route and made them verbalise at least those features that they saw as important for following the route. Consequently, the results do highlight those landmarks that are potentially effective in route directions, but they should be regarded as an explorative rather than a comprehensive set of all the prominent landmarks.

Technical wayfinding support by geospatial applications is most effective when given in real time, landmark by landmark, during the navigation (Rehrl et al. 2010). Therefore, the landmarks in wayfinding maps and navigators should be selected based on the real perception in the environment. The found loss of perceptive differences between conditions in the recall phase highlights the need for perception-based directions in real-time navigation guidance—human conception and memory do not always inherently focus on the most perceptible landmarks. The results from studies like ours in the present paper provide sets of empirically verified, perceptionally prominent landmarks to be employed in geospatial navigation applications.

Conclusions and Future Work

The present study addressed the open question of the perception and recall of landmarks in nature under limited light conditions at night. We approached the issue by means of a thinking-aloud study on a nature trail with participant groups in the day and night conditions and used previously developed natural language processing and sketch map methods for the analysis of the

(continued)

experiments (Kettunen et al. 2013). The study concentrated on the perception of landmarks along a route using a thinking-aloud task, and on the recall of landmarks afterwards using a sketch map task. According to the results, the perceived landmark types differed between day and night due to the absence of ambient light and the visual focus on the spotlight of the headlamp at night. Similar factors have been previously observed to affect the wayfinding with night vision goggles (Kumagai and Tack 2005; Gauthier et al. 2008). However, the recall of landmarks did not differ according to the sketch maps: the drawn landmark groups remained similarly frequent between day and night. The participants' prior ontologies about the important landmarks on maps appeared to influence the sketch map drawing, as many well-recalled landmarks were observed to be left consciously undrawn. Significant differences in the use amounts of landmark groups between the tasks further confirmed these conclusions.

The found feature-specific particularities in the perception of landmarks at night in nature can be applied to the development of adaptation of real-time navigation applications, such as maps or navigators. Such adaptation to time of day would better support the navigation in nature not only during the day but also at night, which is essential for many round-the-clock activities. In particular, water landmarks should be avoided in nightly route directions, whereas salient point-like landmarks, such as rocks, can be more helpful at night than during the day. In addition, the study highlighted illuminated features to be particularly perceptible as distant landmarks at night. However, their use for route directions must be carefully considered concerning their impermanence and homogeneity.

The analysis presented above may be further extended in the future, particularly regarding the study of landmark recall according to the sketch maps, which showed no differences between day and night. Further study of the sketch maps in regard to the omitted landmarks, spatial correctness and individual landmark types might reveal differences between the spatial recall at day and at night. Further synthesising investigation would also be beneficial, considering our previous study between seasons (Kettunen et al. 2013) together with the present study. In general, confirming studies on the use of landmarks in nature at night would be important. The use of more exact behavioural methods than the ones in the present study, such as mobile eye-tracking, could be considered. Moreover, there is still a lack of comprehensive night navigation studies in real urban environments.

Acknowledgements This survey is part of the research project "Ubiquitous Spatial Communication" (UbiMap) in 2009–2012. The UbiMap project was funded by the Academy of Finland (MOTIVE programme) and was carried out in cooperation with the Finnish Geodetic Institute (FGI), Department of Geoinformatics and Cartography and the University of Helsinki, Cognitive Science. The third-party funding sources did not affect the design or conduct of the study. The authors want to thank all the participants for taking part in the study.

References

Adams CJ, Beaton RJ (2000) An investigation of navigation processes in human locomotor behavior. Proc Hum Factors Ergon Soc Annual Meet 44(26):233–236. doi:10.1177/154193120004402625

Anderson MJ (2001) A new method for non-parametric multivariate analysis of variance. Austral Ecol 26(1):32–46. doi:10.1111/j.1442-9993.2001.01070.pp.x

Anderson M, Braak CT (2003) Permutation tests for multi-factorial analysis of variance. J Stat Comput Simul 73(2):85–113. doi:10.1080/00949650215733

Bird S, Edward L, Klein E (2009) Natural language processing with Python. O'Reilly Media, Sebastopol. http://nltk.org/book/. Accessed 26 Mar 2014

Boren MT, Ramey J (2000) Thinking aloud: reconciling theory and practice. IEEE Trans Prof Commun 43(3):261–278. doi:10.1109/47.867942

Brosset D, Claramunt C, Saux E (2008) Wayfinding in natural and urban environments: a comparative study. Cartographica 43(1):21–30. doi:10.3138/carto.43.1.21

Caduff D, Timpf S (2008) On the assessment of landmark salience for human navigation. Cogn Process 9(4):249–267. doi:10.1007/s10339-007-0199-2

Cohen R, Baldwin LM, Sherman RC (1978) Cognitive maps of a naturalistic setting. Child Dev 49(4):1216–1218. doi:10.2307/1128763

Denis M (1997) The description of routes: a cognitive approach to the production of spatial discourse. Cahiers de psychologie cognitive 16(4):409–458. doi:10.1037/0033-295X.87.3.215

Ericsson KA, Simon HA (1980) Verbal reports as data. Psychol Rev 87(3):215–251. doi:10.1037/0033-295X.87.3.215

Ericsson KA, Simon HA (1998) How to study thinking in everyday life: contrasting think-aloud protocols with descriptions and explanations of thinking. Mind Cult Act 5(3):178–186. doi:10.1207/s15327884mca0503_3

Gauthier MS, Parush A, Macuda T, Tang D, Craig G, Jennings S (2008) The impact of night vision goggles on way-finding performance and the acquisition of spatial knowledge. Hum Factors 50(2):311–321. doi:10.1518/001872008X288295

Hegarty M, Richardson AE, Montello DR, Lovelace K, Subbiah I (2002) Development of a self-report measure of environmental spatial ability. Intelligence 30(5):425–447. doi:10.1016/S0160-2896(02)00116-2

Heth C, Cornell EH (1998) Characteristics of travel by persons lost in Albertan wilderness areas. J Environ Psychol 18(3):223–235. doi:10.1006/jevp.1998.0093

Hill KA (2013) Wayfinding and spatial reorientation by Nova Scotia deer hunters. Environ Behav 45(2):267–282. doi:10.1177/0013916511420421

Jones E, Oliphant T, Peterson P et al (2001) SciPy: open source scientific tools for Python. http://www.scipy.org. Accessed 26 Mar 2014

Kaplan R (1976) Way-finding in the natural environment (chap 3). Dowden, Hutchinson & Ross, Stroudsburg, pp 46–57

Kettunen P, Irvankoski K, Krause CM, Sarjakoski LT (2013) Landmarks in nature to support wayfinding: effects of seasons and experimental methods. Cogn Process 14(3):245–253. doi:10.1007/s10339-013-0538-4

Kumagai JK, Tack DW (2005) Alternative display modalities in support of wayfinding. Technical report, Defence Research and Development Canada, Toronto. http://pubs.drdc-rddc.gc.ca/PDFS/unc48/p524736.pdf. Accessed 26 Mar 2014

Lindén K, Silfverberg M, Pirinen T (2009) HFST tools for morphology—an efficient open-source package for construction of morphological analyzers. In: Mahlow C, Piotrowski M (eds) State of the art in computational morphology, communications in computer and information science, vol 41. Springer, Berlin, pp 28–47. doi:10.1007/978-3-642-04131-0_3

Lynch K (1960) The image of the city. MIT, Boston

McLeay H (2003) The relationship between bilingualism and the performance of spatial tasks. Int J Biling Educ Biling 6(6):423–438. doi:10.1080/13670050308667795

Montello DR, Sullivan CN, Pick HLJ (1994) Recall memory for topographic maps and natural terrain: effects of experience and task performance. Cartographica 31(3):18–36. doi:10.3138/W806-5127-7W41-12H8

Okabe A, Aoki K, Hamamoto W (1986) Distance and direction judgment in a large-scale natural environment: effects of a slope and winding trail. Environ Behav 18(6):755–772. doi:10.1177/0013916586186004

Omodei MM, McLennan J (1994) Studying complex decision making in natural settings: using a head-mounted video camera to study competitive orienteering. Percept Mot Skills 79 (3f):1411–1425. doi:10.2466/pms.1994.79.3f.1411

Pick HL, Heinrichs MR, Montello DR, Smith K, Sullivan CN, Thompson WB (1995) Topographic map reading. In: Hancock PA, Flach J, Caird J, Vicente K (eds) Local applications of the ecological approach to human-machine systems. Lawrence Erlbaum Associates, Hillsdale, pp 255–284

Presson CC, Montello DR (1988) Points of reference in spatial cognition: stalking the elusive landmark. Br J Dev Psychol 6(4):378–381. doi:10.1111/j.2044-835X.1988.tb01113.x

R Core Team (2013) R: a language and environment for statistical computing. R Foundation for Statistical Computing, Vienna. http://www.R-project.org. Accessed 26 Mar 2014

Rehrl K, Leitinger S (2008) The SemWay skitouring experiment—how skitourers find their way. In: Gartner G, Leitinger S, Rehrl K (eds) Electronic proceedings of the 5th symposium on location based services, November, Salzburg, pp 64–68

Rehrl K, Leitinger S, Gartner G, Ortag F (2009) An analysis of direction and motion concepts in verbal descriptions of route choices. In: Hornsby KS, Claramunt C, Denis M, Ligozat G (eds) COSIT 2009, LNCS, vol 5756. Springer, Berlin, pp 471–488. doi:10.1007/978-3-642-03832-7_29

Rehrl K, Häusler E, Leitinger S (2010) Comparing the effectiveness of GPS-enhanced voice guidance for pedestrians with metric- and landmark-based instruction sets. In: Fabrikant S, Reichenbacher T, van Kreveld M, Schlieder C (eds) Geographic information science, LNCS, vol 6292. Springer, Berlin, pp 189–203. doi:10.1007/978-3-642-15300-6_14

Sarjakoski LT, Kettunen P, Flink HM, Laakso M, Rönneberg M, Sarjakoski T (2012) Analysis of verbal route descriptions and landmarks for hiking. Pers and Ubiquitous Comput (Special Issue on Extreme Navigation) 16(8):1001–1011. doi:10.1007/s00779-011-0460-7

Sarjakoski LT, Kettunen P, Halkosaari HM, Laakso M, Rönneberg M, Stigmar H, Sarjakoski T (2013) Landmarks and a hiking ontology to support wayfinding in a national park during different seasons. In: Raubal M, Frank AU, Mark DM (eds) Cognitive and linguistic aspects of geographic space, LNGC. Springer, Berlin, pp 99–119. doi:10.1007/978-3-642-34359-9_6

Smagorinsky P (1998) Thinking and speech and protocol analysis. Mind Cult Act 5(3):157–177. doi:10.1207/s15327884mca0503_2

Smith B, Mark DM (2001) Geographical categories: an ontological investigation. Int J Geogr Inf Syst 15(7):591–612. doi:10.1080/13658810110061199

Snowdon C, Kray C (2009) Exploring the use of landmarks for mobile navigation support in natural environments. In: Proceedings of the 11th international conference on human-computer interaction with mobile devices and services, 15–18 September, Bonn. doi:10.1145/1613858.1613875 (article no. 13)

Steck SD, Mallot HA (2000) The role of global and local landmarks in virtual environment navigation. Presence 9(1):69–83. doi:10.1162/105474600566628

Steidle A, Werth L, Hanke EV (2011) You can't see much in the dark. Darkness affects construal level and psychological distance. Soc Psychol 42:174–184. doi:10.1027/1864-9335/a000061

Tolman EA (1948) Cognitive maps in rats and men. Psychol Rev 55(4):189–208. doi:10.1037/h0061626

Waller D, Hunt E, Knapp D (1998) The transfer of spatial knowledge in virtual environment training. Presence: Teleoperators & Virtual Environments 7(2):129–143. doi:10.1162/105474698565631

Whitaker LA, Cuqlock-Knopp G (1992) Navigation in off-road environments: orienteering interviews. Sci J Orienteering 8(2):55–71. http://orienteering.org/wp-content/uploads/2010/12/Scientific-Journal-of-Orienteering-1992-Vol.82.pdf. Accessed 26 Mar 2014
Winter S, Raubal M, Nothegger C (2005) Focalizing measures of salience for wayfinding. In: Meng L, Reichenbacher T, Zipf A (eds) Map-based mobile services. Springer, Berlin, pp 125–139. doi:10.1007/3-540-26982-7_9

Exploring the Influence of Colour Distance and Legend Position on Choropleth Maps Readability

Alžběta Brychtová

Introduction

Main and the most important requirement which must a map fulfilled regardless its audience or purpose is its good readability, which allows quick and accurate acquisition of information. Stigmar and Harrie (2011) list factors which negatively influence the map readability. Among those which can be attributed to the issue of cartographic design belongs the insufficient difference of map symbols. In order to make map symbols easily discriminable and thus prevent their misleading interpretation, it is necessary to maintain a sufficient visual distance among them (Bjorke 1996). The graphical representation of maps can be significantly improved by colours which are clearly distinguishable by human visual perception (Steinrücken and Plümer 2013). Using and reproducing colours in a way that is close to the human perception requires their description in a perceptually uniform color space (CIE 2012). Perceptually uniform means that a change of the same amount in a color value should produce a change of about the same visual importance (Slocum et al. 2008).

Color, among other visual variables, operates in a preponderant way in readability problems (Chesneau 2007; Stigmar 2010). As it was already mentioned, it is important that the difference between two colors, as visual variables representing two different qualities or quantities of map objects, is sufficiently high to allow easy identification of symbols. The most used method of color distance calculation is determining the linear distance in the CIELAB color space called CIE76 (the colour distance metric is called ΔE^*_{ab}) and method called CIEDE2000 (also called and used in the text as ΔE_{00}), which is a refinement of the first mentioned method (Werman 2012). The latter mentioned method was applied in this presented

A. Brychtová (✉)
Department of Geoinformatics, Palacký University, Olomouc, Czech Republic
e-mail: alzbeta.brychtova@gmail.com

© Springer International Publishing Switzerland 2015
J. Brus et al. (eds.), *Modern Trends in Cartography*, Lecture Notes in
Geoinformation and Cartography, DOI 10.1007/978-3-319-07926-4_23

research. Specification of computational procedures and formulas of both methods could be found for example in Kuehni (2003).

This article focuses on the issue of choropleth maps readability. It aims to investigate how is the efficient visual distance between classes in the colour scheme, which allows map users its easy interpretation. Many scientific articles concerning design of choropleth maps have been written on topics such as colour perception, rate estimation, optimal classification techniques, and the impact of enumeration unit geography (Harrower 2007). Contemporary, much attention is paid to customization of colour schemes for map users with various color vision deficiencies (e.g. Kröger et al. 2013; Culp 2012; Jenny and Kelso 2007; Olson and Brewer 1997). Another studies are focused on evaluating the efficiency of sequential color schemes for various purposes (Schiewe and Weninger 2013; Chesneau 2007; Mersey 1990; Gilmartin and Shelton 1989; Kimerling 1985, etc.). Harrower (2007) was focused on comparing the efficiency of classed versus unclassed choropleth maps. Gilmartin and Shelton (1989) tested how many classes of choropleth map are optimal. They also found that humans perform better while reading choropleth maps with grey lightness continuum, than green or purple. The great contribution has brought a research activity of C. Brewer, who was developing colour schemes for visualization both qualitative and quantitative data (e.g. Brewer et al. 2003). Quantification of the appropriate colour distance between classes of choropleth maps is still apparently unexplored part of cartography.

Eventhough colour (and other visual variables) related issues belongs to those frequently discussed, but also more subtle effects may affect the final map usability. It is believed that an overall map design could distinctly involve the pleasantness of map reading. Positioning major design elements should results in a visually balanced map importance (Slocum et al. 2008). This issue was preliminary evaluated by Brychtová et al. (2012). However no extensive map users' oriented research on how the design can influence the usability of map wasn't probably done so far.

Both mentioned gaps in the research of maps usability—the effective colour distance and pleasant map design—are the subject of this presented research.

The Experiment

The goal of the experiment was to find out whether particular placement of the legend could positively or negatively affect the success to search for the information and to determine which level of colour distance between classes of choropleth maps allows the most accurate results. Respondents of the experiment were given a simple task to determine a value of a depicted area in the choropleth map by matching its colour with the legend. Results of the experiment were obtained by analysing the accuracy of respondents' answers and sum of fixation duration in defined areas of interest.

Experimental Design, Conditions and Participants

The experiment consists of 46 stimuli which are characterized by 6 different legend positions I to VI (see Fig. 1) and 8 colour schemes with different colour distance step between classes. Five of these schemes have equal colour distance steps between their classes—the lightness of their colour decreases by the same amount of colour distance ($\Delta E_{00} = 2, 4, 6, 8$ or 10) through the whole scheme (see Table 1a). Next three schemes were designed and tested based on the first phase of evaluation of experiment, which is described further. Those schemes are designed so the steps between classes are not homologous. The sequence of colour distance steps conform to values ΔE_{00} 4-8-10-8-4, 6-8-10-8-6 and 4-6-8-10-12 (see Table 1b). Each choropleth colour scheme is compounded by 6 classes A to F (A refers to the lightest, F to the darkest shade of the scheme).

All colour schemes are made of lightness continuum of green. In order to keep the endurable length of the experiment no other hue continua were tested.

The experiment was carried out under controlled conditions in the laboratory at Department of Geoinformatics, Faculty of Science, Palacký University in Olomouc, which is equipped with a low-frequency contactless eye-tracker SMI RED 250 with a sampling frequency 120 Hz. Stimuli were projected on 23 "LG Flatron monitor IPS231P". Stimulus size was $1,920 \times 1,080$ px. Design of the experiment was prepared in SMI Experiment Center™ (SensoMotoric Instruments 2013). Detection of fixations was performed through the SMI BeGaze™ (SensoMotoric Instruments 2013) using ID-T (dispersion threshold algorithm). Dispersion threshold was set to 50 px and a minimum length of 80 ms.

The experiment was attended by 35 respondents [19–35 years; 19 males and 16 females; 18 experts in the field of cartography and 17 novices (based on self-evaluation)]. Prior to projecting experimental stimuli, participants were subjected

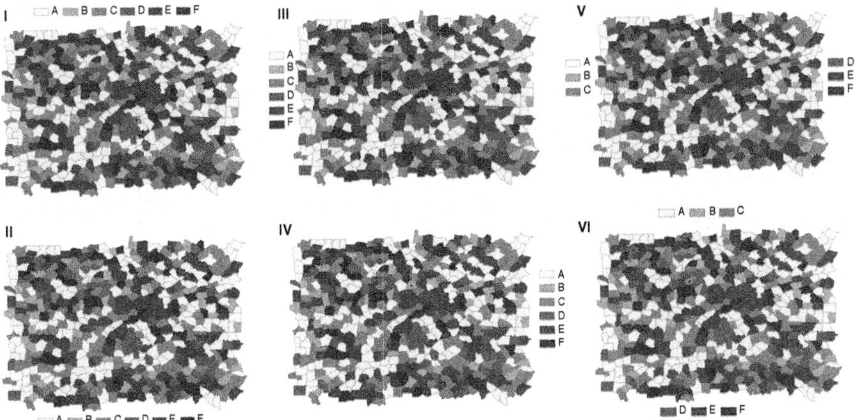

Fig. 1 Six groups of stimuli with different position of the map legend: (*I*) up; (*II*) down; (*III*) left; (*IV*) right; (*V*) vertically split; (*VI*) horizontally split

Table 1 Specification of colour schemes with (a) equal and (b) unequal colour distance steps given in CIE Lab and RGB values

ΔE_{00}	L	a	b	R	G	B
a)						
10	94.80	−30.00	30.00	201	255	179
	79.45	−30.00	30.00	159	211	138
	66.10	−30.00	30.00	123	174	104
	54.67	−30.00	30.00	93	143	76
	44.67	−30.00	30.00	68	117	52
	33.12	−30.00	30.00	39	89	26
8	94.80	−30.00	30.00	201	255	179
	82.39	−30.00	30.00	167	219	146
	71.22	−30.00	30.00	136	188	117
	61.34	−30.00	30.00	110	161	92
	52.63	−30.00	30.00	88	138	71
	44.58	−30.00	30.00	67	117	52
6	94.80	−30.00	30.00	201	255	179
	85.40	−30.00	30.00	175	228	154
	76.63	−30.00	30.00	151	203	131
	68.63	−30.00	30.00	129	181	110
	61.34	−30.00	30.00	110	161	92
	54.71	−30.00	30.00	93	143	76
4	94.80	−30.00	30.00	201	255	179
	88.51	−30.00	30.00	184	237	162
	82.40	−30.00	30.00	167	219	146
	76.65	−30.00	30.00	151	203	131
	71.24	−30.00	30.00	136	188	117
	66.15	−30.00	30.00	123	174	104
2	94.80	−30.00	30.00	201	255	179
	91.70	−30.00	30.00	192	246	171
	88.50	−30.00	30.00	183	237	162
	85.40	−30.00	30.00	175	228	154
	82.39	−30.00	30.00	167	219	146
	79.47	−30.00	30.00	159	211	138
b)						
4	94.80	−30.00	30.00	201	255	179
8	88.32	−30.00	30.00	183	236	162
10	76.47	−30.00	30.00	150	203	131
8	63.54	−30.00	30.00	116	167	98
4	54.57	−30.00	30.00	93	143	76
	50.50	−30.00	30.00	82	132	66

(continued)

Table 1 (continued)

ΔE_{00}	L	a	b	R	G	B
6	94.80	−30.00	30.00	201	255	179
8	85.22	−30.00	30.00	174	227	153
10	73.72	−30.00	30.00	143	195	123
8	61.18	−30.00	30.00	110	161	92
6	52.49	−30.00	30.00	87	138	71
	46.48	−30.00	30.00	72	122	57
4	94.80	−30.00	30.00	201	255	179
6	88.32	−30.00	30.00	183	236	162
8	79.31	−30.00	30.00	158	211	138
10	68.49	−30.00	30.00	129	180	110
12	56.71	−30.00	30.00	98	149	81
	44.69	−30.00	30.00	68	118	53

by the test of colour vision with pseudoisochromatic plates. All participants succeeded in this test.

Respondents were asked to find an area in the map depicted with a dot, match it with one of the six classes of the legend and mark their answer via questionnaire which was projected after the experimental stimulus.

Time to answer the question was not limited. The order of stimuli was randomized for each single participant in order to avoid the effect of the eye adaptation to handle such tasks. It was assumed that at the beginning of the experiment participants might have more problems to recognize colour differences, while at the end of the experiment they might show better results (Holmqvist et al. 2011).

In the last part of the experiment the self-evaluation of users' strategy was carried out. Volunteers were asked to describe their strategy during solving the given task and visual searching for the answer.

Results

The experiment was analysed in two phases. In the first step only colour schemes with equal colour distance step were examined. Based on preliminary results of 11 participants, which indicated strong influence of order of classes within the colour scheme on their correct value identification, three more colour schemes were added to the experiment and tested. New colour schemes setting were optimized in the sense of making visual steps between classes unequal. Specification of these schemes (Table 1b) was done deterministically. It was believed, that colour schemes with growing colour distance between classes in the middle (e.g. B, C, D and E) will cause more efficient solving a spatial task with use of choropleth map.

Results are based on analysing the accuracy of answers and eye-movements regarding specified areas of interest.

Analysing the Accuracy of Answers

Monitored accuracy of answers was analysed separately for groups of stimuli with the same position of legend, colour distance steps between classes and order of class within the colour scheme, which corresponded to the correct answer to the experimental task.

The less problematic composition of the map sheet was observed for the group of stimuli, where the legend was placed on the right to the map sheet (no. IV). However the observed 79 % accuracy does not exceed others groups of different compositions obviously. Groups of stimuli with composition II, III, V and VI showed the accuracy of 70–71 %. The lowest ratio of correct answers was observed for the composition, where the legend is above the map sheet (I). To compare accuracy of all groups see the Fig. 2a.

Analysing the accuracy according to the colour distance steps between classes of colour scheme brought prima facie evidence that the increasing colour distance has a positive impact on the choropleth map readability (see Fig. 2b). The lowest overall accuracy (54 %) was observed for maps with the smallest colour distance between areas ($\Delta E_{00} = 2$). The best accuracy (82 %) was recorded for groups of stimuli with $\Delta E_{00} = 10$.

The last experimental task was to find out how accurate are map-users while matching different shades with the corresponding class in the legend. Particularly problematic was to match the correct answer with classes in the 'middle' of the colour scheme. Less trouble causing was matching the lightest shades—the observed accuracy for class A (corresponding to the very first class of the legend and thus the most lightest) was higher than 90 % (see Fig. 2c). The second class (B) was also quiet easy to match (80 % accuracy). The most difficult was to match correctly classes C and D (58 %, resp. 60 % accuracy). The 71 % accuracy for the class F was rather surprising. It was assumed, that matching the extreme classes will be easier, because by fast visual search one can quickly decide whether they can see darker (or lighter) area than the one intended. However matching the lightest class was much easier than to match the darkest class.

By deeper investigation of observed phenomenon it was found that low accuracy observed for class F was caused by colour schemes with the small colour distance between classes ($\Delta E_{00} = 2$ and 4). Other colour schemes confirmed given assumption—the extreme shades (classes A and F) are the most easier to match with the legend (see Fig. 2d).

Based on results above, it is apparently inappropriate to apply colour schemes with the equal colour distance step between classes. Equal colour distance steps, regardless their ΔE_{00} value, are not ideal to allow easy discrimination between 'inner' classes of the colour scheme.

In order to find out better solution three more colour schemes were designed and tested. Optimized schemes are characterized by unequal colour distance steps with higher ΔE_{00} values differentiating middle classes. For evaluation of optimized

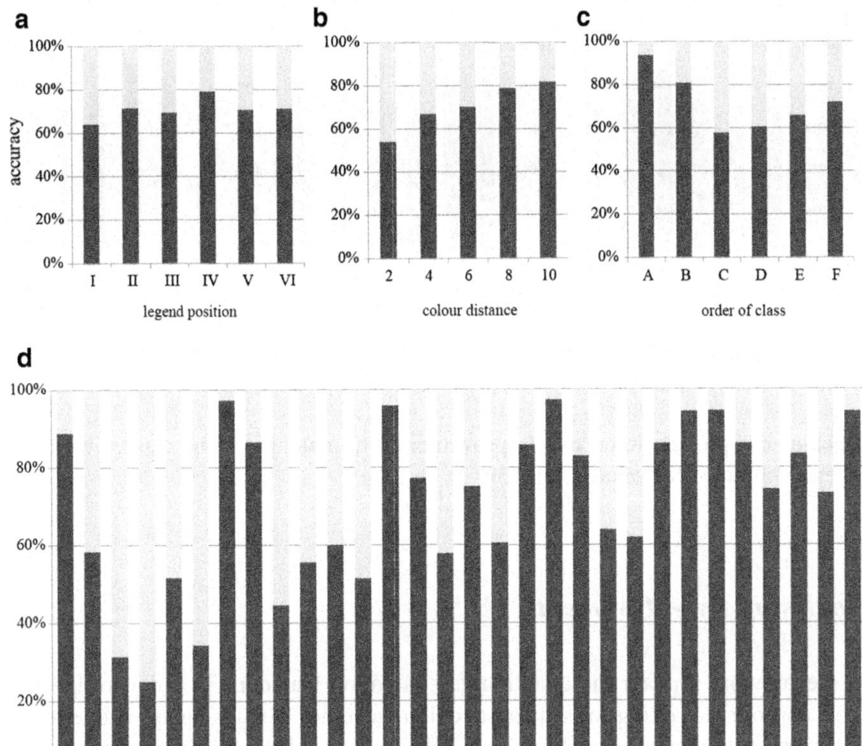

Fig. 2 (**a**) Accuracy of answers monitored for groups of stimuli with particular legend position (I–VI). The highest accuracy (79 %) was observed for the design IV. (**b**) Accuracy of answers monitored for groups of stimuli with particular colour distance between colour scheme classes (2–10). The accuracy increases with increasing colour distance. The highest accuracy (82 %) was observed for groups of stimuli with $\Delta E_{00} = 10$. (**c**) Accuracy of answers monitored for groups of stimuli with particular class to match (A–F). It is evident that the most troublemaking are classes in the middle of the colour scheme (C and D). (**d**) Accuracy of answers according to the colour distance (2–10) and class of the colour scheme (A–F). Regardless the colour distance the lightest class (A) is always easier to match. Lower accuracy on 'central' classes is evident in all schemes

colour schemes, 18 more stimuli were prepared: 6 for each of 4-8-10-8-4, 6-8-10-8-6 and 2-4-6-8-10-12 schemes (numbers corresponds to the colour distance between classes from the lightest to the darkest). Examination was done only with the composition IV. Technical description of colour schemes was already given above (Table 1b).

The best accuracy was observed for the colour scheme 4-8-10-8-4 (see the Fig. 3a). The average accuracy was in this case 93 %. Next two optimised colour

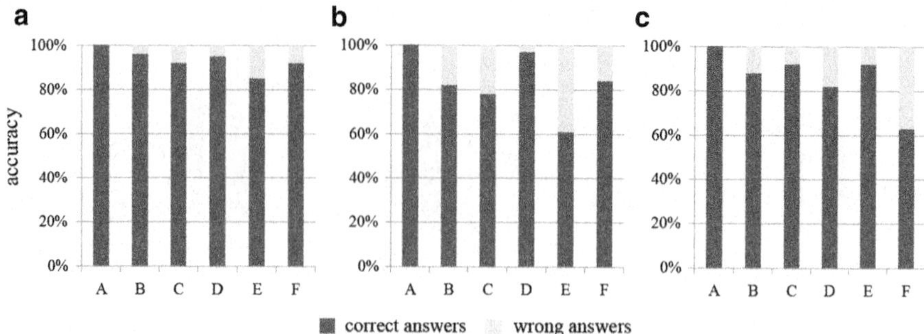

Fig. 3 Accuracy of answers monitored for stimuli with different classes to match (A–F) of colour schemes (**a**) 4-8-10-8-4, (**b**) 2-4-6-8-10-12 and (**c**) 6-8-10-8-6

schemes did proven noticeable improvement, but matching some classes with the legend was still problematic (Fig. 3b, c).

Analysing Eye-Movements

In the previous chapter the quantitative evaluation was done in order to find the most appropriate setting for sequential colour schemes. Analysing eye-movements can bring more knowledge about what behaviour is behind the process of reading maps. As it was already mentioned, at the end of the experiment, participants were asked to recall the procedure of their searching for the information and describe it. In all cases participants were seeking over the stimulus trying to find lighter or darker areas and keeping in their memory the order of the one depicted area. In few cases participants stated, that they matched the depicted area directly with the legend, without comparing it with other areas in the map.

To find out how different levels of colour distance, or order of classes within the colour scheme influences the map-users strategy the analyses of eye-movements regarding specified areas of interest (AOI) was done.

In all experimental stimuli four different AOIs were defined. These AOIs refer to a region just around the depicted area to be matched with the legend (AOI 1), more distant region around the depicted area (AOI 2), rest of the map (AOI 3) and the legend (AOI 4). The proportion of all listed AOIs was the same in all experimental stimuli.

The hypothesis was, that participants will spend more time in AOI 2 and AOI 3 if the task is difficult to solve. The more difficult task was given, the more time had to be spend in order to compare depicted area with others in the map.

In order to find out the interest of experiment participants sum of all fixations durations over AOIs were analysed. Longer fixation duration could mean difficulty in extracting information, or the object is more engaging in some way (Eastman 1985; Poole and Ball 2005).

Dataset containing all participants' fixations duration observed at specified AOIs was analysed with Kruskal-Wallis test. Pairwise comparison of data related to the colour distance and the class order was done in order to find out statistically significant differences of median location. Analyses were done separately for each AOI. In the case of equally-stepped schemes only imperceptible number of compared pairs was found significantly different, thus only elementary visual interpretation of results can be done based on boxplots (Fig. 4).

The sum of fixations spent in AOI 1 is generally the highest among other AOIs. Higher medial values of fixations duration can be observed at classes B–F, which correspond to the previous findings. Looking at boxplots for AOI 2 (close region to the depicted area) it is clear that lightest and darkest classes (A and F) in most cases shows the smallest, even zero, duration of fixations. Observations for AOI 3 are more or less balanced, which means that participants spend the same (obviously short) time in this AOI regardless the colour distance or class. Values at AOI corresponding to the legend (AOI 4) are tracing values of AOI 1; even observed fixations durations are much shorter.

The same analysis was done for the 'winning' optimised colour schemes (4-8-10-8-4). Kruskal-Wallis test did not prove any significant differences as well as for schemes with equal colour distance step. However in the case of all AOIs the medial values are by visual control more equable (see Fig. 5), than in original colour schemes.

Conclusion and Discussion

Results of the experiment didn't prove any influence of legend position on the map sheet on the effectiveness of the map reading process. The effect of the colour distance on the accuracy of reading the choropleth maps was confirmed. The higher is the colour distance between classes of the colour scheme the more accurate are map-users while estimating the value of concrete area comparing it with the map legend. Another interesting finding is that equal colour distance steps are not optimal solution in designing sequential schemes. Map users perform particularly better when determining minimal and maximal values, which are plotted by the lightest, respectively darkest colour shade. Based on these findings new colour schemes of unequal visual steps between classes were designed and evaluated. The most efficient colour scheme was the one with colour distance steps 4-8-10-8-4. Observed accuracy was for this particular scale balanced regardless the class order and the overall rate of correct answers achieved 93 %.

In the presented experiment only lightness continuum of six green shades was evaluated. It is known that the human eye maximal sensitivity is in the green region at 555 nm. Even though in theory the human eye can distinguish millions of colours, in cartography would be unreasonable to ask map users to distinguish more than 12 classes (Dent et al. 2009). It is assumed, that the

(continued)

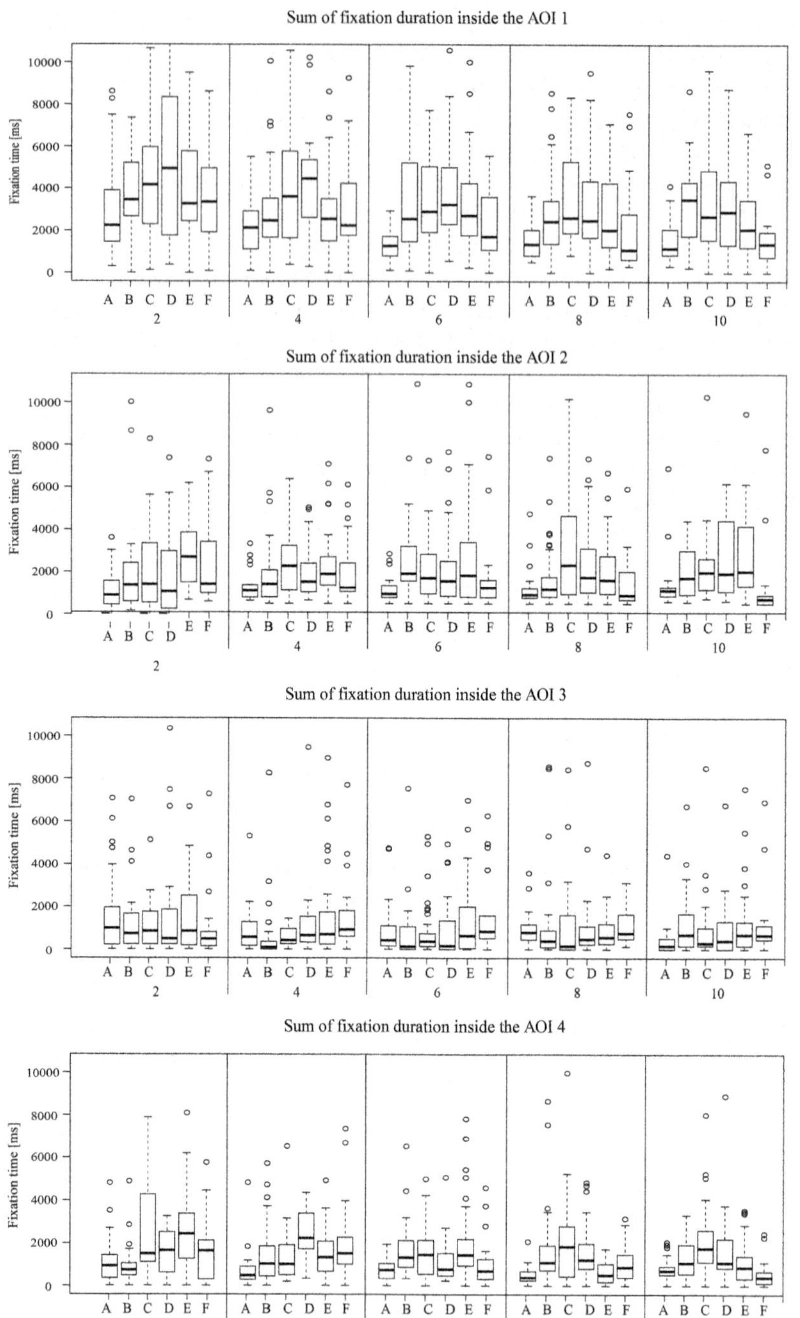

Fig. 4 Summed fixations duration inside AOI 1, 2, 3 and 4 observed at examined colour schemes with equal visual steps between classes. Results are shown separately for groups of stimuli with particular colour distance, order of the class and regardless the legend position

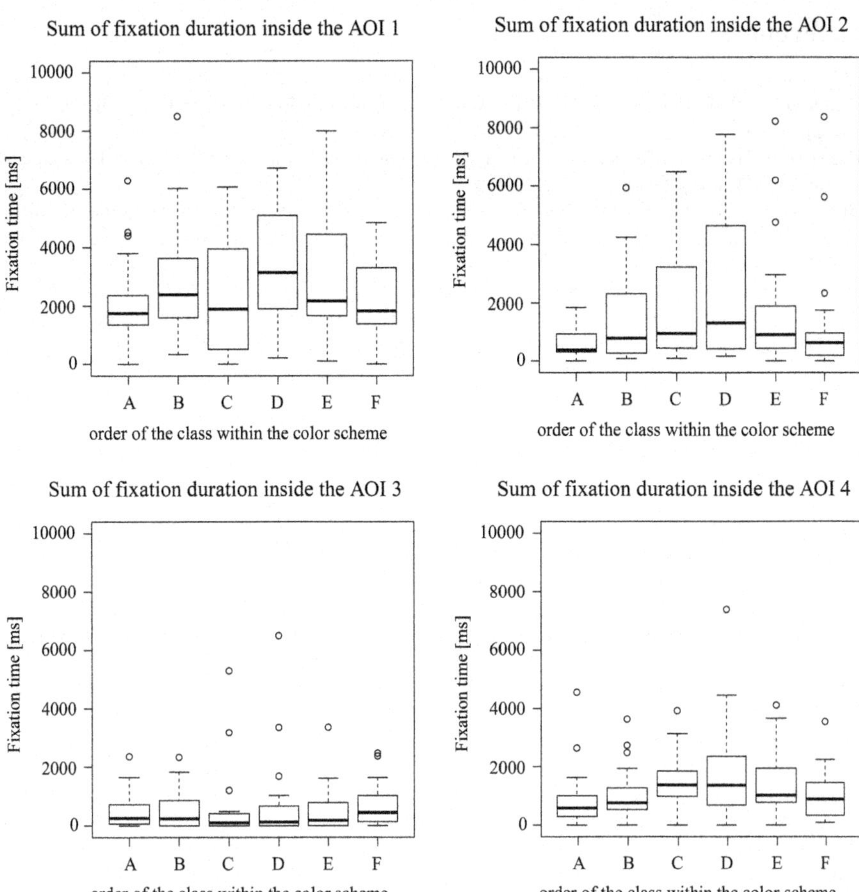

Fig. 5 Summed fixations duration inside AOI 1, 2, 3 and 4 observed at optimised colour scheme (4-8-10-8-4), which was evaluated as the most efficient (the accuracy of answers was the highest while solving the task where this scheme was applied)

change of experimental setting would significantly change presented results. Thus, the future plan of the author is to evaluate colour schemes varying in the number of the classes and colour hue and make the experimental results more universal for purposes of cartography.

Acknowledgments The article was created within the project CZ.1.07/2.3.00/20.0170, supported by the European Social Fund and the state budget of the Czech Republic.

References

Bjorke JT (1996) Framework for entropy-based map evaluation. Cartogr Geogr Inf Syst 23 (2):78–95

Brewer CA, Hatchard GW, Harrower MA (2003) ColorBrewer in print: a catalog of color schemes for maps. Cartogr Geogr Inf Soc 30(1):5–32

Brychtová A, Popelka S, Dobešová Z (2012) Eye-tracking methods for investigation of cartographic principles. In: 12th international multidisciplinary scientific geoconference (SGEM 2012), vol II, pp 1041–1048

Chesneau E (2007) Improvement of colour contrasts in maps: application to risk maps. In: Proceedings of 10th AGILE international conference on geographic information science, pp 1–14

CIE (2012) Termlist of International Commission on Illumination. Retrieved from http://eilv.cie.co.at/

Culp GM (2012) Increasing accessibility for map readers with acquired and inherited colour vision deficiencies: a re-colouring algorithm for maps. Cartogr J 49(4):302–311

Dent BD, Torguson JS, Hodler TW (2009) Cartography: thematic map design, 6th edn. Thomas Timp, p 336

Eastman JR (1985) Cognitive models and cartographic design research. Cartogr J 22(2):95–101

Gilmartin P, Shelton E (1989) Choropleth maps on high resolution CRTs: the effect of number of classes and hue on communication. Cartographica 26(2):40–52

Harrower M (2007) Unclassed animated choropleth maps. Cartogr J 44(4):313–320

Holmqvist K, Nyström M, Andersson R, Dewhurst R, Jarodzka H, Van de Weijer J (2011) Eye tracking: a comprehensive guide to methods and measures, 1st edn. Oxford University Press, Oxford, p 560

Jenny B, Kelso NV (2007) Designing maps for the colour-vision impaired. Bull Soc Cartogr SoC 41:9–12

Kimerling JA (1985) The comparison of equal-value gray scales. Am Cartogr 12(2):132–142

Kröger J, Schiewe J, Weninger B (2013) Analysis and improvement of the open-streetmap street color scheme for users with color vision deficiencies. In: Proceedings of the 26th international cartographic conference, Dresden, p 17

Kuehni RG (2003) Color space and its divisions: color order from antiquity to the present. Wiley, p 434

Mersey JE (1990) Colour and thematic map design: the role of colour scheme and map complexity in choropleth map communication. Cartogr: Int J Geogr Inf Geovis 27(3)

Olson JM, Brewer CA (1997) An evaluation of color selections to accommodate map users with color-vision impairments. Ann Assoc Am Geogr 87(1):103–134

Poole A, Ball LJ (2005) Eye tracking in human-computer interaction and usability research: current status and future. In: Ghaoui C (ed) Encyclopedia of human-computer interaction. Idea Group, Inc., Pennsylvania, p 13

Schiewe J, Weninger B (2013) Visual encoding of acoustic parameters – framework and application to noise mapping. Cartogr J 50(4):332–344

SensoMotoric Instruments (2013) SMI experiment CenterTM. Retrieved from http://www.smivision.com/en/gaze-and-eye-tracking-systems/products/experiment-center-software.html

Slocum TA, McMaster RB, Kessler FC, Howard HH (2008) Thematic cartography and geovisualization, 3rd edn. Prentice Hall, p 576

Steinrücken J, Plümer L (2013) Identification of optimal colours for maps from the web. Cartogr J 50(1):19–32

Stigmar H (2010) Making 21st century maps legible – methods for measuring and improving the legibility and usability of real-time maps. Dissertation thesis, Lund University

Stigmar H, Harrie L (2011) Evaluation of analytical measures of map legibility. Cartogr J 48 (1):41–53

Werman M (2012) Improving perceptual color difference using basic color terms. Computer Research Repository, pp 1–14

Physiological and Cognitive Aspects of Sound Maps for Representing Quantitative Data and Changes in Data

Jochen Schiewe

Introduction

Map graphics is without doubt the central form for presenting geospatial information, as it ensures a space-saving and a geo-referenced display. However, other coding formats like animations, videos or acoustic representations are conceivable which can be combined to multimedia (better: multimodal) maps on demand.

With respect to the *acoustic coding of geospatial information* Edler et al. (2013) give an overview of applications fields and specific functions of sounds. So far, only a limited number of implementations (and with that, little experiences) can be observed. Exceptions are for example the work by Krygier (1993; with the topic of AIDS in the United States), Fisher (1994) or Lodha et al. (1996; the latter two with the topic of data uncertainty) or Brauen (2006; elections in Canada). Focusing even more on the acoustic coding of *quantitative* geospatial information—a topic widely neglected in literature in the past (with exceptions like Bearman and Lovett 2010; Lodha et al. 1999)—several disadvantages become obvious: Because one or many sounds should not be displayed all the time, an interactive release is necessary and an actually simultaneous evaluation of multiple carriers of features (like counties in a thematic map) is not possible. With that, the analysis of complex patterns, anomalies, etc. becomes rather difficult, at least in comparison to the visual counterpart method. And finally, the recognition of absolute values through sound parameters like loudness or pitch is hardly possible for humans.

Despite of these disadvantages, there are several *arguments for using sound in maps*. Besides the general motivational aspect of a new and additional coding format as well as the proven improvement of an active learning process, a multicodal representation follows the general idea of reducing the amount of

J. Schiewe (✉)
Lab for Geoinformatics and Geovisualization (g2lab), HafenCity University Hamburg, Überseeallee 16, 20457 Hamburg, Germany
e-mail: jochen.schiewe@hcu-hamburg.de

© Springer International Publishing Switzerland 2015
J. Brus et al. (eds.), *Modern Trends in Cartography*, Lecture Notes in Geoinformation and Cartography, DOI 10.1007/978-3-319-07926-4_24

information which has to be processed by each perception channel. Along with this, negative recognition phenomena might be reduced; for example the split attention effect (i.e. the limited ability to discriminate between objects that are spatially or graphically apart from each other; Harrower 2007), the change blindness (i.e. missing changes in image sequences that are too fast; Simons and Ambinder 2005), or the visual clutter (i.e. the graphical overload; Brewster et al. 1994). The latter aspect is critical in multivariate maps where, for example, two graphical variables have already been introduced for the first two topics and more themes have still to be covered. In particular, the usage of graphics within tiny areas becomes difficult due to space limitations, consequently leading to cognition problems. With that, acoustic coding can be seen as an application dependent, complementary option in a multimodal map. Finally, sound maps have still to be considered and further developed for the 300 million visually impaired and blind people worldwide. In this context, the availability and functionality of modern mobile devices leads to several advantages of multimodal maps in comparison to conventional haptic maps (Jacobson 2007; Delogu et al. 2010; Koch 2012).

Based on this background, *goal of this contribution* is to investigate acoustic options for depicting *quantitative data* and to derive recommendations for practical usage depending on given data and purpose. In particular, representations for a *single point in time* (second section) and for *time series* (third section) will be considered.

This detailed and systematic analytic review goes beyond the current literature, in particular the work done by Krygier (1994) who assembled the sound variables (location, loudness, pitch, register, timbre, duration, rate of change, order, attack/ decay) in order to describe the transformation of spatial data properties into abstract acoustic (i.e. neither verbal nor musical) properties (*sonification*). Our investigation will consider mainly physiological and cognitive, but also practical map imple- mentation factors.

Based on this analytical review we will briefly describe an own empirical study as well as demands for future tests (fourth section).

Acoustic Coding of Quantitative Data for Single Point of Time

Goal of the following analytical review is the identification of suitable abstract sound variables for representing quantitative information for a single point of time. This review is based on the above mentioned list of sound variables by Krygier (1994). Whereas Krygier evaluated the suitability for data at nominal and ordinal scales, an explicit systematic check for data at cardinal scale has not been performed so far. For this the following suitability criteria are taken into account:

- There should be an analogy between metric scale and sound variable; in partic- ular, not only a sequence of values (from low to large values) but also—in contrast to ordinal scale—it should be possible to map a proportional and measurable discrimination (*mapping issue*);

- Effective and efficient perceptibility of sound variables (*physiological and cognitive issue*);
- Simple design and implementation of the respective sound variable (*implementation issue*).

Concerning the *mapping issue* we know from the graphical domain that the variables form, filling (except for hatching), hue and direction (except for the indirect angle specification) are not or hardly suited for the representation of quantitative data. In analogy to this we can state for an acoustic coding that the following variables are not suited: timbre (i.e. the dominant characteristic or quality of a sound), register (i.e. the relative position of a pitch in a given range of pitches), sequence of sounds, location (like sound coming from right hand side) as well as attack and decay of sounds.

Considering the *physiological and cognitive issue* we can rely on studies and empirical values that have mainly been derived in the domains of psycho-acoustics and music psychology. Focusing only on the remaining variables after applying the mapping criterion from above, we have to look at

- the *total interval*, as well as
- the *resolution*

that can be perceived by humans. In this context, the following can be stated concerning their perceptibility for humans:

- *Loudness*: Expressing loudness by the sound level, a human is able to distinguish differences of 2 to 5 dB in optimal settings, while differences between 5 and 10 dB are recognized as clear, and differences between 10 and 20 dB as quite large (Lercher 1998). Considering the well perceivable range between the thresholds of hearing and pain (e.g. between 45 and 65 dB), we end up with a rather small number of classes (e.g. with five classes at a level width of 5 dB). While this is generally sufficient for data an ordinal scale representing a limited number of classes, this range is generally not suited for a generally larger number of quantitative values or levels.
- *Pitch*: The human being is able to recognize frequencies between 20 and 20,000 Hz. The frequency resolution is proportional to the frequency of the standard sound; for example a 1,000 Hz tone allows for a perceivable resolution of about 3 Hz. Cutting off extreme values, we are typically able to differentiate between 48 and 60 pitches (which corresponds to four to five octaves; Yeung 1980). For comparison purposes: Transferring this number to the graphical counterpart, we would need for example a combination of eight to ten different color hues and six intensities for each hue. While the human ear is quite well suited for the description of *relative* pitches, an exact acquisition of *absolute* frequencies is only possible for quite a few musical people.
- *Duration*: Depending on the total interval and the required resolution an acoustic coding that maps a value to the length of a sound might last very long. For

example, if a resolution of two seconds between two (classified) values and a mapping of ten different values are required; one ends up with a 20 s sound display in the worst case. But as experience teaches, attention significantly decreases after a few seconds so that a relative comparison of two (or more) values released one after the other becomes very difficult. With that, duration is not a suitable option for depicting quantitative data.

- *Rate of change*: The discriminability of different tempi depends on a variety of factors, like the duration and the rest between sounds. Empirical values from music state a relative discriminability of ca. 4 beats per minute; however, this also requires a certain total duration. Like with the aforementioned parameter duration an exact or reliable determination of absolute tempo values is not possible for the human.
- *Number*: The number or repetition of a sound is variable that is not proposed by Krygier (1994). But in analogy to the graphical method of symbol repetition also the playback of a certain number of sounds is suited in order to represent absolute values. However, depending on the actual number of the values (either unclassified or classified) this playback can last quite a long time.

From these considerations we can conclude that

- for the determination of *absolute values* only the use of respective number of sound displays seems to be appropriate; however, as described above, this method is strongly limited to the total amount of values;
- for the representation of *classified values* with the aim of a relative comparison pitch can be recommended as best choice: Pitch is very often denoted as "brightness" of a tone, which very well describes the analogy to graphical coding. In analogy to the graphical bi-polar coding (using two colors and varying brightness) we can also apply two timbres in combination with varying pitches: For example, a high flute sound represents an increase, whereas a lower tuba sound stands for a decrease of values with respect to a certain zero, mean or standard value (which might be encoded by an additional base sound).

With respect to *implementation issues* the general problem remains that sounds have to be presented in a sequential and time-dependent manner, thus requiring an additional *interactive release* for each use case. This can be realized through pressing on loudspeaker symbols associated to objects (like county polygons) or simply by mouse-over movements over those objects. In any case, an additional explanation for this interaction is necessary.

Like with graphics, also acoustic coding requires a *suitable legend*. Again, the sound legend needs a certain interaction concept which finally leads to a longer processing time compared to the graphical counterpart and with that to a reduced usability.

Finally, as a consequence of the sequential display, a simultaneous recognition of values of more than one object, and with that a quick comparison "at a glance" becomes impossible. This again reduces the effectiveness and efficiency of nearly all comparison tasks.

Table 1 Sound variables and their suitability for representing data at different scales ("– –": not suited; "–": poorly suited; "+": suited with limitations; "++": suited)

Sound variable	Suitability depending on data scale		
	Nominal	Ordinal	Cardinal
Location	–	–	– –
Loudness	– –	++	+
Pitch	– –	++	++
Register	– –	+	–
Timbre	++	– –	– –
Duration	– –	–	–
Rate of change	– –	+	–
Order	– –	–	– –
Attack/decay	– –	–	– –
Number	– –	+	+

Summarizing this analytical review we can modify and expand the aforementioned list of Krygier (1994) by assessing the suitability of sound variables also for the representation of data at a cardinal scale (Table 1). In particular, it considers differences between data at ordinal and cardinal scales according to the above made statements.

Acoustic Coding of Quantitative Data for Time Series

For depicting time series (i.e. describing more than one point of time) diagrams are the standard choice in cartography. Turning over to an acoustic representation, we can of course transfer the aforementioned statements (second section). However, there are further acoustic variable combinations and design options which should be taken into consideration.

For the representation of a *bi-temporal change* the following options are feasible (Fig. 1).

- the attack or decay (crescendo/de-crescendo) of a sound can represent an increase or a decrease, resp.;
- two different pitches are used to encode increase or decrease, resp. (a third pitch might be used for the case "no change");
- each of the three change cases are represented by a series of two tones (for example, a second higher, a second lower, same tone).

Also for *multi-temporal series* of data values only little implementations and studies have been published so far. One exception is the work done by Mezrich et al. (1984) who propose the display of multivariate time series in a dynamic manner with support of audio.

More conceptually, we can encode multi-temporal series of data again through pitch (Fig. 2) which can be changed

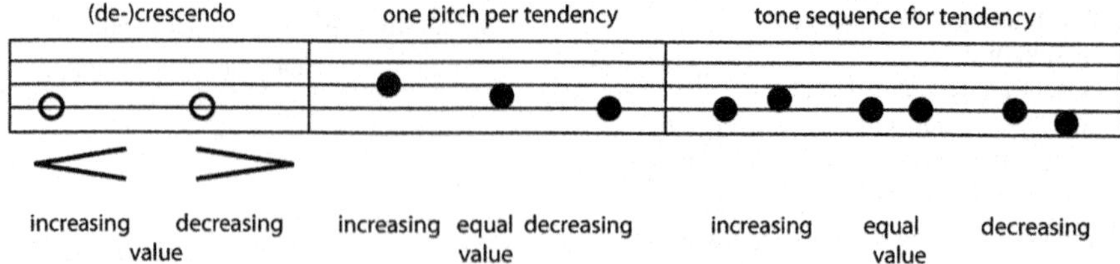

Fig. 1 Coding bi-temporal changes with variations of pitch

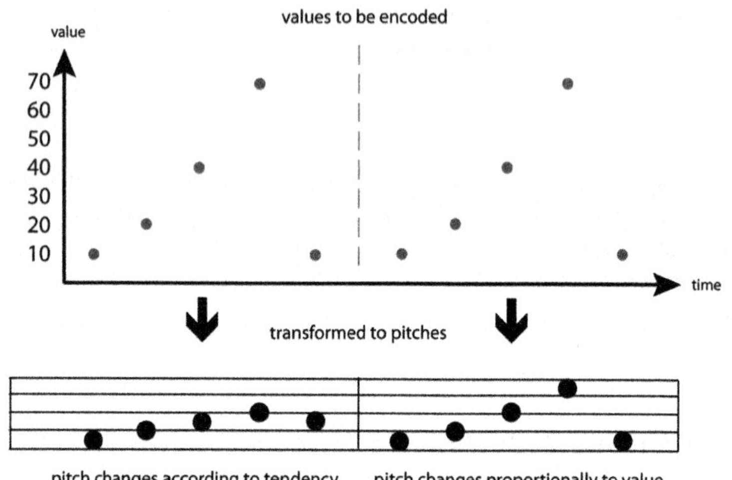

Fig. 2 Coding multi-temporal changes with variations of pitch

- either proportional to the change of values, or
- using always a constant pitch difference (e.g. a second up for an increase and a second down for a decrease), no matter how large the value difference is.

The latter case can be perceived much easier by the human; however, this goes along with a loss of information. If pitch differences are required, our experiences show that using standard differences from the musical scale (i.e. second, third, etc.) are more convenient to the human ear in order to represent equidistant class values compared to a strict transformation of value differences into frequencies (i.e. using the unit Hertz). Finally, it has to be mentioned that no studies are known that reveal usability recommendations concerning the total number of sounds (i.e. the total number of different values that can be represented), the suitable tone interval, and the total number of time stamps that can be perceived and reliably recognized by the user.

Empirical Investigation of Sound Map Usability

Existing Studies

Based on the aforementioned analytical review it is still necessary to perform user studies aiming at an optimization of different representation methods and respective parameter settings. There are quite a few examples from other disciplines like scientific visualization (Bly 1982) or statistics (Mezrich et al. 1984); however, an explicit and profound application on cartographical representations has not been performed to our knowledge.

A small user study that picked out selected aspects for the representation of quantitative data for a single point of time was described in an earlier publication of our research group (Schiewe and Weninger 2012). General aim of this study was to assess the overall usability through a comparison of acoustic versus conventional graphical coding methods. Each test person received four maps that used color, hatching, sound and a double coding (color/sound) for the representation of one value. Two more maps deployed combinations of color/hatching and color/sound for the display of bi-variate issues. In each map the lowest or largest value had to be determined. Effectiveness was evaluated via the correctness of the answers, while efficiency was expressed by the ratio of correct answers and used processing time.

Because goal of the study was to identify basic problems of acoustic encoding and differences compared to the graphical counterpart, a sample size of 18 test persons was seen as being large enough.

The key findings of this study can be summarized as follows:

- Not surprisingly, color coding has been proven the most usable method for extracting quantitative information, while sound coding showed worst results. On the other hand, a double coding of color and sound was clearly superior against the hatching option with respect to effectiveness (while efficiency was slightly worse). An additional value of the double coding approach was seen through the fact that sound was a good alternative for discriminating between very similar values which is due to the better resolution of the sound variable pitch as mentioned before.
- Considering the display of two map topics, the combination of color/hatching showed a better usability compared to color/sound. There was, however, an important application case for the color/sound combination: If a couple of small object or reference areas exist, sonification becomes meaningful because hatching is not appropriate due to space limitations.
- Observing the heterogeneous and partly not comprehensible interaction patterns in sound maps, we conclude that a significant portion of the worse usability effectiveness and efficiency is due to the unfamiliar coding method. Vice versa, we also observed a significant learning effect throughout the study, despite of an increasing complexity of tasks. These findings correspond very well with results

of Bly (1982) who could even quantify an increase in effectiveness through repetitions of similar experiments.

Further details of this study are described in Schiewe and Weninger (2012).

Further Studies

For both cases, visualizing data for one time stamp as well as time series, more detailed user studies have to be performed in order to create an additional value out of an acoustic coding in maps.

In particular, *detailed parameter settings* have to be investigated. For example, with time series coding a meaningful number of points of time, tone differences and total length of the acoustic series needs more attention. Of course, such an acoustic coding has to be compared to the graphical counterpart (i.e. a diagram); however, also specific scenarios (like missing graphical space) and tasks have to be considered. But not only the pure hearing capabilities of different users have to be considered but also the interplay between visual and acoustic perception that takes place while using a map.

Furthermore, *interaction design* aspects are of major importance. For example, a mouse-over can ensure a quick release of a sound (and with that, reducing the time span between multiple releases for comparison purposes). The disadvantage is that sounds are played back unintentionally when an area is just passed during mouse movements. Again, empirical investigations using different map layouts have to be performed in order to gain better insight and better usability.

Also, different types of *sound legends* and different ways of positioning them need to be investigated. As mentioned before, we observed a significant *learning effect* throughout the study. It would be of interest how this learning curve develops and which measures can be taken in order to support a quick customization.

Finally, in the context of *maps for blind and visually impaired people* also the question arises how "hybrid maps" have to be designed that are useful for seeing people at the same time in order to allow for an immediate exchange of information.

Conclusions

Our analytical review has revealed that the sound variable *pitch*—and with strong limitations also the variable *loudness*—are suited for representing quantitative data in maps for comparison purposes. For an absolute determination of values neither pitch nor loudness are really suited because here a very well trained musical hearing is required. Instead of this, presenting a certain number of sounds according to the value to be described is a possible, but also not satisfying solution due to the limited maximum number of sounds that is well recognizable. Summarizing this analytical review we could

(continued)

modify and expand the aforementioned list of Krygier (1994) which looked only at data at nominal and ordinal scales.

Although there are strong practical disadvantages with using sound in maps, there are a couple of specific scenarios and users where acoustic coding can produce a significant additional value. On the other hand, only little experiments have been conducted so far in order to understand and to optimize the use of sound. To fill this gap, more usability research will be necessary in the next few years.

References

Bearman N, Lovett A (2010) Using sound to represent positional accuracy of address locations. Cartogr J 47(4):308–314

Bly S (1982) Presenting information in sound. In: CHI '82 conference on human factors in computer systems, pp 371–375

Brauen G (2006) Designing interactive sound maps: using scalable vector graphics. Cartographica 41(1):59–71

Brewster SA, Wright PC, Edwards ADN (1994) The design and evaluation of an auditory-enhanced scrollbar. In: Adelson B, Dumais ST, Olson JD (eds) Conference on human factors in computing systems, 24–28 April 1994, Boston, pp 173–179

Delogu F, Palmiero M, Federici S, Plaisant C, Zhao H, Belardinelly O (2010) Non-visual exploration of geographic maps: does sonification help? Disabil Rehabil Assist Technol 5 (3):164–174

Edler D, Lammert-Siepmann N, Dodt J (2013) Die akustische Dimension in der Kartographie – eine Übersicht. Kartographische Nachrichten 62(4):185–194

Fisher PF (1994) Visualization of the reliability in classified remotely sensed images. Photogr Eng Remote Sensing 60(7):905–910

Harrower M (2007) The cognitive limits of animated maps. Cartographica 42(4):269–277

Jacobson D (2007) The future of tactile cartography: from static raised lines to multimodal dynamic portable computer interfaces. In: Abstracts of Papers. XXIII international cartographic conference, 4–10 August, Moscow, p 358

Koch WG (2012) State of the art of tactile maps for visually impaired people. In: Buchroithner M (ed) True-3D in cartography: autostereoscopic and solid visualisation of geodata. Lecture Notes in Geoinformation and Cartography. Springer, Berlin pp 137–151

Krygier JB (1993) Sound and cartographic design. Videotape. Department of Geography, Pennsylvania State University

Krygier JB (1994) Sound and Geographic visualization. In: MacEachren AM, Taylor DRF (eds) Visualization in modern cartography. Pergamon, pp 149–166

Lercher P (1998) Medizinisch-hygienische Grundlagen der Lärmbeurteilung. In: Kalivoda MT, Steiner JW (eds) Taschenbuch der Angewandten Psychoakustik. Springer, Wien, pp 42–102

Lodha SK et al (1996) Visualizing geometric uncertainty of surface interpolants. Graphics Interface, pp 238–245

Lodha SK, Joseph AJ, Renteria JC (1999) Audio-visual data mapping for GIS-based data: an experimental evaluation. In: Proceedings of the 1999 workshop on new paradigms in information visualization and manipulation in conjunction with the eighth ACM international conference on information and knowledge management, Kansas City, pp 41–48

Mezrich J, Frysinger S, Slivjanovski R (1984) Dynamic representation of multivariate time series data. J Am Stat Assoc 79(385):34–40

Schiewe J, Weninger B (2012) Akustische Kodierung quantitativer Information in Karten – Ergebnisse einer Studie zum Vergleich mit klassischen Darstellungsformen. Kartographische Nachrichten 62(3):126–135
Simons DJ, Ambinder MS (2005) Change blindness. Curr Dir Psychol Sci 14(1):44–48
Yeung ES (1980) Pattern recognition by audio representation of multivariate analytical data. Anal Chem 52(7):1120–1123

The User Centered Framework for Visualization of Spatial Data Quality

Jan Brus and Vilém Pechanec

Visualization of uncertainty of spatial data includes a wide range of issues relating to data quality of the individual components, their definition, nature, database design, graphic methods, selection and implementation, dynamic monitoring and description of the error propagation in GIS operations, and many more. It follows that it is necessary to explore, extend and integrate existing research in uncertainty visualization of spatial data. The notion of uncertainty is defined the first part of paper, there is also described the issue of data quality as a variable which is used for visualization of uncertainty in spatial data. The principals of uncertainty visualization are discussed further more with emphasis on basic theories that have major impact on shaping the contemporary concept of uncertainty visualization. The attention is focused primarily on issues related to communication models, cognition, perception and semiotics. It is emphasized the need to perform research not only in particular visualizations, but also to carry out research which develops specific frameworks and concepts of visualization of spatial data quality. Proposed frameworks should incorporate users as a key factor which influences whole visualization. In this study the widen user centered framework of uncertainty visualization is proposed. This framework is based on results from user centered research on uncertainty visualization and on the basis of obtained findings and results from testing. This framework is able to reflect the various hierarchical natures of spatial data and integrate the various parameters of quality of spatial data as a basis for visualization.

J. Brus (✉) • V. Pechanec
Department of Geoinformatics, Palacký University in Olomouc, Olomouc, Czech Republic
e-mail: jan.brus@upol.cz; vilem.pechanec@upol.cz

© Springer International Publishing Switzerland 2015

J. Brus et al. (eds.), *Modern Trends in Cartography*, Lecture Notes in
Geoinformation and Cartography, DOI 10.1007/978-3-319-07926-4_25

Uncertainty

The notion *uncertainty* is mentioned quite often in both foreign and Czech literature and in the last twenty years has become the subject of focus of more and more professionals. The current development of technology only supports this fact. In the field of geoinformatics, the situation is no different, quite to the contrary. Quick development of implementations of individual theories working with this approach makes uncertainty an integral part of most operations with spatial data. Despite the stated fact, understanding uncertainty in the field of geoinformatics is still limited by problematic definition of this notion in individual science disciplines.

In literature, there are quite a few definitions of uncertainty. Great number of definitions reflects the complexity of the issue hidden under this general phenomenon. For instance, the difference between *accuracy* and *uncertainty* is debated very often. In cartography, the two terms are understood as very similar, and sometimes, they are even considered to be synonyms (Boukhelifa and Duke 2007). In connection with uncertainty, we can talk about accuracy, reliability, clearness, distinctiveness, etc. (Duckham et al. 2001). More generally, uncertainty can be defined as imperfection of the users' knowledge concerning data, processes or results. Thus, *uncertainty* is a multi-faceted notion including many types. They include inaccuracy, indeterminacy, imperfect knowledge, inconsistency, missing information, buzz, ambiguity, lack of reliability, doubt and others. Definitions of uncertainty are thus very rich and reflect its various properties. Thanks to numerous types of uncertainty, there is no clear agreement in the used terminology or a generally accepted meaning (MacEachren et al. 2005). The ambiguity of translation from various languages, in particular from English, has a distinctive impact on this situation too, and moreover, every scientific discipline operating with uncertainty, uses different terminology. Even professional literature states notions like uncertainty, indeterminacy, haziness, generality, bilinguality and others often as synonyms. From this view, the effort to follow authoritative foreign sources is surely praiseworthy, however, it is important to take into account also various specifics of the environment (professional, territorial, linguistic) in which these authorities formulate their standpoints (Kalouda 2010). To define and use individual terms correctly in geoinformatics, it is necessary to know how to work with the notions. The main aim of the paper is to discuss and suggest user centered framework for spatial data quality visualisations. In the first part of the paper is described spatial data quality, then continuously mention visualisation issues. Finally all aspect of new framework are discussed and described with min aim to facilitate the whole process.

Data Quality

This part deals with the quality of spatial data as a variable used to visualize uncertainty in spatial data. Therefore, in the beginning of this chapter, we must define notions and mutual relationships that are used in the text and thus illustrate problematic and ambiguous view of data and information with geographic or spatial aspects. The definition of the notion spatial data itself is difficult, and there are often problems with terminology and correct understanding of the meaning of this expression. In practice, there is not only this notion, but a whole spectrum of notions bearing similar and often confused meaning.

To determine and define the quality of spatial data, the actual meaning of the word *quality* must be understood. Moreover, the notion quality must be used with great discretion, because the notions uncertainty and quality of spatial data are not synonyms, although, in many cases, they are very close to each other. They have similar categories, deal with similar domains and in many cases, quality is confused with uncertainty and vice versa (Brus et al. 2013). Uncertainty in this meaning is considered a property allowing to evaluate how good (of what quality) are the available data, and uncertainty distinctively influences the quality of spatial data. Basic and distinctive difference is the fact that uncertainty can be introduced in any phase of the production of mapping data and GIS analyses (monitoring, conceptual modelling, measuring, analysis, visualization etc.) including perception by the end users, and as such, it concerns not only the quality of the spatial data, but the whole process of transferring geographic information.

The notion of quality and its contents has been undergoing historical development and transformation. From this point of view, it must be deciphered not only as part of the given context and discourse of the scientific field, but also with regard to its genetic and semantic interpretation. Thus, quality in the meaning of this paper can be most conveniently defined as each property or quality that belongs to something, or is connected to something, as the thing that modifies the thing to fulfil its purpose for which it was created or founded (Brugger 1994), and quality means the quality of the acquired information. From this point of view, a high-quality piece of information can be considered such piece of information that is useful and helpful for its "consumer", which, in connection with traditional definition of information means that the information quality is very dependant not only on basic data, but on the users themselves. High-quality information can then be defined as "accurate, trustworthy and sufficient for the decision-making of the user". In connection with the possibility to interpret data and information in many ways, also above-mentioned knowledge (tacit, explicit) and experience of the originator and recipient of the information enter the process. Therefore, it is very interesting to consider also the efficient quality of information—actual usefulness where an important factor is the skills of the information user. For users without sufficient skills, information is not of high-quality (useful), but, quite to the contrary, the quality of the presented information can even decrease. This fact is also reflected by the used definition of quality "convenience of use" (Beard et al. 1991; Chrisman 1984; Hunter 1999). In industrial sense, quality is basically the level of fulfilment of the production process, it is i.e. a function of non-substantial qualities such as completeness and consistency influenced by production process (Veregin 1998). Based on the definition of quality derived from encyclopaedic dictionary (Weber and Buttenfield 1993), quality is viewed as a distinguishing property allowing to distinguish whether the given property has positive or negative character. Based on this division, two other terms can be derived—accuracy and mistake. If quality has a positive character, it could be measured by the likeliness to the chosen model or reality, and the result is the already mentioned term accuracy. On the other hand, negative approach is measured by discrepancy, i.e. mistake. From the point of view of geoinformatics, there is a number of other views at the definition of quality of spatial data, because quality is always related to a range of data characteristics. Individual definitions thus usually reflect various views of the terminology of the notion itself and differ from author to author.

From the beginning of the development of quality of spatial data, this field has been a key subdiscipline of geoinformatics. It is considered to be of great importance both in the field of academic workplaces and in governmental and industrial organizations. The quality of spatial data has been the topic of many conferences and seminars and international activities (e.g. AGILE, GEOIDE, NCGIA and others). However, the whole discipline, with regard to uncertainty, deals with the above-mentioned problem, i.e. the disunion of used definitions and inaccurate terminology. The stated can be demonstrated by an example where representatives of governmental institutions and scientists working in academic environment are asked to define "data quality" and "uncertainty". It is very likely, that differing and ambiguous answers will be given. The problem lies in the community of people dealing with the quality of spatial data and uncertainty. In scientific articles, professional reports and studies, it is very problematic to identify various concepts that are called the same, or same concepts bearing differing names. As stated in the preceding chapters, notions like uncertainty, mistake, accuracy, ambiguity and their interpretation always reflect the scientific field from which the author using the terminology is coming from. From the point of view of quality and uncertainty of spatial data, in this interpretation, a division into two different communities of people dealing primarily only with quality or uncertainty can be traced. Each of these groups goes to their "own" conferences such as International Symposium on Spatial Data Quality or International Symposium on Spatial Accuracy Assessment in Natural Resources and Environmental Sciences. However, information and opinions are usually not interchanged between these two groups (Devillers et al. 2010).

Visualization of Data Quality

Recently, requests to visualize data quality occur more and more often. This fact was distinctively influenced by the acceptance of uncertainty as an important part of data (Schneiderman and Pang 2005). Despite the researching efforts in this field, there is still no defined way to visualize uncertainty making this activity user-friendly for wider group of users. Traditionally, information about uncertainty is visualized using very limited methods and to acquire final result, combination of various software is necessary. To find convenient methods and implement them, user testing of created representations must be still performed. In general, research in the field of spatial data visualization is primarily based on the definition of uncertainty itself, development of methods for graphic depicting of the information, creation of connections between spatial parts of data quality and visualization methods. User evaluation of thus created visualization is what matters, and still there are many questions concerning the impact of the use of specific types of visual representations on decision-making process to be answered.

Cognitive Aspects of Spatial Data Quality Visualization

As stated by Weick (2001), regardless of the field, it is necessary to take into account rather the perception of a specific individual than objective reality. This approach will allow easier understanding of uncertainty and related analyses, modelling and other aspects connected to this issue. In this part of the chapter, we are going to focus in particular on human factors influencing visualization of uncertainty and data quality.

In compliance with this approach, we can talk about cognitive cartography. The history of cognitive research in the field of geography and cartography can be traced back to the 1960s. With the beginning of occurrence and definition of the notion GIScience (Goodchild 1992), cognitive research was applied also as part of this discipline (Montello 2009). Cognitive cartography represents a research field that uses concepts and procedures of cognitive psychology to uncover mental representations and images with the objective to evaluate those images and to acquire geographic information from individual users (Blades and Spencer 1987). As part of this approach, human psyche is viewed as information processing system. Unlike computing approach which rather works with cartographic tradition, cognitive approach emphasizes in particular the user. The objective is to provide visualizations that work in compliance with what is known about human perception and cognition of geographic space and its visual representations (Fabrikant and Buttenfield 2001; Fabrikant 2001).

Leading professional in the field of cognitive cartography Montello (2002) said that the research performed in this field is an approach that can, thanks to technological progress, help to develop cartography as a whole. In his conception, cognition includes inner mental structures and processes that are part of perception, attention, thinking, consideration, learning, memory and linguistic and nonverbal communication, when cognitive cartography itself includes the application of cognitive theories and methods to understand and read maps, mapping and using the map to understand cognition. Cognition includes perception, learning, memory, thinking and solving problems including communications. It also includes external symbolic structures and processes, for instance maps and written instructions to perform formal spatial analyses that help internal recognition (Montello et al. 2004; Montello 1998). From the perspective of cognitive science, as part of information science, which, according to Peuquet (2002) also includes cartography, sensual constructions of individual situations are created by an individual, i.e. an individual is an active agent. From this point of view, an individual is also active in interaction between information structures and own conceptual framework. In the case of cognitive cartography, this interaction is performed between representations of models of geographic information such as maps, and saved knowledge structure of the user. This interaction allows the individual to perform individual adjustments or changes of their knowledge, because a piece of information acquired from the surrounding environment is the element that creates the knowledge itself. We should distinguish between cognition as a process and cognition as a result of this process, i.e. findings and knowledge. Basic approaches to explain cognition are symbolic representation, connectionism and semiotic approach.

Perception becomes an important element of visualization of uncertainty. Perception is an internal feeling created from an impression to our senses, accepting impressions or feelings using one of our senses. It is important to realize that human senses are not perfect and can thus distort reality. In other words, we can observe that perception itself provides more or less imperfect view of reality (Mezias and Starbuck 2003). Some professionals are of the opinion that human personality influences perception of individuals in particular. Others (Fiske and Taylor 2008) are of the opinion that properties that influence perception the most include clarity, legibility and clearness of the grounds. Here, it is convenient to state that each cartographic work can be viewed in three different ways: from the professional (contents of the map), technical (cartographic interpretation of the map contents, print, font, quality of work) and aesthetical (map composition, used colours, etc.) points of view (Voženílek 2004). From the point of view of uncertainty visualization, it can be observed that even though all stated parameters and variables have minimum impact on the correct transfer of information, none of the possibilities can be excluded (or declared insubstantial) for deeper understanding of this issue. By defining uncertainty across the spectrum of fields, other interesting facts that are more or less, directly or indirectly connected to depicting of uncertainty and subjective perception in maps were ascertained. As Bourgeois (1985) described, perception of individuals can differ distinctively, which leads to the perception of relativity.

User-Centered Framework for Uncertainty Visualisation

The preceding chapters summarized individual findings concerning great amount of individual aspects that can have distinctive influence over the visualization of uncertainty. Visualization of uncertainty is, from this point of view, a discipline lying at the borders of a number of different scientific fields. As a completely starting point during study of uncertainty can be considered the properties of uncertainty lying in the basis of natural processes and it is very difficult to describe them. Subsequently, all spatial data acquired from real environment can be analogically affected by a certain form of uncertainty. From the visualization point of view, it is very difficult to visually represent data in a form reflecting their true nature. Moreover from the perspective of the paper and proposed framework is necessary to consider also another frameworks in the same domain which have been already proposed. As the most important frameworks can be considered frameworks done by Voženílek and Kaňok (2011), Buttenfield (2000), Davis and Keller (1997) or Aipperspach (2009).

Uncertainty is an integral part of all operations with spatial data, but also part of all results and cartographic representations. From the point of view of the preceding text, uncertainty was understood as, predominantly, quality of spatial data, as a principal and key element influencing a number of other factors. Visualization of uncertainty also influences the whole process of transfer between the originator and recipient of information, because during cartographic communication, many of problems occur. While defining the concept of visualization of uncertainty, it must be stated that there are more concepts of uncertainty and it was not possible for this paper to deal with all of them.

Individual components of spatial data can be researched from the point of view of individual standards and groups of descriptive parameters of data that can be subsequently used for the visualization of uncertainty can be created. Most of those components are described using metadata. Metadata recording of individual components thus plays a principal role in describing uncertainty. These statements on information bearing data concerning the level of quality of the data however, despite all efforts in the GIScience community still do not include descriptions, instructions or at least recommendations how subsequently represent the information concerning the quality of the data. The situation is further complicated by the fact that as part of research in the field of visualization of uncertainty, there are only few works dealing with the complex issue of the visualization of uncertainty from the cartographic point of view. This fact is an initial presumption for designing a concept for decision-making process concerning the visualization of uncertainty with regard to the user. The design of the concept itself is based on decision-making process concerning the selection of expressing tool as part of topical cartography. As an integral part of individual criteria in the decision-making process during the creation of a map, individual aspects of the visualization of uncertainty are subsequently added in order to create a complete concept. The designed concept also includes many findings acquired during solving the project called Intelligent

System for Interactive Support of the Creation of Topical Maps that was solved at the Department of Computer Science of the Palacký University in Olomouc and as part of which, the issue of decision-making processes and aspects of visualization are studied in detail (Brus et al. 2009, 2010; Dobešová and Brus 2011).

During creation of a topical map, the initial step is the requirement of the visualization itself, given either by the request of a topical map or only by the need to visualize data for better understanding of the studied data and phenomena. In this moment, the author of visualization must decide the subsequent option and individual steps in the process of the creation of the map. There are usually many options to choose from and there are more variants on which the solution of a specific task can be based. The decision-making process concerning specific properties of the resulting representation is subsequently always influenced in particular by already acquired cartographic skills of the creator of the map and also by the knowledge concerning visualization techniques in general. The primary source of information is therefore study of professional literature. In this phase, a professional should gather and classify all generally available information based on which they will learn the basics, in particular in the given domain, in advance. With regard to the fact that almost 95 % of all map outputs is connected with digital representation of data, the needed knowledge must also include the knowledge of computer graphics and related disciplines. From the point of view of data processing, and subsequent visualization, the ability of the author to work with available software must be also mentioned, since knowledge and availability of technological and software equipment principally influences the whole process of the creation of a map. In many studies, this situation must be emphasized, because of the fact that the rules of optimal process of the creation of a map are often broken and the authors of the final visualization are often domain experts themselves. Cooperation of a domain expert with cartographer is rather unusual. The group of users that visualize and process spatial data using programs for the creation of maps concerning for instance land-use planning, environmental protection and other fields often create the maps intuitively and do not respect cartographic rules. This situation changes the traditional relationship between cartographer and user of the map. Therefore, it is necessary to emphasize initial knowledge of the map creator and consider it as key for the designed concept.

For the designed scheme (Fig. 1), initial step is to determine the objective of the map. The main objective of the map should always be defined before visualization itself and subsequently developed into individual objectives in order to ensure that all objectives will be met. An important factor is, in this case, feasibility of individual objectives and the possibilities to fulfil them. The objectives of visualization including uncertainty should be always evaluated critically in advance in connection to subsequently formulated functions of the occurring visualization. Also, it should be based on already defined methods and already created partial visualizations in order to proceed as is usual for the given topic. However, the used methods should be viewed prudently and critically since many visualization methods were only designed but not tested for user aspects. In general, also

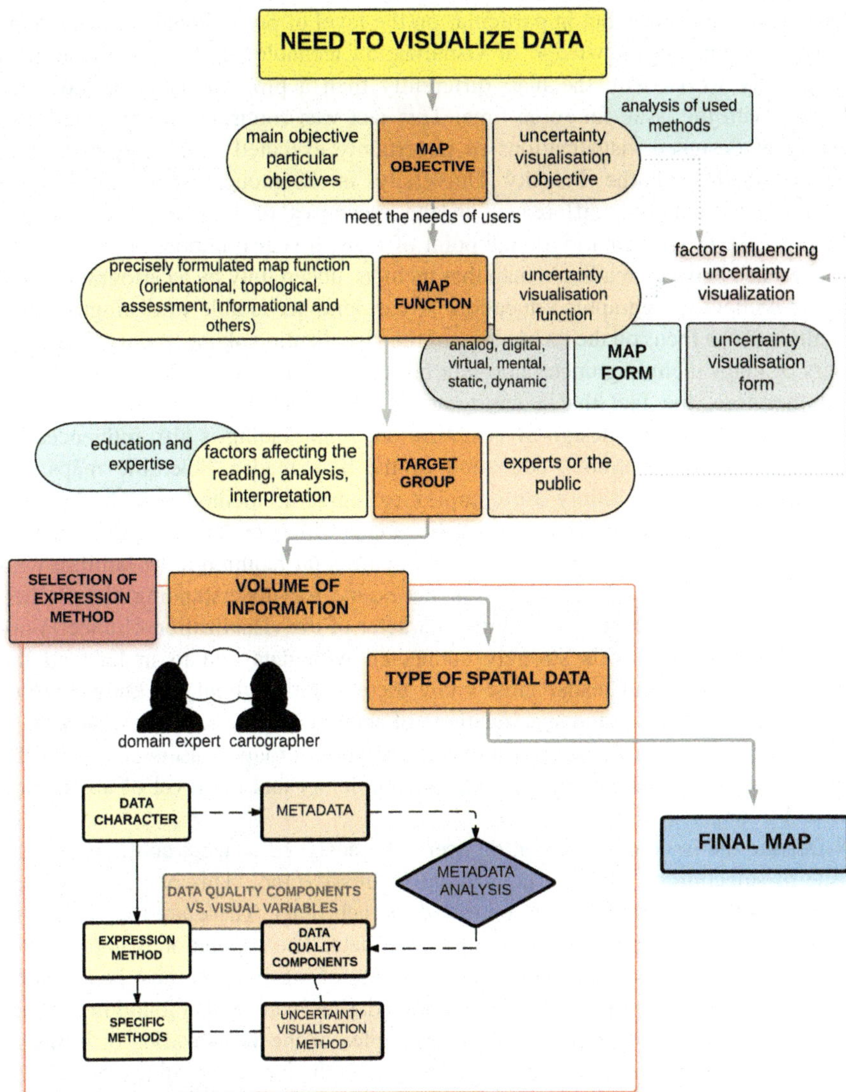

Fig. 1 Proposed framework for user centered spatial data quality visualisation

cartographic rules should be maintained so that visualizations of uncertainty can become a tool that would help to understand reality more comprehensively.

The main contents of the primary and secondary objectives of the map are in particular meeting of the needs of the expected target users of the map. The creator of the visualization who wants to visualize also the quality of spatial data and uncertainty connected to them must realize that individual steps in the process of map reading are influenced by many factors. Communication function of a map depends on education, but in particular on the level of professionalism of the map user. User with no knowledge of visualization technologies for visualization of uncertainty works with the map differently than a professional who has been dealing with the issue for some time. This fact was confirmed by executed user testing of designed visualizations of uncertainty executed at the Department of Geoinformatics of the Palacký University in Olomouc, where there were ascertained distinctive differences in the responses of individual respondents (Brus 2013). From the cartographic point of view, it is also important that there is a difference between reading a map that includes uncertainty by a cartographer and reader without cartographic education. A cartographer is able to distinguish fine details and can focus on the technology of the visualization of uncertainty. Most end users of maps including uncertainty are not cartographers from this point of view and therefore, the fact that a cartographer reads the map differently is of no importance during the design of visualization. Map reading is also influenced by age and already mentioned education of the map reader. Scientific maps and visualizations of uncertainty use complex cartographic methods which are commonly used by specialists, however, reading maps with uncertainty requires the user to be well versed in the field. The user is in particular required to be familiar with specific terminology. Terminology is necessary for understanding the studied domain and, in particular, the specific character of the visualization of uncertainty itself. This uncertainty is strongly connected with data and to understand the visualization, the map reader must know these facts. Without the knowledge of terminology, processes and specifications of specific data sets, the designed maps are often illegible for the user. Therefore, the target group of users is key for the selection of the visualization method, its complexity and the level of abstraction needed to understand it.

Despite the issue of visualization users, the designed concept continues by the issue of selection of the most convenient expressive tool. This part is connected with two other criteria that are connected to each other. These are the volume of conveyed information and the kind of spatial data. The stipulation of the level of complexity of the map and the resulting volume of information must be based on the already established principles. Recommendations can be also found in individual perception theories and mentioned principles of the visualization of information. Subsequently, the contents of a map are distinctively influenced by the objective of the map, which then also defines the selection of methods convenient for visualization. The selection of specific methods must be in particular influenced by the primarily depicted reality and by the amount of information that is to be included in the map to fulfil the main objective of the map and from the point of view of the visualization of uncertainty, to fulfil these objectives. The conclusions from this consideration will significantly influence the overall amount of information and therefore also graphical contents of the map. The selection of method for the visualization of uncertainty is then subsequently directly depending upon the selection of cartographic methods for the depiction of a phenomenon. The selection of method is also determined by the criteria of evaluation of the kind of spatial data, which is executed according to the character of the description of properties of depicted phenomena. In this phase, the creation of visualization is usually influenced by an expert in the field and by a cartographer. The creator of visualizations must, as part of selection of the most convenient expressive tool acquire quite a few pieces of information from the expert, then analyze it and subsequently decide the specific way of visualization. After the decision, the map including uncertainty is created.

As part of selection of the expressive tools, the following questions can be asked for instance:

- What is the character of the depicted data
- Which components of the quality of data will be visualized?
- Is there a metadata record for the data?
- What are the convenient expressive tools for the visualization of uncertainty?
- To depict uncertainty, which method shall be used: internal or external?
- Which specific methods for the visualization of uncertainty shall be selected?

Acquired expert knowledge is based on the education, gathered experience, intelligence and intuition and communication with an expert. Efficient integration of the mentioned parts leads to the creation of a decision and selection of cartographic expression method for the visualization of uncertainty. Necessary presumption is information on visualized data, in particular metadata description of the data. Based on a detailed analysis of individual components of data quality, it is then necessary to select convenient visual variables and be able to achieve by this integration correct transfer of information concerning the quality of the data using a specific visualization method or combination of methods, bearing the end user in mind as main and key element of the whole concept in mind.

If individual recommendations for the visualization of uncertainty are not maintained, the theory of perception is violated, so mistakes and inaccuracies occur in the map outputs, which leads to incorrect interpretation of not only individual parts, but the whole system of depicted data. From the point of view of the demanding character of the visualization of uncertainty, purposeful and correct map can only be created if three professionals (domain expert, cartographer and geoinformation scientist) or a team of scientists cooperate, which is proven by actual situations when the most elaborate visualizations of uncertainty can be found in project intentions and mutual cooperation between teams of professionals.

The whole process of the visualization of uncertainty can be, based on studying individual key aspects of the visualization of uncertainty, divided into a few basic areas. Uncertainty is created in all parts of the process of cartographic communication and from the point of view of possible elimination, this is an element that cannot be removed. According to the occurrence of uncertainty, we can talk about the uncertainty during the presentation of spatial data due to differing quality of data due to gathering, transformation and other factors. Visualization uncertainty (the uncertainty of visualization) occurs due to a number of factors, when basic processes concern the transfer of information into digital form and subsequent presentation to the end users. Even the users themselves represent a source of uncertainty, because perception is affected by a number of personality and other characteristics (Kubíček and Šašinka 2011). It is necessary to emphasize that some factors during the visualization of uncertainty can be removed completely, others can be eliminated partially, and some must be taken as initial and unchanging presumption while designing newly created visualizations.

Discussion

From the point of view of this paper, it is of utmost importance to view the definition of the notion of uncertainty in environmental studies. As was already said in the very beginning of the paper, there is not just one interpretation of uncertainty, different concepts, opinions and definitions can be found. Based on this inconsistency, finding convenient tools and theories that would apply to the whole issue of this scientific discipline is problematic. Due to the complexity of the issue, there is also no unified terminology that would approach the topic globally.

Final and main objective of uncertainty visualizations is to influence the decision-making process of the map reader in way which will help to make more adequate decision based on cartographic symbology. From the point of view of the ability to distinguish individual cartographic signs, the biggest problem and subject of discussion is the lack of software tools for the visualization of uncertainty. Current geographic information systems have almost no implemented tools that can work with uncertainty more comprehensively. It is always only partial functionality of possible ways of visualization. The computing background itself offers program tools, however, only as part of some approaches to modelling of uncertainty. A tool that would allow the user simple and quick display using more methods without the need to convert data between various programs is still missing.

As part of some software, a number of modules and extensions that can work with uncertainty is available, and there is also a considerable number of individual applications, however, these are predominantly the results of research activities, which were not spread extensively or in some cases were not even put into proper operation after testing or termination of financial support. Very often, these tools are specialized for a certain part of the work with uncertainty, so there is no complex solution in this regard. Moreover, these tools are usually not user friendly, and require the user to master deeper theory of work with uncertainty. Therefore, there are rather used as supportive tools, since the user must know where to find them, which again makes their use even more complicated.

From the user point of view, deeper knowledge of the stated issue and also familiarity with GIS products is necessary. Common GIS user is thus not informed by classic software about the uncertainty connected to spatial data. Moreover, the crucial problem with some program environments is cartographic functionality itself, which varies between GIS products considerably. In practice, complicated solutions that prevent spreading between common users must be used. Depicting all aspects of uncertainty in a map is impossible. Basic and limiting elements are cartographic expression tools, limitation of the form of the map and predominantly the users of the data themselves. From this point of view, the most important and most critical condition of meaningfulness of visualization is its purpose and in particular the end user.

References

Aipperspach R (2009) Categorization and toolkit support for uncertainty visualization. http://vis.berkeley.edu/courses/cs294-10-sp06/wiki/images/2/21/AipperspachFinalReport.pdf

Beard MK, Buttenfield BP, Clapham SB (1991) NCGIA research initiative 7: visualization of spatial data quality. Scientific report for the specialist meeting, 8–12 June 1991, Castine, Maine. National Center for Geographic Information and Analysis

Blades M, Spencer C (1987) How do people use maps to navigate through the world? Cartographica 24:64–75

Boukhelifa N, Duke D (2007) The uncertain reality of underground assets. In: Joint workshop on visualization and exploration of geospatial data, ISPRS, ICA, DGFK, June, pp 27–29

Bourgeois LJ (1985) Strategic goals, perceived uncertainty, and economic performance in volatile environments. Acad Manage J 28:548–573

Brugger W (1994) Filosofický slovník. Naše vojsko, Praha

Brus J (2013) Visualization of uncertainty in environmental studies. Dissertation thesis, Palacky University Publishing House, 144 pp

Brus J, Dobešová Z, Kaňok J (2009) Utilization of expert systems in thematic cartography. In: International conference on intelligent networking and collaborative systems (INCOS 2009), pp 285–289

Brus J, Dobešová Z, Kaňok J, Pechanec V (2010) Design of intelligent system in cartography. In: 9th Roedunet IEEE international conference, pp 112–117

Brus J, Pechanec V, Kilianová H (2013) Uncertainty vs. spatial data quality visualisations: a case study on ecotones. In: 13th international multidisciplinary scientific geoconference, v tisku

Buttenfield B (2000) Mapping ecological uncertainty. In: Hunsaker CT, Goodchild MF, Friedl MA, Case TJ (eds) Spatial uncertainty in ecology. Springer, New York, pp s.116–132

Chrisman NR (1984) The role of quality information in the long-term functioning of a geographic information system (Part 2: Issues and problems relating to cartographic data use exchange and transfer). Cartographica 21:79–88

Davis TJ, Keller P (1997) Modelling uncertainty in natural resource analysis using fuzzy sets and Monte Carlo simulation: slope stability prediction. Int J Geogr Inf Sci 11:5

Devillers R, Stein A, Bédard Y, Chrisman N, Fisher P, Shi W (2010) Thirty years of research on spatial data quality: achievements, failures, and opportunities. Trans GIS 14:387–400

Dobešová Z, Brus J (2011) Coping with cartographical ontology. In: Conference proceedings SGEM 2011, 11th international multidisciplinary scientific geoconference. STEF92 Technology Ltd., Sofia, pp 377–384

Duckham M, Mason K, Stell J, Worboys M (2001) A formal approach to imperfection in geographic information. Comput Environ Urban Syst 25:89–103

Fabrikant SI (2001) Evaluating the usability of the scale metaphor for querying semantic spaces. In: Montello DR (ed) Spatial information theory: foundations of geographic information science. Springer, Berlin, pp 156–172

Fabrikant SI, Buttenfield BP (2001) Formalizing semantic spaces for information access. Ann Assoc Am Geogr 91:263–280

Fiske ST, Taylor SE (2008) Social cognition: from brains to culture. McGraw-Hill Higher Education, Boston

Goodchild MF (1992) Geographical information science. Int J Geogr Inf Syst 6:31–45

Hunter GJ (1999) Managing uncertainty in GIS. Geogr Inf Syst 2:633–641

Kalouda F (2010) Riziko a nejistota–příspěvek k precizaci pojmů. In: 5. mezinárodní konference Řízení a modelování finančních rizik. VŠB-TU Ostrava, Ekonomická fakulta, katedra Financí

Kubíček P, Šašinka Č (2011) Thematic uncertainty visualization usability–comparison of basic methods. Ann GIS 17:253–263

MacEachren AM, Robinson A, Hopper S, Gardner S, Murray R, Gahegan M, Hetzler E (2005) Visualizing geospatial information uncertainty: what we know and what we need to know. Cartogr Geogr Inf Sci 32:139–160

Mezias JM, Starbuck WH (2003) Studying the accuracy of managers' perceptions: a research odyssey. Br J Manage 14:3–17

Montello DR (1998) A new framework for understanding the acquisition of spatial knowledge in large-scale environments. In: Egenhofer MJ, Golledge RG (eds) Spatial and temporal reasoning in geographic information systems. Oxford University Press, New York, pp 143–154

Montello DR (2002) Cognitive map-design research in the twentieth century: theoretical and empirical approaches. Cartogr Geogr Inf Sci 29:283–304

Montello DR (2009) Cognitive research in GIScience: recent achievements and future prospects. Geogr Compass 3:1824–1840

Montello DR, Waller D, Hegarty M, Richardson AE (2004) Spatial memory of real environments, virtual environments, and maps. In: Allen G (ed) Human spatial memory: remembering where 251–285. Lawrence Erlbaum, Mahwah

Peuquet DJ (2002) Representations of space and time. Guilford Press, New York

Schneiderman B, Pang AT (2005) Visualizing uncertainty: computer science perspective. In: National Academy of Sciences Workshop on toward improved visualization of uncertain information, 2–4 March 2005

Veregin H (1998) Data quality measurement and assessment. NCGIA Core Curriculum in Geographic Information Science, pp 1–10

Voženílek V (2004) Aplikovaná kartografie I.: tematické mapy. Univerzita Palackého, Přírodovědecká fakulta, Olomouc

Voženílek V, Kaňok J (2011) Metody tematické kartografie: vizualizace prostorových jevů. Univerzita Palackého v Olomouci, Katedra geoinformatiky, Olomouc

Weber CR, Buttenfield BP (1993) A cartographic animation of average yearly surface temperatures for the 48 contiguous United States: 1897–1986. Cartogr Geogr Inf 20:141–150

Weick KE (2001) Making sense of the organization. Blackwell Business, Oxford

Comparing Paper and Digital Topographic Maps Using Eye Tracking

Annelies Incoul, Kristien Ooms, and Philippe De Maeyer

Introduction

Paper Versus Digital Maps

For long, paper was seen as the primary means of communication. Because of its many advantages, it is a universal and dynamic medium: ease of use, transportation and archiving (Johnson et al. 1993). Therefore, in the past it was obvious to choose for the paper medium to depict geographical information. More recently however, maps are not limited anymore to the paper medium, but are also disseminated in a digital format (Peterson 1997). The Internet has played a significant role in the rise of the digital cartographic products: they are accessible by a wide range of users, can be easily updated and manipulated. Examples of leading projects in digital cartography are Google Maps and OpenStreetMaps. Projects like these introduced the geographical information (visualised through digital cartographic products) to the general public. This is closely linked with the world wide use of Global Positioning Systems (GPS), which also caused a considerable rise in the number of available maps (Hurst and Clough 2013; Pederson et al. 2005).

These digital maps also come with a number of drawbacks, mainly related to the limitations of the screen on which they are depicted: resolution, colour range and dimensions (e.g. Kraak and Brown 2001; Peterson 2003). The resolution of the screens on which digital maps are depicted are typically much lower than this of paper documents (e.g. 92 ppi versus 1,200 dpi respectively). What is more, different screens have different colour ranges. Maps that are distributed through the Internet are visualised on different screens, which could cause a shift in how the colour is represented. Finally, the dimensions of the screens can vary significantly: 7 in.,

A. Incoul • K. Ooms (✉) • P. De Maeyer
Department of Geography, Ghent University, Krijgslaan 281 (S8), 9000 Ghent, Belgium
e-mail: kristien.ooms@ugent.be

© Springer International Publishing Switzerland 2015

J. Brus et al. (eds.), *Modern Trends in Cartography*, Lecture Notes in
Geoinformation and Cartography, DOI 10.1007/978-3-319-07926-4_26

21 in., 50 in., etc. Mostly, the dimensions are rather limited. These limited screen sizes are compensated by zooming and panning tools that are nowadays available on most web maps (e.g. Cartwright 2012; Kraak and Brown 2001; Peterson 2003; van Elzakker and Griffin 2013). Despite these tools, it is more difficult to have an overview of the depicted region, in combination with details of a selected sub-region.

Since we are living in a digital era, one might wonder why paper documents still haven't disappeared. Bondarenko and Janssen (2009) indicate that the 'paperless office' is still an ambition: despite the technological advancements, not all paper documents have been replaced by digital ones. Paper documents have a number of functions and characteristics that cannot be replaced digitally: their tangibility, they are used to reduce the cognitive load, obtain an overview of the tasks at hand, to indicate the state of a task, as visual cues, because they are easier to read, to have an overview over the complete document, etc. The same arguments can be used to explain why paper maps still have not disappeared (Hurst and Clough 2013).

Previous Research

Because both media are still frequently used, it is interesting to study their relationship to the map users: what do they prefer, what is more efficient, what is more effective and under what conditions (application, user characteristics, etc.). However, not many studies have been devoted to this subject. Hurst and Clough (2013) used a web-based questionnaire and a task-based user study to investigate the future role of paper maps in the digital era. They found that the preference for a certain medium depended on the task at hand: digital maps for route planning and locating points of interest; paper maps for navigation on foot. Expert map users also preferred paper maps for car navigation, whereas novices preferred the digital navigation aid. The research indicated that paper maps still possess enough qualities which makes them survive in the foreseeable future. Regarding reading on paper and on a screen, several authors found a different user behaviour (e.g. Liu 2005; O'Hara and Sellens 1997; Sutherland-Smith 2002). Nevertheless, reading full texts is different that a map reading task, which means that these results cannot be applied directly to cartographic visualisations.

Pederson et al. (2005) and Verdi et al. (2003) investigated which of both media (paper or digital) should be used while teaching pupils to read a map. The two studies agreed in their conclusion: the digital format is efficient, but it could not surpass the paper maps. It is however, important to include both formats in the teaching materials.

The present research aims to extent these studies, with a focus on the map users' attentive behaviour. Previous research has investigated the map users' preference, satisfaction, efficiency and effectiveness on both media (Hurst and Clough 2013; Pederson et al. 2005; Verdi et al. 2003). However, when completing a task on a map, the user will need to read the content of the map. The structure and difficulty of this reading process could have an influence on the users' cognitive load (Harrower 2007; Bunch and Lloyd 2006). Due to the different characteristics of screens and paper, the same information is displayed differently: resolution, colour display, etc. This could cause differences in how the users read the information on the map. These differences could have an influence on the users' perceived level of satisfaction and thus preference towards a certain medium. Therefore, it is important to study these processes to be able to explain the findings in previous research.

Extending the Research: Eye Tracking

Processing a visual scene, such as a map, is based on the visual input on the one hand (bottom-up processing) and the user's background knowledge (top-down processing) on the other hand (e.g. MacEachren 1995). Wolfe (1994) described a model (or set of rules) that predict where a user will focus his attention during a visual search. Recent findings in the field of change blindness also stressed the importance of attention to be able to interpret a visual scene (Rensink 2002; Simons and Ambinder 2005).

Using eye tracking, the users' eye movements are registered: where a user is looking (Point of Regard, POR), but also how long and how often. These latter metrics can give insights in the user's cognitive processes while trying to interpret the visual content. When studying the interpretation process, fixations are most important. These define the time intervals when and locations where a user is processing the visual content. Saccades are fast eye movements between fixations during which no information is processed. More fixations at a certain location indicates that the user's attention is attracted by something at this location. Longer fixations can indicate that the user finds it difficult to interpret the content. These and many other eye movement metrics have been studied in previous research, including their potential link to the user's cognitive processes (Holmqvist et al. 2011). It is beyond the scope of this paper to discuss them all. A good overview can be found in a number of books and journal articles: Duchowski (2007), Holmqvist et al. (2011), Goldberg et al. (2002), Jacob and Karn (2003), Poole and Ball (2006), Rayner (1998), among others.

Paper maps have already been included in eye tracking studies in the 1970s (e.g. Dobson 1977; Jenks 1973) and the 1980s (e.g. Castner and Eastman 1984, 1985; Steinke 1987). The stimuli at that time were mainly dot maps. The authors concluded that, although the method was applicable, no new knowledge could be derived. Consequently, the method almost disappeared in cartographic research afterwards. In the beginning of 2000 however, a mobile (head mounted) eye tracker was successfully used by Brodersen et al. (2001) to study the effectiveness of a new symbology for the printed topographic maps in Denmark. Nevertheless, no comparison was made with digital maps.

Recently, a rise is noticed in the application of the method in cartographic user research (e.g. Çöltekin et al. 2009; Fabrikant et al. 2008; Kiefer et al. 2014; Ooms et al. 2012, 2014; Popelka and Brychtova 2013). These studies focus on digital cartographic products, possibly with animations (Çöltekin et al. 2009, 2010), or navigation tasks (Kiefer et al. 2014). This renewed interest is closely linked to the recent need to gain better understanding of the cognitive processes (and limits) of map users while working with highly dynamic, interactive, animated screen maps (Cartwright 2012; Fabrikant and Lobben 2009; Montello 2009).

Hegarty et al. (2010), for example, studied expert and novice map users' eye movements while executing a task on (digital) weather maps. They discovered that eye movements were mainly guided by top-down factors and that a good design facilitated the processing of task relevant visual features. Ooms et al. (2014) found that both expert and novice users tend to focus their attention on a reference frame in the map image (main structuring elements) in order to store the map content in memory.

Previous research using eye tracking used paper or digital stimuli. However, up till now both media have not been combined and compared in a single user study. The aim of this research is to investigate whether the attentive behaviour is different when the same content (topographic map) is presented on paper or on a screen (with the same dimensions), in a controlled environment. Therefore, the users' attentive behaviour will be analysed and compared in a controlled environment, by registering the participants' eye movements on both digital and paper stimuli. The research question is as follows: Can users find labels more efficiently on a paper map compared to the digital medium? More details and justifications regarding the study design (participants, stimuli, task, etc.) are described in the next section.

Study Design

Participants

In total, 32 participants took part in the user study. All participants were students (M.Sc.) or researchers of the Department of Geography (Ghent University). This group was selected to ensure that all participants would have similar domain knowledge in geography and cartography. They are familiar with the design of the Belgian topographic maps (see section "Stimuli") and have experience in the use of both digital and paper maps. In a post-study questionnaire, these characteristics were also verified. These participants were divided into two groups, each with similar user characteristics. This grouping operation is necessary to avoid that a participant would see the same regions twice (once on screen and once on paper) which would distort the measurements due to a learning effect. The participants received a code that corresponded to the order in which they participated. Participants with an odd code number belong to the first group and participants with an even code number belong to the second group.

Due to some problems with registering the participants' eye movements on a large screen (at close viewing distance), not all trials could be included in the analyses (missing data, large deviations, etc.). Still, sufficient data was gathered (at least 20 persons per trial) to be able to conduct statistical analyses. The selection of suitable data is discussed in detail in section "Creating the Gridded Visualisation".

Stimuli

Twelve different stimuli were used in the user study, based on six different map sheets of topographic maps on 1:10,000 of the Belgian national mapping agency (IGN/NGI Belgium): Assesse 54/1 N, Ciney (Est) 45/2 S, Daussois 52/8 S, Leignon 54/6 N, Onhaye 53/7 N and Walcourt 52/8 N. These six map sheets (see Fig. 1a) were visualised on a screen (digital medium) or on paper. Each participant will perform a task (see section "Task") on three digital and three paper maps. Using two similar user groups with different stimuli (switching between paper and digital map sheets) avoids a learning effect and keeps the duration of the user test limited (as this could result in distorted measurements due to a diminished level of concentration or fatigue by the participant).

The selection of these six map sheets was based on a number of criteria. First, it should be unlikely that the participants knew the region by heart. All participants lived in the northern part of Belgium and, therefore, the selected regions are located in the southern part. Furthermore, the participants had to indicated in a post-study questionnaire if they were familiar with one of the depicted regions (which would then be removed from the results for that participant). Second, the type and number of map objects should be limited and similarly distributed across the six map sheets. As a consequence, the selected map sheets consist mainly of some major roads, a small number of villages surrounded by meadows, fields and forests. Third, the number of toponyms, their hierarchical classifications and their distributions should be similar for all stimuli.

The aim of the study is to investigate the potential difference in the users' attentive behaviour while performing a task on two maps, between which the medium (screen or paper) is the only difference. Therefore, all stimuli are presented with exactly the same dimensions and on exactly the same location. Furthermore, only the map page is presented to the participant, without the title, legend, scale bar, meta data, etc. as this could distract the user. The dimensions correspond with these of the original map sheet (82.5×52.7 cm). The digital maps have a resolution of 125 dpi and were visualised on a TV screen (see section "Apparatus and Set-Up" for more details). The corresponding paper maps were printed at a resolution of 250 dpi.

In order to be able to compare the results between the two user groups, the six regions were displayed to all participants. The only difference was the medium with which the maps was displayed. While one saw a stimulus on the paper medium, the

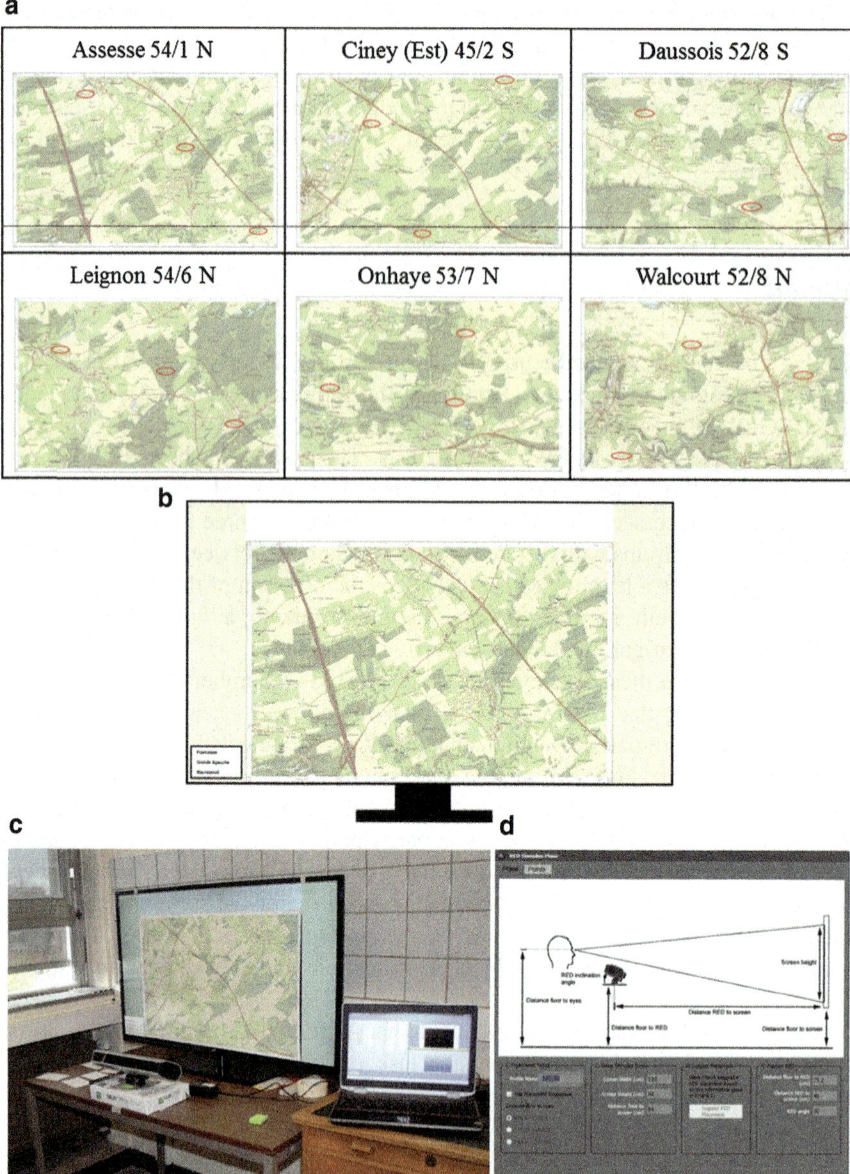

Fig. 1 Selected map sheets with indication of the labels that had to be located (**a**), TV screen, with a stimulus and the list with labels (**b**), experiment set-up (**c**) and settings (**d**)

Table 1 Order of the stimuli (and medium) for each user group

Trial number	Group 1 (odd)	Group 2 (even)
1	Assesse (digital)	Ciney (digital)
2	Ciney (paper)	Assesse (paper)
3	Daussois (digital)	Leignon (digital)
4	Leignon (paper)	Daussois (paper)
5	Onhaye (digital)	Walcourt (digital)
6	Walcourt (paper)	Onhaye (paper)

other group saw the same stimulus on digital medium and vice versa. Both groups started with a digital map and the next five maps were displayed alternately between paper and digital. An overview of which participants saw which stimuli and in what order is given in Table 1.

Task

Each participant had to conduct a visual search on the six stimuli. The visual search task was selected because this is often executed 'in real life' on a map: finding one or multiple point(s) of interest. The participants were instructed that they had to locate three labels in the map image. These labels were listed in the lower left corner of the screen on which both the digital and paper stimuli were presented (see Fig. 1b and section "Apparatus and Set-Up" for the set-up). The same list was used for the corresponding paper and digital stimuli. The participants could choose in which order they wanted to search for the names. Forcing the participants to search the names, for example, in the same order as they are listed could distort the participant's natural search behaviour and thus attentive behaviour. Nevertheless, the participants had to indicate their search strategy in the post-study questionnaire.

After the participants had found all three labels on a certain stimuli, they had to indicate this to the moderator. If a paper map was depicted, the moderator would detach it from the screen and present the next map digitally on the screen. If a digital map was depicted, the moderator would depict a black screen and attach the next paper map on the correct location to the screen (see also section "Apparatus and Set-Up"). The participants eye movements were recorded during the visual search.

After completing the six trials, the participants had to fill in a questionnaire to obtain background information (age, sex, education), but also regarding the familiarity with the depicted regions, their search strategies, etc.

Apparatus and Set-Up

The user study was conducted in the Eye Tracking Laboratory of the Department of Geography, Ghent University. This laboratory is equipped with an SMI RED system, which registers the users' eye movements at 120 Hz. This remote eye tracker was not used in its standard set-up, attached below a 22 in. screen, because the stimuli with much larger dimensions had to be depicted. In order allow objective comparison between digital and paper maps, the digital maps had to be displayed at exactly the same scale, size and position as the paper maps. Consequently, the eye tracker was used in a stand-alone mode and placed in front of a large television screen. The TV is a JVC LT-50HW45U with a diagonal of 50 in. and a resolution of $1,920 \times 1,080$ pixels (or 44 ppi). During the experiment, the moderator could follow the participants viewing behaviour on a separate laptop. This set-up is displayed in Fig. 1c.

The paper maps were attached to the screen with tape. A black rectangle was depicted on the screen, which indicated the correct position for the paper map. Furthermore, the black colour avoided that too much light would be transmitted through the paper map. In the set-up in Fig. 1c, a paper map is displayed.

Using this stand-alone mode, a number of parameters had to be specified in the eye tracking software (SMI Experiment Center), such as the viewing distance, dimensions of the screen, position of the eye tracker relative to the screen, etc. Dependent on the height of the participant, the inclination angle of the eye tracker had to be adapted. Consequently, this parameter is different for each participant. These settings are displayed in Fig. 1d.

Methodology

To set-up the user study and subsequently analyse the data, a combination of different software packages was used. This consists of software related to the eye tracker (SMI), existing open source software (such as OGAMA), specialised software (such as SPSS) and tools we have written ourselves, for example, to be able to convert the data between the different programmes.

The study is 'programmed' in SMI's Experiment Builder: specifying the stimuli, their order, the parameters for the calibration, the parameters for the stand-alone set-up, the questionnaire, etc. These settings were then used in combination with SMI's iViewX software, which drives the eye tracker during the tests. The recorded data can be loaded into BeGaze (SMI). This software contains tools to do (mainly visual) analyses with the data and export the raw or processed (fixations and saccades) data in ASCII format.

In this experiment, it was decided to use the open source software OGAMA (Open Gaze and Mouse Analyzer). This is an open source software, that can visualise and analyse mouse and eye movement data, developed at the Freie

Universität Berlin. Raw data (in ASCII) from different types of eye trackers can be imported and processed based on user defined settings regarding the duration and dispersion of a fixation. The tools available in OGAMA to visualise and analyse the eye tracking data are similar as in BeGaze, but the software has the advantage that it does not require a licence key.

The structure of the raw eye tracking data exported by BeGaze does not correspond to the requested structure of that data by OGAMA. Consequently, the authors have written a tool in JAVA that can do this conversion automatically: *SMI2OGAMA*. The statistical analyses are executed in a specialised software package: SPSS. Because the (difference in) users' attentive behaviour is the main interest of this user test, a fixation report will be exported from OGAMA. This text file lists all fixation from all participant underneath each other, with its characteristics (e.g. duration, x-position, y-position) on the same line.

In order to be able to visualise the values of the fixations in a grid of AOIs (see section "Creating the Gridded Visualisation"), the authors have written a second tool in JAVA, *fr2grid*, that takes the fixation report as input, places the fixations in the corresponding cell in the grid (based on its x and y position) and calculates the total number of fixations and total duration of fixations in each grid cells. This is done separately for the twelve stimuli. The resulting grids are visualised in (Microsoft Office) Excel, based on the 'conditional formatting' tool.

However, before analysing the recorded eye movements, the data had to be verified and subsequently selected. This selection was based on the accuracy of the calibration, the tracking ratio of the eye tracker and a visual verification of the location of the fixations. This is explained in more detail in the next section.

Data Selection

Not all data could be included in the analyses due to some problems in the calibration phase, which could result in a too low accuracy, too low tracking rate, or a shift in the data.

Calibration Accuracy and Tracking Ratio

When using the eye tracker in a standard set-up (attached to the 22 in. screen), the acceptable accuracy is typically limited to $0.5°$. Because this study places the eye tracking device in a stand-alone setting, in front of a much larger screen and with different relative distances, higher deviations in the calibration will still be accepted. For this study the limit is placed on $1°$. The justification of this values takes the viewing distance and the task into consideration. The distance between the user and the screen was 116 cm. A deviation of $1°$ in the participant's calibration would consequently result in a deviation of 2.02 cm on the screen. As the task is to locate a label on the map, this can be verified within a 2.02 cm range. All deviation

Table 2 Deviation (in x and y) and tracking ratio per test person

Part. code	Deviation X (°)	Deviation Y (°)	Tracking ratio (%)	Part. code	Deviation X (°)	Deviation Y (°)	Tracking ratio (%)
P01	0.5	0.9	93.8	P19	1.2	0.9	91.4
P02	1.1	1.9	91.0	P20	0.6	0.6	94.7
P04	0.2	1.3	91.4	P21	0.5	1.1	93.6
P05	0.4	0.6	95.1	P22	0.8	0.8	80.1
P06	0.6	2.4	66.7	P23	0.6	1.4	92.2
P07	1.2	1.1	96.4	P24	0.5	0.8	94.9
P08	0.8	1.3	96.4	P25	0.8	0.1	93.8
P09	0.7	0.5	96.3	P27	0.6	0.6	73.0
P10	0.8	0.6	88.8	P28	1.0	1.5	96.0
P11	0.4	0.5	95.1	P29	0.5	0.8	95.9
P13	0.8	0.7	96.7	P30	0.6	0.5	93.9
P14	0.7	0.9	92.3	P31	0.3	0.8	93.7
P15	0.9	0.8	96.2	P32	1.0	0.3	90.9
P16	0.7	1.6	95.6	P33	0.3	0.4	96.6
P17	0.3	0.6	93.6	P34	0.7	0.6	94.1
P18	0.6	0.2	95.5	P36	0.6	0.4	95.6

above 1° deviation are therefore deemed unacceptable. An overview of the participants' deviations in X and Y is presented in Table 2.

For several reasons (such as interference of other light sources, participants' movements, the range of the eye tracker, etc.) the eye tracker can lose the position of the eye (pupil and/or registered reflection). No measurements are recorded in these cases. Since the eye tracker was not used in its standard setting, it was of utmost importance to check this ratio (% in time that the eye tracker could track the position of the eye and thus record the participant's POR). The result is listed in Table 2 (last column) and shows that overall the tracking ratio is rather high. Only three participants show a tracking ratio below 85 %, which is considered unacceptable for this study.

Visual Verification and Shift Correction

The parameters discussed previously do not give insights in the acceptability of the recordings during the individual trials. Therefore, the eye movement data for every trial (and participant) is inspected in OGAMA. The fixation module in OGAMA allows visualising the fixations as dots on top of the corresponding stimuli. During this visual inspection, it was noticed that the location of the fixations could be shifted in the X or Y direction. The fixation module in OGAMA has a tool with which the fixations can be translated over a certain distance in case of a shift. As a consequence, the shifted measurements can be corrected and still be used in the analyses. Other recordings could not be corrected.

Table 3 Selected data: participants and stimuli

Part.	1D	2P	3D	4P	5D	6P	Part.	1P	2D	3P	4D	5P	6D
P01	x	x	x	x	x	x	P10		x	x	x	x	x
P05	x	x	x	x	x	x	P14	x	x		x	x	x
P07	x	x	x		x		P16	x	x	x	x	x	x
P09	x	x	x	x	x	x	P18	x	x	x	x	x	x
P11	x	x	x	x	x	x	P20	x	x	x	x	x	x
P13	x	x	x	x	x	x	P22			x	x	x	x
P15	x	x	x	x	x	x	P24	x	x	x	x	x	x
P17	x	x	x	x	x	x	P28	x		x	x	x	x
P21	x		x	x	x	x	P30	x	x	x	x	x	x
P25	x	x	x	x	x	x	P32	x	x	x		x	x
P27	x			x	x	x	P34	x	x	x	x	x	x
P29	x	x	x	x	x	x	P36	x	x	x	x	x	x
P33	x	x	x	x		x							
TOT.	13	11	12	12	12	12	TOT.	10	10	11	11	12	12

The recordings that have been selected are listed in Table 3. The recordings of some participants have been removed completely, whereas for other participants the recordings of a limited number of trials was considered unacceptable. From this table it can be derived that, after selection, a sample size of at least 10 individuals is still available for each stimulus.

Creating the Gridded Visualisation

The registered data will be visualised in a grid AOIs. The method has already been used by Brodersen et al. (2001) and Ooms et al. (2014). Since the aim of the paper is to study the users' attentive behaviour, only the fixations will be considered. The number of fixations on a specific location might indicate where interesting elements are located. The duration of the fixations can indicate whether the participants find it difficult to interpret the visual content (at a certain location) (e.g. Duchowski 2007; Holmqvist et al. 2011).

Heatmaps (which are actually density maps) are often used to visualise the distribution of the fixation densities of a certain participant across the stimuli (Spakov and Miniotas 2007). Typically, a green-yellow-red colour ramp is used, although this is not cartographically correct (e.g. Bertin 1967). In OGAMA it is possible to adapt this colour ramp, and use only one hue (which is more suitable). However, it is not possible to compare the different heatmaps objectively as the colour ramp is calculated separately (and thus differently) for each stimulus. For example, the maximal total fixation duration (or dwell time) is 11,595 ms for the paper version of 'Assesse' and 9,194 ms for its digital version. The same colour ramp (0–100 %) will be applied to both stimuli. Consequently, the same colour value will correspond to a different fixation duration. This problem is also described in more detail in Ooms et al. (2014).

To overcome this problem, a grid of Areas of Interests (AOIs) will be used to visualise the fixations: their count and total duration in each cell. This allows comparing the overall fixation counts and durations between the two mediums in a more objective way. The selected colour ramps to visualise the data will not be based on one stimulus, but on the values of all stimuli.

The dimensions of the grid cells is based on the size of the stimulus (which is equal for all stimuli) and the accuracy of the eye tracker. The original screen had a resolution of $1{,}920 \times 1{,}080$ pixels. However, the actual map image was displayed on $1{,}400 \times 880$ pixels. The accuracy that could be reached by the eye tracker, when considering the $1°$ deviation limit, was 2.02 cm. Based on these measurements it was decided to use square sized AOIs of 40×40 pixels (or 2.30 cm on the screen). This results in a grid of 35×22 cells. As was mentioned in section "Methodology", the grid with the corresponding values (fixation counts and durations) was obtained by applying the JAVA-tool *fr2grid* (which converts the exported fixation report from OGAMA to a grid). The resulting grid was visualised in Excel using the 'conditional formatting' options. The results are discussed in the next section.

Result

Search Times

During each trial the participants had to locate three names in the map image. The registered search times can indicate the efficiency with which these names were located (Nielsen 1993; Rubin and Chisnell 2008). The mean search time over all test persons and stimuli (paper and digital) is 146.842 s (SD $= 97.367$ s). The mean search time for paper maps (M $= 147.306$ s, SD $= 114.500$ s, N $= 68$) is somewhat higher than for digital maps (M $= 146.391$ s, SD $= 78.054$ s, N $= 70$). However, statistical analyses (t-test) indicate that no significant difference could be found between both mediums (P $= 0.956 > .05$). The mean search times and associated standard deviations for each stimulus are listed in Table 4. No trend can be detected in this table.

Fixation Count

The mean fixation count (number of fixations per second) over all participants and stimuli is 3.139 fix/s (SD $= 0.409$ fix/s). The mean fixation count is lower on paper stimuli (M $= 2.981$ fix/s, SD $= 0.402$ fix/s, N $= 68$) than on their digital counterparts (M $= 3.293$ fix/s, SD $= 0.356$ fix/s, N $= 70$). The statistical t-test indicates that this

Table 4 Mean search times with standard deviations for the twelve stimuli

	Paper		Digital	
	Mean (s)	SD (s)	Mean (s)	SD (s)
1	134.583	137.745	113.537	45.887
2	146.422	117.637	178.752	119.999
3	198.151	145.550	149.879	64.901
4	162.910	127.633	180.556	93.384
5	123.033	76.006	136.053	71.293
6	120.782	77.553	130.544	55.821

Table 5 Mean fixation counts (with standard deviations) for the twelve stimuli

	Paper		Digital	
	Mean (fix/s)	SD (fix/s)	Mean (fix/s)	SD (fix/s)
1	3.004	0.440	3.364	0.377
2	2.940	0.505	3.273	0.420
3	2.958	0.407	3.262	0.415
4	2.928	0.442	3.329	0.309
5	3.069	0.351	3.259	0.293
6	2.985	0.331	3.264	0.373

difference is very highly significant ($P < 0.000$). The results for all stimuli are listed in Table 5 and this shows that the fixation count is always highest on the digital medium.

In order to verify whether the distributions of the fixations are also different across the map image, the gridded visualisation were created for the twelve stimuli. Taking into account the different maximum values for each stimuli, the colour ramp proposed in Fig. 2a is applied on the grids and the corresponding maps can as such be compared. These results are depicted in Fig. 2a.

Fixation Duration

The fixation duration is closely linked with the number of fixations. In order to be able to interpret the findings correctly and objectively, it is good practice to study both metrics (e.g. Duchowski 2007; Holmqvist et al. 2011; Ooms et al. 2012). The mean fixation duration over all participants and stimuli is 268.158 ms ($SD = 186.321$ ms). The mean fixation duration is longer for paper stimuli ($M = 265.652$ ms, $SD = 40.160$ ms, $N = 68$) than for digital stimuli ($M = 257.594$ ms, $SD = 34.827$ ms, $N = 70$). However, the statistical t-test shows that this difference is not significantly different ($P = 0.210 > 0.05$). Table 6 shows the variation of the mean fixation duration over all stimuli.

In order to verify whether the distributions of the fixations and their corresponding durations are also different across the map image, the gridded visualisation are created for the twelve stimuli. Taking into account the different

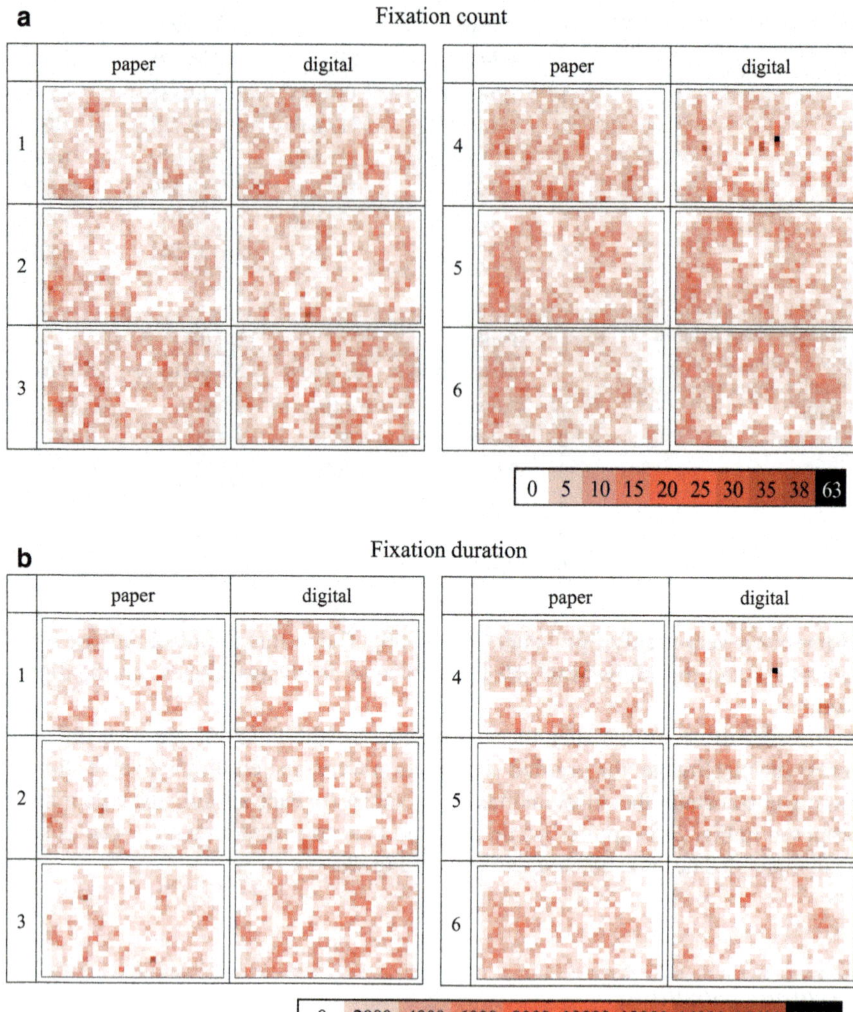

Fig. 2 Gridded visualisation: distribution of the users' fixation counts (**a**) and fixation duration (**b**) on the twelve stimuli

Table 6 Mean fixation duration (with standard deviation) for the twelve stimuli

	Paper		Digital	
	Mean (ms)	SD (ms)	Mean (ms)	SD (ms)
1	225.701	87.085	254.941	32.264
2	275.485	40.275	248.760	36.465
3	261.438	48.541	271.314	46.700
4	273.458	37.491	248.316	32.265
5	258.505	34.533	272.113	29.772
6	274.327	34.863	248.099	25.925

maximum values for each stimuli, the colour ramp proposed in Fig. 2b is applied on the grids and the corresponding maps can as such be compared. These results are depicted in Fig. 2b. This visualisation corresponds to the distribution of the participants' dwell times on the map image.

Discussion

The obtained data was analysed descriptively, statistically and visually for several components to detect how efficient and effective a user can perform a certain task related on map use. Does the medium on which the maps are displayed have an influence on the user's cognitive processes? First, the search times, the needed time to find the three names in the map image, show no significant difference between the two mediums. Moreover, no trends could be detected in the graphs that visualised the mean search times and associated deviations for each stimulus. This indicates that the digital medium is not more or less efficient when a user has to search for a points of interest (indicated by their label) on it.

Second, the mean fixation count (number of fixations per second) was analysed for the two mediums. This mean is closely related to the durations of the fixations which may indicate the degree of difficulty experienced by the user when interpreting a certain visual content (e.g. Duchowski 2007; Holmqvist et al. 2011). Longer fixations or a lower fixation count can signify that the user struggles with interpreting the visual input and needs more time to process the information. This is closely linked with a higher cognitive load (Harrower 2007). It can also indicate that the user considers the visual input more interesting, however, this explanation can be omitted since in this study only topographic maps with the same visual qualities were applied and it could thus be assumed that the users look at all the stimuli with the same level of interest (Ooms et al. 2012, 2014). The fixation count also depends on the length of the saccades: the shorter the saccades, the higher the fixation count. Table 5 clearly shows that the digital stimuli are linked with more fixations per second than the paper stimuli. The statistical t-test established the difference is very high significant. This result can indicate that the digital maps were less difficult to read and thus interpret than the paper maps. It could thus be derived that the lower resolution of the screen (digital maps) is likely to have no negative influence on the user's visual behaviour.

Third and last, the fixation duration was calculated. In contrast to the fixation count, the differences in mean fixation duration for the two mediums are, accordingly the statistical t-test, not significantly different for paper en digital stimuli. From this we can conclude that none of the mediums is easier to interpret than the other. When linking this result to this of the fixations counts, it can be concluded that the users used shorter saccades to study the digital map's content: the distance between subsequent fixations is shorter.

The distribution of these eye movement metrics is visualised in a grid of Areas of Interests (AOIs): their count and total duration (or dwell time). The applied colour ramps are based on the values of all stimuli for both components. For every stimulus the levels of intensity and their positions are compared for the two formats. In general, a similar pattern is noticed on every stimulus independent of the medium and the metric that is visualised. This (fixation) pattern reflects the general map structure and its similarity indicates a similar search behaviour for the users' regardless of the medium. This general structure can be seen as a reference frame that the map users need to study (and structure) the map's content. It consists mainly out of map objects with linear structures, such as roads and rivers (Ooms et al. 2014). Furthermore, the cluster of grid cells with the highest level of intensity generally corresponds to the locations of the labels that the users have to look for (Fig. 1a). This can be explained by the higher level of interest in those areas, which is a result of the task the user had to execute. Because each user has to focus on the given labels a longer dwell time is registered on their locations. Besides the locations with the longest dwell times, the overall hue of the gridded visualisations can be compared. But there is no medium which has consistently a darker background then the other medium.

The results of the statistical test and the gridded visualisations of the fixation durations and the gridded visualisations of the fixation counts do not confirm the outcomes of the statistical test of the fixation count. In this case, no unidirectional conclusions can be drawn. So, the lower resolution of the screen where the digital maps are depicted on does not seem to have an influence on the user's attentive behaviour. This is mainly guided by the map's content and the task at hand. These findings are in correspondence with previous research on the user's attentive behaviour on (digital) maps (Hegarty et al. 2010; Ooms et al. 2014).

Conclusion and Future Work

In contrast to previous research, the users' attentive behaviour is studied when searching on a digital and a paper map. To investigate the influence of (only) the medium, a controlled study design was used. The obtained results indicate that the users could interpret the content of the digital map somewhat more efficiently, but this did not result in an overall more efficient task execution. The distribution of the users' attentive behaviour was similar on both media, but the results indicate a difference in the length of the saccades: the distance between subsequent fixations was shorter on the digital medium.

However, because the study was executed in a controlled set-up, the results are restricted and may not reflect real-life situations. Typically, digital maps are presented on smaller screens than paper maps, where the user has tools to zoom and pan on the map. Furthermore, participants should be allowed to adjust the contrast, brightness and/or sharpness of the screen which could also influence user performance in the real-life situations. Therefore, further research is necessary in this line of work, taking into account the

(continued)

different screen sizes on which the maps are presented (smart phone, tablet, PC, etc.), the interaction tools and the specific design for maps presented on a digital medium.

References

Bertin J (1967) La sémiologie graphique. Mouton-Gauthier-Villars, Paris

Bondarenko O, Janssen R (2009) Connecting visual cues to semantic judgments in the context of the office environment. J Am Soc Inf Sci Technol 60(5):933–952

Brodersen L, Andersen JHK, Weber S (2001) Applying the eye-movement tracking for the study of map perception and map design. In: Kort & Matrikelstyrelsen. National Survey and Cadastre, Copenhagen, p 98

Bunch RL, Lloyd RE (2006) The cognitive load of geographic information. Prof Geogr 58(2): 209–220

Cartwright W (2012) Neocartography: opportunities, issues, and prospects. S Afr J Geomatics 1(1):14–31

Castner HW, Eastman JR (1984) Eye-movement parameters and perceived map complexity I. Am Cartogr 11(2):107–117

Castner HW, Eastman JR (1985) Eye-movement parameters and perceived map complexity II. Am Cartogr 12(1):29–40

Çöltekin A, Heil B, Garlandini S, Fabrikant SI (2009) Evaluating the effectiveness of interactive map interface designs: a case study integrating usability metrics with eye-movement analysis. Cartogr Geogr Inf Sci 36(1):5–17

Çöltekin A, Fabrikant SI, Lacayo M (2010) Exploring the efficiency of users' visual analytics strategies based on sequence analysis of eye movement recordings. Int J Geogr Inf Sci 24(10): 1559–1575

Dobson MW (1977) Eye movement parameters and map reading. Cartogr Geogr Inf Sci 4(1): 39–58

Duchowski AT (2007) Eye tracking methodology – theory and practice. Springer, London

Fabrikant SI, Lobben A (2009) Introduction: cognitive issues in geographic information visualization. Cartographica 44(3):139–143

Fabrikant SI, Rebich-Hespanha S, Andrienko N, Andrienko G, Montello DR (2008) Novel method to measure inference affordance in static small-multiple map displays representing dynamic processes. Cartogr J 45(3):201–215

Goldberg JH, Stimson MJ, Lewenstein M, Scott N, Wichansky AM (2002) Eye tracking in web search tasks: design implications. In: Proceedings of the eye tracking research and applications symposium, New York

Harrower M (2007) The cognitive limits of animated maps. Cartographica 42(4):349–357

Hegarty M, Canham M, Fabrikant SI (2010) Thinking about the weather: how display salience and knowledge affect performance in a graphic inference task. J Exp Psychol Learn Mem Cogn 36(1):37–53

Holmqvist K, Nyström M, Andersson R, Dewhurst R, Jarodzka H, Van De Weijer J (2011) Eye tracking: a comprehensive guide to methods and measures. Oxford University Press, Oxford

Hurst P, Clough P (2013) Will we be lost without paper maps in the digital age? J Inf Sci 39(1): 48–60

Jacob R, Karn K (2003) Eye tracking in human-computer interaction and usability research: ready to deliver the promises. In: Radach R, Hyona J, Deubel H (eds) The mind's eye: cognitive and applied aspects of eye movement research. Elsevier, Amsterdam, pp 593–660

Jenks GF (1973) Visual integration in thematic mapping: fact or fiction? Int Yearbook Cartogr 13:112–127

Johnson W, Jellinek HD, Klotz L, Rao R, Card SK (1993) Bridging the paper and electronic worlds: the paper user interface. In: Proceedings of INTERCHI '93, pp 507–512

Kiefer P, Giannopoulos I, Raubal M (2014) Where am I? Investigating map matching during self-localisation with mobile eye tracking in an urban environment. Trans GIS 18(5):660–686

Kraak MJ, Brown A (2001) Web cartography: developments and prospects. Taylor & Francis, London

Liu Z (2005) Reading behavior in the digital environment: changes in reading behavior over the past ten years. J Doc 61(6):700–712

MacEachren AM (1995) How maps work: representation, visualization, and design. Guilford Press, New York

Montello DR (2009) Cognitive research in GIScience: recent achievements and future prospects. Geogr Compass 3(5):1824–1840

Nielsen J (1993) Usability engineering. Morgan Kaufmann, San Francisco

O'Hara K, Sellen A (1997) A comparison of reading paper and on-line documents. In Proceedings of the ACM SIGCHI conference on human factors in computing systems. ACM, pp 335–342

Ooms K, De Maeyer P, Fack V, Van Assche E, Witlox F (2012) Interpreting maps through the eye of expert and novice users. Int J Geogr Inf Sci 26(10):1773–1788

Ooms K, De Maeyer P, Fack V (2014) Study of the attentive behavior of novice and expert map users using eye tracking. Cartogr Geogr Inf Sci 41(1):37–54

Pederson P, Farrell P, McPhee E (2005) Paper versus pixel: effectiveness of paper versus electronic maps to teach map reading skills in an introductory physical geography course. J Geogr 104(5):195–202

Peterson MP (1997) Trends in Internet map use. In: Proceedings of the 18th International Cartographic Conference, vol 3, pp 1635–1642

Peterson MP (2003) Maps and the Internet: an introduction. In: Peterson MP (ed) Maps and the Internet. Elsevier Science, Oxford, pp 1–16

Poole A, Ball LJ (2006) Eye tracking in human computer interaction and usability research: current status and future prospects. In: Ghaoui C (ed) Encyclopedia of human computer interaction. Idea Group, Pennsylvania, pp 211–219

Popelka S, Brychtova A (2013) Eye-tracking study on different perception of 2D and 3D visualisation. Cartogr J 50(3):240–246

Rayner K (1998) Eye movement in reading and information processing: 20 years of research. Psychol Bull 124(3):372–422

Rensink RA (2002) Change detection. Annu Rev Psychol 53(1):245–277

Rubin J, Chisnell D (2008) Handbook of usability testing. How to plan, design and conduct effective tests, 2nd edn. Wiley, Indianapolis

Simons DJ, Ambinder MS (2005) Change blindness: theory and consequences. Curr Dir Psychol Sci 14(1):44–48

Spakov O, Miniotas D (2007) Visualization of eye gaze data using heat maps. Int J Electrical Electron Eng Res 74(2)

Steinke TR (1987) Eye movement studies in cartography and related fields. Cartographica 24(2): 40–73

Sutherland-Smith W (2002) Weaving the literacy Web: changes in reading from page to screen. Read Teach 55(7):662–669

van Elzakker CPJM, Griffin AL (2013) Focus on geoinformation users: cognitive and use/user issues in contemporary cartography. GIM Int 27(8):20–23

Verdi MP, Crooks SM, White DR (2003) Learning effects of print and digital geographic maps. J Res Technol Educ 35(2):290–302

Wolfe JM (1994) Guided search 2.0: a revised model of visual search. Psychon Bull Rev 1(2): 202–238

Non-photorealistic 3D Visualization in City Maps: An Eye-Tracking Study

Stanislav Popelka and Jitka Doležalová

Introduction

There are several perspectives on the term "3D" in cartography. Kraak (1988) in one of the first articles focused on 3D maps states that image will be considered 3D if it contains those stimuli or depth cues which make it being perceived as 3D.

Wood et al. (2005) described the role of three dimensionality used in both process of visualization and representation of 3D objects and space. To clarify the term "3D", they considered the model of a general visualization pipeline. According to Upson et al. (1989) and Haber and McNabb (1990) it is possible to distinguish five levels of dimensionality, which correspond to the various stages within a visualization process (data management, data assembly, visual mapping, rendering, display). Wood et al. (2005) associates the term 3D with the phase of "visual representations of data", but it can also be part of the outputs from all other stages of the process.

Non-photorealistic Models of the Cities

According to Cartwright et al. (2007), two basic concepts of 3D cartography exist—photorealistic and non-photorealistic. Durand (2002) emphasizes that non-photorealistic visualization provides "extensive control over expressivity, clarity and aesthetics", but Jedlicka et al. (2013) mentions also the limits (generalisation, simplification, non-perspective projections, distortions etc.)

This general fact could be applied also for the visualization of cities. Photorealistic visualization of cities became popular due to the ubiquity of Google

S. Popelka (✉) • J. Doležalová
Department of Geoinformatics, Palacký University, Olomouc, Czech Republic
e-mail: standa.popelka@gmail.com; jitka.dolezalova@upol.cz

© Springer International Publishing Switzerland 2015

J. Brus et al. (eds.), *Modern Trends in Cartography*, Lecture Notes in Geoinformation and Cartography, DOI 10.1007/978-3-319-07926-4_27

Earth and similar applications, where models of large areas are created with automatic or semi-automatic acquisition method. Models are created in high detail, and aerial photo is used as a base map. Because of the lack of the clarity, it is difficult to use photorealistic visualization as the map of larger areas. Non-photorealistic models influence the map complexity too, but not so significantly.

Generally characteristics of non-photorealistic rendering techniques include the ability to sketch geometric objects and scenes, to reduce visual complexity of images, as well as to imitate and extend classical depiction techniques known from scientific and cartographic illustrations (Döllner and Buchholz 2005). Standardization of 3D maps was discussed in Herman and Reznik (2013).

Modern internet map portals use non-photorealistic 3D models in different levels of abstraction as an enhancement of the map, especially in large cities. Maps used as stimuli in the experiment are described in chapter "Stimuli" in more detail.

Evaluation of 3D Maps

Three dimensional non-photorealistic 3D visualization is used by an increasing number of applications. However, there is a still little known how 3D can be used in visualization most efficiently. As Konečný et al. (2011) highlights, the creation of the usability tests for different types of maps and visualizations is quite a challenge.

There exist few studies, focused on evaluation of 3D in maps. Most of them use the questionnaire as the main investigation method. Savage et al. (2004) and Petrovic and Masera (2006) analysed user's preferences on 2D and 3D maps. Schobesberger and Patterson (2008) investigate differences between 2D and 3D map of the Zion National Park in Utah. Haeberling (2004) evaluated design variables for 3D maps.

In few studies, eye-tracking was used for the evaluation of 3D maps. Fuhrmann et al. (2009) analysed differences between perception of 2D map and its holographic equivalent. Irvankoski et al. (2012) investigates visualization of elevation information on maps. Interaction with a 3D geo-browser under time pressure was evaluated by Wilkening and Fabrikant (2013). Possibilities of eye-tracking evaluation in cartography were discussed in Popelka et al. (2012) and Popelka and Voženílek (2012).

Different perception of 2D and 3D terrain maps was investigated in Popelka and Brychtová (2013). In this study, two eye-tracking tests were used for observing the user perception of the pair of maps representing the terrain. On one map, the terrain was represented by contour lines. Second map contained the perspective view of the same data as the first map.

The purpose of the paper is to analyse the user perception of two types of 3D visualization in maps of the cities.

Case Study

Equipment

For the case study, an eye-tracking device was used. As Ooms et al. (2014) states, eye-tracking is a direct method to study users' cognitive processes. Eye-tracker is situated in the special room—eye-tracking laboratory. Windows are covered with non transparent foil, to unify the lighting conditions.

SMI RED 250 eye-tracker was used within the study. This device is capable of recording eye-movements with the frequency of 120 Hz. Eye positions are recorded every 8 ms. Eye-tracker is supplemented by web camera, which records participant during the experiment. This video helps to reveal the cause of missing data, respondents' reactions to the stimuli and their comments to the particular maps.

For data visualization and analyses, three different applications were used. First one is SMI BeGaze, which is software developed by the manufacturer of the device. Open source software OGAMA and CommonGIS developed at Fraunhofer Institute in Germany were used for visual analytics of eye-tracking data.

Participants

Total of 40 participants (24 females, 16 males) attended the eye-tracking experiment. Most of them were representatives of academic staff and students. Respondents were originated from different fields. Some of them were cartographers, some of them were not. Majority of participants were 20–25 years old. Participants were not paid for the testing.

Before the experiment, respondents filled out the short questionnaire with personal information. Apart from elementary information, like age or sex, they had to answer the question, how often are they using the internet map portals like Google Maps or OpenStreetMap. Most of the participants use the web map portals every day.

Experiment Design

At the beginning of the experiment, respondents fill out a short questionnaire. Then, the 9-point calibration was performed. Eye-movement recordings with deviation smaller than 1° were included in the experiment.

After the calibration, welcome screen with instructions was displayed. The instructions included also the sample question. Respondents' task was to find out one particular point symbol in the map as fast as possible and mark it with the mouse click.

Study was performed in within-subject design. The experiment contained 18 static stimuli with 2D and 3D maps of cities. The stimuli with maps were presented in random order. To unite the starting point of the eye-movement trajectories, the fixation cross was displayed for 500 ms before the stimulus. Respondents had maximum time of 30 s to find the target, but for the most of the tasks, the time was fully sufficient.

Stimuli

The experiment contained screenshots of different internet map portals. These maps were complemented with point symbols. Two types of maps were used—the first one was standard map with buildings represented by polygons, second contained 3D (2,5 D) visualization of buildings.

Maps from three different sources were used. The first one is well-known Google map (stimuli 1–5). In bigger cities, the Google map in zoom 17 and higher contains 3D representation of block of buildings. These screenshots were compared with 2D maps, which were captured in the different part of the city (where 3D coverage was not used) or with the use of merging and scaling down of the map of the same area in zoom 16. The maps were styled (with the use of Gmaps wizard), because the original map available at maps.google.com contains a large amount of symbols and labels. For the purpose of the experiment, the fictitiously placed point symbols (designed according to the original ones from Google Maps) were used. For each pair, the same number and set of symbols was used.

Second type of maps contained maps from OpenStreetMap.org (stimuli 6–8). In the default version, there exists no option to display 3D block of buildings. However, thanks to the free availability of OSM data, there exist some possibilities, how to display 3D content. Well-known is project osmbuildings.org, which is an additional layer to existing web maps. It is currently working with LeafletJS and OpenLayers.

Last stimulus (map 9) also uses OpenStreetMap data through project F4Map. This project is only in beta version, but it automatically creates the 3D variant of cities all over the world. The map is enhanced by ground elevation, animated water, dynamic shadows, urban and natural details. It is possible to switch between 2D and 3D version of the map, but also to change the camera angle and rotation.

The example of each type of stimuli (Google Maps, OSMbuildings and F4map) is in the Fig. 1.

Fig. 1 The pair of the stimuli. 2D (*left*) and 3D (*right*), where the buildings are represented with 3D blocks. *Up*: Google Maps (*source*: http://maps.google.com), *Middle*: OSMbuildings (*source*: http://osmbuildings.org), *Bottom*: F4map (*source*: http://map.f4-group.com)

Results

Analysis of Questionnaire

Part of the experiment was the questionnaire, focused on participants' personal opinion about presented maps. The questionnaire was presented after all stimuli, and it contained only two questions. First one concerned the suitability of the map. Respondents were asked to answer, which variant of the map was more suitable for finding the answer. Second question was focused on the aesthetic factor—which variant did they like more. In both questions, three options were available—"2D"; "3D" and "Depends on the specific map".

Participants found 2D map more suitable for answering the question (finding the point symbol in the map) than 3D map. Majority of them (24 from 40) preferred 2D variant of the map. Relatively high number of respondents (13 from 40) chose answer "Depends on the specific map".

Participants were also asked, which type of map like they more. The distribution of the preferences between 2D and 3D maps was almost balanced (19 for 2D vs. 13 for 3D).

Fixation Detection

One of the most important issues in eye-tracking data analysis is event detection of recorded data. For almost all analyses, the fixations and saccades are needed. Eye-tracking data were recorded with sample frequency of 120 Hz, so the dispersion algorithm (I-DT), which is more appropriate for the low-frequency data, was used.

I-DT takes into account the close spatial proximity of the eye position points in the eye movement trace (Salvucci and Goldberg 2000). The algorithm defines a temporal window which moves one point at a time, and the spatial dispersion created by the points within this window is compared against the threshold. If such dispersion is below the threshold, the points within the temporal window are classified as a part of fixation; otherwise, the window is moved by one sample, and the first sample of the previous window is classified as a saccade (Komogortsev and Khan 2009).

The threshold values were set to 80 ms (duration) and 50 px (dispersion). These values were selected based on the author's unpublished study, which compares four settings, used in cartographic papers and identified the thresholds, which fits to the recorded raw data.

For data analysis, open-source software OGAMA was also used. Most important parameters are "Maximum distance" and "Minimum number of samples", which corresponds to dispersion and duration in BeGaze. For optimizing the event detection parameters in OGAMA, the image of scanpath from BeGaze was used in OGAMA instead of SlideResource image. The fixations in OGAMA were plotted over the image of the BeGaze fixations. Event detection parameters in OGAMA were changed until the scanpath was very similar to the scanpath from BeGaze.

Statistical Analysis

For statistical analysis of eye-movement data, several eye-tracking metrics were calculated. For all metrics, median values for 40 respondents were calculated. Median was used instead of mean because the data had not normal distribution and median also filter out the extreme values. Data was analysed with the use of the Wilcoxon rank sum test and statistically significant difference between 2D and 3D maps was observed on the significance level $\alpha = 0.05$ in all cases.

Analysed metrics were Time to Answer (click), Fixation Count, Fixation Duration Median and Scanpath Length.

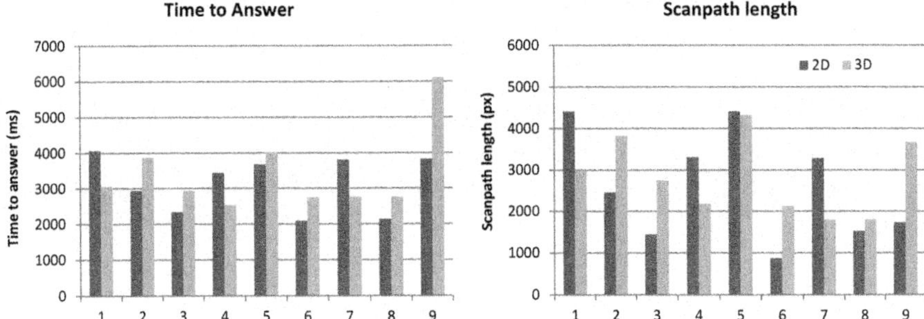

Fig. 2 Graph of median time to answer (*left*) and scanpath length (*right*) values for each map in the experiment

Table 1 Wilcoxon test of differences between fixation count for 2D and 3D variant of the map

Fixation count	Alpha	W	p-value	Statement
Map 1	0.05	1,070	0.009296	Rejecting H0
Map 2	0.05	565	0.02381	Rejecting H0
Map 3	0.05	505	0.004499	Rejecting H0
Map 4	0.05	1,011	0.04238	Rejecting H0
Map 5	0.05	745.5	0.6027	Failed to reject
Map 6	0.05	498.5	0.003702	Rejecting H0
Map 7	0.05	1,035.5	0.02351	Rejecting H0
Map 8	0.05	722.5	0.4575	Failed to reject
Map 9	0.05	482.5	0.002266	Rejecting H0

In case of Time to Answer metric (Fig. 2, left), highest difference between 2D and 3D variant was observed for map 9. This result was expected, because the 3D map no. 9 is tilted and orientation in this map is harder. Second highest value of Time to Answer was recorded in case of 3D variant of map 5. This map, which displays the downtown of New York with many 3D skyscrapers, is the most complex one from the set of "Google maps (map 1–5)". It is surprising that the difference between 2D and 3D variant is so small in this case.

Value of Time to Answer is interlinked with the Fixation Count metric (see Table 1), where statistically significant difference was observed for 7 from 9 maps. In contrast to Fixation Count, statistically significant difference for Fixation Duration Median was observed only in three cases (Map 5, 7 and 9).

According to Goldberg et al. (2002), a longer scanpath indicates less efficient searching. Statistically significant difference was found in 5 cases from 9 (Fig. 2, right). For maps 1 and 7, higher median scanpath length was recorded for 2D variant. For maps 3, 6 and 9, higher values were recorded for 3D variant. This fact suggests that the scanpath length was dependent on other variable than 2D or 3D visualization method.

Table 2 Wilcoxon test of differences for 2D and 3D maps for whole dataset

Metric	Alpha	W	p-value	Statement
Trial duration	0.05	49,054.5	0.3591	Failed to reject
Time to answer	0.05	48,686.5	0.2826	Failed to reject
Fixation count	0.05	48,965	0.3387	Failed to reject
Fix. duration med.	0.05	47,549.5	0.1183	Failed to reject
Scanpath length	0.05	114,566	0.8873	Failed to reject

Apart from the analyses for particular maps; also the whole dataset for all maps was analysed (see Table 2). It was found that results for map 9 influenced all results. If the map 9 was included in the dataset, statistically significant differences were found for Time to Answer and Fixation Duration Median. The values were higher for the 3D map. When the map 9 data were omitted, with the Wilcoxon rank sum test on the significance level $\alpha = 0.05$, no differences between 2D and 3D variant were found for any of the metrics.

Statistical analysis showed that there are statistically significant differences in eye-tracking metrics between 2D and 3D variant of a particular map, but the results did not indicate that one of the variants is better than the other. Within the analysis of the entire dataset as a sum of all maps, no statistically significant differences were found for any of the studied eye-tracking metric.

Visual Analytics of Data

Visual analytics, the science of analytical reasoning facilitated by interactive visual interfaces is an important tool for investigation of a large amount of data. For the visual analytics of recorded eye-tracking data, software CommonGIS developed at the Fraunhofer Institute IAIS was used. For data conversion from BeGaze software to CommonGIS environment, the conversion tool created by Kristien Ooms was used. Fixations from BeGaze software were transformed into the trajectories, which are represented as lines in CommonGIS.

For data analyses, two methods introduced by Andrienko et al. (2012) were used. First method, Flow Map, represent results of discrete spatial and spatio-temporal aggregation of trajectories. Arrows represent multiple movement of gaze from one location to another. The thickness of arrows is derived from variable Number of moves between defined voronoi polygons. Only arrows representing more than three moves are displayed.

Second used method is Temporal View of Trajectories. The horizontal dimension of the graph represents time, and the colour of the lines displays the distance between current gaze position and the target in pixels.

In Fig. 3, three map pairs are shown. First image from the top shows the situation, when more of cumulative gaze trajectories were observed in case of 3D variant of the map (right part of the image). The Temporal View shows that

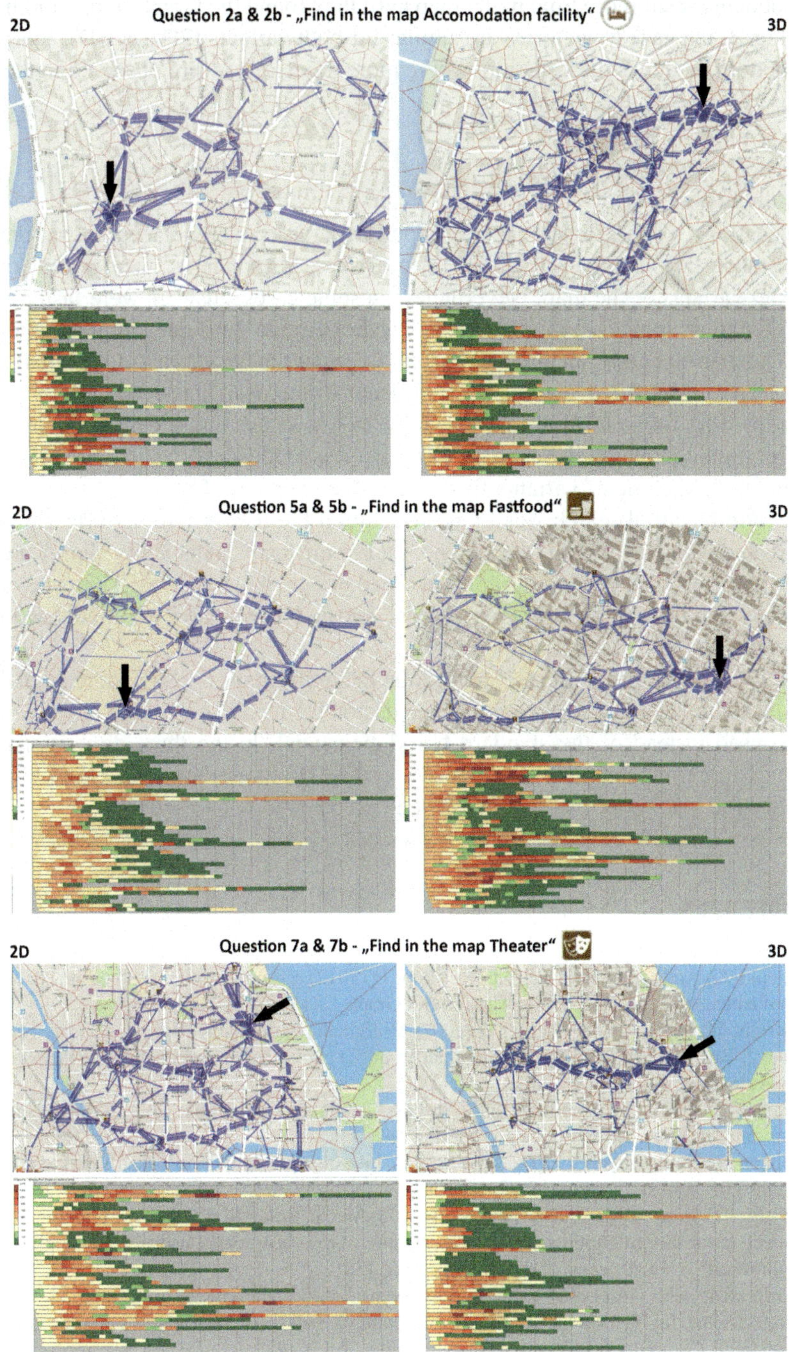

Fig. 3 Visual analytics of selected tasks in CommonGIS. *Black arrows* point to the location of correct answer. 2D variant is on the *left*, 3D on the *right*. Figure in full resolution is available at www.eyetracking.upol.cz/Research_images/Cities.jpg

respondents spent more time in the map until they found the target. In the middle of the Fig. 3, the gaze trajectories are similar for both variants (2D and 3D). Image at the bottom of shows the situation, where more trajectories were observed in the 2D map.

Conclusion

Locations of the targets in the experiment tasks were placed in similar distance from the centre of the image (where respondents gaze starts) for both variants of the map. Nevertheless in some cases, answers were faster in 2D map, in some cases in the 3D one. Data from the short questionnaire after the experiments shows that respondents consider the 2D variant more suitable for answering the question. No significant differences between 2D and 3D maps were found for four metrics (Time to Answer, Fixation Count, Fixation Duration Median and Scanpath Length). Respondents also did not clearly incline to the one of variants from the aesthetics point of view.

Point symbol search was more difficult on the map where 3D effect was created with use of map tilt (map no. 9). For this map, statistically significant differences were observed for all recorded eye-tracking metrics. This type of maps should not be used very often, because users have problems with orientation in this map.

On the other hand, results for all other stimuli (maps no 1–8) indicate that in situations when it is reasonable and desirable, the 3D map of the city could be used instead of the standard two-dimensional. In the three-dimensional map, more information is contained and the 3D representation did not influence the reading of the map and its comprehensibility.

References

Andrienko G, Andrienko N, Burch M, Weiskopf D (2012) Visual analytics methodology for eye movement studies. IEEE Trans Vis Comput Graph 18(12):2889–2898

Cartwright W, Peterson MP, Gartner G (2007) Multimedia cartography. Springer, Heidelberg

Döllner J, Buchholz H (2005) Non-photorealism in 3D geovirtual environments. In: Proceedings of AutoCarto, ACSM, Las Vegas, pp 1–14

Durand F (2002) An invitation to discuss computer depiction. In: Symposium on non-photorealistic animation and rendering (NPAR)

Fuhrmann S, Komogortsev O, Tamir D (2009) Investigating hologram-based route planning. Trans GIS 2009:177–196

Goldberg JH, Stimson MJ, Lewenstein M, Scott N, Wichansky AM (2002) Eye tracking in web search tasks: design implications. In: Proceedings of the eye tracking research and applications symposium, New Orleans, 25–27 March, pp 51–58

Haber RB, McNabb DA (1990) Visualization idioms: a conceptual model for scientific visualization systems. In: Nielson G, Shriver B, Rosenblum L (eds) Visualization in scientific computing. IEEE Computer Society Press, Los Alamitos, pp 74–93

Haeberling C (2004) Topografische 3D-Karten ñ Thesen für kartografische Ge-staltungsgrundsaetze. Dissertation, Institut für Kartographie der ETH Zuerich, Zuerich

Herman L, Reznik T (2013) Web 3D visualization of noise mapping for extended INSPIRE buildings model. In: Hrebicek J et al (eds) ISESS 2013, IFIP AICT 413, pp 414–424

Irvankoski K, Torniainen J, Krause CM (2012) Visualisation of elevation information on maps: an eye movement study. In: Conference proceedings, The Scandinavian Workshop on Applied Eye Tracking (SWAET), Karolinska Instituet, Stockholm, 56 p

Jedlicka K, Cerba O, Hajek P (2013) Creation of information-rich 3D model in geographic information system – case study at the Castle Kozel. In: Conference proceedings, informatics, geoinformatics and remote sensing, vol I. STEF92 Technology Ltd., Sofia, pp 685–692

Komogortsev O, Khan J (2009) Predictive compression for real time multimedia communication using eye movement analysis. ACM Transactions on Multimedia Computing, Communications and Applications 6

Konečný M, Kubíček P, Stachoň Z, Šašinka C (2011) The usability of selected base maps for crises management – users' perspectives. Appl Geomatics 3(4):189–198

Kraak MJ (1988) Computer-assisted cartographic three-dimensional imaging techniques. In: Raper J (ed) Three dimensional applications in geographical information systems. Taylor & Francis, London, pp 99–114

Ooms K, De Maeyer P, Fack B (2014) Study of the attentive behavior of novice and expert map users using eye tracking. Cartogr Geogr Inf Sci 41(1):37–54

Petrovic D, Masera P (2006) Analysis of user's response on 3D cartographic presentations. In: Proceedings of 5th ICA mountain cartography workshop, Bohinj

Popelka S, Brychtová A (2013) Eye-tracking study on different perception of 2D and 3D terrain visualisation. Cartogr J 50(3):240–246s. ISSN: 1743–2774 (Maney Publishing)

Popelka S, Voženílek V (2012) Specifying of requirements for spatio-temporal data in map by eye-tracking and space-time-cube. In: Proceedings of international conference on graphic and image processing, Singapore, 5 p

Popelka S, Brychtová A, Voženílek V (2012) Eye-tracking a jeho využití při hodnocení map. Geografický časopis Geografický ústav SAV, pp 71–87

Salvucci DD, Goldberg JH (2000) Identifying fixations and saccades in eye tracking protocols. Eye tracking research and applications symposium. ACM, New York

Savage DM, Wiebe EN, Devine HA (2004) Performance of 2D versus 3D topographic representations for different task types. Proc Hum Factors Ergon Soc Annu Meet 48(16):1793–1797

Schobesberger D, Patterson T (2008) Exploring of effectiveness of 2D vs. 3D trail-head maps. In: Proceedings of 6th ICA mountain cartography workshop, Lenk im Simmental

Upson C, Faulhaber T, Kamins D, Schlegel D, Laidlaw D, Vroom J, Gurwitz R, van Dam A (1989) The application visualization system: a computational environment for scientific visualization. IEEE Comput Graph Appl 9(4):30–42

Wilkening J, Fabrikant SI (2013) How users interact with a 3D geo-browser under time pressure. Cartogr Geogr Inf Sci 40(1):40–52

Wood J, Kirschenbauer S, Döllner J, Lopes A, Bodum L (2005) Using 3D in visualization. In: Dykes J, MacEachren AM, Kraak M-J (eds) Exploring geovisualization. Pergamon, Oxford, pp 295–312

Generating Cartographic Representations of Volunteered Environmental Noise Data from Mobile Phones

Petr Duda

Introduction

If one is solving a targeted mapping task, he/she usually firstly looks for a method that would capture the phenomenon as objectively as possible, and only then look for ways to physically realize such a method. But today this trend is reversed in some cases. When designing a research method, we begin with study of the normal human behavior and willingness to voluntarily collect data during other human activities.

As a result of the rapid development of information and communication technology, particularly miniaturization, data collection using handheld mobile devices (especially smart phones) is increasingly prominent also in the field of geoinformatics. These devices have sufficient computing power for managing relatively big data and also have some kind of GNSS receiver (e.g. GPS).

This combination came to be the basis for various kinds of either voluntary or commercial *in situ* data collection tasks, conducted through sensors connected to the mobile phone. One of these data types is sound pressure level. Almost every mobile phone has a built-in microphone for voice communication. This microphone can be successfully used for measuring environmental noise. Currently there are some applications (e.g. NoiseTube: D'Hondt et al. 2013, NoiseSpy: Kanjo 2010, or NoiseDrone: Nüst 2013), which allow collecting and sharing noise data from mobile phones, and also perform basic data processing, including creating map outputs.

The aim of this article is to outline basic characteristics, methods of processing and interpretation of volunteered data from mobile phones with an example of noise data obtained mainly during informal user activity, and to propose methods of cartographic representation of that data. Because this data is not nearly as accurate

P. Duda (✉)
Masaryk University, Brno, Czech Republic
e-mail: 147169@mail.muni.cz

© Springer International Publishing Switzerland 2015 369
J. Brus et al. (eds.), *Modern Trends in Cartography*, Lecture Notes in
Geoinformation and Cartography, DOI 10.1007/978-3-319-07926-4_28

as professionally collected data, a relatively significant component of uncertainty enters into the measurement chain. For this reason, it is necessary to firstly disassemble the nature of voluntary noise data and compare it with the standard noise data that are commonly processed today.

Volunteered Input Data Quality and Their Dependence on User Behavior

The quality of data obtained during voluntary measurements is entirely dependent on volunteer behavior during measurement. Generally, it is possible to distinguish between two kinds of data obtained by this so called crowdsourcing:

- Data obtained through dedicated activity in the volunteer's free time (macrowork). In the case of environmental noise measurement, this means getting near to the standards and methodologies for measuring noise as possible, and therefore performing relatively long static measurements. Resulting data is standard in nature and certified procedures can be used for processing and interpretation. The only major difference from volunteered data compared with that obtained by a certified device is a lower accuracy due to use of less accurate sensor. Feick and Roche (2013) calls it quasi-scientific data, but the question here is "what is scientific and what not?".
- Data obtained during normal (informal) human activities (microwork). Using this method it is possible to collect much more data than in the previously described case, since it does not require any special volunteer time (if the mobile collecting app is properly designed and requires only a minimum of user operations). When a sufficient number of measurements are collected and the user's attention on the measurement process is sufficient, this method also collects fairly reliable data on environmental conditions to which the volunteer is exposed in his/her real life.

It is problematic to combine these types of data, or it may confuse the map's reader. The second type of data is much more prone to interpretation errors due to its higher degree of its uncertainty and therefore graphical presentation of this data requires a more complex post-processing. It is thus necessary to apply the theory of uncertainty, which may greatly affect the resulting cartographic representation.

Measuring Environmental Noise Data with Mobile Phones

Mobile Phones as a Sensor Network

Sensor data collection, which uses mobile phones as sensors, is technologically based on mobile sensor networks. The difference between conventional cellular networks and specialized sensor networks lies more or less on communication technology that is used. The main advantage of sensor networks, consisting of mobile phones, is their wide accessibility. Smart mobile phones have become the easiest way, how to engage public in environmental data collection. Applications enabling collecting of various sensor data can be distributed through centralized servers and services, which offer applications designed for the particular smart mobile phone's operating system [e.g. Apple 2013 (iTunes), Google 2013 (Google Play), etc.].

In environmental noise mapping, semi-mobile sensor network technology has been successfully tested on a relatively large area of Madrid (Manvell et al. 2004). Technical feasibility of environmental noise mapping using mobile phones was tested e.g. at Trinity College in Dublin (McDonald et al. 2008). Projects aimed to measure noise in an urban environment (D'Hondt et al. 2013; Kanjo 2010; Nüst 2013) have also report generally positive results.

Mobile Phones as Noise Dosimeters

The main difference between noise mapping using mobile phones and classical mapping approaches is the style of noise source classification. Current applications for measuring noise with mobile phones use a person-centralized approach in which the mobile phone is used as a noise dose meter, which indicates the overall noise experienced by one particular person (equivalent to radiation dosimeters). The noise level is, simultaneously with time and possibly position, recorded every second (which corresponds to the "slow" regime of a standard sound meter).

This approach is considerably simpler in comparison with conventional mapping of the spatial distribution of environmental noise. First of all, different noise sources (e.g. transport, industry) are not differentiated, including noise created by user himself. Also information about position is just secondary, does not have to be provided, and is used rather for orientation purposes (e.g. in which environment was this value measured). For this reason, NoiseTube application (Stevens 2012) provides post-processing tool, which automatically assigns position (if measured) to the nearest point on the nearest street. For information purposes such position manipulations are adequate, but when studying the spatial distribution of noise, a somewhat more sensitive approach should be chosen.

Despite these shortcomings, we can say that the dosimetry data, if collected with sufficient accuracy, can be used to construct maps of the spatial distribution of noise or even be combined with data from conventional measurements.

Cartographic Processing of Voluntarily Measured Environmental Noise Data

Currently there is not too much scientific literature that would address cartographic visualization of environmental noise data. Technical standards (old ISO 1996-2:1987) and best practice literature (European Commission 2006) present only overview maps with contours, or choropleth maps. The area method prevails in large scale and detailed maps. Virtually all maps give information only on the average noise in the area, but not (except for maps at a very large scale) on the effects of noise pollution on humans (this information must be deduced), and are usually result of calculations. Such maps can be constructed from measurements if there is a sufficiently dense network of sensors which produces representative data. The number of required measurements, however, is directly proportional to size of the reference area and the variability of noise sources. Therefore, in most cases of using volunteered data collected by mobile phones, choropleth or contour methods are not applicable, because not enough of the space is sampled.

Authors of the first attempts at environmental noise measurement using mobile phones (D'Hondt et al. 2013; Kanjo 2010; Nüst 2013) divided the area of interest into regular cells (usually square). The measured point noise data are then assigned to overlaying cell. If multiple points fall within one cell, their values are simply averaged.

Older maps, which depict traffic noise measurement and calculation, use the method of linear symbols. The noise level of surface transport at a standardized distance from source (e.g. railways or roads), or in some cases the noise level at the facades (e.g. Senatsverwaltung für Gesundheit, Umwelt und Verbracherschutz 2008) is typically what is represented in these maps. In both cases it is necessary generalize measured values into cells (or lines).

Generalization of Regions with Presupposed Homogeneous Noise Distributions

The basic visualization parameters of a volunteered noise map, formed from a large volume of values, measured at various positions, are (1) shape, (2) size and (3) arrangement of visualization cells. In most common cases, the cells are square and are arranged to a square grid because processing of that configuration is the most simple, but other shapes and arrangements, e.g. hexagonal, may be more

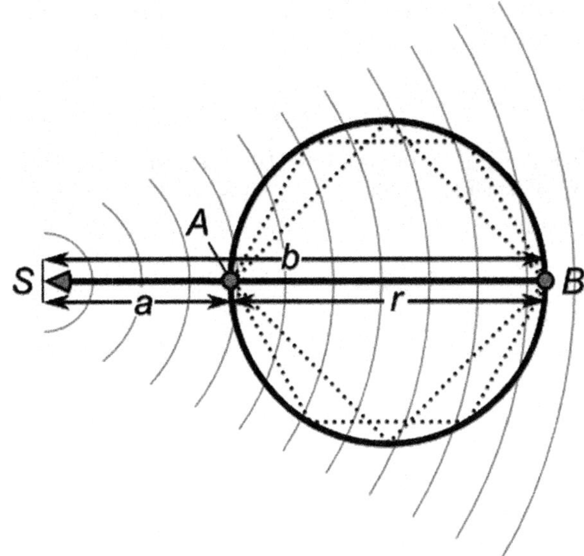

Fig. 1 Determination of appropriate size of visualization cells. S is noise source, A the closest measurement point in the cell, B most distant possible measurement point in the cell, r cell diameter, a distance from noise source to closest measurement point, b distance from the noise source to most distant measurement point. *The dotted square* and *hexagon* suggest a possible approximation of ideal circular-cell shape

suitable. The largest possible size of one cell must be estimated on the basis of dependence of noise level change with distance and the average speed of a pedestrian or cycler (volunteer). According to ISO 1996-1:2003, to produce reliable data, it is desirable to have at least three samples per cell from one pedestrian or cycler pass, preferably five.

If we take into account that the average speed of a pedestrian is about 6 km/h (ca 1.7 m/s), the distance between the outer boundary of the cell at a standard sample rate of 1 Hz (*slow* mode) from its center results in about a 5 m cell for 3 samples, or 8.5 m for 5 samples. If the measured noise level at a distance of 15 m from the source of noise is 72 dB [e.g. current modern passenger car at a speed of about 90 km/h, according to Finley and Miles (2007)], the level measured at a distance of 20 m from the source (i.e. at the opposite boundary of the cell with a diameter of 5 m, see Fig. 1) should be about 70 dB; while with distance of 23.5 m, it should be about 68 dB. In the first case, difference between these values is within the accuracy of measurement itself (± 5 dB, see section "Accuracy of Noise Measurement Conducted by Mobile Phone Microphones") and regular human perception (± 3 dB), in the second case it is just slightly above the threshold of both accuracy and perception.

If the noise level is 60 dB at a distance of about 3 m from source (vehicle speed of 30 km/h—a typical noise exposure situation of pedestrians on the sidewalk in a residential urban area of Europe), noise at 8 m will be about 52 dB. In this case, the difference between these two extremes in a given cell is about 8 dB. This difference is already well perceptible by hearing. But the level of 60 dB is also loudness of normal human walk (see section "Data Aging") while a noise level of 52 dB is a typical city noise background value and it is on edge of being distinguishable by a

Table 1 Typical urban noise situations from a road traffic, letters in parentheses correspond to the labeling in Fig. 2

Situation [car speed (km/h)]	Cell diameter (r) (m)	Closer measurement point (A)		Most distant measurement point (B)	
		Distance from source (a) (m)	Noise level (dB)	Distance from source (b) (m)	Noise level (dB)
Passenger car, 30	5	3	60	8	52
	7			10	49
	8.5			11.5	48
Passenger car, 50	5	3	70	8	62
	7			10	59
	8.5			11.5	58
Passenger car, 65	5	15	64	20	62
	7			22	61
	8.5			23.5	60
Passenger car, 105	5	15	70	20	68
	7			22	67
	8.5			23.5	66

person. Localities with both noise levels 52 and 60 dB can be still considered as quiet (in urban environments, Kang (2007)). The most acceptable universal cell size for noise in the urban environment is an area with a diameter of 7 m. The data for common distances and speeds is summarized in Table 1.

In case of GPS position measurement, reducing the cell size under 5 m will not lead to better spatial accuracy, because, according to Modsching et al. (2006), the spatial error of typical GPS receiver can be well one order of magnitude above.

However, in such a homogeneous space layout, it was necessary to perform a large number of measurements, even in the case when sufficient noise measurement accuracy is secured. Based on the observations of the noise situation in cities, it is clear that (especially in the case of linear noise sources, e.g. from road traffic), the average noise level values along certain segments of this source (e.g. road segments) in a given measurement period is virtually the same. (As long as the source is in the same condition. For road traffic the surface condition is considered to be the same if the same number of vehicles passes through individual measurement positions at the same speed.) The noise situation in a direction transverse to the linear source is also predictable, and is possible to generalize its behavior.

Therefore, in the case of noise from surface traffic, which passes through city streets, it is appropriate to adjust cell shape and orientation to the street directions and generalize the measured data into some form of linear map which indicates the average noise level on each homogeneous street segment. The method of allocating street segments to homogenous units is shown in Fig. 2.

Usually the pedestrian moves on the street, on Fig. 2(a) is volunteer's path noted as a *blue line*. Error in position recorded by GPS device is then expressed by *violet vectors*. Measured point values of noise level can be in map then represented by

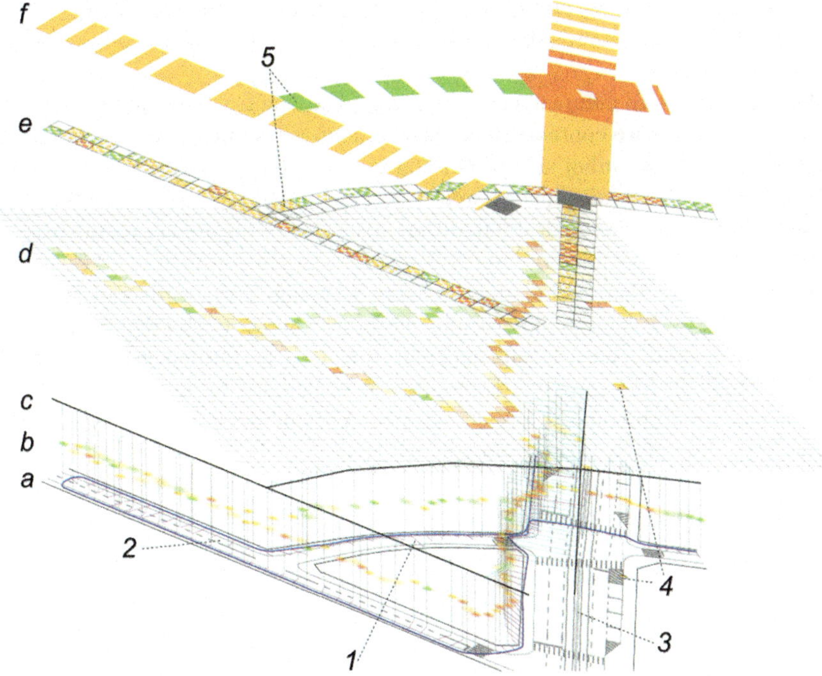

Fig. 2 Derivation of urban traffic noise map from the string of noise point data collected by mobile phone with GPS. Description in text

color hue and color value, and measurement uncertainty [on Fig. 2(b) excluding positional uncertainty] can be represented by blurred edges.

It should be noted, that difference between original position of measured points and model street lines [depicted by *cyan vectors* on Fig. 2(c)] may not be correlated with GPS measurement error.

Today, a relatively large effort has been directed to uncertainty visualization and interpretation in the map. A large number of methods have been developed (see MacEachren 1995). However, only a few of these methods have been tested for (1) how they really influence the perception of the user, and (2) which of those methods are most effective.

Evans (1997) compared four basic methods for displaying uncertainties on choropleth maps. She found that best results of uncertainty interpretation are still achieved on bivariate maps and on maps with user uncontrolled animations. According to her, users do not use uncertainty information from the map, located next to the map of a particular phenomenon, because such comparison are too laborious.

Normal processing of noise map is done on a regular cell network [grid, see Fig. 2(d)]. For sufficiently accurate processing, this raster must have relatively large number of cells. Cartographic visualization of the overall measurement uncertainty on the grid is usually represented by transparency, but, on a light background, transparency acts as a lightening and, according to Schweizer and Goodchild (1992) and Leitner and Buttnefield (2000), affects perception of both noise level value and uncertainty.

Due to fragmented nature of environmental noise data from the mobile mapping, it is not possible to use contours (until observations are made on the vast part of the area of interest) or symbol size (according to Senaratne et al. 2012), because the size of acoustically homogeneous areas is not equal and there may be a misinterpretation based on either inability to differentiate size of the area or on the basis of overlapping symbols.

If the cells are tied to the street network [see Fig. 2(e)], cells can be larger because the noise is measured perpendicular to the source. Cartographic visualization of the overall measurement uncertainty can be performed using the cell's texture. This will prevent misinterpretation of noise level by the map reader due to transparency lightening. Cells also can be on both sides of street line, which allows recording of asymmetric noise situations.

Because volunteers usually move along fixed routes (sidewalks, streets) when measuring, cartographic visualization of noise level average values can be done by using a line symbology. *Line hue* (or saturation) can then illustrate noise level, *line structure* depict uncertainty [see Fig. 2(f)].

Default and ideal for mapping by direct assignment of measured values to the street line is a narrow street without lanes (Fig. 2, 1). In narrow road with lanes is traffic in different directions already reflected in the sidewalk's noise levels (Fig. 2, 2). For a wider street, which has more lanes, parking space for cars, intersections with asymmetrically placed crossings, tramway lanes, etc. (depicted on Fig. 2, 3), it is useful to map each side completely separate.

Using uncertainty is possible to combine noise measurement data from static measurement station with long-term measurements and calibrated noise meter. On a typical raster map long term measurement is projected in one pixel, in the case of an oriented cell it is possible to extrapolate its value (see Fig. 2, 4). Short sections that cannot to be recognized on a smaller scale maps may be generalized (see Fig. 2, 5).

Adjustment of the Measured Data and Evaluation of Their Uncertainties

During environmental noise measurement it is practically very difficult to determine the function of quantities of individual error sources. As a consequence, standards ISO 3745:2012 and ISO 1996-2:2007 identify some of the most important error sources: (1) Error in the measurement chain (sound level meter in the simplest case, further denoted as δ_{slm}), (2) error due to the difference from the source ideal operating conditions(δ_{sou}), (3) error due to meteorological conditions different from the ideal(δ_{met}), (4) error of sensor position(δ_{loc}), and (5) error on residual noise level (δ_{res}). Their overall contribution can be written as follows:

$$L_{Aeq,true} = L_{Aeq,m} + \delta_{slm} + \delta_{sou} + \delta_{met} + \delta_{loc} + \delta_{res} + \varepsilon \tag{1}$$

where $L_{Aeq,true}$ is the true equivalent sound pressure level adjusted with weighting filter A (weighting filter A is defined in IEC 61672:2002 standard), $L_{Aeq,m}$ is the sound pressure level measured by a noise meter and adjusted with the weighting filter A, and ε the residual error. $\delta_{sou} + \delta_{met}$ are often obtained directly by measurement in the field during static noise measurements. But when measuring using a mobile phone in motion, meteorological data is often only available from remote weather stations, which are insufficient due to their remoteness.

The uncertainty of each measurement can then be expressed according to the error propagation law as:

$$u^2\left(L_{Aeq,m}\right) = u_{slm}{}^2 + u_{sou}{}^2 + u_{met}{}^2 + u_{loc}{}^2 + \left(c_{res} \cdot u_{res}\right)^2 + \varepsilon^2 \tag{2}$$

where c is a sensitivity coefficient (all sensitivity coefficients are equal to 1.0 for a certified static measurement, except for residual noise c_{res}).

For measuring environmental noise using mobile phones, the largest sources of uncertainty lay in the following areas: (1) Accuracy of measurement, (2) accuracy of positioning, (3) temporal variability of noise level during different time periods (hour/day/week/year), (4) dependence of measurements on weather conditions, and (5) User knowledge, motivation and behavior.

Accuracy of Noise Measurement Conducted by Mobile Phone Microphones

It is obvious that the accuracy of the sensors (microphones) in mobile phones is lower than the accuracy of a certified sound meter. However, this disadvantage can be almost completely eliminated by using an accurate calibration, as shown in various studies (Weber 2013; Stevens 2012). In the case of artificially generated sound ranging from 35 to 110 dB in an anechoic chamber, the measurement error is within ±1 dB (Weber 2013). Simultaneous field measurements against class 2 noise meters (according to IEC 61672-1) shows, that during measurements of standard urban noise situations, there are deviations ±1 dB from the value measured by the noise meter (Stevens 2012). For solely class 2 sound level meters, uncertainty was ±2 dB (Douglas and Erik 2005), therefore it can be said that the overall measurement accuracy of the instrument does not exceed ±5 dB (at 95 % confidence level). This is sufficient to create informative maps that record the average (or percentile) value for a certain period.

Inaccuracies in the Positioning of Mobile Phones and Their Impact on Data Quality

A mobile phone's GPS is frequently affected by high inaccuracies in positioning, especially in urban areas. The main cause is a reflected GPS radio signal, which leads to an incorrect calculation of transmitter-receiver distance and subsequently, incorrect positioning. The resulting inaccuracy in an urban environment lies usually between 3 and 40 m (Modsching et al. 2006). However, the noise level decreases with the square of the distance. If the positioning system indicates, that the distance from the noise source is 2 m, but the actual distance of the measuring device is only 1 m, the measured value should be about 6 dB higher than it would be for indicated distance, which makes moderately loud noise sources (e.g. industrial installations) hardly mappable.

Positioning using trilateration from GSM base stations has an error of one order of magnitude higher, than GPS (Pent et al. 1997) and therefore cannot be recommended at all.

One of the possible solutions may be to link positional measurements up to map data. The position information of each measurement point can be assigned to the nearest street [a similar principle has been tested in the project NoiseTube (Stevens 2012)]. In the case of a chain of sample points (obtained, for example, while the user walks or rides a bike), the corresponding section of this sample chain may be assigned to that street segment. However, this means that detailed digital maps, which also include sidewalks outside of the main street layout, must be available for post-processing. For particularly wide or asymmetric streets is necessary to consider on which side user is.

If the position signal is lost or user is outside of paths, stored in the reference geodatabase, it is advisable to create an opportunity to additionally georeference measured data (both in the field and during post-processing). If the user does not have a GPS signal, he/she should be alerted, and given the option to record map points or nodal points of his/her path (such as road intersections, stops, etc). Then the map application could identify those points with places on the map and the measured values would be uniformly spatially distributed between these points.

Effect of Background Noise and User Behavior on Noise Measurements

When measuring the general noise situation, residual noise from distant sources does not constitute a significant problem because it has to be included in each measurement and it need not to be separated like for the evaluation of one type of noise source's properties.

But in the case of measurement conducted by a moving pedestrians, background noise loudness of his/her own footsteps is undesirable. The noise level of the walk depends mainly on the type of footwear (shoe sole), the nature of the surface, on which the walker goes, and frequency of his/her steps. For shoes with soft to medium-hard soles, when walking on hard surfaces (asphalt, concrete, or bricks) and when holding the microphone horizontally at waist height, volume levels about 55–60 dB were recorded. In the case of shoes with a slim hard heel measured values reach up to 70 dB. Smaller contributions can also be caused by rustle of pants' fabric. Therefore, it is appropriate that each user, who intends to measure the noise during his/her walk, will measure (or set) the noise level, which he/she emits while walking on different surfaces (user's footwear noise profiles should be created).

Even though the noise level of ca. 60 dB forms a relatively high threshold, below which there is virtually no measurement resolution during walk, this threshold is still sufficient for measuring noise in cities. If the level of residual noise is more than 10 dB below the level of the phenomenon, residual noise is traditionally seen as irrelevant.

When measuring quiet locations, user can make regular stops, and measure noise during these stops. If each stop lasted longer than ca. 5 s, it should be possible to reliably separate data collected during walking from the data measured during stops.

Effects of Mobile Phone on Noise Measurements

Operation of the mobile phone itself may also have a significant effect on measured noise levels. This includes activities associated with calling, sending data, or the loudness of hardware components. For example, pressing hardware buttons can greatly distort measurement results (measured noise level can be around 80 dB), so it is necessary to exclude samples recorded during other user activity with the phone. This may not apply to clicks on the touch screen if the user is careful enough.

Data Aging

Noise data become gradually outdated, like any other environment data. Noise situation changes occur as a result of new infrastructure construction, as well as changes in traffic flows, individual devices (e.g. cars) loudness, etc. Infrastructure changes can be detected with base spatial data updates, or they can be detected also by volunteers themselves. However, it is more difficult to determine which sites are affected by this change and that therefore what data is already outdated. This situation can be verified by observation, or, better, by direct measurement. At places, where was no fundamental change, environmental noise data are considered as obsolete 5 years after acquisition (European Commission 2006).

Volunteers Motivation

It takes a huge effort to ensure, that the collected data correctly predicts the real situation, which implies a need for sufficiently motivated participants. If the project is based on volunteers, it is necessary to rely on their good feeling to do socially useful activities. It is convenient to create a community of users.

The willingness of ordinary residents to spontaneously organize larger sensing events is limited. Spontaneous measurements are rather performed by individuals, who tend to focus on certain areas, in which they typically move (Stevens 2012; Martí et al. 2012) Therefore, it is appropriate to proceed to more sophisticated methods of motivation. One of them is gamification, which allows creating a community of players based on their competitiveness and efforts to compare their achievements with others. Such a community can then measure the noise also in areas not covered by regular contributors (Martí et al. 2012).

According to Zichermann and Cunningham (2011), it is appropriate that such games should be divided into several levels, where players gradually go through on the basis of their efforts from the lightest to more difficult. New game elements should be unlocked by players, based on their activities and contribution to the game. To motivate players to longer contribution, it is recommended to transfer certain powers to the user, who reached high game levels. Likewise a gamer's inactivity should slowly decreases his/her evaluation and capabilities. The main stimulant of such game is however a system of rewards and free gifts. These gifts are attractive for players because they materialize players' differences. A significant stimulus may be also involving in the wider context of social networks, such as FourSquare (2013) (a successful social game created for commercial purposes, which crosses the boundaries between the virtual and the real world by providing physical rewards to its users).

Such an application for noise observation is suggested by Martí et al. (2012). A player's game progress is based on the number of measurements and thereby increasing his or her game level. The game begins with measurements on sites where the player spends time most often, and continues to farther sites. Obtaining measurements from different areas and completing various other tasks of the game narrative opens uncommon tasks that can give some unusual abilities to the player. Using this strategy, not only number of measurements is evaluated, but also their quality. This concept is currently being tested (Universitat Jaume I 2013). The greatest danger of this technique is the possibility of proliferation of such games.

As follows from the previous text, also the cartographic visualization must be designed to motivate users to supply additional data and improve their quality. Therefore, it should be appropriate to create a visualization that supports the concept of gamification. The map should remind players of the game plan and the visualization should support the user's playfulness and game narratives.

Conclusions

Complex processing of uncertainties for noise maps created by information volunteered from mobile phones is the result of efforts to enable adequate interpretation of the map with significant lack of data. It is clear that creating of such noise maps for standard European city is very lengthy process and already stored data may become obsolete. Therefore it is not assumed that complete maps of all time periods and with sufficient accuracy would be available. It is also necessary to prevent the reader impression that person-centered data can be used interchangeably with data obtained by objective measurements.

Both of these problems can be resolved by displaying noise measurement uncertainty. However, introduction of uncertainty theory requires a fairly extensive theoretical base. The presented analysis revealed that measurement noise using mobile phones is facing following main groups of problems:

(1) Accuracy of environmental noise measurement—this problem is solved to a certain extent and can be quantified in terms of uncertainty; further refinement depends on quality of calibration. Visualization of the overall uncertainty reflects different accuracy of different types of devices. (2) Accuracy of positioning—quality of position data can be improved by adequate post processing techniques, but additional data is necessary. Due to the double error (inaccuracy to actual position measurement usually disagrees with lines in the street network model), it is therefore still necessary manually determine locations on map sometimes. (3) Temporal variability of noise levels during different time periods (hour/day/week/year)—short-term measurements are often just a random sample, so it is necessary to combine different measurements. Uncertainty is estimated comparatively reliably, at least in traffic noise. It depends mainly on knowing the source of the noise (4) The dependence of measurements on weather conditions. These data can be only indirectly estimated from meteorological records and uncertainty cannot be quantified directly, only cataloged. So it depends mainly on how users in the field evaluate the situation. (5) User knowledge, motivation and behavior—users can be motivated by gamification, both by fully morphing data acquisition into interactive games and simpler forms suitable for serious contributors, similar to applications such as Foursquare.

An adequate number of measurements remains as the most pressing problem. If enough data is available, we would expect to counterbalance random errors as different characteristics of devices (sensors), operator error, spatial error and random environmental effects. Such data would also monitor changes during the day, because different types of days or periods during the day are not comparable. Collecting a sufficient amount of relevant data to produce adequately accurate maps is mainly dependent on user motivation.

(continued)

Further research should therefore focus on methods of position refining, for example by geotagging or other technologies (e.g. based on Wi-Fi), on methods of user control and activity increase (such as applications of automated systems for recognition of sound sources in post processing in the first case and gamification or support from public budgets in the second case) and on the techniques of estimating changes in noise levels during the day from the low number of short measurements. A separate, very complicated task then is the cartographic visualization of the data, including relevant contextual metadata and uncertainties.

References

Apple (2013) iTunes. http://www.apple.com/itunes/charts/free-apps/. Accessed 31 Dec 2013

D'Hondt E, Stevens M, Jacobs A (2013) Participatory noise mapping works! Pervasive Mobile Comput 9(5):681–694. doi:10.1016/j.pmcj.2012.09.002

Douglas M, Erik A (2005) Uncertainties in environmental noise assessments – ISO 1996, effects of instrument class and residual sound. Paper presented at Forum Acousticum, Budapest, 29 August–2 September 2005

European Commission. Working Group Assessment of Exposure to Noise (2006) Good practice guide for strategic noise mapping and the production of associated data on noise exposure. European Commission, Bruxelles. http://ec.europa.eu/environment/noise/pdf/wg_aen.pdf. Accessed 31 Dec 2013

Evans B (1997) Dynamic display of spatial data reliability: does it benefit the map user? Comput Geosci 23:409–422

Feick R, Roche S (2013) Understanding the value of VGI. In: Sui D, Elwood S, Goodchild M (eds) Crowdsourcing geographic knowledge, 1st edn. Springer, Heidelberg, pp 15–29

Finley M, Miles J (2007) Exterior noise created by vehicles traveling over rumble strips. Paper presented at the Transportation Research Board 86th annual meeting, Washington, 21–25 January 2007

Foursquare (2013) Foursquare. https://foursquare.com/. Accessed 31 Dec 2013

Google (2013) Google Play. https://play.google.com/. Accessed 31 Dec 2013

IEC 61672-1:2002 (2002) Electroacoustics – Sound level meters – Part 1: Specifications

ISO 1996-1:2003 (2003) Acoustics – Description, measurement and assessment of environmental noise – Part 1: Basic quantities and assessment procedures

ISO 1996-2:1987 (1987) Acoustics – Description and measurement of environmental noise – Part 2: Acquisition of data pertinent to land use

ISO 1996-2:2007 (2007) Acoustics – Description and measurement of environmental noise – Part 2: Determination of environmental noise levels

ISO 3745:2012 (2012) Acoustics – Determination of sound power levels and sound energy levels of noise sources using sound pressure – Precision methods for anechoic rooms and hemi-anechoic rooms

Kang J (2007) Urban sound environment. Taylor & Francis, London

Kanjo E (2010) NoiseSPY: a real-time mobile phone platform for urban noise monitoring and mapping. Mobile Netw Appl 15(4):562–574. doi:10.1007/s11036-009-0217-y

Leitner M, Buttnefield B (2000) Guidelines for the display of attribute certainty. Cartogr Geogr Inf Sci 27(1):3–14

MacEachren A (1995) How maps work: representation, visualization and design. Guilford Press, New York

Manvell D, Ballarin Marcos L, Stapelfeldt H, Sanz R (2004) SADMAM – Combining measurements and calculations to map noise in Madrid. Paper presented at the 33rd international congress and exposition on noise control engineering (inter-noise 2004), Prague, 22–25 August 2004

Martí I, Rodríguez L, Benedito M (2012) Mobile application for noise pollution monitoring through gamification techniques. In: Proceedings of 11th international conference on entertainment computing (ICEC 2012), Bremen, 26–29 September 2012. doi:10.1007/978-3-642-33542-6_74

McDonald P, Geraghty D, Humphreys I, Farrel S (2008) Assessing environmental impact of transport noise with wireless sensor networks. In: National Research Council (ed) Environment and Energy 2008. Transportation Research Record is 2058. Transportation Research Board, Washington, pp 133–139

Modsching M, Kramer R, Hagen K (2006) Field trial on GPS accuracy in a medium size city: the influence of built-up. Paper presented at the 3rd workshop on positioning, navigation and communication, Hannover, 16 March 2006

Nüst D (2013) OpenNoiseMap. https://wiki.52north.org/bin/view/SensorWeb/OpenNoiseMap. Accessed 31 Dec 2013

Pent M, Spirito M, Turco E (1997) Method for positioning GSM mobile stations using absolute time delay measurements. Electron Lett 33(24):2019–2020. doi:10.1049/el:19971375

Schweizer D, Goodchild M (1992) Data quality and choropleth maps: an experiment with the use of color. Proc GIS/LIS 2:686–699

Senaratne H, Gerharz L, Pebesma E, Schweing A (2012) Usability of spatio-temporal uncertainty visualisation methods. In: Gensel J, Josselin D, Vandernbroucke (eds) Bridging the geographic information sciences. International AGILE'2012 conference, Avignon, 24–27 April 2012.

Senatsverwaltung für Gesundheit, Umwelt und Verbracherschutz (2008) Umweltatlas Berlin. Strategische Lärmkarten. http://stadtentwicklung.berlin.de/umwelt/laerm/laermminderungsplanung/de/laermkarten/index.shtml. Accessed 31 Dec 2013

Stevens M (2012) Community memories for sustainable societies – the case of environmental noise. Dissertation, Vrije Universiteit, Brussel

Universitat Jaume I (2013) GeoGaming: NoiseBattle. http://www.geotec.uji.es/geogaming-noisebattle/. Accessed 31 Dec 2013

Weber D (2013) A review of 30 sound (noise) measurement apps. http://www.safetyawakenings.com/safety-app-of-the-week-42/. Accessed 30 Aug 2013

Zichermann G, Cunningham C (2011) Gamification by design: implementing game mechanics in web and mobile apps. O'Reilly Media, Sebastopol

Multivariate Data Visualization and Usability: Preliminary Notes

Petr Kubíček, Radka Báčová, and Zdeněk Stachoň

Introduction

Effective and understandable communication of geographic information is currently under discussion by the scientific community in general and cartographers in particular. New data sources as well new visualization methods are re-opening the discussion about the user's ability to read the map and interpret its content in a coherent way.

Multivariate mapping techniques used for representation of three or more variables are even more challenging for individuals with a limited cartographic background. Several types of visualization techniques exist for representing multivariate spatial patterns on the same map. These include trivariate choropleth maps (Brewer 1994), multivariate dot maps (Rogers and Groop 1981), or multivariate point symbol maps (glyphs) (Chernoff and Rizvi 1975; Dorling 1994). Another possibility is to compare three or more maps at the same time. Such visualization is termed small multiples (Tufte 1990; Kousoulakou and Kraak 1992; Slocum et al. 2009). A more complex discussion about multivariate visualization techniques can be found in Guo et al. (2005) or Slocum et al (2009).

With respect to the data that can be represented, several authors have found that multivariate depictions can be used successfully for representation of both temporal and non temporal data with a wide variety of data nature (nominal, ordinal, interval, and ratio). However, only preliminary discussion with users has been held regarding their ability to use such representations. In this empirical study, it was decided to compare the effectiveness and efficiency of two multivariate mapping methods—ring maps and small multiples—for both spatial and temporal data. The choice of small multiples is justified by the following points. Small multiples are traditionally

P. Kubíček (✉) • R. Báčová • Z. Stachoň
Faculty of Science, Department of Geography, Masaryk University, Kotlářská 2, 611 37 Brno, Czech Republic
e-mail: kubicek@geogr.muni.cz; rada.ba@mail.muni.cz; zstachon@geogr.muni.cz

© Springer International Publishing Switzerland 2015
J. Brus et al. (eds.), *Modern Trends in Cartography*, Lecture Notes in
Geoinformation and Cartography, DOI 10.1007/978-3-319-07926-4_29

used for multivariate epidemiological data as described by several authors (Battersby et al. 2011; Stewart et al. 2011). Small multiples use areal symbols for depiction of spatial or temporal variables and can serve as a comparable representation with ring maps. Other multivariate visualization methods use different graphical variables and are not easily comparable with ring maps. Moreover, experimental methods such as trivariate dot maps are not traditionally used and their use by Maršík (1998) caused an extensive discussion about the usability of such maps among the health community. The main goal was to validate the suitability of a relatively new visualization method (ring maps) for multivariate data analysis and to compare it with the conventionally used method (small multiples). We adopted the hypothesis, that ring maps are a more useful tool for a complex analytical task rather than for simple map reading.

The Use of Ring Maps

The pursuit toward reaching the aforementioned goal was initiated in several studies of multivariate visualization using ring maps (see below) and introducing this method as: "…positive and encouraging in terms of continued production of the maps for (epidemiological) data analysis" (Battersby et al. 2011). Huang et al. (2008) and Zhao et al. (2008) constructed and described the ring map as a depiction of multiple attribute data sets as separate rings of information surrounding a base map of a particular geographic region of interest. Battersby et al. (2011) utilized ring maps for visualizing spatial and temporal relationships in multivariate epidemiological datasets and demonstrated a spectrum of possible uses. Stewart et al. (2011) deployed the ring map method for non temporal (spatial attribute) data and also discussed the disadvantages of ring maps, citing the complete loss of topology in the first place followed by the limited ability to display a large number of enumeration units.

The other authors proposed two future research directions for studies evaluating the advantages and disadvantages of ring maps. The internal direction is focused on the method itself and revelation of possible limits—number of aggregation units, data types and predetermination of data classes, number of rings and their optimal shape, and the effect of graphical variables used for symbolization (hue, saturation, point symbols and texture). The external research is directed towards a comparison of the ring map visualization method with other multivariate visualization methods and also with method usability for a different map use (exploration, analysis, synthesis, and presentation). Both research directions can be further divided into the static and dynamic visualization branches. Table 1 compares several existing studies from the internal research point of view.

Explanation of terms is given on Fig. 1.

Chan et al. (2013) developed a Python tool enabling the creation of a simple ring map within the ArcGIS environment and thus opening the possibility of more extensive use of this visualization method. This tool allows its users to design

Table 1 Examples of existing studies and particular construction methods and data types used

Author (s)	No of rings	Continuous rings (vertical/horizontal)	Symbolization	No of spatial units	Data type
Huang et al. (2008)	12–24	N/N	Hue	25	Spatial/ temporal
Zhao et al. (2008)	Up to 96	Y/Y	Hue	47	Spatial/ temporal
Battersby et al. (2011)	2–5	Y/N	Saturation/ shade	46	Spatial
Fede et al. (2011)	4–5	N/N	Saturation/ texture	46	Spatial
Stewart et al. (2011)	2–5	Y/N	Saturation	46	Spatial

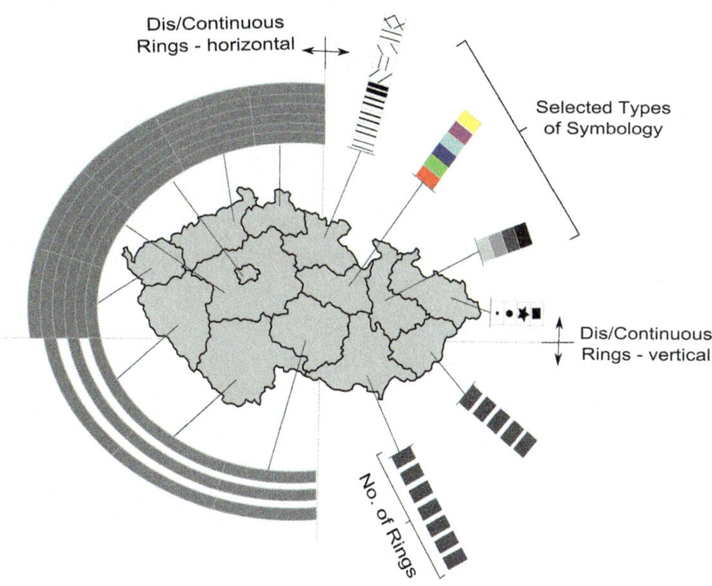

Fig. 1 Internal aspects of ring maps construction

only horizontal and vertical continuous rings for a selected geographic region, which is a different concept of making ring maps from that mentioned in the original studies (discussed above).

Based on the varied options in cartographic visualization of ring maps, we decided to take the first step in the external research direction by starting with a comparison of the ring map with other existing multivariate visualization methods. Evaluation of the cartographic methods and products is a broad issue with different approaches—from qualitative (e.g. user questioning) to quantitative (e.g. multicriterion evaluation)—both for exploratory geovisualization (Koua et al. 2006) and for mobile applications (van Elzakker et al. 2008). Various papers can be mentioned as examples: Sedlák et al. (2011) focused on evaluation of map portals, Štěrba et al. (2011) focused on map symbols, Popelka and Brychtová (2013) dealt with 2D and 3D visualizations, and Fabrikant et al. (2008) focused on small multiples and eye tracking. Rogers and Groop (1981) made one of the first empirical evaluations in the field of multivariate data visualization. They compared the multivariate dot map with several univariate dot map methods and evaluated map reading tasks. Rogers and Groop proposed that the multivariate method was more effective for perceptual tasks, but also pointed out the limited legibility of the multivariate method for more than three categories.

Kaspar et al. (2011) performed a study comparing contiguous cartograms and choropleth maps for spatial inference making. They confirmed significant differences in the user's performance according to the visualization type and the dominance of choropleth maps over cartograms. Usability of the visualizations was also dependant on the complexity of the map use tasks and on the shape of the enumeration units. van den Elzen and van Wijk (2013) compared small multiples and large singles for both the effectiveness and efficiency of visual exploration. They discovered no advantages of small multiples in efficiency (execution time) or in the number of errors (correctness of answers). However, they found that users were more satisfied and preferred exploration methods using small multiples. All the studies generally confirmed that overall efficiency and effectiveness depends not only on the visualization method itself, but also on the complexity of the tasks (map reading, information exploration, and information inference) and the visual variables used.

Data Used and Preparation of the Test

To empirically evaluate the difference between particular types of multivariate representation, we designed and performed an empirical experiment. Two different types of visualization—namely ring maps and small multiples—were used for three types of tasks.

Health data, which are typical representatives of large and multivariate data sets, were used for map preparation. We created two different base maps for the empirical experiment using two cartographic visualization techniques—attribute, and temporal characteristics.

The first base map (attribute) contains the following data:

- Main characteristic—Number of heart attack deaths per 10,000 people—(shown on a central choropleth map in ring maps, or on one of the choropleth maps in small multiples) was classified into five categories.
- Secondary characteristics—Crude mortality rate, Percent of urban population, GDP per capita, Unemployment rate, and Pharmacy profits from supplements in thousands of CZK (shown on rings in ring maps, or on the remaining choropleth

maps in small multiples) were classified into five similar categories classified verbally from very high to very low.

The second base map (spatio-temporal changes) includes:

- Main characteristic—Percentage increase in the incidence rate of malignant neoplasms between 2000 and 2004—(shown on the central choropleth map in ring maps, or on one of the choropleth maps in small multiples) was classified into five classes.
- Secondary characteristics—Incidence rate of malignant neoplasms in each survey period 2000–2004—(shown on rings in ring maps, or on the five remaining choropleth maps in small multiples) were classified into five similar value categories.

No commercial or open tools for automatic ring map development were available at the time of test preparation. Therefore, we decided to use ESRI ArcMap for creating both types of cartographic visualizations. Small multiples were designed on one printable page to allow for better comparison of related displayed variables. Ring maps were created manually, while we designed a shapefile-based template connected with selectable database attributes (spatial or temporal).

The Multivariate Testing Program (MuTeP), developed by the Department of Geography, Masaryk University, was chosen for evaluation of the usability of the multivariate data visualization methods. MuTeP enables the creation of testing batteries and the collection of qualitative and quantitative data about the test participants and their performance on presented cartographic materials. Important advantages of the MuTeP tool for this study are the possibility of maintaining test batteries in Internet Explorer, which the participants are familiar with, the possibility of simultaneous recording of time consumed on a particular section of the test battery and the correctness of the answers. MuTeP software is described in more detail in Štěrba et al. (2011).

Study Design

Stimuli

Two different types of visualization were in our empirical study, namely ring map and small multiples (see Fig. 2). Both cartographic visualization methods displayed similar thematic data mentioned in previous chapter. All the colour templates used in the study were defined using ColorBrewer 2.0 (2013).

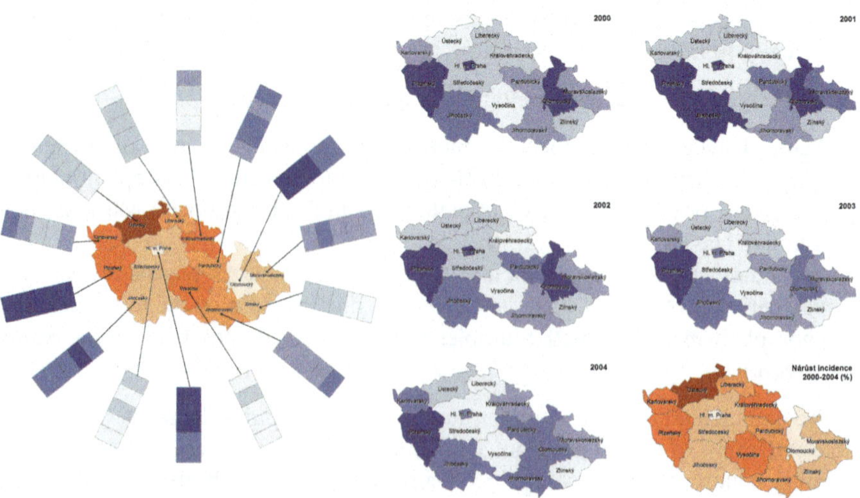

Fig. 2 Example of used visualizations: ring map (*left*) and small multiples (*right*)

Structure

The testing battery was designed for the research of two parallel groups. There were two independent tests: Test A using ring map visualization and Test B using small multiples visualization of equivalent characteristics. Each test contains four main sections. Correctness of answers and the time consumed by particular tasks were recorded.

- At the beginning of the test there was a personal questionnaire which focused on the personal characteristics (e.g. age, map use experience, gender, computer use experience etc.) of the particular participant followed by an example of the cartographic visualization method used in the relevant test. The following sections contained tasks on relevant visualization methods.
- The second section of the test contained ten tasks focused on basic identification of one particular value from the map or from the map legend (map reading). The first five tasks involved using six different characteristics, followed by five tasks visualizing two characteristics which one was temporal. All the tasks offered a choice of four possible answers.
- The third section of the test entailed visualizing six characteristics and increased in terms of the complexity of the tasks. It began with the requirement of comparing one particular characteristic in the map legend to the map and proceeded up to the comparison of three different characteristics.
- The fourth section was similar to the third section using two characteristics where one represented temporal development.

Participants

The participants of the test were 52 students of the Department of Geography, Masaryk University, with advanced cartographic knowledge. The participants were randomly split into two groups (A—ring maps, B—small multiples). Both groups were homogeneous with an average age of 23.5 in Group A and 23.7 in Group B. The ratio of male/female was also similar (55 % males/45 % females). The respondents were asked about additional qualitative parameters—frequency of computer and map use, and vision defects. Only two participants declared a relevant vision defect—reduced colour perception. All the respondents participated voluntarily and the experiment was carried out under identically monitored conditions.

Recordings

The results of the test contain personal information about the participants and also qualitative information about task solving time (in seconds for each task) and task completion correctness. We tried to evaluate the results considering differences between cartographic visualization methods and also the distinctions between the user groups, e.g. male/female.

Even a simple graphical comparison of the cartographic visualization methods showed differences between ring maps and small multiples in task solving time. An important fact is that participants performing on ring maps achieved a faster response time in completing most of the tasks. This fact is more apparent in the third and fourth sections of the test where the ratio was 9/1 (see Table 2). The results were due to the low number of the participants tested only by a nonparametric U test, but no statistically significant differences were discovered on 5 % significance

Table 2 Average response time (s) for tasks (T) performed on ring map (RM) and small multiples (SM)

		Section 2									
		T1	T2	T3	T4	T5	T6	T7	T8	T9	T10
Time (s)	RM	46.8	23.5	22.7	45.1	39.6	36.5	50.7	30.8	20.3	16.6
Time (s)	SM	44.0	24.8	20.7	15.9	38.3	37.1	44.4	22.2	20.2	12.0
		Section 3									
		T1	T2	T3	T4	T5	T6	T7	T8	T9	T10
Time (s)	RM	33.1	41.3	27.7	42.7	61.6	33.9	62.5	73.1	61.3	137.0
Time (s)	SM	31.5	33.0	30.6	43.3	35.1	39.7	49.1	88.0	69.1	122.2
		Section 4									
		T1	T2	T3	T4	T5	T6	T7	T8	T9	T10
Time (s)	RM	66.0	39.9	153.8	52.2	27.0	60.2	63.1	111.5	38.2	54.9
Time (s)	SM	60.7	37.8	124.8	51.9	24.6	45.5	52.6	102.1	44.9	68.7

level. Because of its dependence on the sample size (N; estimated N will be about 50 participants in each group of our experiment), there was a calculate effect size of (ω^2). It is a characteristic of the strength of a phenomenon and basically means the proportion of explained variance (Hayes 1963). The resultant $\omega^2 = 0.06$ explains only a 6 % effect of change in the cartographic visualization method to the average time of correct answers. We identified a higher efficiency in female participants working with ring maps. It was approximately 6 s per person per task. However, this difference must be approached with caution due to the low number of female participants and the possibility of coincidental results. The rate of response in the small multiples group was equivalent in both users' subgroups and comparable with the female subgroup using ring maps.

There was not an unambiguous trend in the error rate between ring maps and small multiples. The accuracy of answers was closely connected with particular tasks and varied through all sections of the test. The overall error rate achieved 11 % for both types of visualization in the whole group of respondents. Ring maps reached higher effectiveness in spatial information inference, and small multiples were more effective in map reading tasks. Overall results were not statistically significant, because of the small test-participants sample size.

Discussion

The presented paper investigates how users read and infer basic information from two different multivariate spatial data visualization methods/techniques—ring maps and small multiples. Preliminary results based on a limited group of users show that the ring map depiction achieved faster (more efficient) responses with a comparable error rate (effectiveness) for both the map reading and spatial information inference. This result is surprising taking into account the common use of choropleth depiction used for small multiples.

The achieved results cannot be generalized. The tested students are experienced users with extensive experience in map use and are familiar with both methods as well as with the background health statistical data. The overall results can also be influenced by the spatial units used. Both the spatial patterns and topology of the basic administrative units (14) in the Czech Republic are generally known to all tested students and can bias the users' performance.

Future Work

In order to get generally valid results we will expand the group for empirical testing and also include a more heterogeneous group of users, including non-experts. This quantitative testing will also be enhanced with qualitative eye tracing testing in order to analyse differences in personal strategies of task completion for both small multiples and ring maps depiction.

Preliminary results provide stimuli for further investigation of multivariate methods' effectiveness and efficiency. The proposition that the more unfamiliar depiction method (ring maps) can be comparable or even more usable than the commonly used method (small multiples choropleth maps) needs more extensive research and more robust confirmation. However, these results corroborate the findings of Kaspar et al. (2011) that non traditional methods (cartograms in the case of Kaspar et al. and ring maps in our case) deserve attention and a detailed analysis of use. Both research directions outlined at the beginning of our study should be combined in order to find user-friendly and acceptable map design for particular types of tasks.

References

Battersby S, Stewart J, Fede A, Remington K, Mayfield-Smith K (2011) Ring maps for spatial visualization of multivariate epidemiological data. J Maps 7(1):564–572

Brewer C (1994) Color use guidelines for mapping and visualization. In: MacEachren AM, Taylor DR (eds) Visualization in modern cartography. Pergamon, Oxford, pp 123–147

Chan TCH, Wang CHM, Lee YM (2013) Looking at temporal changes – use this Python tool for creating ring maps. http://www.esri.com/esri-news/arcuser/fall-2013/looking-at-temporal-changes?WT.mc_id=EmailCampaignh10401. Accessed 10 Dec 2013

Chernoff H, Rizvi MH (1975) Effect on classification error of random permutations of features in representing multivariate data by faces. J Am Stat Assoc 70:548–554

ColorBrewer 2.0 (2013) Color advice for maps. http://colorbrewer2.org. Accessed 10 Dec 2013

Dorling D (1994) Cartograms for visualizing human geography. In: Unwin D, Hearnshaw H (eds) Visualization and GIS. Belhaven Press, London, pp 85–102

Fabrikant SI, Rebich-Hespanha S, Andrienko N, Andrienko G (2008) Novel method to measure inference affordance in static small-multiple map displays representing dynamic processes. Cartogr J 45(3):201–215

Fede A, Stewart J, Hardin J, Mayfield-Smith K, Sudduth D (2011) Spatial visualization of multivariate datasets: an analysis of STD and HIV/AIDS diagnosis rates and socioeconomic context using ring maps. Public Health Rep 126(Suppl 3):115–126

Guo D, Gahegan M, MacEachren AM, Zhou B (2005) Multivariate analysis and geovisualization with an integrated geographic knowledge discovery approach. Cartogr Geogr Inf Sci 32 (2):113–133

Hayes WL (1963) Statistics for psychologists. Holt, Rinehart and Winston, New York

Huang G, Govoni S, Choi J, Hartley DM, Wilson JM (2008) Geovisualizing data with ring maps. ArcUser 11(1):54–55

Kaspar S, Fabrikant SI, Freckmann P (2011) Empirical study of cartograms. In: 25th international cartographic conference, Paris, 03 July 2011–08 July 2011

Koua EL, MacEachren AM, Kraak M-J (2006) Evaluating the usability of visualisation methods in an exploratory geovisualization environment. Int J Geogr Inf Sci 20:425–448

Kousoulakou A, Kraak MJ (1992) Spatio-temporal maps and cartographic communication. Cartogr J 29(2):101–108

Maršík V (1998) Atlas výskytu zhoubných nádorů v České republice 1978–1994. MOÚ, Brno, 47 s. ISBN: 80-238-1718-3

Popelka S, Brychtová A (2013) Eye-tracking study on different perception of 2D and 3D terrain visualisation. Cartogr J 240–246. ISSN: 1743–2774 (Maney Publishing)

Rogers J, Groop R (1981) Regional portrayal with multi-pattern color dot maps. Cartographica 18 (4):51–64

Sedlák P, Hub MJK, Komárková J (2011) Preference uživatelů internetových mapových aplikací zjištěné na základě hodnocení použitelnosti. 19. Kartografická konferencia Kartografia a geoinformatika vo svetle dneška. Kartografická spoločnosť Slovenskej republiky, 2011, Bratislava, s131–139. ISBN: 978-80-89060-19-1

Slocum TA, McMaster RB, Kessler FC, Howard HH (2009) Thematic cartography and geovisualization, 3rd edn. Pearson/Prentice Hall, Upper Saddle River

Štěrba Z, Stachoň Z, Šašinka Č, Zbořil J, Březinová Š, Talhofer V (2011) Evaluace kartografických znakových sad v kontextu osobnosti uživatele, Geodetický a kartografický obzor

Stewart J, Battersby S, Fede A, Remington K, Hardin J, Mayfield-Smith K (2011) Diabetes and the socioeconomic and built environment: geovisualization of disease prevalence and potential contextual associations using ring maps. Int J Health Geogr 10:18

Tufte ER (1990) Envisioning information. Graphics Press, Cheshire

van den Elzen S, van Wijk JJ (2013) Small multiples, large singles: a new approach for visual data exploration. Comput Graph Forum 32:191–200 (Wiley Online Library)

van Elzakker C, Delikostidis I, van Oosterom PJM (2008) Field-based usability evaluation methodology for mobile geo-applications. Cartogr J 45(2):139–149 (Use and Users Special Issue 2008, The British Cartographic Society)

Zhao J, Forer P, Harvey AS (2008) Activities, ringmaps and geovisualization of large human movement fields. Inf Vis 7:198–209

Part IV
Cartography in Practice and Research

On Shape Metrics in Cartographic Generalization: A Case Study of the Building Footprint Geometry

Vít Pászto, Alžběta Brychtová, and Lukáš Marek

Introduction

Cartographic generalization is a key point in the process of mapmaking. It can be defined as the process of selecting and simplifying the detail of representation appropriate to the scale and purpose of a map (Savino 2011). The process of generalization extracts and reduces information from reality or source maps and portrays it to represent a specific theme at a smaller scale, while meeting cartographic specifications and maintaining the representative integrity of the mapped area (ESRI 1996). The degree of generalization is proportional to a feature's spatial detail and inversely proportional to map scale (Kimerling et al. 2011).

Cartographic generalization comprises two distinct processes: (a) reducing the shape complexity of mapped features and even displacing linework or symbols in order to improve legibility, (b) reducing the amount of information shown in a map. The first process influences the quantitative accuracy of a map; the second affects the completeness of it (Maling 1988). The generalization becomes more specific in the GIS environment. For this reason two kinds of generalization can be distinguished: (a) cartographic generalization and (b) model-oriented generalization. Cartographic generalization concerns only the visualization of geospatial information and occurs at the graphic level. Yet, none of available GIS software products

V. Pászto (✉)
Faculty of Science, Department of Geoinformatics, Palacký University in Olomouc,
Tř. Svobody 26, 771 46 Olomouc, Czech Republic

Department of Informatics and Applied Mathematics, Moravian University College Olomouc,
tř. Kosmonautů 1288/1, 779 00 Olomouc, Czech Republic
e-mail: vit.paszto@gmail.com

A. Brychtová • L. Marek
Faculty of Science, Department of Geoinformatics, Palacký University in Olomouc,
Tř. Svobody 26, 771 46 Olomouc, Czech Republic
e-mail: alzbeta.brychtova@upol.cz; lukas.marek@upol.cz

© Springer International Publishing Switzerland 2015
J. Brus et al. (eds.), *Modern Trends in Cartography*, Lecture Notes in
Geoinformation and Cartography, DOI 10.1007/978-3-319-07926-4_30

provide tools for graphic-oriented tasks but rather tools for data model (non-graphic) generalization. Model-oriented generalization takes place within the scope of an internal representation of a map and pursues reduction of the information density in a database (Filippovska et al. 2008). Contemporary trend of research is focused on automated methods that enable the creation and display of geographic information at multiple levels of detail (Cheng and Li 2013).

The greatest need of the generalization feel organizations that operate large volumes of spatial data (such as National Mapping Agencies). Their requirements for automated generalization are still not directly available. The reason for these missing requirements are that: (a) currently no formalism has proven to be adequate for fully capturing the specifications of a map, (b) not all requirements are easily to be formalized and (c) much knowledge on generalization requirements and processes still needs to be revealed (Stoter et al. 2009).

Although principles and guidelines of the generalization can be found in cartographic literature and among mapping organizations, there has not existed a set of universal rules that explicitly and completely defines how the generalization should be performed (ESRI 1996).

The challenge of the generalization is to preserve the geographical meaning and relations; it is necessary to recognize between essential and unimportant information (Bard and Ruas 2005). The generalization is a subjective process with an accent to knowledge and experience of the cartographer. The computational geometry makes a process of the simplification less dependent on a subjective view of the cartographer, however it is not an easy task to find and set a geometric criterion that should be satisfied by a simplified element (Bayer 2009).

Due to this fact few authors were searching for approaches or tools how to evaluate results of generalization using various methods (e. g. Girres 2011; Filippovska et al. 2008). Shape metrics may serve as such a tool or method for quantitative evaluation of generalization methods and their outputs.

Shape metrics were originally applied in the landscape ecology in order to quantify characteristics of landscape patches. Landscape patches represent homogenous, further indivisible units of a landscape (according to the scale—forests, humid areas, specific habitats, urban fabric, etc.). These patches are represented as polygons in GIS and therefore are objects of cartographic generalization. Applications of shape metrics in cartography with the focus on polygons can be found in Peter (2001), Schmid (2008), Agent (1999), Burghardt and Steiniger (2005) or Gao et al. (2012). Complexity measures of generalized line features are thoroughly described in Jasinski (1990), Bernhardt (1992), or Skopeliti and Lysandros (2001).

Generally, shape metrics serve as a quantitative description of any planar object (e.g. ground projection of a building) in order to measure its shape complexity, roughness or irregularity. Shape metrics are fundamentally based on an area of a shape and its perimeter (these two characteristics are themselves considered as shape metrics and are very easy to obtain), but most of the metrics are more complicated to calculate and are treated as shape indexes. Nevertheless, it is worth to mention, why it is useful to calculate shape metrics. Since shape metrics take into account only geometric properties of the patch (polygon), it is possible to eliminate expert subjectivity in cartographic generalization process. There is no doubt that expert skills are crucial in cartographic generalization, but shape metrics could serve them as a "statement of fact" to support their expert knowledge. Thus, shape metrics are suitable to help cartographers with decisions about the level of generalization.

In this paper, we applied shape metrics calculation on 22 levels of cartographic generalization in the sense of reducing the complexity of mapped features to obtain quantitative characteristics of generalized shapes of four building ground planes. Our idea was to evaluate results of cartographic generalization in the sense of reducing the polygon complexity by examining the shape metrics values of each generalization level. We believe that this approach could show deviations or mistakes in the process of semi-automated generalization. This pilot study showed great potential for quantitative description of generalization outputs. In addition, there are some recommendations about particular shape metric usability.

Data and Methods

Cartographic geometry generalization and shape metrics calculations were performed on four buildings ground plans. First, the ground plans were vectorized in GIS upon ground plans of four architectonically different buildings with distinctive shapes:

- Petronas Towers (Kuala Lumpur, Malaysia)
- Lund University main building (Lund, Sweden)
- Houses of Parliament (London, United Kingdom)
- Church of St. Maurice (Olomouc, Czech Republic)

Buildings' ground plans are depicted in Fig. 1a. All ground plans were adjusted to uniform size of the X-axis (thus their shapes do not corresponds to their real size and therefore the units in shape area and length indexes are not relevant although being calculated in square meters, and meters, respectively) in order to ensure comparable shape simplification in particular generalization levels.

Then, the ground plans were generalized using Simplify Building tool available in ESRI ArcGIS for Desktop. The algorithm hidden behind this tool reduces details in the boundaries of buildings, while maintaining the essential shape and size of the buildings. The simplification process preserves and enhances orthogonality of buildings. Building simplification is for applications at large scales, where buildings are still represented individually. It works on an entire building boundary, not on selected segment of it. The number of vertices will be reduced, but the measured area will remain roughly the same as the original. The maximum degree of simplification is reached when a building is reduced to a rectangle (Lee and Hardy 2006).

The generalization was done in 22 levels corresponding following simplification tolerances: $t = 2, 4, 6, 8, 10, 15, 20, 25, 30, 35, 40, 45, 50, 60, 80, 100, 200,$

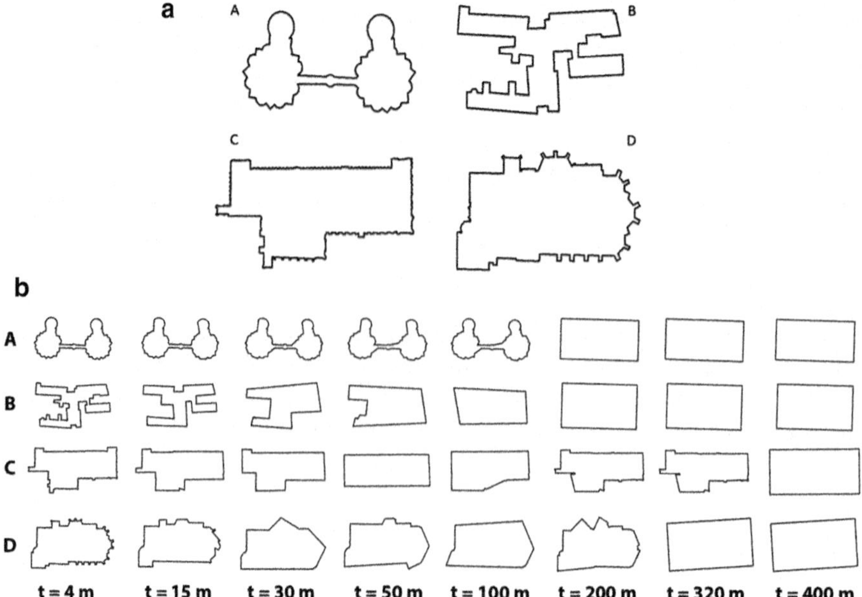

Fig. 1 (**a**) Original ground plans of selected buildings, *A*—Petronas Towers, *B*—Lund University, *C*—Houses of Parliament, *D*—St. Maurice, (**b**) Preview of selected generalization levels with simplification tolerance t = 4, 15, 30, 50, 100, 200, 320 and 400 m

300, 320, 350, 400 and 500 m. These steps were selected in order to cover all the process of polygon simplification—from the most complex to the point in which greater simplification tolerance would not cause any further change of a shape and resulted into a rectangular shape of buildings. Preview of selected generalization steps is shown in Fig. 1b.

Lastly, 15 shape metrics were calculated (Table 1), basic shape metrics (shape area and length) were automatically calculated during the generalization process, Perimeter-area index and Fractal dimension index were calculated separately outside GIS. For other shape metrics, Shape Metrics Toolbox for ESRI ArcGIS for Desktop (Parent 2014) was used. Basic description of used shape metrics is in Table 1. In this case study normalized versions of shape metrics from the Shape Metrics Toolbox were used as they are not affected by a shape area. Metrics were normalized using the Equal Area Circle (Parent 2014). For more details about shape metrics, see Parent (2014), McGarigal (2013), Mesev (2007) or Forman (1995). Shape metrics were calculated for buildings' ground plans in every particular level of generalization.

Normalized shape metrics values range between 0 and 1. According to Angel et al. (2010), a circle is the most compact (or ideal) shape possible for a given area. The higher the value, the more compact the shape. Values equal to 1 indicates that the shape is a circle. In our case, the finite shape of the generalization is the

Table 1 Shape metric used for geometry generalization evaluation

Shape metrics	Description
Shape area	Refers about shape area
Shape length (perimeter)	Refers about total shape length, i.e. shape perimeter
Normalized Cohesion index	Average distance between all pair of interior points
Normalized Dispersion index	Average distance from the centroid to all points on the shape perimeter
Normalized Depth index	Average distance from the shape's interior points to the nearest point on the perimeter
Normalized Detour index	Perimeter of the shape's convex hull
Normalized Exchanged index	Area of the polygon within the Equal Area Circle
Normalized Girth index	Radius of the largest circle inscribed in the shape
Normalized Perimeter index	Normalized perimeter of the shape
Normalized Proximity index	Average Euclidean distance from all interior points to the centroid
Normalized Range index	Distance between the furthest most points on the polygon perimeter
Normalized Spin index	Average of the square of the Euclidean distances between all interior points and the centroid
Normalized Traversal index	Average distance of the shortest paths connecting any two points on the shape perimeter
Perimeter-area index	Ratio of the perimeter and area of the shape
Fractal dimension index	Refers about overall shape complexity and irregularity

Shape metrics description is according to Parent (2014)

rectangle with shape metrics values around 0.8. Shape area and length are in absolute numbers. Nevertheless, when the shape is becoming more compact the shape length (perimeter) is decreasing. Relationship between shape area and length (perimeter) is reflected by the Perimeter-area index. Fractal dimension index describes overall complexity and irregularity of the shape. The "smoother" the shape, the lower the Fractal dimension index value; and vice versa.

In total, there were 22 generalization levels performed upon 4 shapes (building footprints), and for each generalization level and building, 11 shape metrics were calculated in the Shape Metrics Toolbox. In sum, 1,320 values were achieved. The calculation took approximately 183 min. Other shape metrics calculations were not that time consuming and were performed outside the GIS.

Results: Case Study

In this chapter, we highlight some typical cases, where geometry generalization was not satisfactory according to the shape metrics measurement. We evaluated geometry generalization levels by individual shape metrics values complemented with visual comparison in particular level. We also show examples of generalized ground plans together with shape metrics graphs for better comprehension.

Shape metrics were calculated in order to find out if the geometry generalization process is tuning the output polygon properly. Ideally, the geometry generalization method should simplify the polygon more and more with every single subsequent step. In fact, the opposite is true in some cases and generalization levels. Shape metrics values fluctuations during generalization are depicted and commented. Nevertheless, general trend of the polygon simplification was preserved in the sense of shape metrics values.

One of the easiest metric to obtain and interpret is the shape area. Depending on the original shape of the building, the shape area should not decrease during the generalization process. According to Lee and Hardy (2006), algorithm we used should keep the area constant in any generalization level. Figure 2a is depicting shape area values course over generalization levels. It can be clearly seen that there is an increase in the shape area of all four buildings. A steep increase is at the last stages of generalization due to the generalization algorithm. There are also significant local peaks around simplification tolerance value of 30 m, 50 m (Lund University) and 35 m (Houses of Parliament). Subsequently, the shape area decreases, which is not desirable by means of polygon simplification for an output map. Thus, one can identify levels of generalization that are not simplifying the shape gradually, as expected.

Second easily calculable and interpretable metric is the shape length (Fig. 2b). On the contrary to the previous metric, shape length/perimeter should decrease as the generalization continues. Again, the trend is confirming the assumption except the last stages of generalization. Level of generalization at simplification tolerance value of 200 m is too large and preserves undesirable details of ground plans. In the case of the St. Maurice (Fig. 2g) there is opposite trend (slight increase) from simplification tolerance value of 25 m with the exception at values from 60 to 100 m.

Fractal dimension refers about overall complexity of the shape. It is supposed that the higher level of generalization, the lower Fractal dimension value. In Fig. 2c the assumption is confirmed. Except the tail of the graph, in the case of St. Maurice and Houses of Parliament, where Fractal dimension value arose, therefore overall complexity of these two shapes arose too (Fig. 2i). In the last three levels (simplification tolerance values of 350, 400 and 500 m) all the ground plans' Fractal dimension value was more or less the same and constant.

Following shape metrics are more difficult to interpret and are thoroughly described by Parent (2014). Nevertheless, brief explanation is in Table 1. Normalized Detour index describes shape properties using convex hull, i.e. the convex

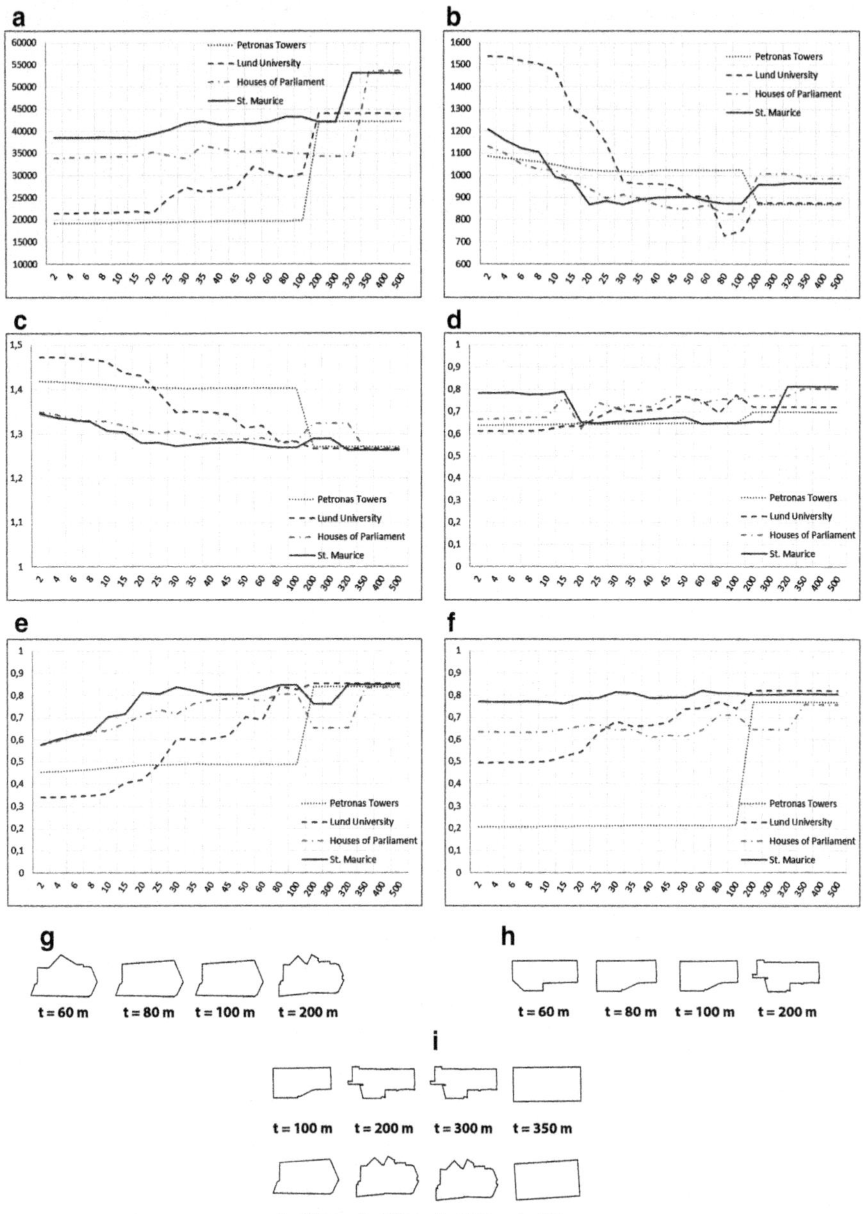

Fig. 2 Graphs of the shape metrics for four ground plans at individual generalization levels (X-axis)—(**a**) Shape area, (**b**) Shape length, (**c**) Fractal dimension, (**d**) Normalized Detour index, (**e**) Normalized Perimeter index, (**f**) Normalized Spin index; Evolution of the ground plans shapes at selected generalization levels—(**g**) St. Maurice church, (**h**) Houses of Parliament, (**i**) Houses of Parliament and St. Maurice church

polygon with the shortest possible perimeter that fully encompasses it (Parent 2014). In other words, the index refers about the shape compactness. The more extremities (e.g. spikes) the shape has, the lower the index value. Ideally, in the case of a circle, the index value is equal to 1. Output values of Normalized Detour index (Fig. 2d) are relatively high with no major extremes, in this case. However, interesting trapezoidal graph has St. Maurice church. This is because the generalization algorithm preserves only a few big spikes at simplification tolerance ranging from 20 m to 300 m. This is desirable to keep typical features of the shape on one hand, but on the other, these sudden changes may confuse the cartographer. Whenever big spikes are eliminated, the shape becomes more compact and Normalized Detour index arise. Similar interpretation could be done with the other building ground plans. In general, the higher level of generalization, the higher Normalized Detour index value.

Normalized Perimeter index transforms shape perimeter value into relative number using Equal Area Circle. It is the ratio of shape perimeter to perimeter of Equal Area Circle. This index indicates how much the shape differs from the ideal circular shape. Lower the value, the less compact the shape. In the generalization process, this index should gradually increase. This presumption is proven in Fig. 2e. Also in this case, there are some exceptions, especially in the case of Houses of Parliament and St. Maurice church at the tail of the graph. Level of generalization at simplification tolerance values of 200 m, 300 m (Houses of Parliament) and 320 m (St. Maurice) affects the shape and preserves higher complexity of the shape (Fig. 2i), instead of simplifying the shape. This should be stressed, because geometric generalization procedures should preserve core geometry characteristics. This requirement was violated in this particular case.

Normalized Spin index is appropriate for measuring compactness when focus is on shape extremities (Parent 2014). Again, the higher the index value the more compact the shape (close to the circularity). Abrupt and steep increase of Normalized Spin index values of Petronas Towers (Fig. 2f) underlined the specific character of its ground plan. In most cases, Petronas Towers holds its shape metrics values constant until the level of generalization at simplification tolerance value of 200 m is reached. At this level, the shape is suddenly transformed into the rectangle with no intermediate stage (see Fig. 1b). Looking at the Lund University graph, for example, it is possible to see that it loses extremities of its shape more or less constantly. In this case, geometry generalization is the most effective between 10 and 80 m simplification tolerance values. Shape metrics values fluctuate the most in case of Houses of Parliament ground plan generalization. Unexpected increase of its shape extremities can be best identified at simplification tolerance values ranging 60 to 200 m. Generalization algorithm simplified the shape at values 80 and 100 m, but afterwards it resulted into more complicated shape (Fig. 2h).

Rest of the calculated shape metrics shows analogous findings. Their graphs are depicting shape metrics values progress at various generalization levels revealing other nuances but not so significantly as presented ones. Moreover, according to Parent (2014), normalized metrics tend to be correlated with each other, therefore it is redundant to present all shape metrics graphs in this article.

Discussion

In the article, we showed the deployment of shape metrics on generalized ground plans of four buildings in order to evaluate geometry generalization quantitatively. In most cases, the basic assumption that the shape is being simplified during the generalization was in most cases confirmed. Thus, general trend of increasing/ decreasing shape metrics values (depending on what shape metric is under the investigation) as the generalization proceeds was also confirmed. Nevertheless, via calculating shape metrics it is possible to detect discrepancies at particular generalization levels. When looking at shape metrics graph, steep increase/decrease between two generalization levels could indicate major changes in the shape. In other words generalization algorithm significantly reshapes the geometry. If this change follows the general trend, we can claim that the shape was simplified properly at particular generalization level. On the contrary, if the change of shape metrics values does not follow the overall trend, the shape was not simplified. Thus, the generalization algorithm made the shape even more complicated than at previous generalization level. It is also important to check the graph for peaks. These local maximums/minimums indicate that generalization either oversimplify, or does not simplify the shape at particular level. Suggesting that generalization algorithm performs as it was programmed, we can identify its suitability for given geometry type as well as generalization level.

Conclusions

Concerning ground plans shapes, here are some findings in connection with shape metrics. Interesting graph of shape metrics values has the Petronas Towers. Generalization of its shape, according to the both shape metrics values and visual interpretation, did not significantly simplify it until the simplification tolerance values of 200 m, where the geometry turned into a rectangle. Thus, chosen generalization algorithm did not have desired effect. On the other hand, the ground plan of Lund University was gradually and most significantly simplified during the whole generalization process, of course with some exceptions. Nevertheless, generalization of Lund University ground plan could be considered as the most effective one. Both Houses of Parliament and St. Maurice ground plans were generalized gradually too but within a smaller range in terms of shape metrics values. In the first generalization steps only small spikes were eliminated and then the overall shape was simplified. Unexpectedly, in several generalization levels their shape was not simplify but rather vice versa.

According to the case study, we can recommend which shape metrics are the most useful for quantitative evaluation of generalization. Basic shape metrics, such as the shape area and length are easy to calculate and are showing fundamental properties of the shape generalization. These metrics

(continued)

are also clearly interpretable. Fractal dimension and Perimeter-area index uses the same inputs (shape length and area) with the difference that Fractal dimension index applies logarithm in calculations. Most of indexes from Shape Metrics Toolbox are more difficult to interpret. Nevertheless, we suggest using shape metrics whose graphs we showed in the paper, i.e. Normalized Detour index, Normalized Perimeter index and Normalized Spin index. Since some of shape metrics are correlated (Parent 2014), it is possible to substitute one metric for another, e.g. Normalized Cohesion index for Normalized Proximity index, and Normalized Exchange index for Normalized Dispersion index. Other indexes, such as Normalized Girth index, Normalized Depth index, Normalized Range index, Normalized Traversal index, can be used regarding on what the user wants to show. We also recommend using only normalized versions of metrics from Shape Metrics Toolbox, because some indexes, when calculated in absolute numbers, do not reflect extremities of the shape.

Shape metrics seems to be a useful tool for quantitative evaluation of cartographic generalization along with expert knowledge. They can help to reveal discrepancies in the process of semi-automated generalization and identify such stages of generalization that bring unsuitable results.

Acknowledgments The article was created within the project CZ.1.07/2.3.00/20.0170, supported by the European Social Fund and the state budget of the Czech Republic.

References

Agent (1999) Selection of basic measures. Technical report C1, Agent Consortium, 81 p

Angel S, Parent J, Civco DL (2010) Ten compactness properties of circles: measuring shape in geography. Can Geogr 54(4):441–461. doi:10.1111/j.1541-0064.2009.00304.x

Bard S, Ruas A (2005) Why and how evaluating generalised data? In: Fisher PF (ed) Developments in spatial data handling. Springer, Heidelberg, pp 327–342

Bayer T (2009) Automated building simplification using a recursive approach. In: ICA symposium on cartography for Central and Eastern Europe, LNG&C. Springer, Vienna, pp 121–145. ISBN: 978-3-642-03293-6

Bernhardt MC (1992) Quantitative characterization of cartographic lines for generalization. Report no. 425, Department of Geodetic Science and Surveying, The Ohio State University, Columbus, 142 p

Burghardt D, Steiniger S (2005) Usage of principal component analysis in the process of automated generalisation. In: Proceedings of 22nd international cartographic conference, pp 9–16

Cheng T, Li Z (2013) Effect of generalization on area features: a comparative study of two strategies. Cartogr J 43(2):157–170

ESRI (1996) Automation of map generalization: the cutting-edge technology. ESRI White Paper Series, 12 p

Filippovska Y et al (2008) Quality evaluation of generalization algorithms. In: The International Archives of the Photogrammetry, Remote Sensing and Spatial Information Sciences, Beijing, pp 799–804

Forman RTT (1995) Land mosaics: the ecology of landscape and region. Cambridge University Press, Cambridge, 656 p

Gao W, Gong J, Yang L, Jiang X, Wu X (2012) Detecting geometric conflicts for generalisation of polygonal maps. Cartogr J 49(1):21–29

Girres JF (2011) An evaluation of the impact of cartographic generalisation on length measurement computed from linear vector databases. In: Proceedings of the 25th international cartographic conference, Paris, 11 p

Jasinski MJ (1990) The comparison of complexity measures for cartographic lines. NCGIA Report 90-1, NCGIA, Santa Barbara, 73 p

Kimerling AJ et al (2011) Map use: reading, analysis, interpretation, 7th edn. ESRI Press Academic, 620 p

Lee D, Hardy P (2006) Design and experience of generalization tools. In: Proceedings of AutoCarto 2006, Vancouver, 25 June

Maling DH (1988) Measurements from maps: principles and methods of cartometry. Pergamon, 577 p

McGarigal K (2013) FRAGSTATS help. University of Massachusetts, 168 p. http://www.umass.edu/landeco/research/fragstats/documents/fragstats_documents.html

Mesev V (2007) Integration of GIS and remote sensing, 1st edn. Wiley, 312 p

Parent J (2014) Shape metrics tool overview. University of Connecticut. http://clear.uconn.edu/tools/Shape_Metrics/index.htm

Peter B (2001) Measures for the generalization of polygonal maps with categorical data. In: Fourth ICA workshop on progress in automated map generalization, 2–4 August 2001, Beijing, 21 p

Savino S (2011) A solution to the problem of the generalization of the Italian geographical databases from large to medium scale: approach definition, process design and operators implementation. Dissertation thesis, Universita di Padova, 141 p

Schmid S (2008) Automated constraint-based evaluation of cartographic generalization solutions. Doctoral dissertation, Geographisches Institut der Universität Zürich, 163 p

Skopeliti A, Lysandros T (2001) A methodology for the assessment of generalization quality. In: Fourth ICA workshop on progress in automated map generalization, 2–4 August 2001, Beijing, 13 p

Stoter J et al (2009) Specifying map requirements for automated generalization of topographic data. Cartogr J 46(3):214–227

Analysis of Basic Relations Within Insights of Spatio-Temporal Analysis

Andreas Hall and Paula Ahonen-Rainio

Introduction

Visual analytics (VA), defined as "the science of analytical reasoning facilitated by interactive visual interfaces" (Thomas and Cook 2005, p. 4), can be seen as an effort to combine human intelligence and computing in an optimal way and have two lines of research: developing representations and tools that support human reasoning by computational, visual, and interactive means, and studying human reasoning in its interaction with VA tools. This research positions itself in the second line of research and in the context of exploratory spatio-temporal analysis.

Spatio-temporal analysis operates in three dimensions, with spatial (geographic), temporal, and thematic (attributive) components of data. The purpose of VA tools is to facilitate insights by enabling the analyst to find the meaningful relations inside and between these three dimensions. Of these dimensions, space and time are given and governed by nature. They are predictable and invariant. The thematic dimension, on the other hand, is not predictable and its contents can vary infinitely, depending on the data and purpose of the analysis.

The goal of this research is to gain a better understanding of the role that the basic spatial and temporal relations play in the analytical reasoning process. The research question is: How do the basic spatio-temporal relations translate into human qualitative categorical reasoning? The hypothesis is that these relations, being the invariants of spatio-temporal analysis, are of key importance to our analytical reasoning. By gaining a better understanding of them and the role they play in the reasoning process we can develop VA representations and tools that better support the reasoning process of the analyst. The basic relations in space and time are distance, direction, and topology. These cannot be broken down to simpler

A. Hall (✉) • P. Ahonen-Rainio
Aalto University School of Engineering, Espoo, Finland
e-mail: andreas.hall@aalto.fi

© Springer International Publishing Switzerland 2015
J. Brus et al. (eds.), *Modern Trends in Cartography*, Lecture Notes in
Geoinformation and Cartography, DOI 10.1007/978-3-319-07926-4_31

relations and they do not duplicate the meaning of any other relation, as has already been pointed out by Nystuen (1968).

Spatio-temporal relations do not exist in the real world but in representations of the real world (Mark and Frank 1989), and, depending on the type of the representation, they relate somewhat different entities to each other. Cognitively, the basic entities are objects which can be decomposed into parts or aggregated into configurations, while geometrically, the basic entities are points that can be aggregated to lines, areas, objects, and more complex structures (Freksa 2013). The approach taken in this research considers the basic relations as being both inherent in the data and something that the analyst perceives and processes cognitively.

One of the original recommendations in the VA research and development agenda (Thomas and Cook 2005) was that we need to refine our understanding of the reasoning processes in VA. As Arias-Hernandez et al. (2012) observed, most of the developments in the field have focused on the design of computer-based visualizations, and there is a lack of research on the reasoning process. This study is specifically about the reasoning processes and thus contributes to filling this gap. Bhatt and Wallgrün (2014) argue that we need a transdisciplinary scientific perspective that brings together geography, artificial intelligence (AI), and cognitive science. The basic relations in space and time, which are central to this research, can help in uniting geography, AI, and cognitive science.

VA implies human-computer interaction. The strength of computers lies in their capacity for fast and accurate quantitative processing; they are good with metric values. Human reasoning, on the contrary, is mainly qualitative, and in a spatial context it works by qualitative categorical terms such as 'near' and 'far' instead of metric values (Kosslyn et al. 1992; Egenhofer and Mark 1995; Mark 1999; Renz and Nebel 2007; Galton 2009). This means that computers and humans speak different languages. The reasons why a human analyst finds certain relations important, e.g., that two points are near each other, cannot easily be translated into a language understandable by computers because the metric value of, e.g., 'near' is unclear and depends on the context of the analysis. Research in the field of computer science has addressed this problem by creating various formalizations, or spatial calculi, enabling computers to reason qualitatively (Galton 2009). The issue has also been studied from a GI Science viewpoint, e.g., as naive geography (Egenhofer and Mark 1995).

Although there is a lack of research on analytical reasoning in the VA community, some previous research can be found. One approach that has been taken involves examining analysts' interaction logs with a VA tool in order to identify the analysts' strategies, methods, and findings (Dou et al. 2009). There have also been attempts at developing a human cognitive model for VA (Green et al. 2008). However, these and other research initiatives mentioned by Ribarsky et al. (2009) have mainly resulted in design guidelines for VA tools rather than a deeper understanding of analytical reasoning. A novel approach to capturing reasoning processes, building on protocol analysis, is pair analytics, a method that generates verbal data about the thought processes of two humans using a VA tool together (Arias-Hernandez et al. 2011). Hall et al. (2014) studied reasoning in an analysis process and considered how to make the analysis results transparent by visualizing the reasoning and the knowledge used. There is also a lot of research that has been done that focuses on how to support the analytical reasoning process (e.g. Jankun-Kelly et al. 2007; Shrinivasan and van Wijk 2008; Gotz and Zhou 2009). Of these only Hall et al. (2013) studied reasoning in a spatial context, but without paying attention to the role of the basic spatial and temporal relations.

This study explored the link between the basic spatio-temporal relations and human qualitative reasoning through a case study. In the case study four insights gained by an analyst while watching a simple animated map were dissected and broken down into chains of reasoning arising from the basic relations. The study looked at how the basic relations translate into qualitative human reasoning. The study also identified other factors that were of importance when the analyst gained an insight. The research method applied in the case study was reverse engineering of insights through introspection, supported by a thorough literature review.

The rest of the paper is structured as follows. First, the results of the literature review are presented in the next section. After this, the case study is introduced, followed by the results section, which is divided into two parts. The first part introduces a framework of spatio-temporal cognitive concepts and the second part deals with the dissection of the insights. The paper concludes with a discussion of the results.

Theoretical Background

This section presents the theoretical background to the study. First, reasoning and knowledge, as they are understood in this paper, are introduced, and, after this, the focus moves to cognitive theories of spatio-temporal reasoning. The last part of this section introduces work done in computer science to make computers able to deal intelligently with spatio-temporal relations.

Reasoning and Knowledge Reasoning is the process by which, through a set of mental processes, we derive inferences or conclusions from a set of premises (Samarapungavan 2009). There are two main types of reasoning: deduction and induction (Johnson-Laird 2006). Deductive reasoning is based only on the information given in the premises and on logic. It is the process of establishing that a conclusion is a valid inference from certain premises. Any other type of reasoning is based on induction, which is reasoning that goes beyond the given information and rules out more possibilities than the premises do (Johnson-Laird 2006). This is possible because inductive reasoning depends on knowledge that is not given in the premises. By going beyond the given information, by using existing knowledge, inductive reasoning results in new knowledge.

In this respect, knowledge is defined as information which has been cognitively processed and integrated into an existing human knowledge structure, information as data given meaning through interpretation, and data as symbols or isolated and non-interpreted facts (as in Keller and Tergan 2005). Thus, knowledge can only exist inside our brains; data visualized in a VA environment becomes information and, when interpreted by an analyst, this information becomes knowledge.

Inductive reasoning is the cornerstone of human reasoning because we are only moderately good at deductive logic and make only moderate use of it (Arthur 1994). We use the meaning of premises and our knowledge to construct mental models which contain less information than the models needed for deductive reasoning and thus impose fewer demands on our working memory (Johnson-Laird 2006). The reasoning of spatio-temporal analysis is no exception; it is mainly inductive and builds on knowledge that has already been acquired. More on reasoning from a VA perspective can be found in Hall et al. (2013).

Reasoning about Space and Time Spatio-temporal reasoning is largely reasoning about spatial and temporal relations. Goodwin and Johnson-Laird (2005) argue that this kind of reasoning is based on spatial mental models. These models are iconic, meaning that they are based on the concept of resemblance. The parts and relations of the models (their structure) correspond to those of the situation they represent. They explain that the clever thing about these kinds of models is that they can yield novel conclusions that do not correspond to any of the premises used in their construction. That means that in the same way as maps implicitly convey innumerable spatial relations that were not explicitly added during their making, the spatial relations that follow from the premises can emerge from these models. However, these models are not images (Goodwin and Johnson-Laird 2005; Knauff 2009); rather, they are abstract topological structures.

Reasoning about space is one of the most common and basic forms of human intelligence (Egenhofer and Mark 1995; Levinson 2003). The abundance of spatial metaphors in every language ever studied tells us something about how central spatial reasoning is to our thinking (Pinker 2009). We create these metaphors not just to co-opt words but to co-opt their inferential machinery. They allow us to lend our inferential machinery for space and motion to other concepts, such as possession, circumstances, and time. Pinker (2009) states that these spatial concepts about places, paths, and motions, along with a few other concepts about agency and causation, appear to be the vocabulary and syntax of mentalese, the language of thought.

According to Piaget's theory of spatial development (Piaget and Inhelder 1956), children's comprehension of spatial relations mirrors the general stages of their cognitive development. The theory states that children learn to understand topological spatial properties first (when well under one year of age), followed by projective spatial properties (direction) and finally metric spatial properties. Freundschuh and Blades (2013) presented evidence in support of the theory through their finding that there are differences between age groups when it comes to the understanding of the locative terms 'next', 'near', 'far', and 'away'. Also Tversky (2014) have presented research in favor of Piaget's theory.

We perceive our surroundings through our senses, with vision being the most important for perceiving space. Vision works by means of two major cognitive systems, of which one encodes object properties and the other encodes spatial properties (Kosslyn et al. 1992; Knauff 2009), called the 'what' and 'where' systems. Both systems can be further divided into subsystems and Kosslyn et al. (1992) argue that the system for spatial properties is divided into at least different systems for categorical and coordinate representations of spatial relations and provide evidence in favor of this claim from neural network simulations. They argue that the coordinate, or metrical, representations are used for guiding action (e.g., moving the eyes, reaching) and that (qualitative) categorical representations, which assign a range of positions to an equivalence class (e.g., above/below), are used for identifying an object or scene. Furthermore, the coordinate representations seem to be encapsulated and not accessible to the system used to categorize information.

Knowledge Representation In computer science the field knowledge representation (KR), a subfield of AI, originated when it became apparent that in order to make a computer intelligent it was not enough to give it the capacity for pure reasoning; it also needed knowledge (which enables reasoning to take place) about the world with which it interacts (Galton 2009). Thus, KR is about the quest for explicit symbolic representations of knowledge, suitable for use by computers. Spatial and temporal knowledge is an essential part of this quest.

These symbolic representations go under many different names: qualitative reasoning (Cohn and Hazarika 2001; Renz and Nebel 2007), formal models (Mark 1999), spatial (and temporal) calculi (Freksa 2013), naive geography (Egenhofer and Mark 1995), and qualitative calculi (Klippel et al. 2013). For time, the most influential is Interval Calculus (Allen 1984), which identifies thirteen distinct qualitative relations between two temporal intervals. This system is 'jointly exhaustive and pairwise disjoint' (JEPD), meaning that for any two intervals, exactly one of the relations must hold (Galton 2009).

For space, the development of symbolic representations has lagged behind that for time by about a decade, as space is more difficult to model than time (Galton 2009). The two most influential spatial calculi are Region Connection Calculus (RCC) (Randell et al. 1992) and the 9-intersection model (Egenhofer 1991). In RCC, the key notions are *region* and *connection* and in the 9-intersection model the key notions are *interior*, *boundary*, and *exterior*. See, for example, Galton (2009) for more on these. The RCC and 9-intersection models are topological calculi. Other types of calculi that have been developed include measurement calculi, orientation and direction calculi, position calculi, and calculi for size and shape (Galton 2009; Freksa 2013).

These different calculi are often claimed to reflect the way people think and reason about space and time (Cohn and Hazarika 2001; Klippel 2012) and there is a widely-held belief that topology is the cognitively most important spatial relation (Klippel et al. 2013), or in other words, that topology matters and metrics refine (Egenhofer and Mark 1995; Mark 1999). It has also been argued that our most basic notions of temporal relations are essentially qualitative, that the topology of events, e.g., before/after relationships, is more important than the number of hours that separate them (Galton 2009). These beliefs are more often based on the intuition of the researchers than on empirical data (Knauff et al. 1997). However, there is some evidence both in favor of and against this belief. Knauff et al. (1997) found that topological relationships are a relevant and dominant factor in judging spatial configurations, while others have found that topology was not always the main criterion for conceptualizing spatial information (Klippel et al. 2013) and that the salience of topological relations changes, depending on the semantic domain (Klippel 2012).

Case Study

The case study looked at four insights gained by an analyst (the first author) watching a case of a simple animated map. The goal was to dissect the insights into chains of reasoning arising from the basic relations in the spatial and temporal dimensions through introspection and reverse engineering. Reverse engineering (Pinker 2009) is to take something apart, e.g., a machine, and try to figure out what the parts are for and what they do to make the machine work. Focus was laid on the qualitative categorical representations that the basic relations were given once they had been processed cognitively and on the role that the thematic information and the analyst's knowledge played in the reasoning. The dissected insights are presented in the form of diagrams in the results section.

The approach taken here is that of distributed cognition, where cognition is seen as an emergent property of interactions between an individual and the environment (Liu et al. 2008; Arias-Hernandez et al. 2012). Central to this theory is the concept of cognitive artifacts (Norman 1993), which are external aids that help us overcome the limits imposed on our memory, thought, and reasoning (Arias-Hernandez et al. 2012). Graphic inventions of all sorts, such as maps, are one class of cognitive artifacts.

In the literature, the word 'insight' has been used with two parallel meanings: as a term meaning a moment of enlightenment (mainly in cognition science) and as a term meaning an advance in knowledge (mainly in the visualization community) (Chang et al. 2009). Here, the term is used with the second meaning: an "individual observation about the data" being "a unit of discovery" (Saraiya et al. 2005, p. 444).

Humans are essentially storytellers (Fisher 1984) and they understand their world through causal relationships (Sloman 2009). In order to make sense of and organize their knowledge, people create causally structured narratives that guide the organization of insights into meaningful structures and patterns (Eccles et al. 2008). In order to understand the insights and the knowledge used in the reasoning, we thus also need the analyst's narratives. They are not directly based on the basic relations in the data but they reveal the background knowledge used by the analyst.

The animated map used in the case study is a multimedia artwork named 1945–1998 (Hashimoto 2003) showing the time and location of every known nuclear

Fig. 1 A screenshot from
the animated map; in the
upper right-hand corner
time is displayed down to
the accuracy of a month and
in the *lower right-hand
corner* the cumulative
number of detonations
performed can be found; the
numbers next to the flags
represent the cumulative
numbers of detonations
sorted by country
(Hashimoto 2003)

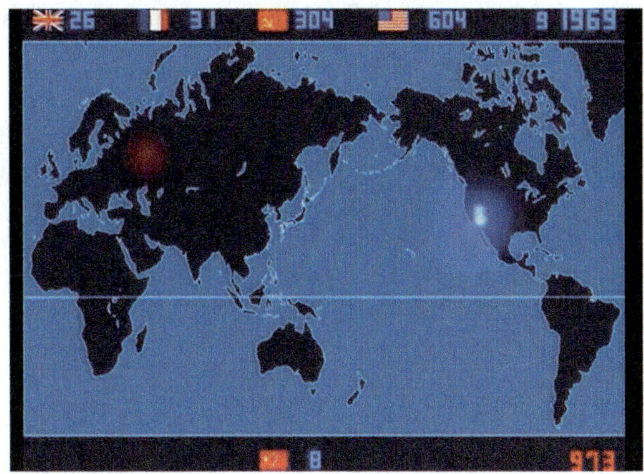

detonation (except those announced by North Korea). The animation was chosen
because it is well-known and freely available on the web (see references for the
internet address). The spatial aspect is communicated through a world map and the
temporal aspect through the fact that it is an animation (every second is one month),
through beeps (one beep for every month, a different beep for the start of every
year), and through textual information. As time goes by, the detonations appear and
disappear as points on the map. The points are color-coded to show to which
country they belong. The total and country-wise cumulative number of detonations
are also given at every moment in time. Figure 1 is a screenshot of the animation.

In this study distance, direction and topology in space, and distance and topology
in time are considered as basic relations. Distance and direction are continuous
quantities defined in relation to some frame of reference. Topology, on the other
hand, is defined as those properties of geometrical figures that are invariant under
continuous deformation (McDonnell and Kemp 1996). As time is one-dimensional,
direction in time is redundant information which can be read from positive and
negative distances in time. Therefore, direction in time is excluded here.

Results

The reverse engineering that led to the results of this study was not a straightforward process but rather iterative, the final diagrams being the result of many
changes. During the work a conceptual framework describing the building blocks
of the insights evolved. The conceptual framework gave a common structure to the
dissection of the insights, and it helps to provide an understanding of the idea
behind the dissection of the insights. It is not a fully evolved theory of insight
making in spatio-temporal analysis, though. This framework is presented below,
followed by the four dissected insights.

The Framework of Spatio-Temporal Cognitive Concepts

The framework builds on the basic relations in space and time and consists of three levels of spatio-temporal cognitive concepts. The cognitive concepts are the building blocks of the insights. In general, they are spatio-temporal properties of the data perceived and cognitively processed consciously or unconsciously by the analyst. As building blocks in a specific insight, they are the properties of some specific data that have been given a categorical mental representation and play an important role in the creation of that insight.

The three levels of the framework are made up of basic, compound, and labeled cognitive concepts. Basic concepts in space are distance, direction, and topology, and in time distance and topology. These are the direct equivalences to the basic relations. On the next level are the compound concepts. Contrary to the basic concepts, they need more than two points in space or time to apply and they are, in a way, derived from, and dividable into, basic cognitive concepts. These are concepts such as intensity, size, shape, and distribution. The third level is called labeled concepts. These are basic and compound concepts enriched with thematic information or knowledge not inherent in the data. Figure 2 illustrates the hierarchies of the framework.

Insights

The dissected insights (labeled I–IV) are presented through a set of diagrams which all have the same kind of vertical structure (Figs. 3, 4, 5, and 6). The basic relations that are relevant in each insight are to be found in the bottom row, in boxes with a dashed border. The cognitive constructs built from these basic relations are above them, in boxes colored in three different shades (light for basic constructs, medium for compound constructs, and dark for labeled constructs). The arrows depict part-of relationships. Boxes with an unbroken border and white background represent information originating from a reference frame. Each diagram presents the final insight, and accompanying narrative, as plain text in a balloon next to the diagram.

The vertical structure reflects the top-down and bottom-up (goal-directed and stimulus-driven) nature of attention and perception (Egeth and Yantis 1997). The basic relations are the stimulus of the process of creating the insight, while the topmost cognitive constructs are closely related to goal-directed processes. The insights are here presented in a bottom-up fashion. This approach is taken mainly to give the descriptions a common structure, but does not indicate that the cognitive processes behind the insights were only structured in a bottom-up manner.

As the concepts are cognitively processed, they are given a qualitative categorical representation. These categorical representations are included in the diagrams as underlined words, while the actual cognitive construct is indicated by bold

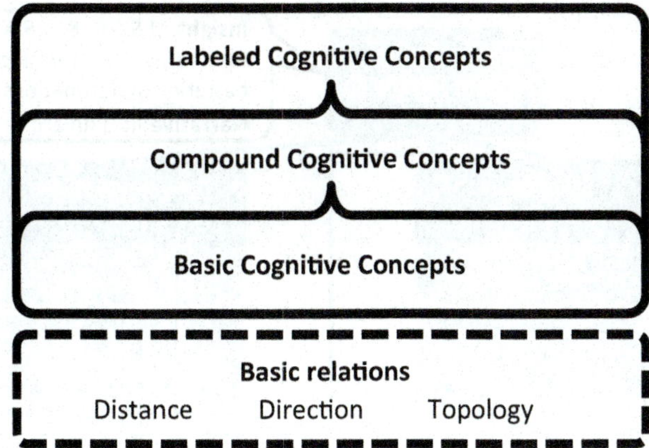

Fig. 2 The framework of spatio-temporal cognitive concepts builds on the basic relations in space and time and consists of three levels

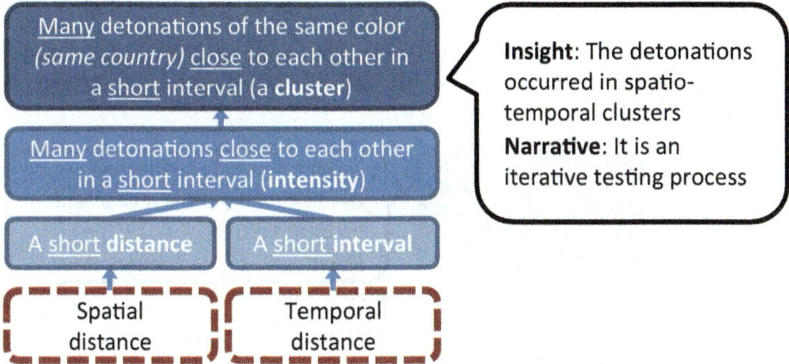

Fig. 3 Insight I is here dissected into the basic concepts distance and interval, the compound concept intensity, and the labeled concept clusters

words. Thematic information included in the labeled constructs is indicated by words in italics.

Insight I This insight was the discovery that during a certain time interval in the animation the detonations occurred in spatio-temporal clusters. It is dissected in Fig. 3. The basic relations that this insight builds upon are spatial and temporal distance. The analyst noticed anomalous spatial and temporal distances between certain detonations which he categorized as short (in relation to the other distances in the data). In this case, it happened that the detonations close to each other in space were also close to each other in time. Hence, as these detonations were quite numerous, the analyst became aware of an anomalous intensity in space and time. Further inspection still revealed that these detonations were thematically the same, i.e., they belonged to the same country, thus creating a cluster. Using knowledge about, among other concepts, product development, the analyst formed a narrative

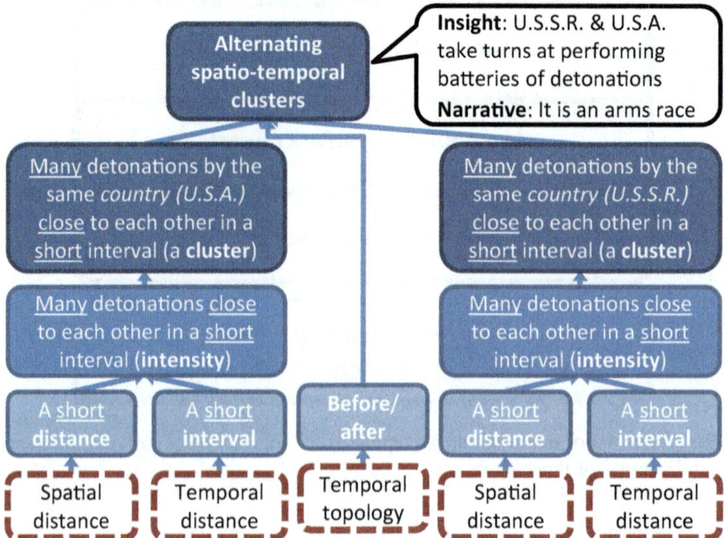

Fig. 4 Insight II is here dissected into the basic concepts distance, interval, and before/after, the compound concept intensity, and the labeled concepts cluster and alternating clusters

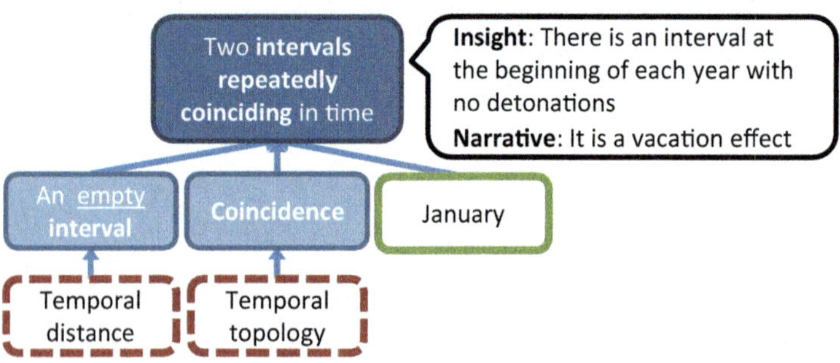

Fig. 5 Insight III is here dissected into the basic concepts interval and coincidence, information originating from a reference frame (January), and the labeled concept repeatedly coinciding intervals

that explained that once a country had performed a battery of detonations, it had to go back to the 'drawing board' to analyze the results and plan new detonations before it could continue detonating bombs and therefore the detonations occurred in clusters.

Insight II This insight was the discovery that the spatio-temporal clusters of Insight I, being the U.S.S.R. and U.S.A. clusters, appeared in an alternating fashion, i.e., the U.S.S.R. and U.S.A. took turns at performing batteries of detonations. This insight is dissected in Fig. 4. The basic relations that this insight builds upon are spatial and temporal distance and, additionally, temporal topology. The analyst noticed a specific before/after pattern among the clusters, so that before a

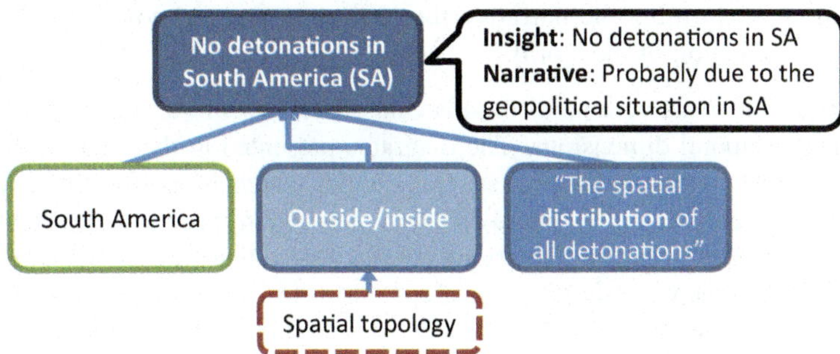

Fig. 6 Insight IV is here dissected into the basic concept outside/inside, information originating from a reference frame (South America), the compound concept distribution, and the labeled concept no detonations in South America

U.S.A. cluster there was a U.S.S.R. cluster, and vice versa. Using knowledge about concepts such as the Cold War and action and reaction, the analyst formed a narrative that explained how this pattern was an indication of the arms race between the U.S.S.R. and U.S.A. that was taking place at that time.

Insight III This insight was the discovery that during one time period in the data there is an empty interval at the beginning of each year. It is dissected in Fig. 5. This insight was gained at a later period in time in the animation, when detonations occurred continuously, contrary to the case of Insights I and II, when detonations occurred in clear temporal clusters. Insight III builds on the relations temporal topology and distance. The analyst repeatedly identified an anomaly in the temporal pattern in the shape of an unusually long interval where there were no detonations occurring. These empty intervals all happened during January. Consequently, there are two temporal intervals, one interval being seen as anomalous because it contains no detonations, and one being defined through the calendar system (January), which repeatedly coincides in time. Using knowledge about the custom of being on vacation during Christmas and New Year's Eve, the analyst formed a narrative that explained it as a kind of vacation effect. When people were on vacation no detonations were prepared at the end of December and therefore there is a break in January.

Insight IV This insight was gained at the end of the animation and it was the discovery that there had not been any detonations in South America. It is dissected in Fig. 6. This insight builds solely on spatial topology and the spatial reference system in use (in this case, a map). After watching the animation, the analyst realized that the spatial distribution of the detonations was located entirely outside the border of South America and formed a narrative explaining that this pattern was probably due to the geopolitical situation of South America. The cognitive concept used in this insight is the outside/inside relation and it is applied between all the detonations and an area defined by the spatial reference system. Thus, knowledge about that spatial reference system was of key importance when gaining the insight.

Furthermore, in forming the narrative the analyst also used knowledge about the geopolitical situation of the world.

Summary Under dissection, all the insights broke down to basic relations in the spatial and temporal dimensions. The diagrams presented in this section illustrate how the insights were made up of several smaller insights, or cognitive concepts, which in turn were, in the end, based on a basic relation in space or time. Most of the cognitive concepts were given various context-dependent qualitative categorical representations, such as 'short', 'many', or 'close to each other'. These kinds of representations cannot readily be dealt with by a computer. On the other hand, some of the concepts, such as outside/inside, can be dealt with by computers through, e.g., the 9-intersection model.

As an analyst watches the kind of animation used in the case study, innumerable spatial and temporal relations are conveyed to him or her. The ones that reach his/her consciousness and become a part of an insight, i.e., those that become cognitive concepts, are those that merge together and form the normal behavior and those that pop out in some way, that are anomalous. To judge from this study, the anomalies were those that differed from the rest of the relations of the same type in the scene or that differed from the knowledge about relations of this kind held by the analyst. Other reasons were that they were of a certain sought-for quality (top-down processes), showed a certain clear pattern (bottom-up processes), or were intriguing in some other way, e.g., the first with a certain thematic value.

Discussion and Conclusions
The results support the hypothesis that the basic relations in the spatial and temporal dimensions are of key importance to our analytical reasoning and the dissection showed that insights can be broken down to, and derived from, the level of basic relations (although direction did not play a role in any of the insights dissected here it most probably plays a role in our analytical reasoning). Furthermore, the importance of the reference system and the previous knowledge of the analyst became evident. Of the four insights dissected here, chosen to give a good representation of different kinds of insights, two included information from the reference system as key building blocks, without which the insights would have been impossible. Previous knowledge was of importance, especially when forming the narratives, but also in the reasoning process prior to that. For example, it is very probable that in Insight III knowledge about vacations in December and January made the empty interval pop out more easily. In the same way, knowledge about the Cold War made the analyst sensitive to interaction patterns between the U.S.S.R. and U.S.A. in Insight II.

Also the framework of spatio-temporal cognitive concepts is a significant outcome of this study. The framework gives a starting point to a better

(continued)

understanding of the reasoning processes involved in spatio-temporal analysis. As the framework keeps spatio-temporal properties and thematic properties as separate input, it is in line with the prevailing view that vision works by two major cognitive systems, the 'what' (thematic information) and 'where' (spatio-temporal information) systems.

As this study built on introspection, the results are highly subjective, but, being based on an extensive literature review, they have an objective basis. The data used in this study were very simple, containing only points and one thematic dimension. Furthermore, as this was an animated map, the temporal relations were mapped as true temporal relations and not transformed into some spatial form. Further research, including other types of data than points, will probably complement the results of this study, but not change them significantly. As such, the study should be seen as a proof-of-concept and a first step towards more objectively verified results. It also contributes to gaining a better understanding of the reasoning processes involved in spatio-temporal analysis.

The basic relations have previously been studied in geography, AI, and cognitive science. This study tried to combine findings from all three fields in order to make new discoveries. The basic relations are understood by both humans and computers and might, as such, be a key to better VA representations and tools. The different calculi of KR are a step in this direction as they try to make computers reason more like humans, i.e. more qualitatively.

The findings of this study neither confirm nor contradict the claim that topology is the cognitively most important spatial relation. Although distance relations were a part of three out of the four insights, it is possible, that cognitively these were processed through topological structures. Piaget's theory (Piaget and Inhelder 1956), stating that children learn to understand topological spatial properties first, would support this possibility. Further research includes trying to find objective approaches to the research questions in this study in order to verify its findings. A future goal is to develop a spatio-temporal reasoning model based on the concepts of basic relations in space and time.

References

Allen JF (1984) Towards a general theory of action and time. Artif Intell 23(2):123–154

Arias-Hernandez R, Kaastra LT, Green TM et al (2011) Pair analytics: capturing reasoning processes in collaborative visual analytics. In: 44th Hawaii international conference on system sciences (HICSS)

Arias-Hernandez R, Green TM, Fisher B (2012) From cognitive amplifiers to cognitive prostheses: understandings of the material basis of cognition in visual analytics. Interdiscip Sci Rev 37 (1):4–18. doi:10.1179/0308018812Z.0000000001

Arthur WB (1994) Inductive reasoning and bounded rationality. Am Econ Rev 84(2):406–411

Bhatt M, Wallgrün JO (2014) Geospatial narratives and their spatio-temporal dynamics: common-sense reasoning for high-level analyses in geographic information systems. ISPRS Int J Geoinf 3(1):166–205

Chang R, Ziemkiewicz C, Green TM et al (2009) Defining insight for visual analytics. Comput Graph Appl 29(2):14–17

Cohn AG, Hazarika SM (2001) Qualitative spatial representation and reasoning: an overview. Fundamenta Informaticae 46(1):1–29

Dou W, Jeong DH, Stukes F et al (2009) Recovering reasoning processes from user interactions. Comput Graph Appl 29(3):52–61

Eccles R, Kapler T, Harper R et al (2008) Stories in geotime. Inf Vis 7(1):3–17

Egenhofer MJ (1991) Reasoning about binary topological relations. In: Günther O, Schek H (eds) Proceedings of the second international symposium on advances in spatial databases, 28–30 August 1991. Lecture Notes In Computer Science, vol 525. Springer, London, pp 141–160

Egenhofer MJ, Mark DM (1995) Naive geography. In: Frank AU, Kuhn W (eds) Spatial information theory: a theoretical basis for GIS, Lecture Notes in Computer Sciences No. 988. Springer, Berlin, pp 1–15

Egeth HE, Yantis S (1997) Visual attention: control, representation, and time course. Annu Rev Psychol 48(1):269–297

Fisher WR (1984) Narration as a human communication paradigm: the case of public moral argument. Commun Monogr 51(1):1–22

Freksa C (2013) Spatial computing. In: Raubal M, Mark DM, Frank AU (eds) Cognitive and linguistic aspects of geographic space. Springer, Berlin, pp 23–42

Freundschuh S, Blades M (2013) The cognitive development of the spatial concepts NEXT, NEAR, AWAY and FAR. In: Raubal M, Mark DM, Frank AU (eds) Cognitive and linguistic aspects of geographic space. Springer, Berlin, pp 43–62

Galton A (2009) Spatial and temporal knowledge representation. Earth Sci Inf 2(3):169–187

Goodwin GP, Johnson-Laird P (2005) Reasoning about relations. Psychol Rev 112(2):468

Gotz D, Zhou MX (2009) Characterizing users' visual analytic activity for insight provenance. Inf Vis 8(1):42–55

Green TM, Ribarsky W, Fisher B (2008) Visual analytics for complex concepts using a human cognition model. In: IEEE symposium on visual analytics science and technology, p 91–98

Hall A, Ahonen-Rainio P, Virrantaus K (2014) Knowledge and reasoning in spatial analysis. Trans GIS 18(3):464–476

Hashimoto I (2003) 1945–1998. http://www.ctbto.org/specials/1945-1998-by-isao-hashimoto/. Accessed 28 Oct 2013

Jankun-Kelly T, Ma K, Gertz M (2007) A model and framework for visualization exploration. IEEE Trans Vis Comput Graph 13(2):357–369

Johnson-Laird PN (2006) How we reason. Oxford University Press, Oxford

Keller T, Tergan S (2005) Visualizing knowledge and information: an introduction. In: Tergan SO, Keller T (eds) Knowledge and information visualization. Springer, Heidelberg, pp 1–23

Klippel A (2012) Spatial information theory meets spatial thinking: is topology the Rosetta stone of spatio-temporal cognition? Ann Assoc Am Geogr 102(6):1310–1328

Klippel A, Li R, Yang J et al (2013) The Egenhofer–Cohn hypothesis or, topological relativity? In: Raubal M, Mark DM, Frank AU (eds) Cognitive and linguistic aspects of geographic space. Springer, Berlin, pp 195–215

Knauff M (2009) A neuro-cognitive theory of deductive relational reasoning with mental models and visual images. Spat Cogn Comput 9(2):109–137

Knauff M, Rauh R, Renz J (1997) A cognitive assessment of topological spatial relations: results from an empirical investigation. In: Frank AU, Kuhn W (eds) Spatial information theory: a theoretical basis for GIS. Springer, Berlin, pp 193–206

Kosslyn SM, Chabris CF, Marsolek CJ et al (1992) Categorical versus coordinate spatial relations: computational analyses and computer simulations. J Exp Psychol Hum Percept Perform 18 (2):562

Levinson SC (2003) Space in language and cognition: explorations in cognitive diversity. Cambridge University Press, Cambridge

Liu Z, Nersessian N, Stasko J (2008) Distributed cognition as a theoretical framework for information visualization. IEEE Trans Vis Comput Graph 14(6):1173–1180

Mark DM (1999) Spatial representation: a cognitive view. Geogr Inf Syst Princ Appl 1:81–89

Mark DM, Frank AU (1989) Concepts of space and spatial language. In: Ninth international symposium on computer-assisted cartography (Auto-Carto 9), Baltimore, pp 538–556

McDonnell R, Kemp K (1996) International GIS dictionary. Wiley, New York

Norman DA (1993) Things that make us smart: defending human attributes in the age of the machine. Basic Books, New York

Nystuen JD (1968) Identification of some fundamental spatial concepts. In: Berry B, Marble D (eds) Spatial analysis. A reader in statistical geography. Prentice Hall, Englewood Cliffs, pp 35–41

Piaget J, Inhelder B (1956) The child's conception of space. Routledge, London

Pinker S (2009) How the mind works. W. W. Norton, New York

Randell DA, Cui Z, Cohn AG (1992) A spatial logic based on regions and connection. In: Proceedings of the 3rd international conference on knowledge representation and reasoning, Morgan Kaufmann, pp 165–176

Renz J, Nebel B (2007) Qualitative spatial reasoning using constraint calculi. In: Marco A, Pratt-Hartmann IE, van Benthem JFAK (eds) Handbook of spatial logics. Springer, Berlin, pp 161–215

Ribarsky W, Fisher B, Pottenger WM (2009) Science of analytical reasoning. Inf Vis 8(4):254–262

Samarapungavan A (2009) Reasoning. The Gale Group. http://www.education.com/reference/article/reasoning/. Accessed 26 Nov 2013

Saraiya P, North C, Duca K (2005) An insight-based methodology for evaluating bioinformatics visualizations. IEEE Trans Vis Comput Graph 11(4):443–456

Shrinivasan YB, van Wijk JJ (2008) Supporting the analytical reasoning process in information visualization. In: Proceedings of the SIGCHI conference on human factors in computing systems, pp 1237–1246

Sloman S (2009) Causal models: how people think about the world and its alternatives. Oxford University Press, Oxford

Thomas JJ, Cook KA (eds) (2005) Illuminating the path: the research and development agenda for visual analytics. IEEE CS Press

Tversky B (2014) Visualizing thought. In: Huang W (ed) Handbook of human centric visualization. Springer, New York, pp 3–40

3D Cartography as a Platform for Remindering Important Historical Events: The Example of the Terezín Memorial

Pavel Hájek, Karel Jedlička, Michal Kepka, Radek Fiala, Martina Vichrová, Karel Janečka, and Václav Čada

Creation of 3D Maps

3D maps can be done in several different ways, each way creating a different representation of the landscape and related objects. 3D maps can be photorealistic, where the landscape is created to match the exact reality, using techniques of overlaying ortho-photo images over a 3D model, or can be "symbolic", where the maps are generalized and symbols designed to show object locations and information. An issue with 3D mapping is that there are currently no adequate standards or design principles in place to guide cartographers in creating functional, user friendly 3D maps. For issues regarding of symbol recognition, depth perception (differences in scale) and so on, all current 3D maps are done individually to show the portrayed information in the best way authors can do. Even though standard cartography design principles still apply to this newer form of geographical representation, we need to look past the technical aspects of 3D model creation and focus on the aspects that can enhance a user's understanding and comprehension of a landscape depicted in a three-dimensional view (Pegg 2011). Elmes (2005) is expressed that there are three processes that are common to all maps: reduction (determining scale), selection and abstraction of features, and symbolization.

Pegg (2011) is also described that while orientation is not really an issue for a 3D map, other factors can have a major effect on the way a 3D presentation is perceived. Recognition of symbology and 3D objects for the user is being an issue. Other issues that can affect the usefulness of a 3D map are different levels of detail and abstraction, depth perception and a changing scale.

P. Hájek (✉) • K. Jedlička (✉) • M. Kepka • R. Fiala • M. Vichrová • K. Janečka • V. Čada
Faculty of Applied Sciences, Department of Mathematics, Section of Geomatics, University of West Bohemia, Univerzitní 22, Pilsen, Czech Republic
e-mail: gorin@kma.zcu.cz; smrcek@kma.zcu.cz; mkepka@kma.zcu.cz; fialar@kma.zcu.cz; vichrova@kma.zcu.cz; kjanecka@kma.zcu.cz; cada@kma.zcu.cz

© Springer International Publishing Switzerland 2015
J. Brus et al. (eds.), *Modern Trends in Cartography*, Lecture Notes in Geoinformation and Cartography, DOI 10.1007/978-3-319-07926-4_32

Regarding mentioned issues, there can be derived steps or rules useful for creation of 3D models. Those steps are taken over from Bandrova (2001), which operates with the term of symbol for the 3D cartographic representation of real objects.

1. The symbols, which are represented in 3D map, should be similar to the real objects (according to photorealistic approach).
2. Minimum polygons should be used when a new symbol is built (for fast and effective visualization).
3. The symbols should be created in their real dimensions (also for a reason of photorealistic approach and also in spite of 2D maps, where symbols are often exaggerated).
4. The symbols are designed for different purposes depending on user's needs.

Not every time these rules can be abided, especially the rule number 3. Usually for the purposes of quick visualization, the generalization of the object or the replacement of the object with an object of lower dimension, need to be done. For more detailed information how these rules are applied within this project see chapter *Modeling of buildings of Terezín Fortresses*.

Modeling of 3D Objects of Terezín

The Terezín Memorial is a very important place of historical events of our culture, thus it is very important to commemorate these events even at present. Therefore the main aim of the project is to geographically relate important events occurred in Terezín ghetto or important personages who lived in the ghetto during the Second World War (2nd WW). Technically the paper further focuses on issues connected with a creation of virtual 3D model of the Main Fortress and the Small Fortress of Terezín reflecting the state of the area in the period of the 2nd WW.

According to Blodig (2003), the first chapters in the history of Terezín began to be written at the end of the 18th century when Emperor Joseph II decided about the construction of a fortress. Its purpose was to prevent enemy troops from penetrating into the Bohemian hinterland and also protect the waterway along the river Elbe. The fortress consisted of the Main and Small Fortress, as well as a fortified area between the Old and New Eger. In June 1940 the Small Fortress in Terezín was turned into the Prague Gestapo Police Prison. The town itself—former Main Fortress—was turned in November 1941 into a ghetto, an internment and transit camp for the Jews. Deported to the ghetto were Jews from the then Protectorate of Bohemia and Moravia, later also from Germany, Austria and other countries. At the initiative of the government of the restored Czechoslovak Republic the Memorial of National Suffering, later renamed to Terezín Memorial, was established in the places of wartime suffering of dozens of thousands of people in 1947. More information can be found in Jedlička et al. (2013).

During the creation of the 3D model of the Terezín it was necessary to search for available information, documents and materials about the situation in the Terezín Fortresses during the Second World War. There was a need to select proper information from huge amount of collected materials, which are useable for both 3D geovisualization of the Terezín Fortresses and also for descriptive data about the life in Terezín ghetto during the 2nd WW. All original archive materials were kept in four different archives in Czech and Austria. More information about data mining used in this project can be found in Hájek et al. (2013).

With regards to the concept of the virtual model of the Terezín, the model consists of two main parts: geometry and related descriptive data. The geometric part consists of 3D models which capture how the Terezín ghetto looked like during the 2nd WW. The descriptive part is based upon structured database of collected documents (descriptive data) which describe the historic information of many particular places in Terezín during the period of interest. For these main parts of the model have been used two main approaches. A Content Management System for the descriptive data and a principle called Level of Detail for the geometric spatial part.

Content Management System (CMS) is a tool for handling, together with a relational database, all the descriptive data (because historical documents of varied characters need to be shown such as scans of plans, photographs, books, etc.). Each document in CMS has not only its own unique identifier, but also an identifier of a place and time where the document is referred to. For more information see chapter *Data storage and interactivity of 3D models.*

As mentioned above, there are a huge number of documents related to events and places in Terezín. But on the other hand, the spatial part of the model is also information-rich. Thus there is a need for dealing with this huge amount of spatial 3D data in the sense of quick querying on data, data visualization and serving relevant data to the user. That's where the principle of various Levels of Detail (LOD) is implemented. The principle comes from the CityGML specification (see OGC 2012) and it deals with different models for a building, while each model has its own amount of displayed details of the building and the more detailed model is built up on the base of the less detailed model. For more information see the next chapter.

Modeling of Buildings of Terezín Fortresses

Each LOD of a building is stored in a base of data separately. Using different LODs facilitates so called view-dependent visualization, while the models which intend to be displayed are chosen based on the position, the view-direction and the distance of an observer. Each building has a bounding box, abbreviated as BB (BB = minimum volume that encloses a set of objects or data points), which represents the building for purposes of selecting appropriate LOD of the building. The definition of displaying a particular building in a particular level of detail is based on KML built-in features. Switching of different LOD for a building is based on calculated size of its bounding box. When the size of BB in pixels is between MIN and MAX values of <LodPixels> attribute, the BB with proper LOD of a model is visualized (see Fig. 1a; note—in this figure the illustrations of LODs with bounding boxes are in the same size, the real proportions of these bounding boxes are depicted in the left bottom side of this particular figure). This displayed size of a bounding box is calculated for a position of the observer and then a suitable LOD is selected for displaying accordingly to the calculated displayed size of the BB and the limits embedded in the model. More information about this KML feature can be found in OGC (2008) or Wernecke (2009).

Cartographic Issues Addressed During the Process of Models' Creation

According to the rules mentioned in the chapter *Creation of 3D Maps* the first issue we needed to handle with was finding a balance between the realism of a created model of a building and a degree of generalization for the smooth visualization of the model. Regarding the realism (similarity to the real objects) of the models, we have been working with different documents showing the form, shape and image of particular buildings not only in the period of the 2nd WW, but also with documents with a date of origin even before the building was build (construction plans of buildings). The others used source of data are the data about the current state of the buildings. Unfortunately it is not rare that the documents from different periods show a building in different form or shape. So there was a need to closely cooperate with the historians from the Terezín Memorial to figure out, how likely buildings looked like during the 2nd WW to create such a model.

Regarding to the second rule dealing with the minimization of polygons used for the representation of a building, we created few rules for simplification of an object. It means to choose and use some methods of generalization, e.g. selection, collapsing and change of the graphical representation of an object. Even though for the LOD3, the view position of an observer to a building will not be too close to the building, so it is not necessary, but even not advisable (because of the purpose of visualization) to include small details, i.e. objects on the facade or on the roof which are smaller than a certain limit or which are not significant for the building. For example the window itself (excluding the recessing of the window into the wall) is not modeled in its whole representation of window glass, windows frame and the division of windows. These are reduced into one polygon and the polygon is textured with the image of the window. The same process is applied for example on elements of doors. An example can be seen in Fig. 1b. But this generalization rule violates the third rule claiming that the symbols should be created in their real dimensions. Nevertheless the quick visualization of data is superior to the modeling of each single detail.

The third rule says that symbols should be created in their real dimensions. As mentioned above, not all details of 3D objects are preserved or modeled in their real dimension. This way of modeling was established for quick smooth visualization of

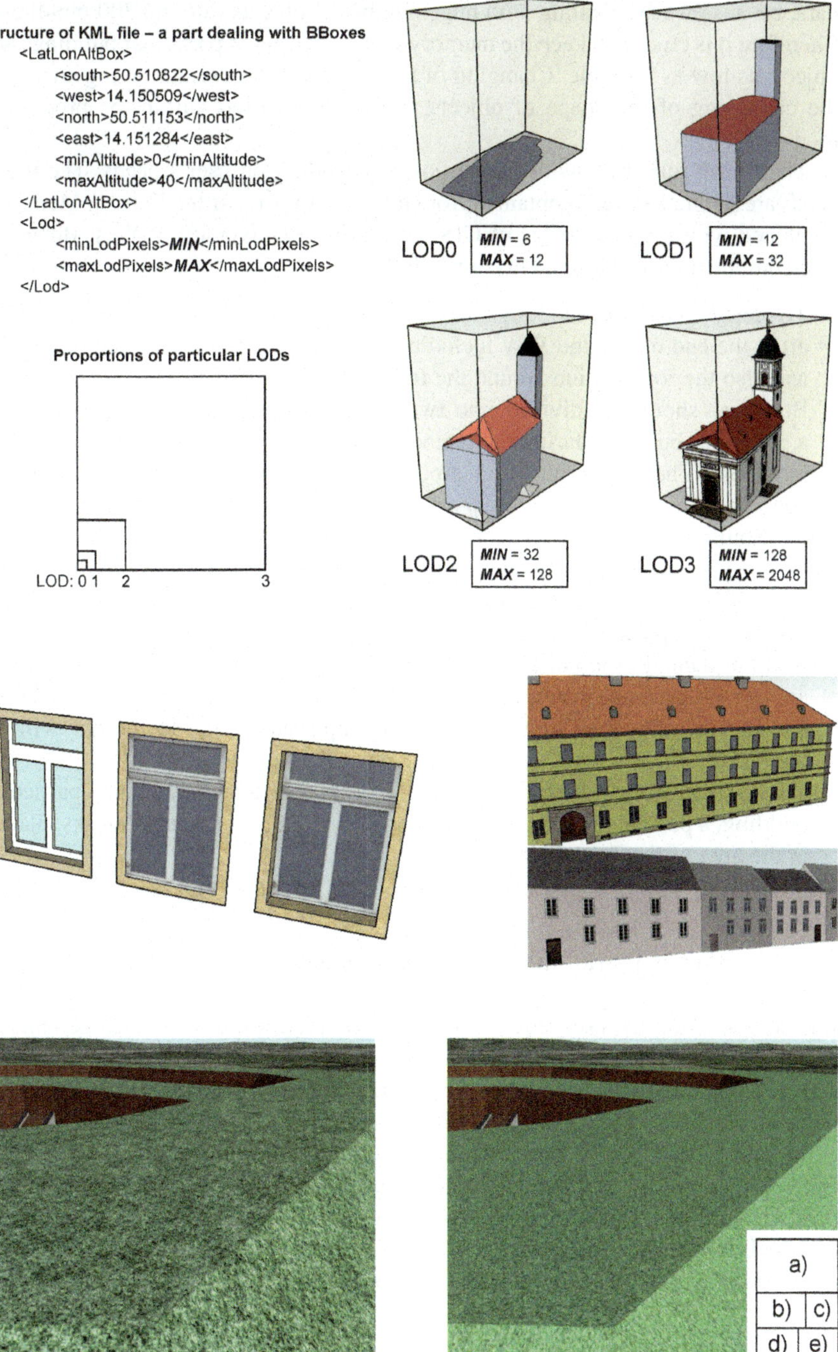

Fig. 1 (**a**) Example of switching LODs based on the pixel size of a bounding box; (**b**) Example of different ways of creating a model (pure vector, vector with flat textures and hybrid); (**c**) Comparison of important building in LOD3 (*upper part*) and unimportant building in LOD2+

data, because we are dealing with huge number of objects (around 300 buildings). Therefore it is crucial to keep the memory size of models of particular buildings and objects as low as possible. Changing of the graphical representation together with the collapsing of the shape of objects, are therefore used within the process of modeling.

The fourth rule deals with the purposes depending on user's needs. The user's needs are defined as needs obtained from historian experts of the Terezín Memorial combined with the knowledge of GIS experts from University of West Bohemia. The combined user requirements are as follows:

- To create a model of the Large and Small Fortress of Terezín in the state of the art at the end of the 2nd WW including buildings inside and outside of the city and also the fortification around the fortress;
- Buildings should be divided into two groups according to the importance of a particular building during the period of interest. Thus there are two groups of buildings. The first group is a group of so called "important" buildings. It consists of buildings of a military purpose (like barracks, hospitals and so on) or of buildings which were generally important (like a church, a water tower or a kitchen). Other buildings are marked as important under the conditions that momentous people lived or were imprisoned in such buildings or remarkable events took a place there (that means e.g. cells, theater, morgue, music pavilion). And the second group of buildings is group of "unimportant" buildings, which are the others. Important objects will be modeled in higher level of detail than the unimportant ones. But even for the important ones, no interior will be modeled;
- To enable visitors of Terezín Memorial to obtain information about a particular building, a person or a place directly from the visualized model at kiosks, placed in the area of the Terezín Memorial.

The Evolution of Works on Creation of 3D Models

It is obvious from the user requirements that we are dealing with the large amount of data which need to be processed and subsequently modeled. It follows that the complexity of the model is great and therefore it is crucial to involve a principle that reduce the amount of data. We have chosen the Level of Detail principle described at the beginning of chapter *Modeling of buildings of Terezín Fortresses*. We have created hierarchical data structure containing three levels: LOD1, LOD2, LOD3 (LOD0 is not necessary for the visualization purposes). In this place it has to be said that the LOD3 for building which are not important isn't the same as it is applied to the important ones, which is derived from CityGML standard. Let's mark the

---

Fig. 1 (continued) (*lower part*); (**d, e**) Comparison of simple (*left side*) and advanced (*right side*) grass texture with identical pixel size (this view represents almost horizontal look over a cavalier)

highest level of detail for unimportant buildings as LOD2+ instead of LOD3, because in this level is the model of an unimportant building just enriched contrast to LOD2 by the textures of windows and doors in a low quality. It can be said, that unimportant buildings are made in a way of the second draft of the model mentioned below, but with a very low quality of textures to minimize the size of the model. See the example of LOD3 of historically important and LOD2+ of unimportant building in Fig. 1c, where the unimportant building is in grey color and an important building is colored.

The main cartographic issue was the handling with the form of 3D cartographic representation according to the rules defined in the end of the chapter *Creation of 3D Maps*. The possibilities of 3D visualization are different in the software chosen for modeling the objects itself (it is Trimble SketchUp in this case—information about why this particular software has been chosen for creation of models can be found in Jedlička et al. 2013) and in the software chosen for the final visualization of the whole model (in this case it is Google Earth, respectively Google Earth plug-in for providing the map online and using the Content Management System.

The Trimble SketchUp can distinguish and visualize the edges between faces, but the Google Earth does not display edges in models, it is based on visualizing textures. This means the only possible way how to visually distinguish among the faces of an object in Google Earth is to use different textures to texture them. There were discussed few possibilities how to handle with 3D cartographic representations of objects, from which has been proposed three drafts of representations.

The first draft (in follow text called as pure vector) for creation of models was designed in a way that all parts of the model would be in a vector form (i.e. walls, roof, windows, doors, etc., including distinction of drawings on the building materials). This way was found not effective. Even though the memory size of a model of a building, which is made completely in vector form, is lower than for the following options, the perception (a cartographic representation) of these models is very flat and it is hard to distinguish the edges of objects while visualizing the model in Google Earth. Another big issue was that creating a model this way would mean to model nearly all parts of a building (of course regarding selected one, see chapter *Cartographic issues addressed during the process of models' creation*) including their small parts, which would significantly increase the number of created polygons (an example can be lattices on the windows) and also the time of production such models would significantly increase.

The second draft (in follow text called as vector with flat textures) proposed to make the shell of a building in vector form and every building parts create as textures on the faces of the shell, but keep the walls flat, without differentiation of embedded windows and doors and raised features on walls like ledges and portals. Such a model takes significantly more memory size than the all vector model, but the perception of a building is still very flat. Thus we have decided to create a hybrid model.

Third draft is a hybrid model where all parts necessary for an impression that a model represents a building were modeled in vector form (like chimneys, embedding of windows and doors, roof overhangs, turrets) and the colorization of the model was done by textures. The impression from such a model is very good, the model constitutes buildings and it is easy to distinguish particular parts of a building.

The all mentioned drafts applied on a window can be seen in Fig. 1b (from left to right, are drafts one to three).

Modeling of Terezín's Fortification

As mentioned above, the main aim of the model is to make a portrait of events taking place in the period of the Second World War. At that time the fortification was no more used for military purposes because it had become obsolete long time before. Nevertheless, the fortification was still useful when the fortress acts as the prison and the ghetto. Today the fortification plays no role in keeping enemies outside or inside. Despite that, the fortification is still essential part of the Terezín city and is considered as a historical and technical heritage.

The fortification did not play any important role in the Second World War and there are no descriptive data related to it. The main reason for modelling the fortification is just to make an impression of a fortified city. So, from the modeling point of view, the fortification falls into the "unimportant" group of objects and it can be considerably generalized.

There are two main sources of spatial data regarding fortification. Plenty of historical documents and plans depicting the fortification system is the first source. They contain information important for modeling (e. g. shape of the fortification, earthworks) but also for our purposes useless information describing invisible underground, casemates and so on. The second source is an actual state of the fortification. In spite of damages to the fortification that has been made so far, it is possible to recover the historical state of the fortification. Of course, some background knowledge of the bastion fortification system is necessary.

A contract for licensing an already-made 3D model of fortification has been assigned. The model is based on historical documents and plans, like in Popelka and Brychtová (2012). Because the original model was too detailed, it was generalized to a reasonable level. In order to keep a consistent appearance with the rest of a city model, only the geometry of the model is used.

Unlike the buildings, the fortification is modeled in one LOD only. There are three reasons for it:

- The fortification is a very large object. It is not a rare situation in which one part of the fortification is very close to an observer (potentially requiring higher LOD), whereas some other part is far from the observer (requiring lower LOD) in the same time. Thus, if the LODs were used, the fortification would have to be divided in several parts for which different LODs could be applied.
- Bastions, together with curtains, form a continuous fortification line around the whole fortress. The profiles of curtains and bastions are linked together and there

are no natural boundaries, where the shape of the fortification can be changed in different LODs.

- The fortification is an "unimportant" object and it just completes the rest of the model. There is no need to display much of details. Making an impression of a fortress is enough.

Because of the problems arising from using multiple LODs and only low requirements on details, the fortification is modelled in one LOD only.

Texturing the Model of the Fortification

The prevailing surfaces in fortification are walls, grass and paving. To be consistent with the rest of the model, not a photorealistic but an artificial textures should be used for them. Up to now, mainly the grass texture has been dealt with, because it is considered to be the most ingenious.

The fortification is modeled in one LOD only, so the grass-like texture is the same for wide ranges of observer distance. This put some special requirements on the texture:

- There must be enough details in the texture so that it looks good when the observer is reasonably close.
- The texture must be at least apparent for a wide range of observer distance. Pouring together all colors in the texture should be avoided for as long distance as possible.
- There should not be visible any texture discontinuities on edges connecting surfaces with the same texture.
- The effect of a repeated pattern should be suppressed as much as possible.

Besides that, the texture should be seamless and its size should be kept small.

There are sophisticated computational methods of creating an artificial texture (for example Chang and Hsiao 1994), but it does not meet our requirements. Because the grass does not create any regular pattern, a simple texture made by random noise added to the green base-color was tested (see Fig. 1d). It looks good for small distances of the observer, where the pixels (or small groups of them) are distinguishable. For farther distances of the observer the pixels pours together and the texture fades out to a non-textured one-color surface only. However, this is the only disadvantage of the simple texture. All other requirements are met.

To avoid pouring colors together for longer distances of observer, an advanced texture had been computed. The advanced texture is based on the random noise also, but the noise is added in multiple scales. The results can be seen in Fig. 1e, where the advanced texture with size of 20×20 m (864×864 pixels) is used. It is obvious that the advanced texture is apparent in much larger distance than the simple texture. There are almost no texture discontinuities on the edges and the effect of repeated pattern is minimized.

Data Storage and Interactivity of the 3D Model

Making models of historical buildings is important and integral part of modern 3D cartography nowadays. We can find models of specific historical object (Angelini et al. 2011; Popelka and Brychtová 2012), of large complexes (Yan and Limin 2011), models of objects still in existence (Hanzalová et al. 2011) or vanished buildings (He 2011). 3D models are created in miscellaneous levels of detail or interactivity. The very important point of modelling is defining the purpose of a model. The purpose of the model determines the approach of creation. We can mention lot of model purposes e.g. for presentation, for virtual reconstruction, as objects of large information system and more other examples. More details can be found in Belai and Jedlička (2012).

There it is possible to distinguish two different purposes of building models created during this particular project: for virtual reconstruction and elements of an information-rich system. 3D models of buildings work here as a front-end of the information system that manage links to historical data about buildings and about important personages and events connected to buildings.

The core of the information system is a central data storage that stores all collected information. This data storage is the source for data presentation in other parts of the system. According to the agreement of project partners the Plone Content Management System solution has been chosen. In general, a CMS is a web content publishing and management system that allows users to create, submit, and publish their content directly within a web site without any development tools or knowledge of HTML (see Plone 2013). This approach allows separation of geometric data creation from attribute data creation. Both types of the data are linked in CMS according to the unique building identifiers used by Germans during the 2nd WW in Terezín ghetto. This unique building identifier consists of building block identifier (capital Latin letter and Roman number—see Fig. 2a) and rectangular street nomenclature (vertical Qs, horizontal Ls—see Fig. 2b) with house orientation number. There is an example in Fig. 2c, showing block DIII and two particular buildings: DIIIL307 (upper bordered building) and DIIIQ408 (bordered on the left side). 3D models are finally visualized in Google Earth plug-in that has been chosen after a discussion within the project partners.

The CMS stores different information types in separate folders like Pictures, Places where KML files and buildings descriptions are stored, Persons where information about important personages are stored, and Tours are for predefined tour through the model of city. The structure of sub-folders for storing models is created according to the unique building identifiers from 2nd WW. Each building folder contains two objects. The first object is a KML file with geometry of a building and the second object is structured text file for the description data related to the building.

Important part of created models is the interactivity that allows discovering the information about individual buildings by clicking on them during visualization. There is a need to create KML files with "clickable" models that allows showing all

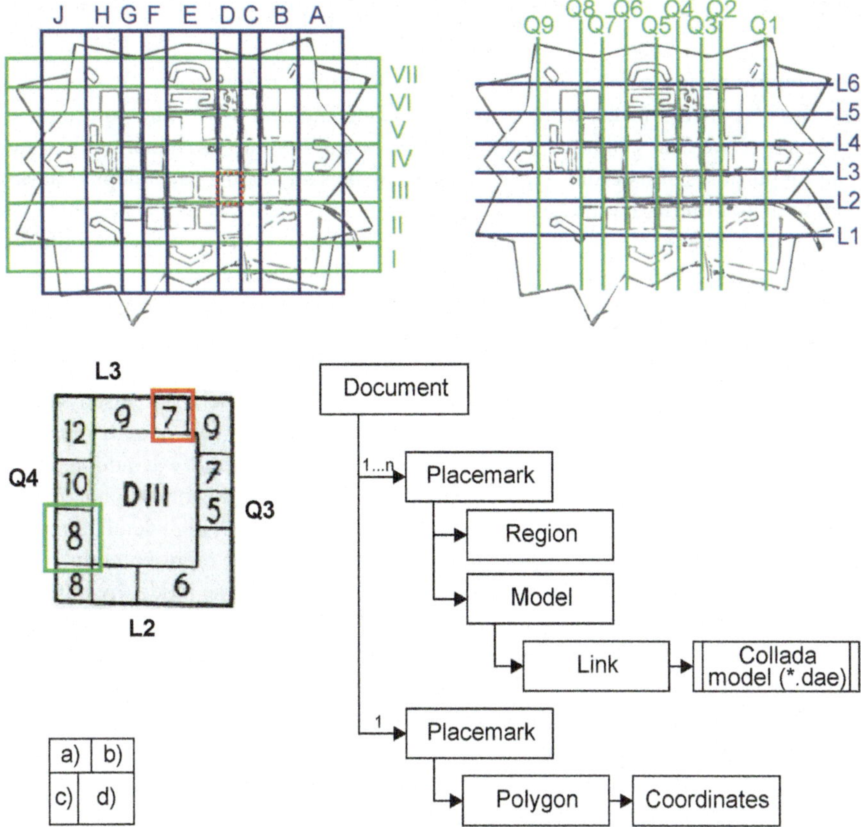

Fig. 2 (**a, b**) System of columns and rows for depicting of blocks (*left side*) and streets (*right side*) in the Main Fortress; (**c**) Example of numbering of particular buildings within the block DIII in the Main Fortress; (**d**) Structure of a combined KML file

related information in the form of a "balloon" by clicking on the model. The "Click on" functionality is attached to simple geometry in Well-known Text (WKT) format but is not implemented for models in Collada format. There was necessary to create simple representation in WKT format of every model the so called "pointer box". The pointer boxes have been created from LOD1 models by ten-percent reduction. This reduction ensures hiding of pointer boxes inside of all LOD models and allowing "click on" functionality to appear information balloon. The structure of such combined KML file (i.e., with real and clickable geometry) can be seen in Fig. 2d.

Summary

This contribution showed a creation of 3D cartographic model of the Terezín Main and Small Fortress in the state of the art during the 2nd WW. The contribution depicts how we deal with a large amount of data and how we use a hierarchy of the data for purposes of its visualization. Thus it can be said that the paper depicts a way how to deal with big hierarchical 3D data.

The first chapter described different approaches of 3D map creation with the stress to useful rules for creation of 3D symbols and models.

The next chapter described 3D model itself which is divided into modeling of buildings and fortification. Buildings are modeled in multiple levels of detail for purposes of dynamic real time visualization. Particular LODs and also the way of details interpretation (vector, flat textures or hybrid approach) are discussed in appropriate chapters of the paper. After that discussion a convenient approach was selected. On the other hand, the shape of the fortification was adopted and adapted from another project (which reconstructed it from historical plans). The adaptation consisted of geometric simplification of the original high detailed model and of proper texturing—to achieve suitable 3D cartographic representation.

Further the contribution narrated about an enrichment of the model by attribute data via Content Management System and publishing the model on the internet. This consists of creation of clickable geometries of buildings (outside the Collada models) and points of interest, collecting documents related to the important places and people living in Terezín during 2nd WW and interconnecting these data (both geometric and attribute) together into CMS.

The final stage is the visualization of the whole model via Google Earth web browser plug-in with an ability to obtain data from CMS via clicking on interactive models and popping up a balloon with the information about the place, related documents, photographs and people. It implies that this paper shows an example of 3D web cartography working with big hierarchical data.

Acknowledgments Authors were supported by projects:

- "Landscape of memory. Dresden and Terezín as places of memories on Shoah", reg. number 100110544, that was approved for financing within the Objective 3—Support of Cross-Border Cooperation 2007–2013 between the Independent State of Saxony and the Czech Republic.
- "NTIS—New Technologies for the Information Society", European Centre of Excellence, CZ. 1.05/1.1.00/02.0090, European Regional Development Fund (ERDF).
- NeoCartoLink—Supporting the creation of a national network of new generation of Cartography, CZ.1.07/2.4.00/31.0010, OPVK.
- EXLIZ—CZ.1.07/2.3.00/30.0013, co-financed by the European Social Fund and the state budget of the Czech Republic.

References

Angelini MG, Costantino D, Milan N (2011) 3D and 2D documentation and visualization of architectural historic Heritage. In: XXIIIth CIPA Symposium, 12–16 September 2011, Prague. http://cipa.icomos.org/fileadmin/template/doc/PRAGUE/008.pdf

Bandrova (2001) Designing of symbol system for 3D city maps. In: Proceedings of the 20th international cartographic conference, Beijing. http://icaci.org/files/documents/ICC_proceedings/ICC2001/icc2001/file/f08012.doc

Belai E, Jedlička K (2012) 3D model of heritage protected area based on a combination of surveying and architectural documentation. Bachelor thesis, University of West Bohemia, Pilsen

Blodig V (2003) Terezín v "konečném řešení židovské otázky" 1941–1945 [Terezín in the "Final Solution of the Jewish question" 1941–1945]. Oswald, Praha, 136 p

Chang RI, Hsiao PY (1994) Artificial texture generation using force directed self-organizing maps. In: IEEE international conference on neural networks, vol 7. IEEE, ISBN: 0-7803-1901-X

Elmes (2005) slide series Geog246: digital cartographic design. http://1-media-cdn.foolz.us/ffuuka/board/tg/image/1366/61/1366617680397.pdf

Hájek P, Jedlička K, Vichrová M, Fiala R (2013) Conceptual approach of information rich 3D model about the Terezín Memorial. Geoinf FCE CTU 11:49–62, ISSN: 1802-2669

Hanzalová K, Čumpelík Z, Mathauserová S, Filípek P, Odvody M, Kostka R (2011) Simplified photogrammetrical documentation and visualization of historic objects in Peru. In: XXIIIth CIPA symposium, 12–16 September 2011, Prague. http://cipa.icomos.org/fileadmin/template/doc/PRAGUE/068.pdf

He Y (2011) Re-relic/Yuanmingyuan: an effective practice in virtual restoration and visual representation of cultural heritage. In: XXIIIth CIPA symposium, 12–16 September 2011, Prague. http://cipa.icomos.org/fileadmin/template/doc/PRAGUE/071.pdf

Jedlička K, Čada V, Fiala R, Hájek P, Janečka K, Ježek J, Roubínek J, Strejcová J, Vichrová M (2013) Techniques used for optimizing 3D geovisualization of Terezín memorial. In: 26th international cartographic conference proceedings. International Cartographic Association, Dresden, s. 686. ISBN: 978-1-907075-06-3

OGC (2008) OGC KML, Open Geospatial Consortium, version 2.2.0, Publication Date: 2008-04-14. http://www.opengeospatial.org/standards/kml

OGC (2012) OGC City Geography Markup Language (CityGML) En-coding Standard, Open Geospatial Consortium, version 2.0.0, Publication Date: 2012-04-04. http://www.opengeospatial.org/standards/citygml

Pegg D (2011) Design issues with 3D maps and the need for 3D cartographic design principles. In: Workshop "Maps for the future: Children, Education and the Internet". http://lazarus.elte.hu/cet/academic/pegg.pdf

Plone (2013) Plone Open Source CMS/WCM, Plone Foundation, 2000–2013. http://plone.org/

Popelka S, Brychtová A (2012) The historical 3D map of lost Olomouc fortress creation. Geography and Geoinformatics: Challenges for practise and education, Masarykova univerzita, s. 147–154, ISBN 978-80-210-5799-9

Wernecke J (2009) The KML handbook – geographic visualization for the Web. Upper Saddle River. ISBN 978-0-321-52559-8

Yan H, Limin C (2011) 3D-GIS application in information management and conservation planning of historic city. In: XXIIIth CIPA symposium, 12–16 September 2011, Prague. http://cipa.icomos.org/fileadmin/template/doc/PRAGUE/157.pdf

Changes in Urban Area Discovered by Analysis of Chosen Places in Old Maps of Liberec

Vojtěch Blažek, Vojtěch Hájek, Ladislav Ličík, Branislav Nižnanský, and Klára Popková

Introduction

This article follows the previous multidisciplinary research of the area of Husova Street, which helped to create the monograph (Nižnanský and Popková 2012). The current research is carried out differently for it is not focus on one particular locality, but more of those, which can characterize the development of Liberec from the centre towards the Jizera Mountains. There are three localities introduced in the article. The multidisciplinary research, which will result in a forthcoming publication, can be characterized with the help of those localities. The localities include: Technical University of Liberec (next TUL) area, the crossroads of streets Masarykova and Vítězná and Králův Háj—old housing development. They have been selected for several reasons. The first reason is the diverse and characteristic development of the areas, which is reflected not only on old maps. The second reason is the contemporary attractiveness of the studied localities, which is connected with their architectural and urban uniqueness. The third reason is the simple identification and availability of relevant secondary sources for historical analysis of the localities.

Liberec itself is one of the most interesting areas for historical–geographical studies. Its history dates back to the Middle Ages when it was founded as a small settlement on a trade route to Zittau (Technik 1967). The expansion of the town was connected with the industrial boom which came during the nineteenth century. This can be seen in the fact that Liberec does not have any city walls which could emphasize its importance in Middle Ages or early modern period (Koňasová, In: Nižnanský and Popková 2012). The industrial revolution and its effects had irreversible impact on the face of the town, which is still visible in the area (Purš 1960). It was as late as in this period when the studied area consisting of the selected

V. Blažek • V. Hájek • L. Ličík • B. Nižnanský • K. Popková (✉)
Technical University of Liberec, Studentska 2, 461 17 Liberec, Czech Republic
e-mail: klara.popkova@tul.cz

© Springer International Publishing Switzerland 2015

J. Brus et al. (eds.), *Modern Trends in Cartography*, Lecture Notes in Geoinformation and Cartography, DOI 10.1007/978-3-319-07926-4_33

localities, nowadays symbolized by a line from Šalda's Square through Dr. Edvard Beneš Square to Soukenné Square (Umlauf et al. 2011), started to develop towards the Jizera Mountains.

Methodology

Specific locations were selected in order to compare the maps. These locations were cut out of particular historical maps and placed next to each other in a chronological line. With the help of this method it is possible to observe the development of map symbols and the way of depicting the areas in time. The comparison of the localities was targeted in six directions: general comparison, street network, housing development, identification of characters using numbers, planned changes and interesting features. The general comparison examined mainly the biggest changes of the graphical representation of a locality between particular maps, i.e. maps with the largest housing development. The level of housing development in the area was deduced by analysing the numbers of symbols, graphical representation of ground plans and position of house symbols. On some maps it was possible to observe the development of functions of important buildings using identification numbers that refer to the map appendix. However, there were problems with old German and its handwriting which in some cases did not allow to clearly clarify the function of a building. The characteristics of street networks, their articulation and graphical representation were examined in a similar way. At the same time, some symbols of public transport were added to the symbols of roads. The planned construction of the localities was depicted only on specific maps and it was mostly the street network. The interesting features include some discrepancies in the use of symbols, their location, ambiguity of their locations etc.

For greater simplicity and clarity, the individual maps in the text are identified only by the year of their origin; more detailed reference is mentioned on the list of the source materials.

The aim of the historical part of the text is to supplement and, if necessary, to verify the cartographic and historical research. For this reason we used, mainly for greater clarity, secondary literature and directories of Liberec.

The Locality of Vítězná and Masarykova Crossroads

Definition of the Locality

Description: Crossroads—museum, policlinic, old exhibition grounds

Nowadays it is one of the most interesting crossroads in Liberec. In the western part of the crossroads there is the building of policlinics, in the northern part there is the building of the North Bohemian Museum and in the southern part there is the former exhibition grounds, recently rebuilt into a leisure centre and a restaurant. In the southern part there is also a newly reconstructed building of former baths where the town gallery is being relocated at the moment. A new modern object has been built in the area of the former baths. It will be used as the gallery depository. There is a tram line going along Masarykova Street, with a tram stop right in the centre of the locality. It is the main street and nowadays it is intersected by a quite busy street called Vítězná.

General Comparison

At the very beginning it is clear that on all the maps (except the one from 1858) those three important buildings are depicted in dark colour. We can also see that the graphic presentations of single maps are not remarkably different from each other. It can be assumed that the most important development of the locality happened between the years when the first and the second map were created i.e. between 1858 and 1899.

Street Network

The axis of this locality is Masarykova Street. In 1858, this area is characterized by only a symbol of a path in fields and meadows without any turnings, intersections or parallel paths. In contrast to this, the map created 41 years later (1899) shows quite developed road infrastructure accompanied by symbols of dense residential development.

Using the map from 1899 we can see that the street network went through remarkable changes, especially to the north, where there are numerous symbols of residential buildings. In contrast to this, the southern direction does not have such visible signs of development in the period between 1858 and 1899. Along the south side of the street there is only a line of symbols of residential houses. This linear housing development does not change very much on later maps.

Parallel with the main street there is a colour line symbol (1899 and 1910—black, 1939—blue white; 1981—red, identification number 3 and 4), which makes us assume that there used to be a tram line there. The only exception are the maps from 1858, when the housing development was not yet presented in the locality and in 1932 and 1943, when the tram line was not displayed due to excessive generalization of the maps.

Housing Development

The most remarkable changes of residential density can be seen on the maps from 1858 to 1899. It can be assumed that this expansion was supported by construction of the tram line.

As early as on the map from 1899 a symbol of the museum with the adjacent park is recognizable. The symbol is the same on all the maps except the maps from 1943 to 1981 where the ground plan does not correspond to the graphical displays on the previous maps. We can assume that the reason is the generalization of map symbols in those years.

The symbol of the former baths ("Stadtbad" on the map from 1899) is not marked with a different dark colour, so we can see that this building was not considered significant. In contrast to this, the map from 1910 displays the symbol of the baths in dark colour. Similarly, the symbol of the building in the western part of the crossroads, which is first displayed on the map from 1899, was a significant building then. In addition to this, the building was extended between 1910 and 1932, which is depicted by the longer ground plan of the symbol on the map from 1932. The building is called "Children's Polyclinic". The newest building—the exhibition grounds—was probably built between 1943 and 1981.

As for the residential development, there were not many remarkable changes. The surrounding areas were fully residentially developed in 1899.

Identification of Symbols with the Help of Numbers (Only Maps 1899, 1910, 1932 and 1939)

Map 1899:

40 ("NordböhmischesGewerbe-Museum"—the North Bohemian Museum of Industry);

37 ("Handels- und Gewerbekammer"—Chamber of Trade and Industry)—the identification number is not legible, but the building can be identified via the orientation square G5.

Map 1910:

27 ("Kaiser-Franz-Josef-Bad"—The Baths of Franz Josef);

38 ("Handels- undGewerbekammer"—Chamber of Trade and Industry);

47 ("NordböhmischesGewerbeMuseum"—the North Bohemian Museum of Industry).

Map 1932:

42 ("Bad d. Reichenberg. Sparkasse"—translation not clear);

60 ("Handels- u. Gewerbekammer"—Chamber of Trade and Industry);

66 ("Nordböh. Gewerbemuseum"—the North Bohemian Museum of Industry);

99 ("Theodor—Körner—Denkm."—memorial of Theodor Körner)

Map 1939 names the buildings as follows:

31 (not mentioned in the legend);

32 ("Bad der ReichenbergerSparkassa"—translation not clear);

33 ("Körner-Denkmal"—memorial of Theodor Körner);

34 ("Nordböhm. Gewerbemuseum"—the North Bohemian Museum of Industry).

Planned Changes

On the map from 1899 there is a symbol of a planned communication connecting streets Husova and Masarykova nowadays. The street is displayed as an already existing communication on the map from 1910.

Interesting Features

The symbols of former exhibition grounds are not equivalent on the maps from 1981 and the present map. The ground plan is more or less the same, but not the location of the building. The map from 1981 names the building "Severočeské výstavy" (North Bohemian Exhibition Grounds). The position of the building between 1981 and today is not the same. We cannot therefore say for sure that the buildings are identical.

Historical Analysis of the Locality

The locality at the crossroads of Masarykova and Vítězná streets is without any doubts one of the most architecturally interesting areas in Liberec. The exhibition grounds is another very interesting place which deserves our full attention. Its present look was given by development after the war, when a large part of the area was constructed. During the period of its biggest boom the exhibition grounds were popular all over then Czechoslovakia (Karpaš 2004). However, from the historical point of view, we have to distinguish two main exhibition areas. The first and older one is the area opposite the museum, in the south-east part of the locality. According to Řeháček (2010), the first exhibition was held there in 1919. This is where pavilion F of the Liberec exhibition grounds was built. Nowadays it is used as a restaurant (Zeman 2011). The second main area of the exhibition grounds is located between the industrial school and houses next to the building of Chamber of Trade and Industry. It was built in 1946 for the exhibition "Budujeme osvobozené kraje" (Technik 2001).

The first building at the crossroads of Masarykova and Vítězná streets is the Chamber of Trade and Industry, built between 1898 and 1900 in the style of German neo-Renaissance. It was used for its original purpose until 1948, then it was the seat of Národní krajský výbor (regional authorities) until 1960, when children's clinic and regional institute of health (Zeman 2011) moved in. The renewed Czechoslovak Chamber of Trade and Industry moved back in 1991 (Technik 1992). Nowadays there are individual private companies and subjects.

Another significant building on Masarykova Street is the town baths, nowadays used as the building of The Regional Art Gallery. It was built by A. Bürger between 1901 and 1902 with the project of P. P. Brang. This neo-Renaissance building is already clearly showing signs of coming Art Nouveau style. The building was reconstructed after World War II, but after the Velvet Revolution it was not used anymore (Technik 1992). Since 2013, the building has been used as the gallery.

The building of the museum is located at the crossroads of Vítězná and Masarykova streets, opposite the gallery. The history of the museum dates back to 1873, when the museum association was founded. But it was not successful until 1882, when Zemský výbor in Prague donated a remarkable financial subvention and thus enabled the construction of the museum building. The building process started with laying the foundation stone from the destroyed old town hall on 24th March 1897. The museum was formally inaugurated on 18th December 1898. After World War II it reopened on 1st September and a year later it made its collections available for visitors of the exhibition (Technik 1992) (Fig. 1).

Locality of TUL Area

Definition of the Locality

Description: Area of TUL

The main axis of the studied locality is Husova Street. In its neighbourhood there is ZŠ Husova (elementary school), the regional hygienic station and between them the dominant campus of Technical University of Liberec. At the moment it is undergoing a remarkable development. New buildings are being built and the whole area is slightly expanding.

General Comparison

The map from 1858 does not display the locality. Thanks to the contour lines on the map from 1899 we can see that Husova Street is located nearly on the ridge and it forks near present building of hygienic station. The development of the main street is not visible as its symbol does not change on individual maps.

Street Network

The main street is a tree-lined avenue on the map from 1899. There is no housing development.

The map from 1932 displays greater development of the area. We can see existing network of streets similar to the plans from 1899 and quite dense housing development. The map from 1910 does not display many differences in the street network compared with the previous map. In contrast to this, the map from 1932 shows the street network remarkably changed and partly copying the plans of streets displayed on the map from 1899. The development of the street network is

Fig. 1 The locality of Vítězná and Masarykova crossroads. Pictures presents changes between 1858 (*top*), 1899, 1932 and 1981 (*bottom*)

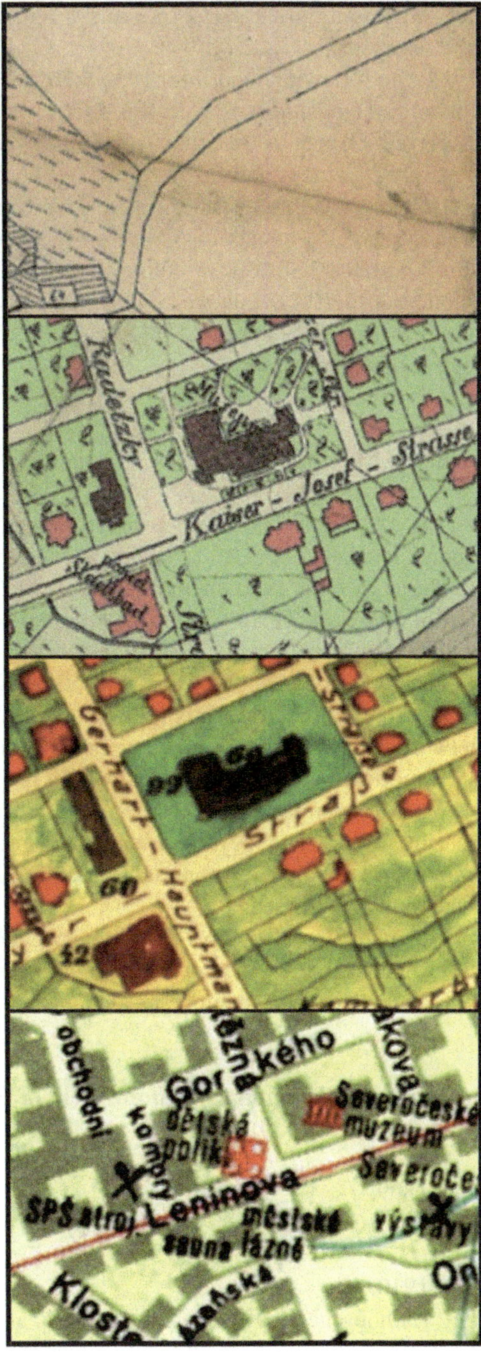

not shown only to the north (nowadays Masarykova Street), but also to the south (the reservoir).

On the following map from 1939 there are no remarkable changes of the street network. The same it is with the difference between 1939 and 1942. The only displayed change is one half of Schütze Strafseis demolished. Nowadays its remaining part is called Boční Street.

How busy and important the streets were can be observed only with the help of the width of the streets displayed on the maps. The map from 1981 highlights Husova Streets with yellow colour, which means that the street is the main street. Public transport is not marked on the maps until 1981, when a red dashed line with numbers 15 and 19 appears, which displays the bus lines in this direction (see the legend of the map).

Housing Development

On the map from 1899 there is only one symbol of an important building in the studied locality. The rest of the area is covered with fields.

The map from 1910 displays symbols of housing development in the western part of the street. There are also two symbols of important buildings planned to be constructed in the locality (see below).

The greatest housing development of the locality is displayed on the map from 1932, where there are symbols of single buildings within the symbols of street network. The map shows three symbols of important buildings as well. However, the important building from 1899 to 1910 maps is missing.

When we compare the maps from 1932 to 1939, we can see slight expansion of housing development as there are symbols of quite remarkable buildings. The map from 1939 displays four important buildings and the building mentioned above is marked as important again.

The maps from 1943 to 1981 are very generalized and that is why the housing development is not recognizable. The only important building on the map from 1981 is the area of contemporary Technical university (acronym VŠST).

Identification of Symbols with the Help of Numbers (Only Maps 1899, 1910, 1932 and 1939):

Map 1899:
50 ("Kath. Gesellenheim"—catholic asylum house)

Map 1910:
55 ("KatholischesGesellenheim"—catholic asylum house).

Map 1932:
9 ("DeutscheStaatsrealschule"—German state secondary school);

8 ("DeutschesStaatsgymnasium, Tschech. Ref. Realgymnasium"—probably German and Czech grammar school);
22 ("Gewerbeförderungsanstalt"—translation not clear);
93 ("Kathol. Gesellenheim"—catholic asylum house displayed as an unimportant building).

Map 1939:
35 ("DeutschesStaats-Gymnasium"—German state grammar school);
36 (not found in the legend);
84 ("Kathol. Gesellenheim"—catholic asylum house)
85 ("DeutscheStaats-Realschule"—German state secondary school).

Planned Changes

The map from 1899 displays a planned rectangular street network grid to the north. The planned streets are depicted by dashed lines. The only street planned in 1899 and finished in 1910 is nowadays Vítězná, connecting Husova and Masarykova streets. The map from 1932 displays the finished street network, but it differs from the designed structure (just the rough concept is the same).

In contrast to this, the map from 1910 displays just a symbol named "proj. Bildergallerie", which is situated in the eastern part of the locality, quite separated from the other buildings. The second symbol of an important building is called "proj.k.k.Staats-Realschule" (contemporary ZŠ Husova). Both buildings were probably designed, which is given by the expression "proj.".

Interesting Features

It is a locality situated on a ridge with the main street as the main axis. On the map from 1899 there is a symbol of a field on that ridge, which is really interesting. This means it must have been a very steep field.

Historical Analysis of the Locality

The locality of Husova Street is the most varied of all the studied localities. The earliest construction activities started in 1904 due to necessary landscape works for the Liberec Czech Germans exhibition in 1906. The earthworks were finished in February. Since the height difference in the area was 45 m, 28,000 m^3 of soil and 7,000 m^3 of stones were excavated. A lot of exhibition premises were built there in this period; they were disassembled later (Melanová 1996). The villas between present Vítězná a Studentská streets were built as permanent residences (Zeman 2011). There were a lot of plans with the empty space, for example a never realized project of a gallery building (Habánová 2010). The land was bought for nowadays

ZŠ Husova (elementary school) in 1909, which was designed by Liberec architect Oskar Rössler, built in 1913 and opened in 1914 (Technik 1993). After this, the area developed in two directions. Residential villas along the reservoir and public buildings in the area of the exhibition, where first contemporary building A of TUL was constructed. It was built in present Čížkova Street between 1917–1920, and it was originally used only as German grammar school, after World War II it was used as Czech grammar school. Present building B of TUL in Čížkova Street was constructed in 1926 as Ústav pro zvelebování živností (Technik 1967). Present building C of TUL was constructed during the 1930s. After the World War II it was used as a special school (Golka and Vild 2003). Next to the buildings mentioned above there was an unused area, where a camp for bomb shelter builders was created during the war (Rous 2009). The camp was later used as a displaced persons camp for people considered to be German at the end of the war (Lozoviuková 2010).

After removal of the camp at the beginning of the 1950s, discussion on how to use the area started. After the university was established in 1953, a decision to build university buildings there was made (Golka and Vild 2003). Building F of TUL at the former football field was constructed because of the lack of student accommodation facilities, drafting rooms and classrooms. It was built between 1957 and 1959 (Technik 1995). The university dining hall was built between the two main buildings of the project in 2000 (Zeman 2011). Building E of TUL is located in the area of the former camp and was constructed between 1956 and 1965. Next to the main building there are still located workshops and laboratories (Golka and Vild 2003). The building of the seat of the rector and information centre was constructed between 2001 and 2004 (Zeman 2011). Building G of TUL is being finalized at the moment.

Residential villa development gradually expanded in the direction from the centre towards the mountains before World War I. It got its present appearance in mid 1930s of the twentieth century. The most important villas include Strossova vila (Stross's villa), the seat of regional hygienic station, or so called Henleinova vila (Henlein's villa), where Konrad Henlein lived during World War II (Halík et al. 2009).

More on the area of Husova Street on a multidisciplinary level can be found in Nižnanský and Popková (2012) (Fig. 2).

Locality Králův Háj 1

Definition of the Locality

Description: Králův Háj—old housing development.

The locality is situated on the top of a hill. In the north it is bordered by woods going down to the reservoir. In the south it is bordered by another group of residential buildings and woods. The axis of the locality is nowadays Březinova

Fig. 2 The locality of the Technical University of Liberec is depicted on maps from 1899 (*top*), 1910, 1932, 1939, 1943 and 1981 (*bottom*)

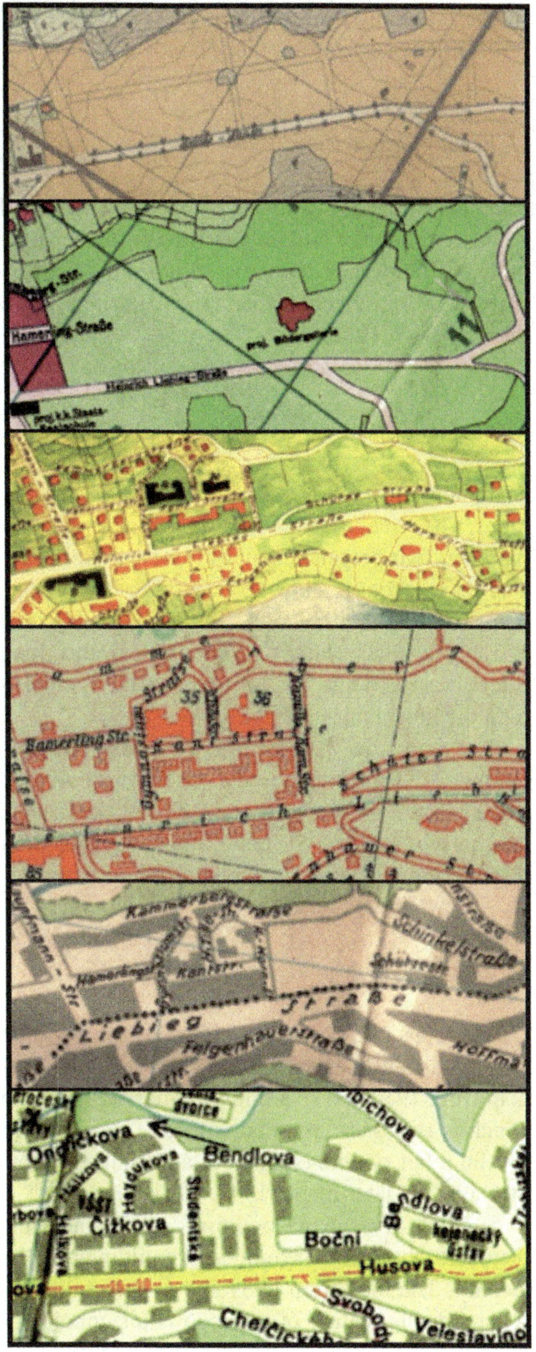

Street. It is a blind part of the main street, closed by a bus turn-table. The street does not continue straight to Starý Harcov.

General Comparison

The map from 1858 does not display this locality. However, it can be assumed that there were not any buildings or streets there. The map from 1899 shows a part of a path as a borderline between woods and a field. Neither this map is complete and that is why it is not clear where the path ends. The whole locality is finally shown on the map from 1932.

Street Network

The main street itself (Březinova Street) is displayed in the same way on all the maps. The street network with branches is displayed on the map from 1932, where there are two perpendicular and one parallel street running out of the main street. The same layout is shown on the other later maps without any major changes. The maps from 1932, 1939, 1943 and 1981 display roads going to the neighbouring part of Liberec (Starý Harcov). In contrast to this, the present map displays Březinova Street closed by the bus turn-table with symbols of paths as the only connection with the neighbouring locality.

The map from 1981 shows a symbol of a red dashed line with identification number 21. This is supposed to be the bus line in the studied locality. The line ends at the point of present turn-table, although it is not displayed on the map from 1981. The present map also displays a bus stop, but from its location we can see that the display of nowadays bus line is different from the map from 1981.

Housing Development

On the maps from 1899 to 1910 there are not any symbols of residences, there are only fields and meadows. However, the symbol for an important building on the map from 1910 shows a building marked as Hochwasser-Behälter (water station), which is still there. At this point on the maps from 1932 to 1939 there is quite a special relief symbol (symbol of a rectangular ground plan) The water station is not displayed on later maps (1943, 1981, present)

Symbols of residence buildings are displayed the map as late as from 1932, when there are about fifty symbols of buildings and the water station in the studied locality. The map from 1939 shows the same situation with a few more symbols. A remarkable feature of the map is a symbol of a line showing the border of the town (see the legends), which is going through the locality and divides it into two asymmetric parts.

The map from 1943 is too generalized and the symbols of residential houses are displayed as monolithic blocks of buildings. In contrast to this, the map from 1981 also generalizes, but only the ground plans of the buildings. We can clearly see the individual buildings there. In the centre of the locality there is a symbol of cutlery, suggesting there used to be a restaurant. The same symbol is located on the border of this a previous locality. There is also a symbol "VB" and an envelope at the same point, so we can assume there was a police station, a post office and a restaurant.

The present map displays these important buildings: garages, kindergarten, elementary school and secondary school of gastronomy.

Identification of Symbols with the Help of Numbers (Only Maps 1899, 1910, 1932 and 1939)

In this locality there are not displayed any numbers identifying map symbols.

Planned Changes

On the map from 1939 we can see a dashed line oval symbol in the western part of the locality, which could symbolize a planned running oval track or a sports ground. This assumption cannot be supported by the map from 1943 due to its generalization. On the map from 1981 and the present map there is a symbol of a sports ground, but it is located further to the west.

Interesting Features

With the map from 1932 it can be assumed that the housing development in the studied locality does not come from the centre of Liberec, but it is the outskirts of Starý Harcov (Alt Harzdorf), which expands towards the centre of Liberec. Except for a few symbols of buildings and slightly more advanced street network, this situation is almost identical to the situation on the map from 1939. Here, however, the "Harcov" hypothesis does not correspond with the city borders, which place the studied locality inside the town of Liberec.

Historical Analysis of the Locality

There are two aspects of the locality of Králův Háj, the youngest, but very varied area. The first aspect is the residential villa area, brick houses and sports ground. According to Karpaš (2004), the residential villa area dates back into mid 1920s of the twentieth century, the sports ground before 1928. The construction from the northern side of the slope reached the top by the end of the 1930s. According to Rous (2011), the housing estate called Heimstätte—Domov was built between the middle of 1939 and 1941. It consists of 15 residential houses no. 325–330 and 332–340, grouped into eight blocks. They were most likely built by the Roma, who were concentrated in several concentration camps in Liberec. An interesting point is connected with the allocation of flats

Fig. 3 The locality of Králův háj old housing development depicted in maps from 1899 (*top*), 1910, 1932, 1939 and 1981 (*bottom*)

and houses during the re-colonization of Liberec in 1945. Thanks to the research of Hájek and Mrázová (2012) we know that those who requested permission to settle in the area had a choice either to get a villa as a worker or to get a flat in a newly built house if they keep their business (Fig. 3).

Conclusion

This article presents a part of the research in selected localities of Liberec. From our point of view, this original approach brings new insights into the way of looking at the particular localities. This will be beneficial not only in the field of science, but for citizens of Liberec as well. Geoinformatical analysis and historical or geographical features will be added into the forthcoming publication. Besides the three studied localities which were selected for the reasons mentioned above, the research will continue in six more localities. The school area in the western part of Masarykova Street, the area of the ZOO and Lidové Sady, the area of Liberec hospital and adjacent buildings, Harcov reservoir, the monastery and Liebigova (Liebig's) villa, Králův Háj—new housing development. They will complement the localities analyzed here and illustrate the development of Liberec from the centre towards the Jizera Mountains. All nine mentioned localities will be thoroughly analyzed and with appropriate multidisciplinarity presented in the forthcoming special publication.

Source Materials

Maps:

1858: Plan der Stadt Reichenberg nach der Reichenberg nach der neuesten Regulirung 730 × 550 mm. In: Anshiringer A (1858) Adressbuch der Stadt Reichenberg. Liberec

1899: Plan von Reichenberg [1:6,000], 510 × 700 mm In: Dressler JF (1899) Adressbuch und Wohnungsanzeiger der Stadt Reichenberg. Liberec

1910: Stadt Reichenberg. Liberec [1:2,880] 1,140 × 1,060 mm (1910)

1932: Tschörner J (1932) Plan der Stadt Reichenberg. Liberec [1: 2,880], 1,490 × 1,080 mm

1939: Bienert R (1939) Stadtplan von Reichenberg. [1:10,000], 390 × 480 mm, Liberec

1943: Plan der Gauhauptstadt Reichenberg; Die innere Stadt. Liberec [1:10,000–1:15,000], 780 × 540 mm (1943)

1981: Liberec—plán města. Kartografia, Praha [1:15,000], 630 × 730 mm

Acknowledgments The article was written in resolving and with the financial support of the Student Grant Competition at Faculty of Sciences, Humanities and Education, Technical University of Liberec in 2013.

References

Golka P, Vild J (eds) (2003) 50 let TUL 1953–2003. Technická univerzita v Liberci, Liberec
Habánová A (2010) Plánovaná novostavba galerijní budovy v Liberci v první polovině 20. století. Technická univerzita v Liberci, Liberec. Fontes Nissae, Prameny Nisy XI/2010 pp 77–111
Hájek V, Mrázová Z (2012) Hodnocení etnicity ZSJ Husova a ZSJ Králův Háj – research review. Technická univerzita v Liberci, Liberec – unpublished
Halík P et al (2009) Slavné vily Libereckého kraje. Foibos, Liberec
Karpaš R (2004) Kniha o Liberci. Dialog, Liberec
Lozoviuková K (2010) Němci v Liberci po druhé světové válce. In: Německy mluvící obyvatelstvo v Československu po roce 1945. Matice Moravská, Brno
Melanová M (1996) Liberecká výstava 1906. Kalendář Liberecka, Liberec
Nižnanský B, Popková K (2012) Využití starých plánů při studiu současného území Liberce – Husova ulice v prostoru a čase. Technická univerzita v Liberci, Liberec
Purš J (1960) Průmyslová revoluce v českých zemích. Státní nakladatelství technické literatury, Praha
Řeháček M (2010) Procházka do Lidových sadů. Pavel Akrman pro Českou besedu v Liberci, Liberec
Rous I (2009) Liberecké podzemí. Kruh autorů Liberecka, Liberec
Rous I (2011) Romové z táborů postavili Králův Háj. Liberecký deník, roč. 13, č. 30, 5. 2. 2011. Liberec
Technik S (1967) Města severních Čech. Severočeské nakladatelství, Liberec
Technik S (1992) Liberecké domy hovoří I. Liberec, Úřad města Liberce, odbor hospodářského rozvoje města ve spolupráci s Českou besedou v Liberci, Liberec
Technik S (1993) Liberecké domy hovoří II. Liberec, Úřad města Liberce, odbor hospodářského rozvoje města ve spolupráci s Českou besedou v Liberci, Liberec
Technik S (1995) Liberecké domy hovoří III. Liberec, Úřad města Liberce, odbor hospodářského rozvoje města ve spolupráci s Českou besedou v Liberci, Liberec
Technik S (2001) Výstava Budujeme osvobozené kraje v Liberci 1946. Česká beseda, Liberec
Umlauf V et al (2011) Střed Liberce v proměnách století. Technická univerzita v Liberci, Liberec
Zeman J (2011) Liberec, urbanismus, architektura, industriál, pomníky, objekty, památky. Knihy 555, Liberec

Alteration of Graphical Part of Local Flood Management Plans Using Geoinformatics: Scottish and Czech Approaches

Petr Vahalík and Přemysl Janata

Introduction

Floods in the Czech Republic are the most relevant extremes found in nature, for other natural catastrophes more destructive in extent, such as large earthquakes, do not occur here, and strong winds do not achieve so devastating effects known from other parts of the world. Floods come as a result of complex influence caused by many factors, namely meteorological (e.g. rainfall), physiogeographical (e.g. properties of surfaces), and anthropogenic ones (changes in the landuse). These factors have a substantial impact on the time and space variability in the flood occurrence rates, on their strength, size and extent of effects. The catastrophic flood in July of 1997 on the upper and central Morava and Odra rivers is still in living memory, not having a parallel in the Czech Republic in the twentieth century as regards the culmination flow rates, the length of duration, the extent of the area affected, the loss in human lives (52 people) and material damage (62.6 billion Czech crowns). Another catastrophic flood followed in eastern Bohemia in July of 1998, with 6 human lives marred and material damage of around 2 billion Czech crowns. The following flood calamity in Bohemia, on the Vltava, the Labe and other rivers in August of 2002 led to the death of 19 people and the material damage amounted to approximately 73 billion Czech crowns (Brázdil et al. 2005).

The flood management strategy includes: (1) pre-flood measures, (2) flood forecasting, and (3) post-flood measures. Pre-flood measures provide the natural, institutional and social infrastructure for the viable management of a flood risk. Strategies for preventive flood management include: technical measures to control and manage the flood (small dams and projects on the retention and stabilization of river banks); regulating measures for land use and the planning of settlements; and

P. Vahalík (✉) • P. Janata
Department of Geoinformation Technologies, Mendel University Brno, Zemědělská 3, 613 00 Brno, Czech Republic
e-mail: petr.vahalik@mendelu.cz; premysl.janata@mendelu.cz

© Springer International Publishing Switzerland 2015
J. Brus et al. (eds.), *Modern Trends in Cartography*, Lecture Notes in
Geoinformation and Cartography, DOI 10.1007/978-3-319-07926-4_34

economic measures for the regulation, promotion and communication (Water Directors 2003).

The synthesis of data and the mapping of the relationships between natural hazard phenomena and the elements at risk require the use of tools such as geographic information systems (GIS). Natural hazards are multi-dimensional phenomena which have a spatial component; thus, the GIS is very well suited in such applications (Zerger 2002). The GIS has evolved in three broad application domains: the first is the use of the GIS as an information database; the second is its use as an analytical tool, a means of specifying logical and mathematical relationships among layers of maps to yield new derivative maps; and the third is its use as a decision support system. To meet a specific objective in a decision support system, several criteria need to be evaluated by means of procedures referred to as "multi-criteria evaluation" or simply "modelling". The integration of multi-criteria analysis (MCA) with a grid-based GIS in order to analyse flood vulnerable areas provides more flexible and accurate decisions (Eastman et al. 1995).

In order to determine flood vulnerability, it is important to identify the sources of danger of a sudden river flood and to locate the highly hazardous areas. The classical approach for the delineation of flood-prone areas with different hazard levels is based on the application of hydrological-hydraulic modelling. However, for ungauged basins, and in areas where expensive and time-consuming hydrological-hydraulic simulations are not possible, the use of an effective tool to delineate the flood-prone areas is essential (Manfreda et al. 2008). Moreover, the identification of areas at high risk of flooding is very important in small hydrological basins where the flood warning time is short (Nektarios et al. 2011). Basic documents dealing with floods at local level are Local flood management plans.

Local flood management plans are primary documents which instruct responsible people in their duties and competence before, during and after a flood event at a local level. Local flood management plans in the Czech Republic (hereinafter CR) are made in accordance with Section 71 of Act No. 254/2001 Coll. on Water and related regulations, in keeping with TNV 75 2931 standard (creation of flood management plans). Under flood management plans for purposes of the act in question we understand documents containing procedures of providing early and reliable information on the flood development, on the possibilities of influencing the drainage mode, on the organization and preparation of safety work. The documents further contain methods of ensuring early activation of flood management authorities, methods of providing flood forecasting service, security guards and protection of buildings, preparation and organization of rescue and salvage operations, ways of providing fundamental functions in buildings and areas damaged by the flood, as well as the authoritative limits for announcing the flood stages.

The flood management plan as a fundamental document of flood protection serves to coordinate activities in the area during a flood event. The document is a summary of organizational and technical measures needed to avert or alleviate the flood damage as far as human lives, the people's and the society's properties, or the environment are concerned.

The content of a flood management plan is divided into:

a) the factual part which includes the data needed to safeguard specific buildings, municipal districts, basins or other regional units against flood events, as well as the authoritative limits for announcing the flood stages;
b) the organizational part which includes lists of names, addresses and ways of connection between the participants in flood protection, tasks for individual participants in flood protection, as well as organization of the warning service and security guards;
c) the graphical part which usually includes maps or plans with the drawings of flood areas, evacuation routes and concentration points, water gauges, and information points.

This article deals with the implementation of geoinformatic technologies mainly into the third, i.e. graphical part of local flood management plans. One of the constituent objectives is the analysis and comparison of organizational and project structures of the flood management documentation at local levels between Scotland and the CR, especially in the extent in which geoinformatics is implemented in those documents. In this respect, Scotland was chosen on purpose since there are flood events caused not only by rainfall (with the average annual amount 1,570 mm/year—MetOffice almost doubling that in the CR 820 mm/year—Czech hydrometeorological institute), but also by high tide, but mostly by the combination of these two factors. Historically, this northern country has been struggling with flood events for long time, and therefore it offers a suitable model for those purposes not only in the frame of Central European countries. The aim of the comparison in the above described sense is the investigation of practical usability of the GIS analyses for the increase of information value contained in the local flood management plans in the CR.

The main objective of this article is to suggest implementation possibilities of the GIS analyses into local flood management plans within the area of the CR. Individual analyses and models are tested on the case study of a flood management plan in the village of Hostašovice.

Methods and Data

The comparison of approaches and GIS implementations in the creation of local flood management plans within the area of the CR was performed in the form of consultation with employees of Czech Hydrometeorological Institute, Odra River Basin Management, Morava River Basin Management and private subjects dealing with the execution of local flood management plans. In Scotland, the following institutions were visited for this purpose: Inverness College—University of Highlands and Islands, Scottish Natural Heritage, The Highland Council and Scottish Environmental Protection Agency.

The case study of a flood management plan in the village of Hostašovice, on the southern fringe of Moravian-Silesian region, was performed in order to put forward a proposal of specific GIS implementations in the local flood management plans.

Characterization of the Locality in Question

The village of Hostašovice is located near the western edge of the Natural Forest Region 39—Silesian-Moravian Foothills. The Natural Forest Region (hereinafter NFR) of Silesian-Moravian Foothills is an autonomic forest area bounded by the state border with Poland in the northeast, neighboring with the Beskids NFR (40) in the south and with the NFR of Low Jesenik (29) in the west. The area lies approximately between $17°55'$ and $18°47'$ of east longitude and between $49°32'$ and $49°57'$ of north latitude (Holuša et al. 2000).

The highest situated point in the village cadastre is the peak of Pohorelec (551 m a.s.l.), median absolute altitude of the village cadastre is 380 m a.s.l., the lowest situated point is 347 m a.s.l.

Following the Quitt classification, a major part of the village is located in a moderately warm area (MW 2 and MW 9), and the southern part is located in a cold area (C 7). The average annual air temperature ranges between 7 and 7.5 °C, with January as the coolest month and July as the warmest month. The average annual rainfall is above 850 mm, with the biggest amount of rainfall in July, and the lowest one in February. As far as winds are concerned, the N–S directions prevail (Demek et al. 2006).

The main European sea-drainage divide runs through the local cadastre of Hostašovice, with 43.6 % of the cadastre land belonging to the Black Sea drainage area, and 56.4 % belonging to the Baltic Sea drainage area. Hydrologically, the area is divided between the Morava basin and the Odra basin. The whole cadastre area is drained off by four streams, namely by the stream called Straník and by Zrzávka, belonging to the Odra basin, and by the streams called Jasenice and Krhova being a part of the Morava basin. With respect to the flood risks in the area, two inflows from the right into the stream Straník are relevant: streams called Hrázkový potok and Bílý potok which flow through the village build-up area and they go through pipes at several points. The hydrological ID numbers of these two streams are ID 10208697 and 10217822 respectively.

Creation of the Local Flood Management Plan in the Willage Hostašovice

The factual and the organizational parts of the plan respect and meet all the requirements of Section 71 of Act No. 254/2001 Sb., on Water. Its graphical part is supplemented with specific geodesic measurements and the follow-up GIS analyses. The geodatabase of DIBAVOD (Digital Database for Water Management managed by T. G. Masaryk Water Research Institute) was the source of the hydrological data. The altitude data come from the ZABAGED data set (Fundamental Base of Geographical Data managed by Czech Office for Surveying, Mapping and Cadastre) data set. On the basis of this group of data, hydrological attributes of the basins in question were determined, such as length, average descent, shape of the basin—expressed by the basin shape coefficient α (De Blij et al. 2006), and type of streaming—expressed by the Froude number Fr (Chanson 2004a, b). The software used is ArcGIS 10.2 (ESRI) and its extensions HEC-RAS and HEC-GeoRAS.

The longitudinal slope, the transversal profiles and parameters of all transversal objects, such as pipes and bridges, were measured for the two streams flowing through the village build-up area with the geodesic GPS using RTK (real time kinematics) of the CZEPOS (the network of permanent GNSS stations in CR) network. This group of data was used to determine the flow rate limits for all transversal objects, two of which were used for creating the water gauges of category C. Flood limits, i.e. state of alert (1), state of emergency (2), state of danger (3), were determined for these water gauges using hydrological calculation.

The drainage from the village build-up area is limited by the maximum flow rates of the piped parts of the streams. This regards 13 piped parts, namely 8 piped parts on the stream Bílý potok and 5 piped parts on the stream Hrázkový potok. All of these passages were geodesically measured together with the longitudinal vertical alignment of the stream bottoms for purposes of hydrological calculation of the maximum flow rates. In front of the entries to specific pipes, the transversal stream profile was measured to determine the Hmax level for the flow rate limits of individual pipes using the Chézy equation and the continuity equation (Landau and Lifshitz 1987). The Hmax value thus gives the maximum depth of the flow before going through the pipe when the limit flow rate (for the pipe given) goes through the streambed. The Manning coefficient of friction 0.033 was used for fully fortified streambeds, the value of 0.04 was taken for unfortified streambeds (Chanson 2004a, b).

With respect to the currently running campaign of laser surface scanning of the Czech Republic, the altitude data for Moravia are not yet available with sufficient precision to model the flood zones in the village. This analysis will be carried out as soon as the 4G data subbase is processed by the Czech Office for Surveying, Mapping and Cadastre—spring of 2014 is the expected date. Selection of the buildings under threat, for the time being, is based on historical experience only.

The western part of the build-up area in the village of Hostašovice is situated in a terrain depression which is surrounded from three sides by slopes with permanent grass growth and forests. In case of burst rainfall events in combination with the saturation of the soil horizon, accumulation of surface water occurs on the slopes in a great measure, which is a threat to some properties during flood events in the village. In case of improper management on the surrounding slopes (planting wide-rowed crops, plowing in lines parallel with the descent, and the like), the soil wash-down is a threat to the village. The intensity of the surface water accumulation and of the soil wash-down is dependent on the slope steepness. The points of contact of the build-up area with the surface wash-down from the adjacent land plots were identified using the GIS analyses of the accumulated surface wash-down (the Spatial Analyst extension).

Results

Structure of Administration and Creation of the Flood Management Documentation in the CR

In the CR, the national flood geodatabase POVIS (Flood Information System) is under creation, which is administered by the Ministry of the Environment (see Fig. 1). This information system serves as a support platform for the communicative, coordinating and decision-making activities at all organizational levels which are bound by the law to manage a flood event. It provides common data structures for the information needed in the flood management plans in the frame of the Czech National Flood Protection Strategy. Local flood management plans, municipal flood management plans with the extended scope, and regional flood management plans

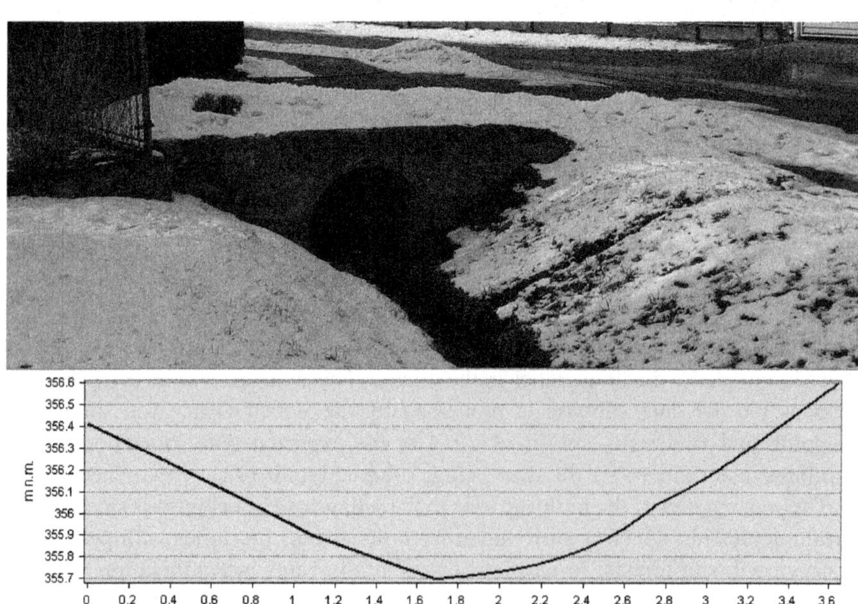

Fig. 1 Transversal object, measured transversal profile and results of hydrological computation (flow limits)

are joined together in the frame of this strategy. Geoinformatical analyses in the POVIS geodatabase are implemented in the flood zones modeling along the main watercourses—they are provided by the T.G. Masaryk Water Research Institute (VÚVTGM). Among other implemented modules of GIS in the national information system there are climate analyses, weather forecasts and flood forecasting service provided by the Czech Hydrometeorological Institute (ČHMÚ).

At lower levels, i.e. in the regions and municipalities with the extended powers, regional and municipal flood management plans are created. They develop the administrative strategy for basins of specific watercourses. Geoinformatics is implemented there by their connection to POVIS.

At the lowest organizational level, local flood management plans are designed. They are produced by physical persons or by design offices. These plans must be in accordance with the superordinate municipal and regional flood management plans—and river administrators, i.e. basin administrators or the state enterprise Forests of the Czech Republic S.E., need to approve them. The extent of implementation of geoinformatics in these plans is quite variable, depending on the specific designer or design office. The GIS analyses are often restricted just to the connection with the POVIS system where one can find just the flood models of the main watercourses—the analyses of the flood zones of smaller streams, and thus the analyses of the buildings under threat, are completely missing. These flood zones are then selected on the basis of historical experience only. Analyses of surface accumulated water wash-down are missing at all.

Structure of Administration and Creation of Flood Management Documentation in Scotland

The process of changes to and authorization of the nation-wide concept of the flood protection strategy in Scotland is carried out by the state-owned organization of Scottish Natural Heritage. This organization handles conflicts at the national level between the flood protection issues and the environment. The strategy of creation of the flood management documentation at all levels is regulated by the Scottish Parliament in the Flood Risk Management (Scotland) Act 2009. This act gives not only the standards and concept-related issues for the creation of flood management documentation, but also the time frame and organizational structure for the preparation of the project documentation at all levels. The national flood protection geodatabase, containing the analytical outputs for all the watercourses flowing through build-up areas, all buildings under threat, all historical flood events, the information on the detour routes or evacuation centers, should be finished and available to the public by 2016.

The creation of the local flood management plans of all the populated areas with potential flood risk is guaranteed by the state organization Scottish Environmental Protection Agency (SEPA), with 24 branch offices throughout all Scotland, and by the Highland Council. The local flood management plans are developed by the individual SEPA branch offices for the region given. The Highland Council designs this type of documentation for the Highlands area (the northwest of Scotland). Both of these organizations keep the same standards, structures and project issues for the flood management plans created. Geoinformatics is implemented in the flood zones modeling for all the watercourses in touch with the build-up area. The models are supported with the lidar data created for this purpose for all populated areas of Scotland. These analyses support the selection of the buildings under threat, not only from the floods caused by rainfall, but also from the floods caused by tidal waves.

GIS Implementation in the Case Study of a Local Flood Management Plan

Hydrological attributes of the two streams flowing through the build-up area of the village were determined using geodesic enquiry and the GIS analyses.

The Stream Hrázkový potok:
Basin area up to the closing profile (until the place where the stream leaves the cadastre): 107.89 ha
  Stream length from the spring to the closing profile: 2,041 m
  Average descent of the stream: 3.1 %
  Basin shape: protracted ($\alpha = 0.143$)
  Type of streaming: torrential (using the Froude number where $Fr = 1.12$)

The Stream Bílý potok:
Basin area up to the closing profile (until the place where the stream leaves the cadastre): 129.14 ha
  Stream length from the spring to the closing profile: 2,223 m
  Average descent of the stream: 4.47 %
  Basin shape: protracted ($\alpha = 0.166$)
  Type of streaming: torrential (using the Froude number where $Fr = 1.19$)

The flow rate limits of all the transversal objects which influence the drainage conditions in the village build-up were further determined (13 piped parts of the streams, and 9 bridges or footbridges). Two of these objects were marked as the water gauges of category C and the flood limits for these gauges were determined on the basis of the above mentioned hydrological computation (see Fig. 1).

The analysis of the accumulated surface wash-down also gives relevant pieces of information in identifying the build-up parts of the village which are under threat of the surface water wash-down from the surrounding plots. This analysis identified four parts in potential danger. The trustworthiness of the analysis is verified by the accounts from the last flood event in the village in 2009 (see Fig. 2). The maps of the buildings under threat from both the flooding of the streams and from the wash-

Fig. 2 Accumulated surface wash-down analyses (*up*), places (contact of analyzed surface wash down and build-up area) shown in the map after the flood event in 2009 (*down*)

down of the adjacent land plots form a part of the graphical outputs. Lastly, the maps of evacuation points and detour routes are also available in the graphical outputs.

Conclusion

This article compares the organizational and project structures of the flood management documentation at local levels between Scotland and the CR, especially the extent of implementation of geoinformatics in the documentation. Both countries approach this issue at local levels in conceptionally different ways. In the CR, the local flood management plans are created by private subjects and the extent of GIS implementation is considerably variable. All the plans do respect the legal and industrial standards for the creation of such documentation, nevertheless, specific ways of implementation of geoinformatics into these documents is not clearly delineated. The POVIS system plays the role of the standard system in the CR. However, this system incorporates the local plans just as the textual and informational sources. The graphical outputs of the local plans would be necessary to get standardized in the way enabling for them to be included in this national geodatabase which

(continued)

so far functions in the flood events management at regional levels (but not at local levels).

The Scottish side works out all the local plans in the frame of the state organizations which keep a unified standard and the extent of GIS implementation for all the localities processed in the country. The resulting geodatabase is thus usable for the management of flood events at all levels, regional as well as local ones. The disadvantage of this approach for the Czech side is the non-existence of a state organization similar to SEPA in size which would ensure the implementation of the local project documentation at a sufficient technological level. This work must therefore be carried out by private subjects.

In the frame of this article, specific GIS analyses were proposed for the change of the design of the local flood management plans, as a part of the case study into the flood management plan of the village of Hostašovice. The specific suggestions for the use of the GIS in the local flood management plans are the following: Determining the flow rate limits for the transversal objects in the streams on the basis of geodesic enquiry and hydrological computation. Similar analysis is used in the exact determination of the flood limits (state of alert, state of emergency, state of danger) in the two proposed water gauges which were made in the village for this purpose. Also, another analysis which can increase the qualitative level of the local flood management plans is the flood zones modeling and the subsequent selection of the properties under threat. And the last suggested GIS implementation into the local flood management plans is the analysis of the accumulated surface wash-down for the identification of the parts of the build-up or the village area which are under threat by the flood from the adjacent land plots. All the above mentioned GIS implementations can qualitatively increase the information level of the respective project documentation without any dramatic increase in the expenses on the project execution.

References

Brázdil R, Dobrovolný P et al (2005) Historické a současné povodně v České republice, 1st edn. Masaryk University in Brno, Praha, Czech Hydrometeorological Institute in Prague, 370 pp. Historie počasí a podnebí v českých zemích, vol. 7. ISBN: 80-210-3864-0

Chanson H (2004a) Hydraulics of open channel flow: an introduction, 2nd edn. Butterworth–Heinemann, Oxford, 650 pp. ISBN: 0-7506-5978-5

Chanson H (2004b) The hydraulics of open channel flow, 2nd edn. Butterworth–Heinemann, Oxford, 630 pp. ISBN: 978 0 7506 5978 9

De Blij H, Harm J, Peter O (2006) Physical geography. Realms, regions and concepts, 12th edn, 560 pp. ISBN: 13 978-0-471-71786-7

Demek J, Mackovčin P et al (2006) Zeměpisný lexikon ČR. Hory a nížiny. 2. ed., Agentura ochrany přírody a krajiny ČR, Brno, ISBN: 80-86064-99-9

Eastman JR, Weigen J, Kyem PA, Toledano J (1995) Raster procedures for multi-criteria/multi objective decisions. Photogramm Eng Remote Sens 61:539–547

Holuša J et al (2000) Oblastní plán rozvoje lesa. Přírodní lesní oblast 40. Moravskoslezské Beskydy, Ústav hospodářské úpravy lesa, branch office Frýdek Místek

Landau LD, Lifshitz EM (1987) Fluid mechanics. Course of theoretical physics, 2nd edn. Pergamon. ISBN: 0-7506-2767-0

Manfreda S, Sole A, Fiorentino M (2008) Can the basin morphology alone provide an insight into flood-plain delineation? WIT Trans Ecol Environ 118:47–56

Nektarios N, Kourgialas G, Karatzas P (2011) Flood management and a GIS modelling method to assess flood-hazard areas – a case study. Hydrol Sci J 56(2)

Quitt E (1971) Klimatické oblasti Československa. Studia Geographica no. 16. Geografický ústav ČSAV, Brno

Water Directors (2003) Core group on flood protection of the water directors (Europe): best practices on flood prevention, protection and mitigation. European initiative on flood prevention, 25.9.2003

Zerger A (2002) Examining GIS decision utility for natural hazard risk modelling. Environ Model Software 17(3):287–294

Mapping a Local Budget Plan: Why and How?

Anja Reinermann-Matatko

Mapping a Local Budget Plan: Why?

Within the project "GeoBudgeting", the local budget plan of the city of Trier is georeferenced and mapped. The aim of the author is to offer the local decision makers (citizens, city councilors and employees of the local authority) decision support by pushing forward the spatial dimension which is by now implicitly hidden in the local budget plan.

In the first part, the study investigates the decision criteria of local decision makers within the process of establishing a local budget plan. As the author knows from her own experience as city councilor, decisions of local politicians are often influenced by various factors, and far from being able to be measured objectively. In this sense, Andrews (2007, p. 161) concludes that "Policy decisions should be rational but sometimes they are not". Special attention in this study is put on the existence of spatial decision criteria in stakeholders' reasoning.

Establishing a local budget plan is the most important decision of a city council. Every project a city wants to realize in the forthcoming years has to be integrated within the local budget plan. Many communities in Germany are short of revenues and therefore have to restrict themselves when taking decisions about investments. During the process of establishing a local budget plan, many discussions take place between the involved actor groups. Since around 15 years, in many cities worldwide not only city councilors and employees of the local authority take part in these discussion processes, but also "normal" citizens: cities offer them the possibility to discuss the local budget plan within the process of establishing a participatory budget plan; which is, in most cases, strongly linked to the whole budget plan process.

A. Reinermann-Matatko (✉)
Universitaet Trier, Kartographie, 54286 Trier, Germany
e-mail: matatko@uni-trier.de

© Springer International Publishing Switzerland 2015
J. Brus et al. (eds.), *Modern Trends in Cartography*, Lecture Notes in
Geoinformation and Cartography, DOI 10.1007/978-3-319-07926-4_35

But the local budget plan is not only the most important, but also one of the most complex decisions in local politics. The complexity of information leads to the fact that city councilors themselves call it incomprehensible. By now, the information of the local budget plan is only presented by the means of tables. The author suggests to use maps as visualization tool for the local budget plan, as maps can summarize the content of a local budget plan and, at the same time explicitly show its spatial impact. One of the hypothesis of this study, generated from the idea that the main interest of local stakeholders should be to control the spatial development of their city, is that the spatial impact is a relevant decision criterion for local city councilors.

Also Wermker and Heil (2003, p. 257), dealing with social inequality in cities, see the necessity to find spatial organization schemes for local budgets. The authors argue that the administrative reform in Germany's local authorities has to be complemented by a spatial dimension, in order to be able to achieve positive developments in poor neighbourhoods. Already in the 1980s, economic scientists deal with the spatial efficiency of public finances (Junkernheinrich and Klemmer 1985). They analysed the labour market areas in North Rhine-Westphalia. Having a look at the local scale, approaches of mapping budgets exist in the participatory budget Berlin-Lichtenberg (Bezirksamt Lichtenberg von Berlin 2006). There, general acceptance of the relevance of the spatial dimension to budget decisions led to first examples of a spatial repartition of the local budget, presented by the means of pie charts. But a completely georeferenced local budget is still missing.

Within the process of establishing a local budget plan, there are several steps which follow the same scheme every year, as they result from administrative rules. The major makes a draft of a local budget plan and presents it to the city council. Then, the city councilors as well as the citizens discuss the draft and try to make changes. Finally, the city council has to decide about the local budget plan, and the approving authority has to agree to the content. All of these steps can be combined with cartographic action fields, as described by Bollmann (2001a). Especially, two of the mentioned action fields can be of relevance within the process of establishing a local budget plan: communication (e.g., when the major presents the content of his draft) and data exploration (e.g., when the city councilors and citizens want to see the spatial impact of certain changes in the draft). Within the research project, both action fields have been taken into account more detailed; finally, they have been divided in even smaller parts of cartographic activity (see Bollmann 2001b). Namely, these are the common cognitive tasks when regarding a map: searching patterns, comparing values of different areas, searching for specific values, and so on.

In order to study the use of maps within the process of establishing the local budget plan, the investments contained in a local budget plan first have to be georeferenced. The aim of the study is then to isolate influence factors of map use within the cartographic action fields.

Georeferencing a Local Budget Plan

The structure of a local budget plans actually contains different kind of plans which in sum, build the local budget plan. The most important parts of the local budget plan which are relevant for georeferencing are the cash flow budget and the profit and loss budget. The first includes all investments (e.g. building new roads, building new houses). The latter one includes all services of the local authority (e.g. pupils' transport, financial support for associations, maintenance of streets and buildings).

When georeferencing the local budget, the question "where is the spatial impact?" has to be answered for every single expenditure. The author constructed a theoretical framework for classifying the spatial impacts. It consists of five groups:

- impact of expenditure can be seen in the living environment, e.g. when building new schools or new streets;
- expenditure goes to an object located at a certain address, but the beneficiaries are not the people living in the surroundings, but the people using the services offered within or through this object; therefore, not the address of the object has to be georeferenced, but the addresses of the beneficiaries; e.g., visitors of a museum or the theatre, members of a sports association;
- the expenditure concerns a line; e.g., maintenance of a certain street;
- the expenditure concerns an area, e.g. the state forest;
- the expenditure has no spatial impact, e.g. annual interests.

For demonstrating the procedure of georeferencing a local budget plan, the local budget plan 2009 from the city of Trier has been used as case study. Similar results of georeferencing can be expected in other cities using the same kind of budget plan structure (which is, actually, the most used one in Germany).

In total, 412 positions of the local budget plan of the case study contain expenditures for investments. Around 20 % of theses investments can not be georeferenced. Around 20 % can be georeferenced as lines or polygons. The rest, around 60 %, can be georeferenced as points: either the investment's impact can be seen directly at the points' coordinates; or, the impact cannot be perceived in the built environment, but users of the establishment which is supported by the investment are the beneficiaries. The latter kind of expenditures is highly linked to data security issues, which could be solved within the research project by signing a data security contract between the author and the local administration. The data were only presented in aggregated units, so no link between particulars and the expenditures can be made.

Especially in the last category, there is a big lack between theoretically possible georeferencing and data availability: from 132 expenditures with indirect spatial impact, only 17 can really be georeferenced by using address data. 94 expenditures can be georeferenced by using the establishment's location instead of the users' addresses. For 21 expenditures, no data is available. The other three categories are easier to georeference.

Mapping a Local Budget Plan

For the eye-tracking study, part of the georeferenced data was used to create different map types (choropleth maps, graduated symbol maps, pie and line chart maps) and variations of legend content (numerical and textual). Table 1 shows the different types of combination of map types and legend content used in this study.

All color schemes used in the study for choropleth maps or pie chart maps have been selected with the Colorbrewer (2013). The maps used in the eye-tracking study can be seen in Fig. 1.

Methods

Within this study, two experimental parts were combined: an online survey and an eye-tracking study. For the online survey, people have been recruited by three ways: local politicians have been contacted by email. All members of the actual city council and the leaders of each city district were included. Second, employees of the local administration who are dealing with financial data have been invited by email. The number of employees working with the local budget is limited, and they can be isolated by checking the organisation chart of the local administration. The third group were the citizens: in Trier, there is a participatory budget process and most of the citizens taking part in this process are registered for a mailing list. This mailing list has been used to sent the invitations to the online survey to the citizens. Within the online survey, the respondents created a personal code (they were asked to create a combination of first letters of their name, their birth place, and so on) which allows to link the online survey's results with the results of the eye-tracking study which was conducted six month later. For showing their interest in the eye-tracking study, people could leave their email-address at the end of the online survey.

Table 1 Variations of map types and legend types used in the eye-tracking study

	Map type 1	Map type 2	Legend	Color scheme	Reference value
Map 1	Pie chart map	Mosaic map	Numerical	None	None
Map 2	Graduated symbol map	None	Numerical	None	None
Map 3	Choropleth map	None	Numerical	Unipolar	Inhabitants
Map 4	Choropleth map	None	Textual	Bipolar	Inhabitants
Map 5	Choropleth map	None	Numerical	Unipolar	Children 6–16 years
Map 6	Line chart map	Graduated symbol map	Numerical	Unipolar	Street meters

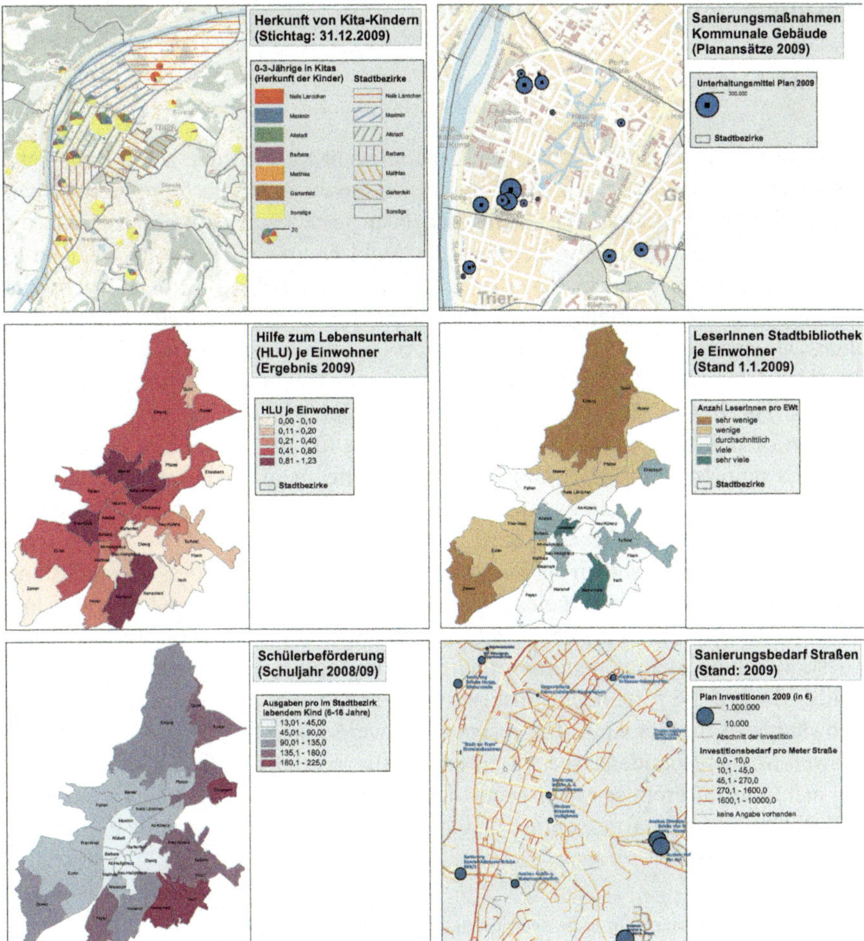

Fig. 1 Maps used in the eye-tracking study (*upper left* = map no. 1, *bottom right* = map no. 6)

The online survey (n = 313) was conducted in order to get information about the decision criteria of local decision makers, as well as an impression about their existing spatial knowledge and their subjective judgement about three different forms of data presentation (table, brochure with text and diagrams, and maps). The survey is established with an xml based tool.

The eye-tracking study was used to compare the gaze behavior of 32 decision makers within different map and legend types. The experiment was performed on two Windows workstations connected via VPN client, one running the Tobii eye-tracking software Clear View, the other one running the web-based questionnaire with integrated maps including logfile registering. The questionnaire was displayed on a 17-in. flat screen at a 1024×768 screen resolution. Eye movements were recorded with a Tobii x50 eye tracker, at a 60 Hz sampling rate.

Results

Online Survey

Concerning the decision criteria which are relevant for the three actor groups, the respondents were asked which of the items shown in Table 2 influences their decision making process. The items had been tested and complemented in the pretest of the online survey. Respondents could choose all items they agreed to and also add additional decision criteria. As Table 2 shows, it was found that spatial thinking plays an important role for a considerable amount of decision makers. The items which explicitly mention "city districts" get the ranks three to five. There are hardly any significant differences in the answer behavior between the three actor groups. The items "the opinion of political parties which I belong to" and "the opinion of political parties which I don't belong to" (which means, the political enemy's opinion) only get 14.6 and 4.4 % of agreement. Seen from the author's perspective as a city councilwomen, the non-responses have been highly influenced by what respondents believe is socially accepted.

Respondents have shown big interest in using maps in the process of establishing a local budget plan. Nevertheless, their spatial knowledge seems to be at a low level. Within the online survey, the respondents should solve answer different questions about spatial patterns of phenomena within different political action fields. Most of the respondents just estimated the answer instead of having a deep knowledge of the domains. The respondents were asked to give reasons for their answer. Some respondents seem to be especially "spatially addicted"; their reasoning mentions different spatial aspects of living quality in a city. Others seem more addicted to non-spatial reasoning, they have been taken into account e.g. financial aspects.

Table 2 Decision criteria of local decision makers in the local budget process

	Percent of respondents (%)
Creating new situations through investments	67.7
Experts' opinion	66.0
The city district for which I think the investment is useful	62.2
The city district where I live	44.2
My personal relationship to a city district	25.5
The opinion of friends and family	18.0
Preserving city districts in the current state	17.0
The opinion of political parties which I belong to	14.6
Other items	10.6
The opinion of political parties which I don't belong to	4.4

Eye-Tracking Study

In eye-tracking literature, the question whether the fixation duration can be interpreted in a way that longer fixation durations mean more intense information processing is not answered yet. Holmqvist et al. (2011), Holŝánová (2008) and Irwin (2004) discuss different aspects of the parameter fixation duration. For the interpretation of the gaze data, the participants in this study were also asked to give reasons for their answering behaviour. The interpretation of all results is a mix of qualitative impressions formulated by the respondents themselves and objective quantitative gaze data.

Comparing the average fixation duration in the case study with the correctness of answers (see Fig. 2), no significant relationship can be found. Respondents with a high amount of correct answers can have a small average fixation duration just as well as a high one.

Within the eye-tracking study, decision makers were asked to execute basic spatial operations (e.g., comparing districts, finding min and max values, getting an idea of spatial repartition of a phenomenon) with the help of the local budget maps. Within the pie chart map (which can be seen in Fig. 1 on the upper left side), respondents were asked to search for the maximum value of investment. They were able to manage this task in a very effective way within the presented map; the interpretation of the gaze maps show that most of the respondents just compared the two biggest dots in the map. It was not necessary to have intense fixations at the

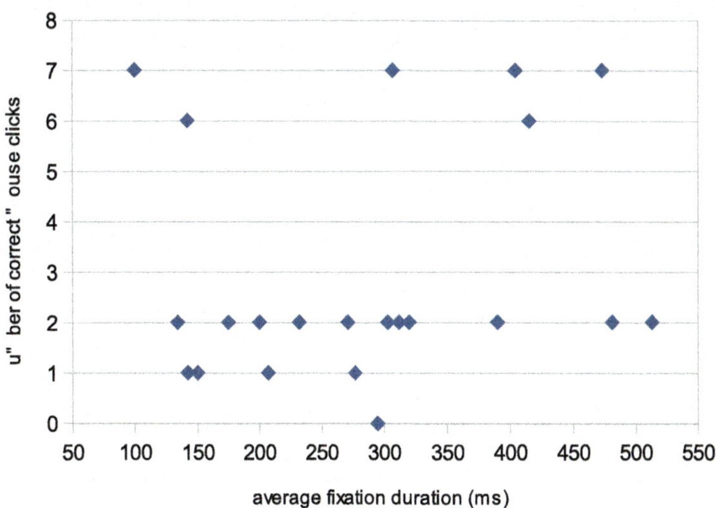

Fig. 2 Relationship between average fixation duration and the correctness of answers

Table 3 Relationship between map complexity, average fixation duration and fixation count for the six maps of the case study

	Complexity hierarchy (1 = highest)	Average fixation duration map	Average fixation duration legend	Fixation count map	Fixation count legend
Map 6	1	273	246	338	109
Map 1	2	286	236	103	75
Map 3	3	286	263	108	34
Map 4	4	280	260	68	28
Map 5	5	235	246	217	55
Map 2	6	279	220	48	27

legend of this map. The choropleth map showing the costs of public pupils' transport (which can be seen in Fig. 1 on the bottom, left side) shows a different eye movement pattern. Respondents were using the legend, and they were looking especially at the areas of the map which had values higher than they expected. This is especially true for the northern part of the city, where they have several schools, but pupils prefer to leave their district and go to school in the city centre or in the neighbour cities instead.

The gaze parameters show that the maps with the most complex information content get the most fixations, but the average fixation duration is not higher than at the maps containing less complex information content (Table 3). It appears that the respondents interest in the presented content influences the average fixation duration instead. The differences between map types are less important than the differences between maps showing interesting content and maps showing less interesting content seen from the decision makers perspective.

Conclusion

The study, especially the online survey in the first part, could prove the relevance of the spatial dimension for the three actor groups within the process of establishing a local budget. All three actor groups see the advantages of using maps as decision support tool. The respondents gave many hints about map content and map design in order to create maps which could support their decisions. Their hints were very different according to the peoples' background and working tasks. Also, they answer that a combination of maps and tables would suit their needs best. To sum it up, the participants favourite local budget plan decision support system would be one where

(continued)

people themselves can choose if they want to see additional textual information, change reference values, interactively add or remove items and get the sums calculated, Further research is necessary to define the role of eye-tracking parameters in a real world cartographic action context; also, the study did not contain measures of the emotional state of the respondents. As the results have shown, this could be helpful when interpreting the results.

References

Andrews CJ (2007) Rationality in policy decision making. In: Fischer F, Miller GJ, Sidney MS (eds) Handbook of public policy analysis. Theory, politics and methods. CRC, Hoboken, pp 161–172

Bezirksamt Lichtenberg von Berlin (2006) Alles, was zählt. http://www.buergerhaushalt.org/wp-content/plugins/wp-publications-archive/openfile.php?action=openfile=2. Accessed 11 Jan 2012

Bollmann J (2001a) kartographische handlungsfelder. In: Bollmann J, Koch WG (eds) Lexikon der Kartographie und Geomatik, vol 2. Karto-Z, Spektrum, pp 23–24

Bollmann J (2001b) kartographische information. In: Bollmann J, Koch WG (eds) Lexikon der Kartographie und Geomatik, vol 2. Karto-Z, Spektrum, pp 25–26

Colorbrewer (2013) Colorbrewer. http://colorbrewer2.org/. Accessed 13 Dec 2013

Holmqvist K, Nyström M, Andersson R, Dewhurst R, Jarodzka H, van de Weijer J (2011) Eye tracking. A comprehensive guide to methods and measures. Oxford University Press, Oxford

Holŝánová J (2008) Discourse, vision and cognition, human cognitive processing, vol 23. John Benjamins Publishing Company

Irwin DE (2004) Fixation location and fixation duration as indices of cognitive processing. In: Henderson JM, Ferreira F (eds) The interface of language, vision and action. Eye movements and the visual world. Psychology Press, New York, pp 105–133

Junkernheinrich M, Klemmer P (1985) Regionalisierung öffentlicher Finanzströme. Eine Analyse zur Zahlungsinzidenz raumwirksamer Mittel in Nordrhein-Westfalen., Beiträge der Akademie für Raumforschung und Landesplanung, vol 89. Vincentz

Wermker K, Heil C (2003) Vom modellprojekt zur regelstruktur. Informationen zur Raumentwicklung 3(4):253–257

Historical Data Processing, Modelling, Reconstruction, Analysis and Visualization of Historical Landscape in the Region of North-West Bohemia

Jan Pacina and Jiří Cajthaml

Introduction

The region of North-West Bohemia (the Czech Republic) has been a victim of very intensive human activity for many decades. Since the beginning of the twentieth century, the landscape has changed from an agriculture region to a heavy industrialized part of the country. The Most brown coal basin contains a huge amount of brown coal located not very deep under the Earth's surface. From the 1930s, the brown coal in this region has been mined using open-pit technology and currently the open-pit mines cover a total area of 3,800 ha. Open-pit mining is a relatively cheap method of coal mining, but has a destructive effect on the surrounding environment. The landscape structure has completely changed—including land-use, hydrological network displacement and settlement extinction. Many towns and villages have disappeared during the last 60 years of brown coal mining. Other types of heavy industries (chemical, power) are closely connected to the mining industry having other impacts on society and the surrounding environment. The other major human activity which influences the landscape in this region are the huge water dams constructed in the Most basin or in the neighboring Ore Mountains. Water dams have caused settlement and landscape change in their surroundings as drinking water dams require specific regulations and treatment. The Ore Mountains were densely inhabited in the past but with respect to the political situations after the Second World War, German inhabitants moved to Germany which resulted in settlement extinction as well.

J. Pacina (✉)
Faculty of Environment, Jan Evangelista Purkyně University, Ústí nad Labem, Czech Republic
e-mail: jan.pacina@ujep.cz

J. Cajthaml
Faculty of Civil Engineering, Czech Technical University in Prague, Prague, Czech Republic

© Springer International Publishing Switzerland 2015

J. Brus et al. (eds.), *Modern Trends in Cartography*, Lecture Notes in Geoinformation and Cartography, DOI 10.1007/978-3-319-07926-4_36

All of these factors have lead to a significant change in the landscape structure (including the georelief), hydrological network and settlement during the past 80 years. Our aim is to collect data, such as old maps, aerial photographs, land-use representation, digital terrain model reconstructions, 3D town reconstructions, old postcards and photographs, history of the places and much more into a comprehensive information system offering all this information to researchers and the public.

In this paper we focus on the newer methods of data collection, processing, analysis and visualization for the purposes of the proposed information system.

The Area of Interest

This project is focused on the whole region of Ústí nad Labem. With respect to limited space within this article, we will introduce the methods used mainly on site 2. The areas 1–6 were processed in detail and the results are available on our server http://mapserver.ujep.cz.[1] A short description of the sites of interest:

- Site 1—The water dam Přísečnice flooded the former town Přísečnice in 1974. The surrounding settlement had to be destroyed due to hygienic regulations. Analysis results presented at (Rekonstrukce zaniklé obce Přísečnice (Reconstruction of vanished town Přísečnice)).
- Site 2—Water dam Nechranice flooded villages in the Ohře valley. The neighboring open-pit mine Nástup Tušimice provides brown coal to the surrounding power plants.
- Site 3—Open-pit mine ČSA is located at the foot of the Ore Mountains. Huge landslides have endangered the surroundings including the castle Jezeří. More about the site in Pacina et al. (2012).
- Site 4—Open-pit mine Vršany has brown coal deposits till the year 2035. Detailed site description and analysis in Škvrna (2013).
- Site 5—Lake Most—a symbol of brown coal mining in this region. The royal town Most has been completely destroyed, brown coal mined and the empty pit turned into a lake. Detailed site description and analysis in Pacina et al. (2013).
- Site 6—The biggest[2] active open-pit mine in this region –Bílina mine and the neighboring large deposit Radovesice. More about the site in Pacina et al. (2011) and Weiss (2011).

---

[1] The links to the resulting web-mapping applications are presented in chapter 4.

[2] Our computations in Pacina et al. (2011) have shown that the material deployed from the mine in-between the years 1938 and 1995 would fill so many railway wagons that length of such a train would be twice the length of the equator.

Data and Methods

Different data and methods are used within this project to cover the variety of proposed tasks. The basis of the information system are well processed old maps and aerial photographs used for further processing and analysis. Other sources of information are used for better understanding of the selected localities—here data collection using standard surveying methods, Kite Aerial Photography (KAP) and Unmanned Aerial Vehicles (UAV) is performed.

Old Maps

Old maps of different scale and age were used for the purposes of this project. We have covered the area of interest with well distributed time-line of the old maps. The oldest processed map is Müller's map of Bohemia (ca. 1720) followed by the 1st, 2nd and 3rd Military survey of the Habsburg empire (1792, 1894, 1938). The Military survey presents the natural development of the area until the heavy industrialization.[3] This time-line is well supported by the Stabile cadaster maps (1842, 1:2,880) containing detailed information about the land-use.

The time period of heavy industrialization is covered by State maps and the State maps derived in the scale 1:5,000 (SMO5) from the years 1953, 1972 and 1981. The map contains two main compounds—hypsography (derived from a Base map 1:10,000) and planimetry (derived from cadastral maps). The hypsography has a 1 m contour interval[4] and thus is well suitable for georelief reconstruction analysis.

Processing (georeferencing) old maps requires an individual approach as each map has to be handled based on its characteristics. Different georeferencing methods were applied to obtain sufficient results for the purposes of our project.

The most problematic was the georeferencing of the 1st Military survey maps as these maps have no geodetic basements. A unique method introduced in Cajthaml (2012) based on polynomial transformation was used for georeferencing the 1st Military survey map sheets preserving the neighboring map sheet continuation.

Georeferencing of the other data sources was based on a large set of identical points (2nd and 3rd Military survey, Stabile cadaster maps) or on the knowledge of the map corner coordinates (SMO5 maps). All maps were georeferenced into the Czech national S-JTSK coordinate system (Georepository) and visualized by the ESRI file geodatabase mosaic dataset which is used to mask the map frame information producing a seamless map.

The processed maps are accessible as a web-mapping application using ArcGIS API for Flex on URL: http://mapserver.ujep.cz/Projekty/Zanikle_obce/.

---

[3] The third Military Survey already shows the beginnings of industrialization and open-pit mining.

[4] The SMO5 from the year 1953 have, in some areas, the contour interval 5–20 m, thus the results are not that accurate. But anyway this map is a unique source of information.

Aerial Photographs

Aerial photographs are a very reasonable data source as they contain a different type of information then old maps. In old aerial photographs there are recognizable different types of land-use, abandoned buildings and other interesting details. The methods of aerial imagery processing are very capable of producing a Digital Surface Model (DSM) based on the input images. Thanks to the location of our area of interest along the border with Germany and thanks to its industrial importance this region has been fully covered by an aerial imagery survey in the year 1938.[5] Different timelines of aerial imagery were tested for the purposes of this project (1938, 1946, 1953, 1987, 1995 and 2008). Finally only three time-lines will be processed for the whole area of interest:

- 1938—the first aerial imagery taken in the whole Czechoslovakia, representing the landscape almost untouched by open-cast mining.
- 1953—precise aerial imagery taken shortly after the Second World War showing the landscape partly affected by heavy industry and open-pit mining activity. This imagery is very well usable for the detection of abandoned houses and villages after the transfer of German citizens from the Czech borderland.
- 2008—imagery showing the up-to-date situation in the region. High quality imagery used mainly for the georelief reconstruction.

Aerial images are processed by the standard ways of photogrammetry using the Leica Photogrammetry Suite environment. The issues of handling old aerial imagery is introduced for example in Collier et al. (2001) and Weiss (2011).

Digital Terrain Modelling

Reconstruction of the original shape of the landscape is a very important part of this project. During the years of intensive open-pit mining the georelief has been transformed in a very dramatic way, which has affected not only the landscape structure but as well the hydrology network and surface water run-off (Bonta et al. 1997; Musilová 2011).

The georelief can be reconstructed from types of elevation data—in this project three data types were used—contour lines, aerial imagery and LIDAR.

- The oldest map with georelief represented by contour lines originates from 1938 (the 3rd Military Survey) when the landscape is almost untouched by mining activity. The consequent time lines (1953, 1972 and 1981) represent the

---

[5] The first test with aerial photogrammetry were within the Czechoslovakia performed (according to Pokusné měření metodou letecké stereofotogrammetrie v Čsl. republice) in 1932. The first photogrammetrical campaign covering a continuous area was performed in 1938.

georelief development. The result of georelief reconstruction is a Digital Terrain Model (DTM) representing the *base ground*—a surface without building, trees, etc.

- Aerial imagery is produced with DSM's from the processed images—a surface containing all the objects on the ground (houses, bridges, trees, etc.). DSM processing is faster than hand digitizing the contour lines, but the resulting model is affected by the quality of the input imagery.
- The latest Digital Terrain Model of the Czech Republic of the 4th generation (DMR 4G) data are used for representing the current state of the georelief—as they represents a picture of natural or human activity modified terrain surface in a digital form as heights of discrete points with X, Y, H coordinates in regular 5×5 m grid. These data were obtained using LIDAR technology (Digital Terrain Model of the Czech Republic of the 4th generation (DMR 4G) and ČÚZK).

The resulting DTMs from the year 1953 and 2012 are shown in Fig. 1, where the original state of the georelief and the current state is introduced—with a large water dam and open-pit mine. This data requires special handling methods and has been tested in Pacina et al. (2012) and Pacina et al. (2013).

Different types of analysis may be performed on the resulting surfaces—here we may perform different analysis to delineate the areas with major georelief changes and further apply volumetric analysis to compute the total volume of the added/removed material from the area. The results of the differential analysis of DTMs are presented in Fig. 1. The dark green represents removed material (open-pit mine) and the orange transported material (deposit, water dam). Detailed results and analysis possibilities are presented in Pacina et al. (2012) and Pacina et al. (2013).

The project results will be offered to the public not only by a web-mapping application but as well as an exhibition showing the results in reality. The reconstructed DTMs (the original and current state) are printed using the ZPrinter® 450 (a 3D printer) in tiles 20×25 cm, in the scale 1:50,000. The resulting tiles will seamlessly cover the Most basin in areas with georelief change. Printed tiles of our area of interest are presented in Fig. 1.

Field Data Collection

A field survey is another reliable data source used for gathering data for our information system. Several methods were used:

- Unmanned Aerial Vehicle (UAV)—so called *drone*, can be used for carrying different kinds of cameras (classic, multispectral, thermal). Results from UAV survey may be orthophotos, DSMs, object 3D models, etc.

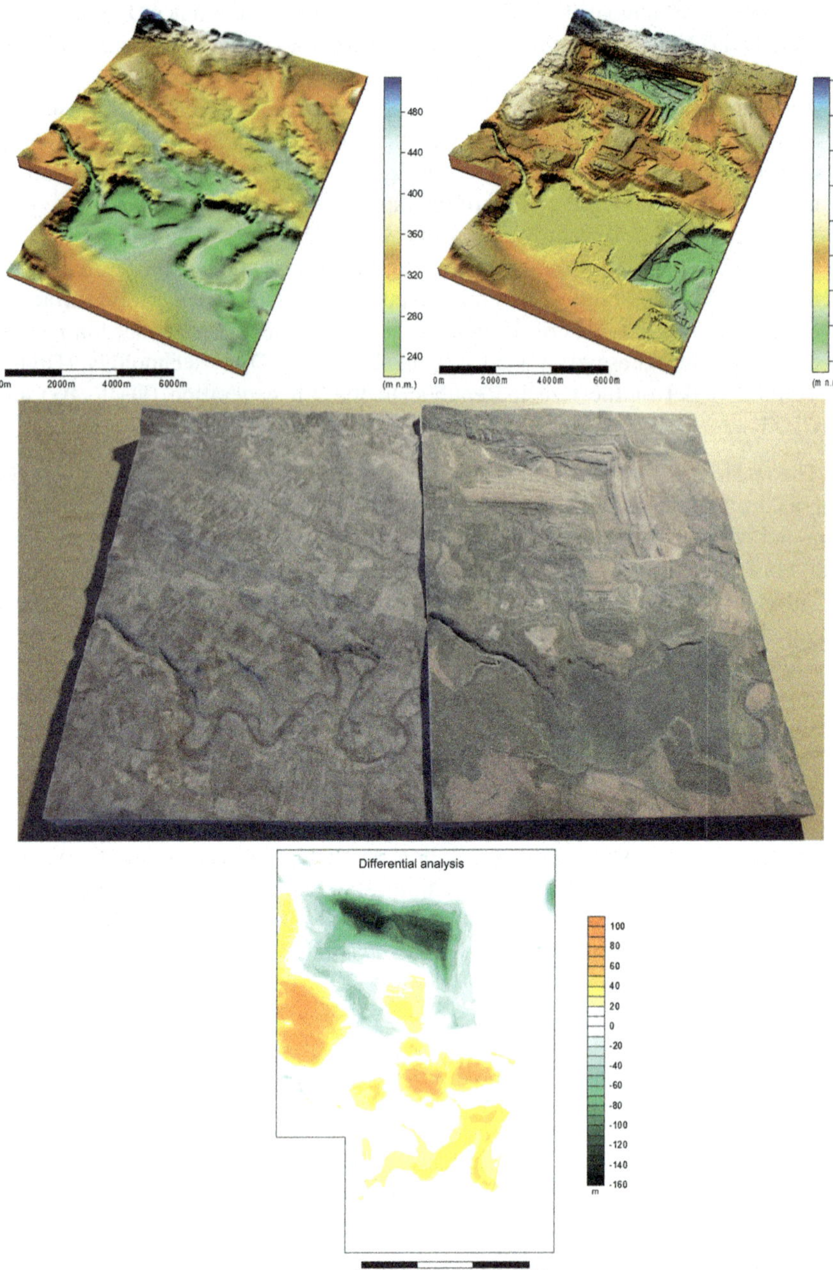

Fig. 1 Georelief reconstruction in the area of interest. Year 1953 (*left*) and 2012 (*right*), result printed on a 3D printer and the differential analysis result

- Kite Aerial Photography (KAP)—used for aerial photography (orthophoto, oblique photography) from ca 200 m altitude in areas where UAV is not suitable.[6]
- GPS survey—for identifying the remains/original locations of extinct settlement and measuring the ground control points for UAV and KAP.

UAV

UAV is a very popular and available method nowadays for collecting spatial data. A UAV can carry different types of sensors like cameras, multi/hyper spectral sensors and thermo sensors, LIDAR or precise GPS. The great advantage of a UAV is the low cost of the equipment compared to manned aircraft, flexibility and possible repeating of the campaigns in different time periods (Rango et al. 2006).

Within this project hexa and octo-copters (drone) are used for taking oblique photography and creating orthophotos of areas of interest. The images are taken using a calibrated camera and further processed using the Agisoft software (Verhoeven 2011).

KAP

KAP is a rather simple way of taking aerial images from ca 200 m altitude. The kite parameters are the following: one string kite Elliot Rhombus Mega Power Sled, size 300 × 170 cm, string length ca 250 m, wind acceptance 2–5 Bft. The kite can carry a Canon PowerShot D10 camera on a special mount (Fig. 2). The images are taken in time-lapse ca every 10 s. The kite has to be operated at least by two people (during take-off/landing, identifying the real camera position in the air). This method is cheaper in comparison with UAV, applicable by less skilled users but dependent on stabile wind. It has been proved that during good weather conditions more than 40 % images are usable. During worse weather conditions the amount of usable images decreases below 20 %. Suitable images are further processed to orthophotos or used for other land-study purposes (Brůna 2013).

KAP was used to study the landscape and for identifying settlement relics in abandoned villages located at the border with Germany. The first stage of this research was the preparation of old maps for use in a field GPS receiver. Based on the information contained in the old maps, the boundaries of the former settlement were identified and staked. The KAP survey was performed in early October as the vegetation was low and the weather conditions up in the mountains (ca. 900 m)

---

[6] Areas unsuitable for UAV may be considered as a high windy area or an area with legislative problems (borders with another country). KAP may be used as well when a person is not licensed to fly an UAV.

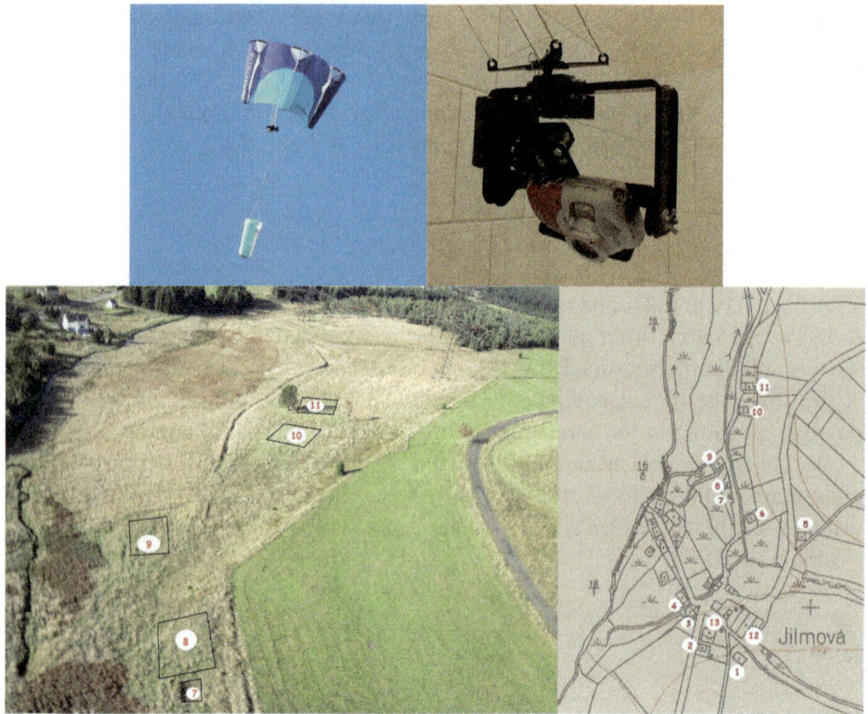

Fig. 2 KAP—the kite and camera mounting

were still suitable. The partial results from the locality Jilmová are presented in Fig. 2.

Information System Front-End

The main purpose of this project is to present a comprehensive information system (IS) offering all the project outputs in one place and in a user friendly way (the scheme of the resulting IS is presented in Fig. 3). The home page of our IS will offer to the user several main themes to choose from.

The main part is dedicated to the presentation of spatial data obtained and processed for the whole area of interest. All the data published uses the ESRI technology and are stored in a file geodatabase (ESRI) and published by ArcGIS for Server 10.2. The data may be accessed in two different ways—as a web mapping application and ArcGIS Rest Services.[7]

---

[7] Only data with a valid license are offered as ArcGIS Rest Services. OGC map services are considered to be used as well.

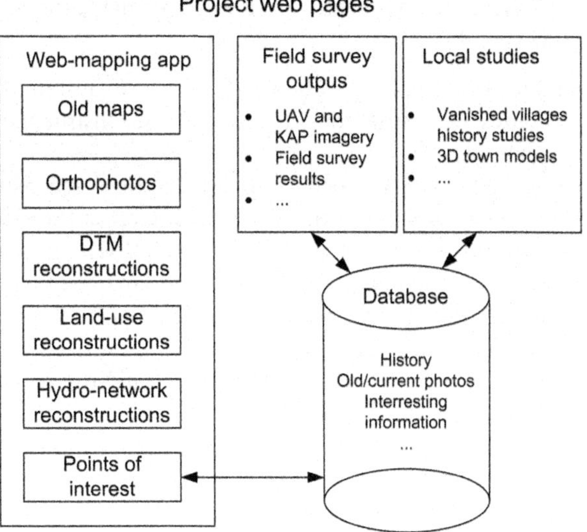

Fig. 3 The information system structure

The *basemap layers* (old maps, orthophotos) are published as tiled (cached) services allowing faster data viewing by the end users. The *operational layers* (vector layers, DTMs) are published as dynamic layers so they are easily transformed into another coordinate system within a desktop GIS. All the published data are protected, so the user is allowed to view, query or print the data, but not to save them to a hard drive. This allows the user to work with a variety of processed maps.

The web mapping application offers the user to view and visually analyze all the processed data. Different types of processed data are currently available for each of the sites of interest as different methodologies for data processing were tested. A comprehensive webpage is under preparation and thus there are currently several isolated web-applications covering all the sites of interest.

The applications are built using the ArcGIS API for Flex and are based on the ArcGIS Viewer for Flex. This technology is very flexible and presents the data in a very effective and feasible way.

Within the web mapping application the user can view the processed maps as the base layers, switch between them and change their transparency. This allows the user to visually compare the presented base maps. Other tools available are implemented for visually comparing the data—for example a swipe tool or a "magnifying glass". The user is allowed to export the currently viewed data to different graphic formats.

The user is offered different types of *operational layers* for each of the sites. Within these layers we can encounter hydrology network reconstructions, digital terrain models and their analysis, land-use interpretations, data obtained from local surveys (KAP, UAV) and other data extracted from old maps and aerial imagery.

A list of web-applications currently available for public containing the following processed data:

- Site 1—complete web-portal available at http://prisecnice.eu/ contains 3D houses reconstructions, town-center reconstruction and land-use development analysis.
- Site 2—georelief reconstruction, land-use reconstruction, hydrology network reconstruction: http://mapserver.ujep.cz/Projekty/NAKI_mapy/Nechranice/ and http://mapserver.ujep.cz/Projekty/NAKI_mapy/Hydrorekonstrukce_CSA/
- Site 3—georelief reconstruction and land-use reconstruction: http://mapserver. ujep.cz/Projekty/NAKI_mapy/Jezeri/   and   http://mapserver.ujep.cz/Projekty/ NAKI_mapy/CSA/
- Site 4—georelief reconstruction and land-use reconstruction: http://mapserver. ujep.cz/Projekty/NAKI_mapy/Vrsany/
- Site 5—detailed analysis of georelief development: http://mapserver.ujep.cz/ Projekty/NAKI_mapy/Jezero_Most/
- Site 6—experimental processing of old aerial imagery: http://mapserver.ujep.cz/ Projekty/NAKI_mapy/Bilinsko/
- More sites—hydrology network analysis for wider area: http://mapserver.ujep. cz/Projekty/NAKI_mapy/Hydrorekonstrukce_CSA/

Conclusions

In this article, new methods for processing, modelling, reconstruction, analysis and visualization of the vanished landscape are implemented. These methods for processing old maps and aerial photographs were tested in a wide range of areas (see Site 1 to Site 6). Each of the processed data sources required very special handling to obtain sufficient results like the right choice of georeferencing method, identical point selection or defined work-flow for processing old aerial imagery.

Modelling, reconstruction and analysis are dependent on the quality of the input data. Digital terrain modelling or digital surface model derivation from the processed aerial imagery is very sensitive to input errors. The resulting georelief reconstructions will be an important part of the information system when printed on a 3D printer, as well a part of an exhibition. A detailed study of the resulting DTMs and DSMs precision is currently under development. Other types of reconstruction (hydrology network, or land-use for example) will be a part of the information system as well.

New methods of vanished landscape visualization were tested within this project—aerial imagery taken from a KAP (Kite Aerial Photography) or a UAV (Unmanned Aerial Vehicle) are bringing a very new point of view on all localities. These methods allow us to obtain orthophotos or oblique

(continued)

photographs of an arbitrary scale (in comparison to traditional photogrammetry), repeated observations, and with the use of a specialized software create orthophotos, DSM's and 3D models objects. All processed data are presented to the public by web mapping applications accessible at http://mapserver.ujep.cz allowing the user to browse and visually analyse the data. When finished, all data will be accessible in one comprehensive application and by ArcGIS Rest Services.

Acknowledgment This work was supported by the Czech Ministry of Culture through the NAKI programme "Landscape Reconstruction and Vanished Municipalities Database for Preserving the Cultural Heritage in the Region of Ústí nad Labem" no. DF12P01OVV043.

References

Bonta JV, Amerman CR, Harlukowicz TJ, Dick WA (1997) Impact of coalsurface mining on three Ohio watersheds: surface water hydrology. J Am Water Res Assoc 33:907–917

Brůna V (2013) Využití KAP (Kite aerial photography) při dokumentaci výzkumu v Abúsíru (The KAP usage for research documentation in Abusir). Pražské egyptologické studie, XI/2013. ISSN 1214–3189

Cajthaml J (2012) Analýza starých map v digitálním prostředí na příkladu Müllerových map Čech a Moravy (Analysis of old maps in the digital environment presented on the Müller's maps of Bohemia and Moravia). Česká technika – nakladatelství ČVUT, Praha. ISBN: 978-80-01-05010-1

Collier P, Inkpen R, Fontana D (2001) The use of historical photography in environmental studies. Cybergeo Eur J Geogr [En ligne], Environment, Nature, Paysage, document 184, mis en ligne le 19 janvier 2001, consulté le 02 décembre 2013. http://cybergeo.revues.org/4019. doi:10.4000/cybergeo.4019

Digital Terrain Model of the Czech Republic of the 4th generation (DMR 4G), *ČÚZK* [online] [cit. 2013-11-03]. http://geoportal.cuzk.cz

ESRI: ArcGIS Desktop 10 Help [online] [cit. 2013-11-30]. http://help.arcgis.com/en/arcgisdesktop/10.0/help/index.htm

Georepository: Geographic 2D CRS used in Europe –Czechoslovakia [online] [cit. 2013-20-11]. http://georepository.com/crs_4156/S-JTSK.html

Musilová L (2011) Effects of mining operations on hydrological regime in the river basins of the Smědá and the Lužická Nisa. Diploma thesis, Masaryk University (Cz)

Pacina J, Novák K, Weiss L (2011) 3D modelling as a tool for landscape restoration and analysis. Environmental software systems: frameworks of Environments. Springer, Heidelberg, s. 16

Pacina J, Novák K, Popelka J (2012) Georelief transfiguration in areas affected by open-cast mining. Trans GIS. ISSN 1361–1682

Pacina J, Novák K, Brůna V, Popelka J (2013) Georelief development analysis in the lake most surroundings. AD ALTA: J Interdis Res 3(2):44–47

Pokusné měření metodou letecké stereofotogrammetrie v Čsl. republice. *Zeměměřičský věstník* [online]. 1932, roč. 20, č. 8, s. 132 [cit. 2012-03-26]. http://archivnimapy.cuzk.cz/zemvest/texty/Rok1932.pdf

Rango A, Laliberte AS, Steele C, Herrick JE, Bestelmeyer B, Schmugge T, Roanhorse A, Jenkins V (2006) Using unmanned aerial vehicles for rangelands: current applications and future potentials. Environ Pract 8:159–168

Rekonstrukce zaniklé obce Přísečnice (Reconstruction of vanished town Přísečnice) [online] [cit. 2013-11-30]. http://prisecnice.eu/

Škvrna D (2013) Povrchová těžba jako významný faktor změny krajiny. Případová studie – velkolom Vršany (Surface mining as an important factor changes the landscape. Case study – quarry Vršany). Bachelor thesis, J. E. Purkyně University in Ústí nad Labem (Cz)

Verhoeven G (2011) Taking computer vision aloft – archaeological three-dimensional reconstructions from aerial photographs with photoscan. Archaeol Prospect 18(1):67–73

Weiss L (2011) *Časoprostorová analýza změn reliéfu Bílinska vlivem důlní činnosti* (Spatiotemporal analysis of relief changes in Bílina region owing to mining activity). Diploma thesis, J. E. Purkyně University in Ústí nad Labem (Cz)

Bayesian Mapping of Medical Data

Lukáš Marek, Vít Pászto, and Pavel Tuček

Introduction

One of the main aims of the spatial analysis of health and medical datasets is to provide additional information to specialized medical research. These analyses can be used for disease mapping; searching for places with a higher intensity and probability of disease occurrence; or an influence assessment of selected natural or artificial phenomena. While the location and space are obviously crucial components within spatial analysis, they represent also a very problematic part in the geospatial research of health data. Medical or health data are usually provided as point data sets—with a direct location, an indirect location, or aggregated frequency data that can be based on administrative units or grids. The latter case prevails and immediately evokes visualization in the form of choropleth maps. However, medical data, as well as other demographical data, require careful manipulation, mapping and subsequent depiction. Suitably selected methods enable proper analysis of these data and the identification of irregularities and deviations of the phenomena in the area of interest. The structure of medical data usually needs to be standardized before comparing different regions. The standardization is based on the population of the administrative unit and/or the age structure of the population. Another procedure is the comparison of cases recorded in the area over time, with the expected number of cases predicted from the known structure of patients in another region or known probability distribution of the disease. The description of the latter is well established in Bayesian statistics, which derives posterior probability as a consequence of prior probability and a probability model for the data to be observed. Geosciences and geomedicine use Bayesian theory for the smoothing of data—to help depict the real spatial pattern and its changeability. Bayesian principles together with the spatial neighbourhood and statistical models are also

L. Marek (✉) • V. Pászto • P. Tuček
Department of Geoinformatics, Palacky University in Olomouc, Olomouc, Czech Republic
e-mail: lukas.marek@upol.cz

© Springer International Publishing Switzerland 2015

J. Brus et al. (eds.), *Modern Trends in Cartography*, Lecture Notes in
Geoinformation and Cartography, DOI 10.1007/978-3-319-07926-4_37

successfully used for the identification of spatial clusters with a significantly higher/ lower risk of incidence of a certain disease. This contribution aims to present the usage of selected Bayesian methods for analyses of infectious disease data of the Czech Republic. The data were provided by the National Institute of Public Health. Both previously mentioned purposes of usage are presented in the paper—the smoothing of prevalence rates and relative risk rates as well as the identification of spatial clusters.

Disease Mapping and Basic Terms in Epidemiology

Geography has a fundamental interest in variations over space and in how things are put together at different scales of thinking about them, which is why they have been used for hundreds of years to describe spatial variations in health and disease (Meade and Emch 2010). One of the first steps in any epidemiological analysis is to visualize the spatial characteristics of a dataset, which enables an appreciation of any patterns that might be present, the identification of obvious errors, and the generation of hypotheses about factors that might influence the observed pattern (Pfeiffer et al. 2008). A comprehensive overview of disease mapping history, from the first usage in Italy in the seventeenth century, through Snow's famous map of cholera in London in the nineteenth century to contemporary mapping, is provided e.g. by Koch (2005). Visualization is also important when providing findings to a wider audience. One has to decide how detailed the description of the methodology should be based on whether the audience is more public or more scientific (Bell et al. 2006).

Spatial disease data are usually provided in two forms, (1) as data points that record every single case of a certain disease or (2) as aggregated data that summarize a group of individual cases into a single value representing the whole group or area, e.g. the total sum or another statistical characteristic. Another data classification can be found in Bailey (2001), where data are sorted with regards to their origin and purpose.

The previously mentioned data types also predefine the main types of visualizations. *Point maps* are the simplest way to visualize disease event information when the locations of events are known, as they provide a means of presenting the data in their 'raw' format, unmodified by any statistical analysis that might be applied to aid or enhance the interpretation (Pfeiffer et al. 2008). The prevailing form of aggregation is the summarizing of disease case frequency in a defined area, which can for example be an administrative unit, ZIP code, regular artificial grid, etc. (Marek et al. 2013). *Choropleth maps* of disease rates (incidence, prevalence, relative risk) then express these sums as a function of the population size in the aggregation area unit, which are subsequently coloured according to the rates. For a complex overview of other mapping techniques and methods, see Waller and Gotway (2004), Koch (2005) or Meade and Emch (2010).

Among the most common rates for disease mapping are the incidence, prevalence and standardized mortality (morbidity) ratio (SMR). *Prevalence* in epidemiology is the total number of cases of a given disease or disorder in a specified population at a specified time regardless of when the illness began (Earickson 2009). Thus, the prevalence of salmonella in the Czech Republic since 2010 includes the cumulative number of citizens suffering from the disease since that year. On the other hand, prevalence is often confused with *incidence*, which refers to only new cases diagnosed during a particular year. *SMR* is the ratio of the number of deaths (or cases) observed in the study group or population to the number that would be expected if the study population had the same specific rates as the standard population, multiplied by 100 and usually expressed as a percentage (Last and Abramson 2001). SMR also represents an estimate of the *relative risk* (Bivand et al. 2008). SMRs should not be directly compared as they are not based on the same standard population, but comparisons of SMRs between geographical areas will be misleading only if the age and sex structure of the populations are extremely disparate, which very rarely occurs in practice (Goldman and Brender 2000; Jarup 2004).

Bayesian Mapping and Smoothing

Presenting disease rates in area units as choropleth maps can inadvertently provide misleading information. This fact is well known mainly in the case of small-area studies that introduce an extra source of variability into the map because of random variation. Typically, sparsely populated areas with few (or zero) cases can generate extreme values of the SMR (and also prevalence), as the variance of the SMR is inversely related to the expected number of cases, and small populations have large variability in the estimated rates (Elliott and Wartenberg 2004), which is why risk estimates and other rates are rather unstable.

Bayesian methods provide a solution for this kind of bias. They use probability models to obtain smoothed estimates consisting of a compromise between the observed rate for each region and an estimate from a larger collection of cases and persons at risk (e.g., the rate observed over the entire study area or over a collection of neighbouring regions) (Waller and Gotway 2004). The basic principle of Bayesian methods is that uncertain data can be strengthened by combining them with prior information (Pfeiffer et al. 2008). In the case of empirical Bayes estimation of spatially-varying disease risk, the posterior risk can be estimated from a weighted combination of the local risk (also called the likelihood) and the risk in surrounding areas, the latter representing the prior information (Clayton and Bernardinelli 1996). Empirical Bayes calculations of disease risk come from the following formula (Bailey and Gatrell 1995):

$$\hat{\theta}_i = w_i r_i + (1 - w_i)\gamma_i \tag{1}$$

where θ_i is the empirical Bayes estimate for area i, w_i are the weights applied to the local and neighbourhood estimates, r_i is the local risk in area i and γ_i is the mean of the prior, and r_i is the local risk in area i.

$$r_i = \frac{y_i}{n_i} \tag{2}$$

where y_i is the number of cases and n_i the population at risk in area i. The weights, w_i, in (1) are estimated as:

$$w_i = \frac{\phi_i}{\left(\phi_i + \frac{y_i}{n_i}\right)} \tag{3}$$

where ϕ_i is the variance of the prior, γ_i is the mean of the prior, and n_i the population at risk in area i. The estimation is made by simplified posterior distributions through likelihood or integral approximations (Lawson et al. 2003). Estimates based on (1) tend to converge towards the global mean. The method is able to perform locally smoothed estimates by employing a local neighbourhood, which allows use of the local mean instead of the global mean. This adaptive smoothing then shrinks unstable risks toward the local mean risk, which means that risks in areas with more information are less smoothed than in areas that exhibit higher sampling variation (typically, those with a low number of cases) (Beale et al. 2008). A comprehensive overview of Bayesian techniques in disease mapping and also in clustering and other spatial analyses can be found in Gelfand et al. (2010), Lawson et al. (2003) or Waller (2005).

Identification of Spatial Clusters

During the study of disease spatial distribution, mainly in the case of aggregated data, it is often suitable to focus on the local variability of disease occurrence or relative risk rather than examine the study area as a whole. This procedure is usually denoted disease cluster detection. A general review of the methodology as well as the usage of spatial clustering methods and its Bayesian enhancements in literature can be found in e.g., Haining (1998, 2004), Lawson (2009) or Waller (2009) etc.

In geosciences, spatial clustering is often encapsulated as an analysis of spatial autocorrelation. Spatial autocorrelation is the correlation among values of a single variable, which is strictly attributable to their relatively close locations on a two-dimensional (2-D) surface, introducing a deviation from the independent observation assumption of classical statistics (Griffith and Arbia 2010). Tobler's first law of geography (Tobler 1979) encapsulates this situation, *"everything is*

related to everything else, but near things are more related than distant things".
Positive spatial autocorrelation refers to patterns where nearby or neighbouring
values are more alike; while negative spatial autocorrelation refers to patterns
where nearby or neighbouring values are dissimilar. One can distinguish two
main types of spatial autocorrelation, which are global and local. While global
clustering methods are used to assess whether clustering is apparent throughout the
study region, local methods of cluster detection define the locations and extent of
clusters (Pfeiffer et al. 2008). The null hypothesis for global clustering methods is
simply that no clustering exists (i.e. random spatial dispersion $\approx$ CSR). These
techniques are collectively denoted as Exploratory Spatial Data Analysis (ESDA)
and Local Indicators of Spatial Association (LISA), which are widespread in
geosciences and GIS software.

Probably the most frequently used method for both global and local analyses of
spatial autocorrelation is Moran's I statistics (together with e.g. Getis-Ord G and
Gearys's C statistics). Moran's I coefficient of autocorrelation is similar to
Pearson's correlation coefficient, and quantifies the similarity of an outcome variable among areas that are defined as spatially related (Moran 1950). Moran's I
statistic is given by:

$$I = \frac{n \sum_i \sum_j W_{ij} (Z_i - \overline{Z})(Z_j - \overline{Z})}{\left(\sum_i \sum_j W_{ij}\right) \sum_k (Z_k - \overline{Z})^2} \tag{4}$$

where Z_i could be the residuals $(O_i - E_i)$ or standardized mortality or morbidity
ratio (SMR) of an area, and W_{ij} is a measure of the closeness of areas i and j. A
weights matrix is used to define the spatial relationships so that regions close in
space are given greater weight when calculating the statistic than those that are
distant (Moran 1950). Local Moran's I is used for the mapping of either similar
(cluster) or dissimilar (outlier) disease frequency values around a given observation
in the space. A comprehensive explanation of the hypothesis and theory is provided
by Anselin (1995) and Scott and Janikas (2010).

The problem with variance instability for rates or proportions, which served as
the motivation for applying smoothing techniques to maps, may also affect the
inference for Moran's I test for spatial autocorrelation (Anselin 2003). The implementation of the adjustment procedure of Assuncao and Reis (1999), which uses a
variable transformation based on the Empirical Bayes principle, may be one of the
solutions. This yields a new variable that has been adjusted for the potentially
biasing effects of variance instability due to differences in the size of the underlying
population at risk (Anselin 2003).

Case Study

The case study in which the previously presented methods are applied deals with the spatial distribution of campylobacteriosis in the Czech Republic between the years 2008 and 2012. There were almost 100 thousand cases of the disease during that period. Using disease rates calculated for the municipal districts in the Czech Republic, we tried to identify areas that are possibly more vulnerable to the disease than their neighbourhood. The 5-year prevalence and the relative risk (SMR) were used as the main disease rates for this study. Subsequently, Local Moran's I was used as the statistic of the local clustering.

Campylobacteriosis is caused by bacteria called Campylobacter jejuni, which is found worldwide in the intestinal tracts of animals. The bacteria are spiral shaped and can cause disease in animals and humans. Most cases of campylobacteriosis are associated with handling or eating raw or undercooked poultry meat or fresh milk. Campylobacteriosis causes gastrointestinal symptoms, such as diarrhoea, cramping, abdominal pain, and fever in domestic animals and humans. Young animals and humans are the most severely affected (CFSPH 2013).

Data

The data set for this study was provided by The National Institute of Public Health of the Czech Republic. The database contains almost 100,000 records of campylobacteriosis occurrence in the period 2008–2012. Names, surnames, identity numbers and sometimes also the full addresses are not included as they are considered sensitive personal data. The data were firstly cleansed of inconsistencies and then the geocoding process was run. Furthermore, individual records were aggregated into municipal districts—administrative units—for clarity of the visualization (Marek et al. 2013). The problem of converting spatial phenomena between different areal or administrative units is well known as MAUP—Modifiable Area Unit Problem (Openshaw 1984). During the calculation of smoothed rates, the population data from the Population and Housing Census of the Czech Republic were used as the main basis for data standardization.

Choropleth Maps of Smoothed Prevalence and Relative Risk

Choropleth maps are probably the most common type of map used to display areal data. These maps use different colour and pattern combinations to depict different values of the attribute variable associated with each area, which is coloured according to the category in which its corresponding attribute value falls (Waller and Gotway 2004). Although choropleth maps do not show continuously distributed values, they often portray densities (Rushton 2003). Viewed in this way, one can consider them as a visual tool for the analysis of spatial distribution of the phenomenon.

All data were aggregated into municipal districts and counts were summarized for all the years. Then these counts were standardized using the population data from the Population and Housing Census of the Czech Republic. This step was processed to allow a comparison of different municipal districts even with dissimilar populations (number and age). Demographic characteristics of the population as a whole were used as the basis for the indirect standardization. Based on the standardized population and observed cases in administrative units, we were also able to calculate the expected number of cases in a municipal district. The standardization itself serves for data smoothing. Furthermore, global empirical Bayesian estimates of the number of cases based on binomial distribution is utilized, as well as local empirical Bayesian estimates of the number of cases based on the first order queen contiguity.

Statistical characteristics that summarize occurrence in the Czech Republic together with standardized and smoothed counts are shown in Table 1. The mean and standard deviation of data does not provide the best estimation of the actual state of the situation as the number of cases in a municipal district is usually closely related to the population and is far from the normal distribution of the data. Moreover, the disease has never been recorded in a significant number of districts. That is why the median and interquartile range describe data better and all the tables contain characteristics that were calculated for both the complete dataset and the reduced dataset, which contained only municipal districts with recorded disease cases.

The first figure (Fig. 1—top) depicts the 5-year prevalence of campylobacter in the population in municipal districts in the Czech Republic. It visualizes the number of cases per 1,000 inhabitants in the area without any standardization or smoothing. The map unintentionally emphasizes mainly areas with rather sparse populations where the disease was recorded (dark areas) on the one hand, and areas with a high number of people and recorded cases on the other. The next choropleth map (Fig. 1—middle part) presents the same original data but this time empirical Bayesian smoothing was applied. The smoothing was based on the global mean in the Czech Republic, which was used to smooth all the data. One can clearly notice that even those areas without any occurrence of the disease now belong to the category with lower prevalence. Also, more than 200 municipal districts that were firstly marked as the highest prevalence areas were dissolved into categories with lower prevalence rates. Finally, the third map (Fig. 1—bottom) shows the smoothed 5-year prevalence of campylobacter, which is obtained by the application of local

Table 1 Statistical characteristics of observed, standardized, globally smoothed and locally smoothed aggregated counts of cases of campylobacteriosis between 2008 and 2012

	Mean		Median		Standard deviation		Interquartile range	
	Complete	Reduced	Complete	Reduced	Complete	Reduced	Complete	Reduced
Observed	6.54	15.99	0.00	4.00	34.81	53.05	3.00	8.00
Standardized	6.54	14.39	1.23	4.11	26.16	39.40	3.14	7.63
EBB	6.60	15.84	0.33	3.69	34.72	52.97	2.12	8.60
LEBQ1	6.64	15.65	0.63	3.56	34.66	52.92	2.52	8.42

Complete stands for observations in all municipal districts of the Czech Republic while *Reduced* refers solely to municipal districts with recorded disease cases
Observed—recorded cases; *Standardized*—population standardized number of cases; *EBB*—global empirical Bayesian estimates of number of cases based on binomial distribution and population; *LEBQ1*—local empirical Bayesian estimates of number of cases based on the first order queen contiguity and population in defined neighbourhood

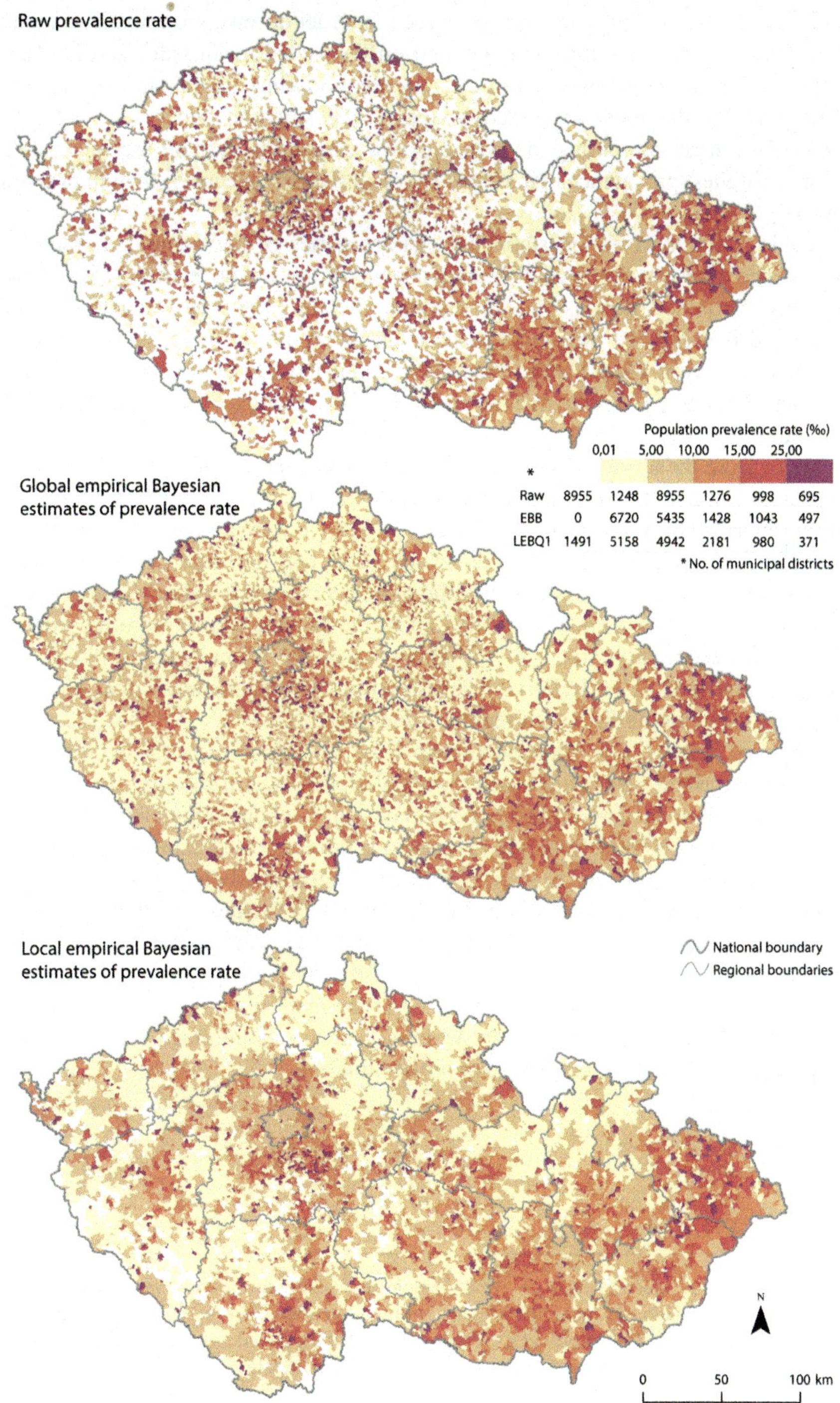

Fig. 1 5-year prevalence of campylobacter in population in municipal districts in the Czech Republic between 2008 and 2012, i.e. division of recorded cases and population (‰) ~ number of

empirical Bayesian smoothing based on the first order queen contiguity. This means that this time the smoothing is not based on the global mean value in the Czech Republic but on the local mean in directly neighbouring municipal districts. This map provides the most smoothed image of the situation and creates spatially continuous areas comparing the previous two maps. Again, there are fewer areas with the highest prevalence that are dissolved into other categories than in previous cases.

Table 2 summarizes the statistical characteristics of the computed smoothed prevalence rates. While the difference between the characteristics of raw and globally smoothed data is not so evident, the difference between the locally smoothed rate and the other two cases is easily visible in the case of mean and standard deviation. On the other hand, the median and IQR confirm the usability of the method, so any possible misinterpretation of the result is on a similar level as in the other two cases.

Apart from the observed number of cases in municipal districts, we also have the expected number of cases available because of the use of standardization and Bayesian smoothing. Thanks to these two values, we were able to find the relative risk (SMR) in administrative areas, which is in fact the ratio of the empirical (observed) number of cases and the theoretical (expected) value expressed as a percentage. This means that only one half of the expected cases were recorded in areas where the relative risk is 50 %, which is why these areas are 'healthier' than they are supposed to be. Relative risk also allows a comparison of areas based on riskiness (or vulnerability) with regard to the diseases.

The first relative risk choropleth map (Fig. 2—top) depicts the relative risk, when the expected values are based on population standardization in the administrative units without any application of Bayesian methods. One can see that the eastern part of the Czech Republic (Moravia) is more affected than the western and central part (Bohemia), and high-risk areas are located mainly in the north-east. The globally smoothed choropleth map (Fig. 2—middle) provides a completely different image of the situation with an extremely smoothed surface, where only a few districts deviate from the global average. Lastly, the locally smoothed values of relative risk were computed and visualized in the map (Fig. 2—bottom). The map shows that Bohemia is riskier than Moravia and Silesia although the prevalence maps tried to claim the completely opposite situation.

Fig. 1 (continued) infection cases per 1,000 people (*top*); Smoothed 5-year prevalence of campylobacter in population in municipal districts in the Czech Republic between 2008 and 2012, which is obtained by global empirical Bayesian estimates of the prevalence rate based on binomial distribution (*middle*); Smoothed 5-year prevalence of campylobacter in population in municipal districts in the Czech Republic between 2008 and 2012, which is obtained by local empirical Bayesian estimates of the prevalence rate based on the first order queen contiguity (*bottom*)

Table 2 Statistical characteristics of raw, globally smoothed and locally smoothed 5-year prevalence of campylobacter in the population (‰) between 2008 and 2012

	Mean		Median		Standard deviation		Interquartile range	
	Complete	Reduced	Complete	Reduced	Complete	Reduced	Complete	Reduced
Raw	7.08	17.33	0.00	9.69	41.33	63.26	7.76	10.16
EBB	8.00	13.26	5.63	9.71	13.80	20.26	5.83	8.89
LEBQ1	15.53	31.20	5.80	8.75	409.23	639.80	6.97	7.19

Complete stands for observations in all municipal districts of the Czech Republic while *Reduced* refers solely to municipal districts with recorded disease cases *Raw*—crude prevalence i.e. division of recorded cases and population (‰); *EBB*—global empirical Bayes estimates of the prevalence rate based on binomial distribution; *LEBQ1*—local empirical Bayes estimates of the prevalence rate based on the first order queen contiguity and rate in defined neighbourhood

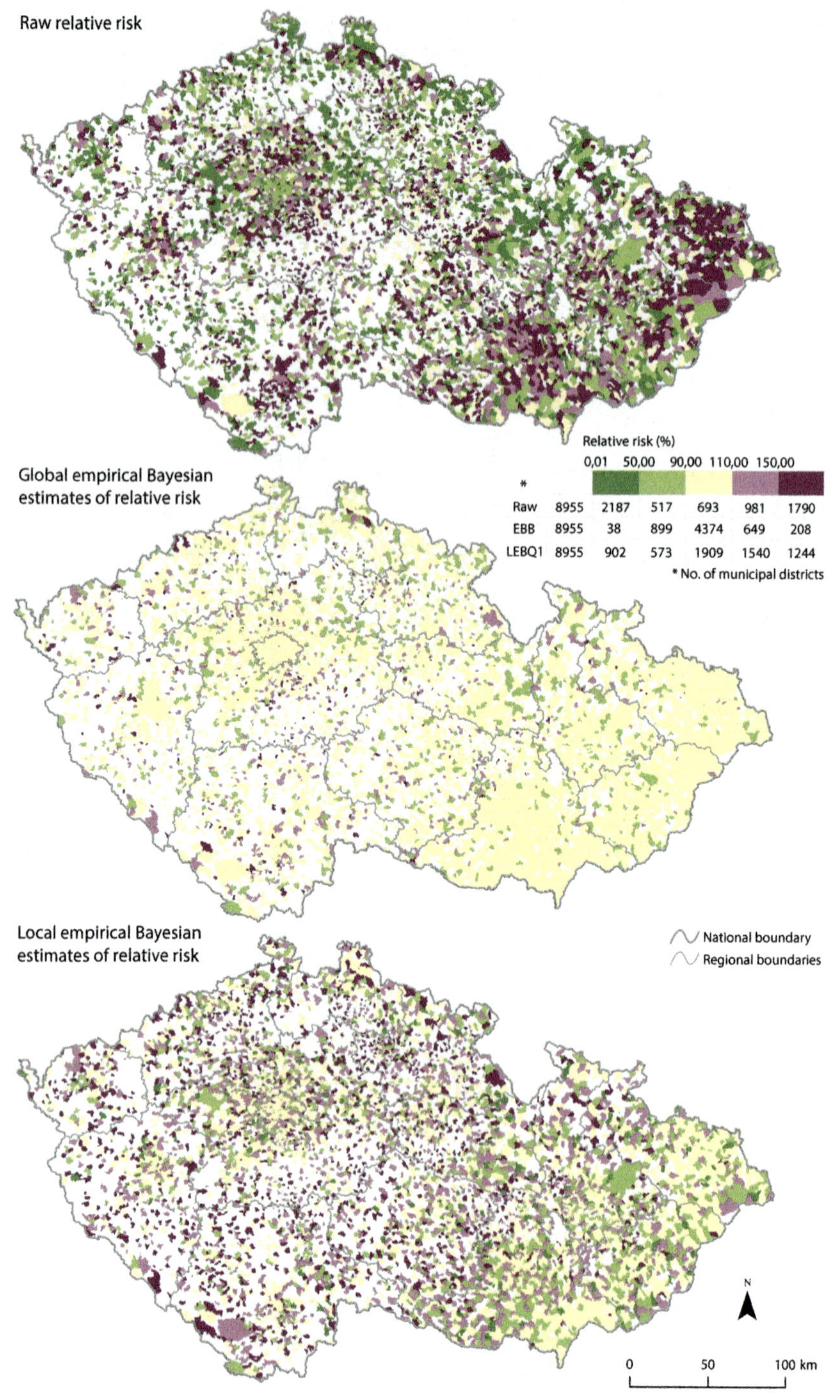

Fig. 2 Relative risk of campylobacteriosis in population in municipal districts in the Czech Republic between 2008 and 2012 (*top*); Relative risk of campylobacteriosis in population in municipal districts in the Czech Republic between 2008 and 2012, which is obtained by global

Table 3 again summarizes the statistical characteristics of the computed relative risk rates. Globally smoothed values evince mostly the lowest characteristics and also variability, which is clearly visible in the middle part of Fig. 2, while the highest statistical characteristics and variability is found in the case of the original relative risk. The advantage of relative risk smoothing is that areas without any recorded cases are preserved as no-risk areas. On the other hand, local differences between methods indicate only limited usability of smoothing in this particular study.

Identification of Clusters

Although it is possible to identify clusters of more affected or vulnerable areas from a choropleth map, it is usually suitable to describe these clusters and quantify them. Moran's I is probably the most widely used method for both global and local analyses of spatial autocorrelation, i.e. estimation of spatial clustering. We mentioned that the eastern part of the Czech Republic (with the core of the cluster in the north-east) is probably more affected than the western part. Upper part of Fig. 3 shows that this situation is in fact true, except for several areas in Bohemia (Prague, Pilsen, etc.). The red in the maps stands for clusters of high values (high number of cases), the blue stands for clusters of low values, and pink and light blue areas are outliers. However, the first computation of local Moran's I is based purely on the number of recorded cases so the analysis is distorted by the population density in all districts.

Lower part of Fig. 3 shows the results of a similar analysis but it is different from the previous one. One can see the changed structure of clusters and the greater number of occurring outliers. This analysis is also based on the recorded number of disease cases but is enhanced by the usage of population density and empirical Bayesian rate with permutations. North-eastern Moravia and part of Silesia is still involved in the cluster of high values but this time the cluster is significantly smaller. Other affected districts also cluster into rather small groups that do not cover large areas.

Fig. 2 (continued) empirical Bayesian estimates of the relative risk based on binomial distribution (*middle*); Relative risk of campylobacteriosis in population in municipal districts in the Czech Republic between 2008 and 2012, which is obtained by local empirical Bayesian estimates of the relative risk based on the first order queen contiguity (*bottom*)

Table 3 Statistical characteristics of relative risk, globally smoothed and locally smoothed relative risk (%) of population to the campylobacter infection between 2008 and 2012

	Mean		Median		Standard deviation		Interquartile range	
	Complete	Reduced	Complete	Reduced	Complete	Reduced	Complete	Reduced
RR	73.32	179.4	0.00	100.27	427.78	654.79	80.32	105.15
EBB	43.09	105.44	0.00	100.01	67.74	68.21	98.95	8.46
LEBQ1	51.52	126.06	0.00	106.25	79.78	78.59	100.06	46.51

Complete stands for observation in all municipal districts of the Czech Republic while *Reduced* refers solely to municipal districts with recorded disease cases *RR*—relative risk i.e. division of recorded cases and expected cases (%); *EBB*—global empirical Bayesian estimates of the relative risk based on binomial distribution; *LEBQ1*—local empirical Bayesian estimates of the relative risk based on the first order queen contiguity

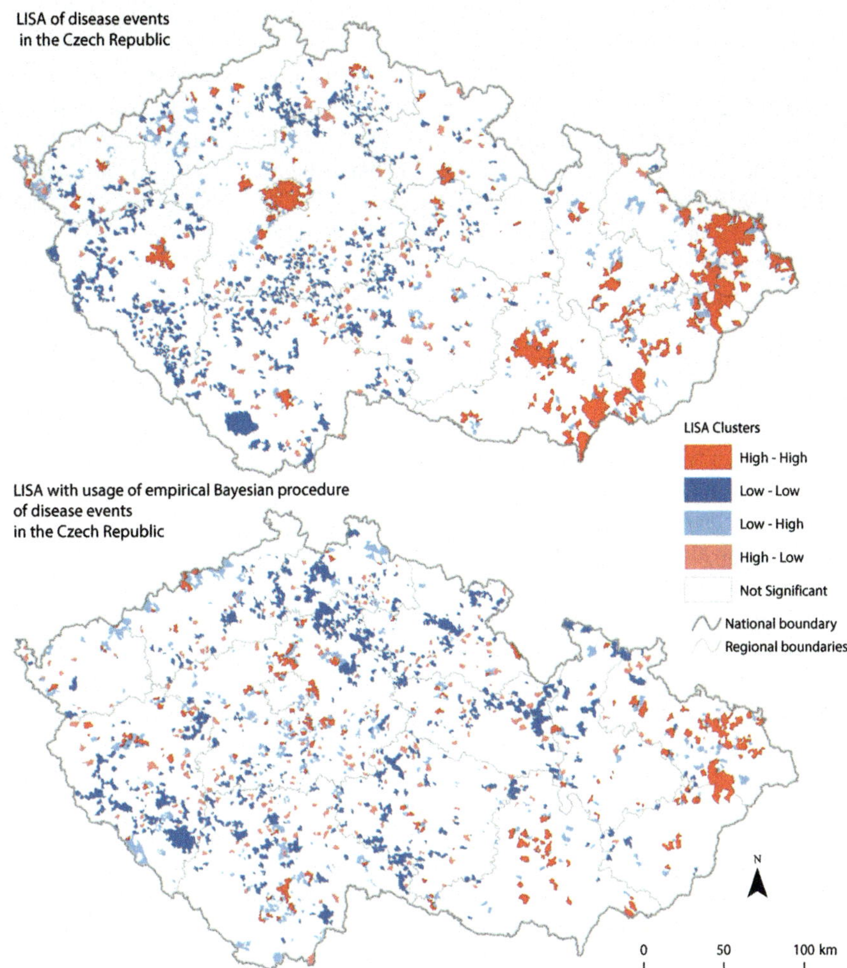

Fig. 3 Clustering of campylobacteriosis cases in municipal districts in the Czech Republic between 2008 and 2012, which resulted from computing local indicators of spatial association (Local Moran's I) based on the first order queen contiguity (*upper*); Clustering of campylobacteriosis cases in municipal districts in the Czech Republic between 2008 and 2012, which resulted from computing local indicators of spatial association (Local Moran's I) with usage of empirical Bayesian rate using two thousand permutations (*lower*)

Discussion and Conclusions

This paper presented the usage of (empirical) Bayesian models for the smoothing of disease rates—5-year prevalence and relative risk. These rates were based not only on the recorded number of cases of infection by campylobacteriosis in the Czech Republic, but also on the population density. The case study showed that smoothing is more suitable for expressing local differences in prevalence than in the case of relative risk.

Choropleth maps pointed out areas that might possibly create spatial clusters based on the similarity of disease occurrence. These clusters were later identified and described by Moran's I enhanced by empirical Bayesian rate with permutations. North-eastern Moravia proved to be most affected by campylobacteriosis.

One has to realize that empirical Bayesian procedures tend to shift values to the mean risk—global or local by incorporating information between areas. The risks in areas with more information (e.g., urban areas) are usually less smoothed than in areas that exhibit higher sampling variation (typically those with a low number of cases), and thus produce more stable estimates of the pattern of underlying disease risk (Richardson et al. 2004). However, although raw risks can produce "noisy" maps that are difficult to interpret, oversmoothed maps may produce a homogeneous risk surface, masking the actual risk distribution (Beale et al. 2008). It is important to mention that all the analyses presented in this paper are heavily dependent on the scale. We chose the scale of municipal districts but results on other scales could show differences.

Acknowledgments The authors gratefully acknowledge the support by the Operational Program Education for Competitiveness—European Social Fund (project CZ.1.07/2.3.00/20.0170 of the Ministry of Education, Youth and Sports of the Czech Republic).

References

Anselin L (1995) Local indicators of spatial association—LISA. Geogr Anal 27:93–115
Anselin L (2003) GeoDaTM 0.9 User's Guide
Assuncao R, Reis E (1999) A new proposal to adjust Moran's I for population density. Stat Med 2162:2147–2162
Bailey TC (2001) Spatial statistical methods in health. Cadernos Saúde Pública 17(5):1083–1098
Bailey TC, Gatrell AC (1995) Interactive spatial data analysis. Longman Scientific & Technical, Essex
Beale L, Abellan JJ, Hodgson S, Jarup L (2008) Methodologic issues and approaches to spatial epidemiology. Environ Health Perspect 116:1105–1110
Bell BS, Hoskins RE, Pickle LW, Wartenberg D (2006) Current practices in spatial analysis of cancer data: mapping health statistics to inform policymakers and the public. Int J Health Geogr 5:49

Bivand RS, Pebesma EJ, Gómez-Rubio V (2008) Applied spatial data analysis with R. Springer, New York

Clayton D, Bernardinelli L (1996) Bayesian methods for mapping disease risk. In: Elliott P, Cuzick J, English D, Stern R (eds) Geographical and environmental epidemiology: methods for small area studies. Oxford University Press, Oxford

Earickson R (2009) Medical geography. In: Kitchin R, Thrift N (eds) International encyclopedia of human geography. Elsevier, Oxford, pp 9–20

Elliott P, Wartenberg D (2004) Spatial epidemiology: current approaches and future challenges. Environ Health Perspect 112:998–1006

Gelfand A, Diggle P, Guttorp P, Fuentes M (2010) Handbook of spatial statistics. CRC, Boca Raton

Goldman D, Brender J (2000) Are standardized mortality ratios valid for public health data analysis? Stat Med 19:1081–1088

Griffith D, Arbia G (2010) Detecting negative spatial autocorrelation in georeferenced random variables. Int J Geogr Inf Sci 24:417–437

Haining R (1998) Spatial statistics and analysis of health data. GIS and health. Taylor and Francis, London, pp 29–47

Haining R (2004) Spatial data analysis: theory and practice. Cambridge University Press, Cambridge

Jarup L (2004) Health and environment information systems for exposure and disease mapping, and risk assessment. Environ Health Perspect 112:995–997

Koch T (2005) Cartographies of disease: maps, mapping and medicine. ESRI Press, Redlands

Last J, Abramson J (2001) A dictionary of epidemiology. Oxford University Press, New York

Lawson AB (2009) Bayesian disease mapping: hierarchical modeling in spatial epidemiology. CRC, Boca Raton

Lawson AB, Browne WJ, Vidal Rodeiro CL (2003) Disease mapping with WinBUGS and MLwiN. Wiley, Chichester

Marek L, Dvorský J, Pászto V, Tuček P (2013) On estimation of spatial clustering: case study of epidemiological data in the Olomouc Region, Czech Republic. VŠB – Technická univerzita Ostrava, Ostrava

Meade MS, Emch M (2010) Medical geography. The Guilford Press, New York

Moran P (1950) Notes on continuous stochastic phenomena. Biometrika 37:17–23

Openshaw S (1984) The modifiable areal unit problem. Geobooks, Norwhich

Pfeiffer D, Stevenson M, Robinson T, Rogers D (2008) Spatial analysis in epidemiology. Oxford University Press, Oxford

Richardson S et al (2004) Interpreting posterior relative risk estimates in disease-mapping studies. Environ Health Perspect 112(9):1016–1025

Rushton G (2003) Public health, GIS, and spatial analytic tools. Annu Rev Public Health 24:43–56

Scott LM, Janikas MV (2010) Spatial statistics in ArcGIS. In: Fischer MM, Getis A (eds) Handbook of applied spatial analysis. Springer, Heidelberg, pp 27–42

The Center for Food Security & Public Health (2013) Campylobacteriosis. http://www.cfsph.iastate.edu/FastFacts/pdfs/campylobacterosis_F.pdf

Tobler WR (1979) Cellular geography. In: Gale S, Olsson G (eds) Philosophy in geography. Reidel, Dordrecht, pp 379–386

Waller L (2005) Bayesian thinking in spatial statistics. Handbook Stat 25:589–618

Waller L (2009) Detection of clustering in spatial data. In: Fotheringham A, Rogerson P (eds) Sage handbook of spatial analysis. Sage, London

Waller LA, Gotway CA (2004) Applied spatial statistics for public health data. Wiley, New York

Spatial-Temporal Evolution of the Unique Preserved Meandering System in Central Europe

Jakub Miřijovský and Monika Šulc-Michalková

Introduction

The channels of alluvial lowland rivers are formed by the action of flowing water on materials that have been deposited by the stream and that can be eroded and transported. The major factors that affect the forms of an alluvial channel include water discharge, sediment characteristics, longitudinal slope, resistance of the bank and bed to erosion, vegetation, geology and human activity. According to Holubová et al. (2005), meandering rivers can be classified as either stable (passive) or actively meandering (active). An actively meandering river has sufficient stream power to deform its channel bends through active bed and bank erosion and the growth of point bars. The result of fluvial processes in a lowland region is an actively meandering river reach. Meanders evolve and respond to floods that have sufficient stream power to mobilise bed and bank sediments. Conversely, a stable meandering (passive) river is one that has insufficient stream power to erode its banks under current conditions.

The objectives of this study are to describe and understand the evolution of lowland meandering streams using the case of the Morava River, a natural meandering river that was modified by the regulation and restoration of oxbow lakes, and to describe the process of cutoff using UAV (Unmanned Aerial Vehicles) technologies. The study area is often suffer from floods. Correct understanding of the development of the river can help protect the surrounding built-up areas.

J. Miřijovský (✉)
Department of Geoinformatics, Palacký University in Olomouc, 17. listopadu 50, 771 46 Olomouc, Czech Republic
e-mail: jakub.mirijovsky@upol.cz

M. Šulc-Michalková
Department of Geography, Masaryk University in Brno, Brno, Czech Republic
e-mail: monika.michalko@gmail.com

© Springer International Publishing Switzerland 2015
J. Brus et al. (eds.), *Modern Trends in Cartography*, Lecture Notes in Geoinformation and Cartography, DOI 10.1007/978-3-319-07926-4_38

Study Area

Case Study Site

Morava is a typical example of a meandering river that has been influenced by regulation (except for the study area of Litovelské pomoraví). Regulation markedly simplified the shape of the river. A direct channel was created from the initially differentiated meandering channel. Floods are now contained within the narrow area between the levees. The natural attenuation of floods by spilling onto the natural floodplain area and thus slowing the downstream progress into lower basin areas does not occur. However, the dyke systems do not provide sufficient flood protection; for example, dams on the Austrian side of the river failed during floods in the spring of 2006. Cutoff lakes, which are isolated outside the dykes, are filling in.

The Kenický Meander National Nature Reserve is situated in the territory of Litovelské Pomoraví PLA (Protected landscape area) at 250.3 km of the Morava River. It is located in Ramena řeky Moravy National Nature Reserve, which includes the main channel and streams of the Morava River. The natural (unregulated) lowland river basin and adjacent bank communities of herbs and wood species, which preserve their original species composition, are protected. The Kenický National Nature Reserve is located on the right bank of the river to the west of the Kenický meander and contains hardwoods (alluvial forests with predominantly hardwood trees such as oak and ash) between the main channel of the Morava Mlynský potok River (occasionally called Střední Morava or Malá voda). This study focuses on a section of the main channel of the Morava River between transverse profiles No. 8 and No. 18 in the Kenický Meander Nature Reserve (Fig. 1).

The flow of the Morava River is unregulated through the Kenický meander and causes natural channel development and geomorphologic activities. In the basin, large numbers of alluvial trees and gravelsand deposits significantly influence the flow and affect its further development. The changing shoals and pools cause changes in the speed of the flow; the river forms alluvial sediments due to short-term losses of energy, and it erodes banks and the channel bottom when the energy increases in faster parts of the river. Due to the quantity of water that flows through the basin, the basin is large and particularly deep (Máčka and Krejčí 2006a). These characteristics allow the basin to efficiently drain floodwaters; however, the groundwater level decreases in times of drought, which can negatively impact the water regime of the surrounding alluvial forests.

A typical feature of water in meanders is its sinuous flow. In the top part of the meander, the streamline flows near the gibbous bank; here, the flow contributes to lateral erosion. If there are several consecutive meanders, the streamline flows towards the next gibbous bank (which is on the opposite side of the river as the previous one). In straight sections, the flow is usually concentrated approximately in the middle of the river. In some cases, the flow is not clear due to the

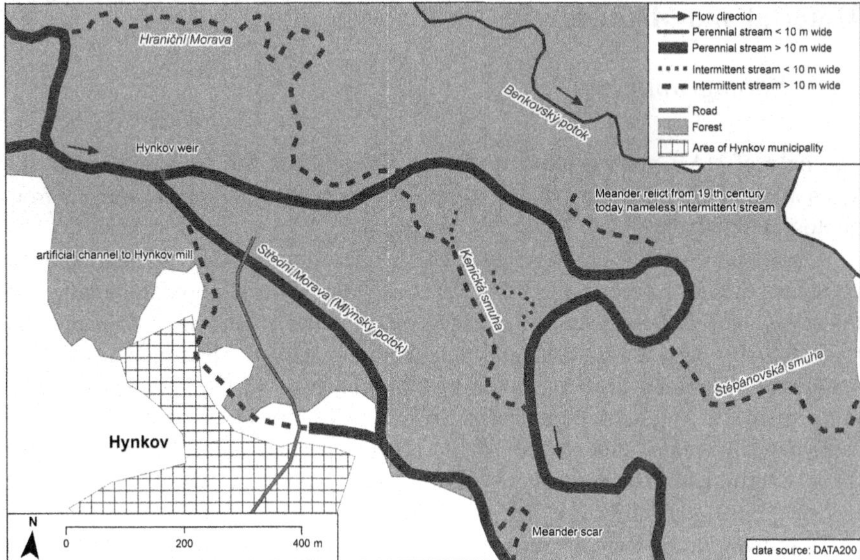

Fig. 1 Area of interest—Kenický meander

compensation of flow dynamics across the flooded profile. Substantially weaker secondary flow occurs close to alluvial banks. Here, the flow energy of the water is very weak, so the water is not able to transport eroded particles. Therefore, side alluvial deposits are formed in these locations. Spiral circulation of water in a meander takes place at the surface towards the gibbous bank and at the bottom towards the alluvial bank. This forms a circulation cell that is divided into two sections in the area of the ford; at the bottom, the water flows towards the centre of the basin, and at the surface it flows towards the banks. The combination of these two components causes the spiral shape of the flow.

The biological value of this region is demonstrated by its inclusion in the list of internationally important wetlands in the Ramsar Convention and in the Natura 2000 system as a European Important Locality and Bird Area. The 3rd oakbeech vegetation degree dominates this region. Hardwood alluvial forests are very well preserved across the entire area, with high species diversity and characteristic alternation of herbal aspects. Wood vegetation is stratified into several levels. As specified by Culek (1995), Morava belongs to the grayling to barbel zones and is characterised by a rich community of fish species and other water animals, such as the critically endangered Fairy Shrimp (*Eubranchipus grubii*) and Tadpole Shrimp (*Lepidurus apus*). Litovelské Pomoraví LPA is also an important migration route and nesting place for approximately 250 species of birds. The region is home to the Little Ringed Plover (*Charadrius dubius*) and Common Kingfisher (*Alcedo atthis*), which are both rare bird species, and rare water animals, such as the European otter (*Lutra lutra*) and Eurasian Beaver (*Castor fiber*), of which a carcass was found during the spring surveying of the cross profiles.

Material and Methods

UAV Photogrammetry

The field of UAV photogrammetry deals with methods and technologies for the acquisition of measurements, maps, digital terrain models and other products from photographic images.

A pilot may not be physically present in the aircraft. UAV photogrammetry combines the advantage of the vertical aerial view of aerial photogrammetry and the advantages of close distances and high image detail of ground photogrammetry (Aber 2010). The basic principle of photogrammetric measurements is the geometric-mathematical reconstruction of the direction of the photographic rays in the image. UAV photogrammetry is a novel measurement tool that works on the same principle and which can be used to obtain geographic data for new applications. Additional information about obtaining data with UAV models can be found in Aber (2010) and Miřijovský and Vávra (2012).

To properly implement aerotriangulation, it is necessary to know the elements of the external and internal orientations. The exterior orientation elements include the X, Y, and Z coordinates of the camera on the platform and the three angles of camera tilt (ω, φ, and κ). These coordinates and angles are relative to the ground coordinate system. The X, Y, and Z coordinates are provided by very precise DGPS systems. The tilts of the camera can easily be measured using a three level system, preferably an INS (Inertial Navigation System) or IMU device (Inertial Measurement Unit). In most cases, the size, weight, and price do not allow the use of UAS in the models, and it is necessary to perform the aerotriangulation without knowing the exterior orientation elements. The most commonly used method is to determine the elements of internal orientation using GCPs (Ground Control Points) with known X, Y, and Z coordinates.

The position accuracy of the GCPs is the most important factor that affects the accuracy of the final aerotriangulation. Measurements using a standard GPS device or subtracting the coordinates from a map can be sufficient when working with small scale images. However, for high accuracy, Small Format Aerial Photography (SFAP), a total station, or GPS device with dual-frequency correction data must be used to focus the GCPs.

The elements of internal orientation can be determined either in special laboratories or by self-calibration. While aerial cameras are usually calibrated by the manufacturer, the calibration protocols for small cameras used in SFAP applications are not known.

In this study, the camera calibration was performed with the PhotoModeler and iWitness software. Several types of calibration were performed, including both a single calibration grid sheet and a multi-grid calibration sheet (Table 1).

Table 1 Interior orientation of the camera. f Focal length (mm), x_0, y_0—Principal point (mm), K_0, K_1, K_2—Radial lens distortion coefficients (mm)		
f		18.4715 mm
x_0		−0.11 mm
y_0		0.15 mm
K_0		-2.75034^{-4}
K_1		5.51047^{-4}
K_2		-1.42621^{-6}

Ground Control Points

Control points are the most essential elements for the correct processing of aerial photographs. Control points serve as a basis for calculating the exterior orientation parameters, and their accurate determination is critical for further processing of the photographs. The accuracy of an AAT (Automatic Aerial Triangulation) calculation is affected by several factors, including the measurement accuracy of the control points, their total number, and their distribution. The aerial triangulation result, or the determination of the outer orientation parameters, will only be correct if all three factors are correctly considered.

The distribution of control points is crucial for the AAT result and its accuracy. Normally, four or five symmetrically-distributed points are used in conventional aerial photography. However, what does "symmetrical" mean? According to commonly used photogrammetry rules, the points should be evenly distributed over the entire area of the photograph. However, there are many ways to distribute the points. The theory of statistical estimates and solvability of linear equations suggests that the error rate of the model increases if there is a linear relation between the points. That is, the accuracy of the result decreases. The addition of equations for a third point that is located on a line connecting two other points will not increase the rank of the matrix; consequently, this point will not increase the solvability of the system of equations. That is, this procedure will not reduce the degrees of freedom. The following example explains this concept. If a system has six unknowns, six equations must be calculated to correctly solve the system. Thus, if we take two points with X and Y coordinates, these points produce four equations (two for X, two for Y). If we add a third point that represents a linear combination of the first two points, then the third point is still a linear combination of the previous points. The resulting system thus adds nothing new, and the entire system cannot be solved.

For a distribution of points on natural terrain, the linear relation means that the points should not lie on lines. In theory, the maximum accuracy can be reached if none of the three points lies on a line. This requirement is clearly difficult to achieve in natural terrain. Natural and artificial obstacles can prevent an ideal distribution of fixed points that can be geodetically surveyed. This situation can occur in photogrammetry images that are collected for fluvial geomorphologic investigations. However, a watercourse does not allow points to be distributed correctly and evenly over the entire photograph.

In UAV photogrammetry, determining the exterior orientation of objects is further complicated because the photographs are usually taken with non-surveying cameras, where the radial distortion can reach up to 200 m. If the values of radial distortion cannot be incorporated in the calculation, only 60 % of the image around its imaginary center should be used. Modern applications can usually calculate the radial distortion in a single step with the AAT calculation. Software for the third sides can potentially be used. However, even if the pattern of the radial distortion is known, using the entire photograph is not recommended, which reduces the potential distribution of the control points.

Based on our experiments, the effect of the correct distribution of control points over the terrain on the correct calculation of the external orientation objects is not negligible. A distribution of points along a straight river or other object is not recommended. It is always best to use an even distribution of points over the surface so that none of the points lies on a line between other points. This corresponds to a random distribution of points.

Two sets of images of the Kenický meander were collected. The images were photogrammetrically processed, and a morphometric analysis of the development of the meander and the woody debris was performed using the processed images (Fig. 2). The parameters of the imaging flight and the image processing are shown in Table 2.

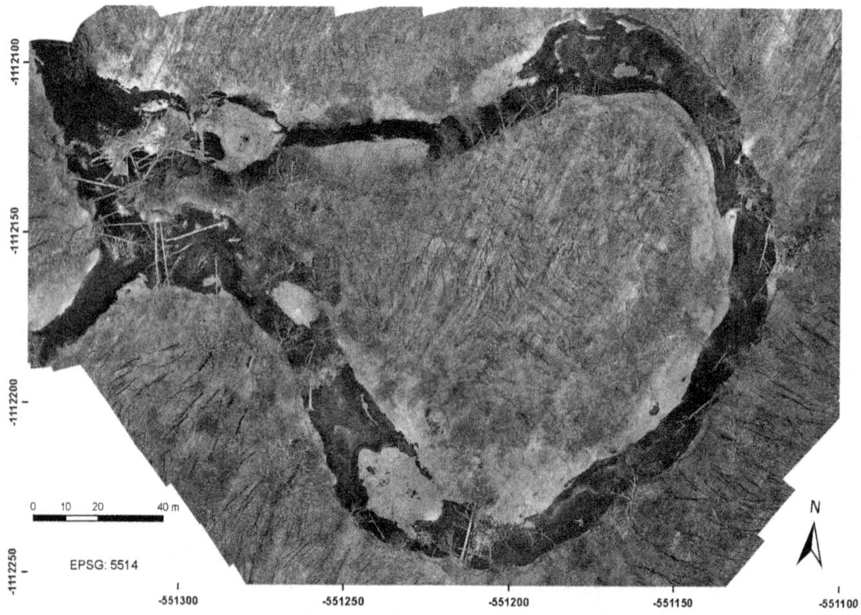

Fig. 2 Orthophoto of the Kenický meander

Table 2 Parameters of the imaging flight and image processing

December 2012	Value
Number of images	47
Flying altitude (m)	106
Ground sample distance (cm)	2.3
Number of Ground control points	17
Number of Tie points	94,365
Error of the image coordinates (pix)	0.59
Point density (points/m^2)	299
The average error in Z (m)	0
The standard deviation (m)	0.093
$RMSE_Z$ (m)	0.093

Results and Discussion

Analysis of the Historical Development of the Morava River Near the Kenický Meander PLA

Anastomosis and the formation of meanders cause continuous changes in the basins of the Morava River. Blind branches covered with soil, former river basins and their parts are clearly visible on detailed maps and in the field. When water stops flowing through the river basin, soil immediately begins covering the former meanders by the accumulation of particles from the surrounding areas after floods and rainfalls and by the collection of organic materials from the surrounding vegetation. One of the methods for determining the historical development of river basins is investigating terrain undulations in the floodplain as well as the ages of soil samples from the former basin, which are always younger than sediments from the same depth outside of the basin. This method can be very precise, but it is both costly and technically demanding. A simpler method is to analyse old maps or aerial photos. This method was used by Grešková (2002) in a morphologic investigation of the lower course of the Morava River. This method allows the development and character of the river course from approximately the first half of 19th century, when the maps with larger scales and different coordinate systems began to be produced, to be determined.

Woody Debris in the River Basin

The impact of woody debris is most important in middle-sized rivers (Máčka and Krejčí 2006b). Small rivers are crossed by large pieces of wood, and the impact of woody debris on large rivers is negligible due to size of the river basin. A tree or part of a tree may fall into a river because of several factors: water erosion, wind, fire, landslides, wood parasites (including beavers), sickness, competition within the phytocoenosis, aging, and human activity (Stevens 1997). The impacts of wood on the flow of water can be divided into morphological (erosion-accumulative processes, transport of wooden debris, basin stability), biological (position and species diversity, oxidisation of water), hydraulic (roughness of the basin, flow direction, flow dissipation) and the circulation of particles (food for animals, collection of mineral sediments, spiraling of nutrients) (Máčka and Krejčí 2006b). Wood is removed naturally in three ways: washing away, covering by sediments and decomposition. Most LWD is washed away during floods, when the energy of the flow is much higher and often washes away entire trees for significant distances. A fallen tree on the gibbous bank can be moved to the alluvial bank due to lateral erosion of the basin, where it is buried by sediments. The decomposition of wood is slower in water than in dry conditions. Even so, wood also often removed by decomposition, but it is more commonly washed away down the stream before it decomposes.

The Kenická Accumulation and Woody Debris

Another unique feature of the study area is the Kénická accumulation, which is one of the largest accumulations of woody debris in the country. It is located north of the former isthmus of the meander. Its age is unknown. Authors (Máčka and Krejčí 2006a) specify that "in 2000, there were only five trees lying diagonally in the basin of the river. Additional material was gradually caught by these key trees." The most important sources of LWD were the withdrawal of the adjacent banks due to narrowing of the isthmus. Smaller pieces of wood and woody debris from the course are captured by large trunks. The accumulation is stabilised on several gravel-sand deposits.

In the autumn of 2011, the accumulation consisted of four parts with different characteristics. The north part consisted of several trunks that have captured medium sized and small pieces of wood as well as large amounts of plastic trash. This part was the highest of the four, and it is higher than the left bank. The central part is the largest and was formed by large and long trunks of different ages that accumulated in a group perpendicular to the flow. This part is the biggest obstacle to flow during the highest water levels. The south part of the accumulation consists of long trunks at different orientations that are separated because most of the flow occurred through this part. A part consisting of SWD (small woody debris) formed directly adjacent to the south bank of the river. SWD was carried by floods and remained on the land after the water subsided. Before the isthmus ruptured, the accumulation divided the flow into three parts: a main part that was directed around the accumulation along the right bank of the river and then crossed the part of the accumulation that consisted of long trunks without woody debris; a middle part (the weakest) where water flowed under the main parts of the accumulation and then merged with the main flow; and the third part where water flowed north along the main part and crossed the accumulation where medium size and incomplete trunks were caught. Live vegetation grows in the top parts of the accumulation. The accumulation is 40×40 m in size, and the height ranges from 1.5 m below the right bank up to 1.5 m above the bank.

A large amount of woody debris was caught near a large fallen tree perpendicular to the basin in the meander between profiles 14 and 15. This unstable accumulation (further transport of material is prevented by this tree) will be eliminated in May. Other woody debris was caught downstream by fallen trees near the former isthmus and formed several small accumulations of wood that were mostly caught by one piece of LWD (large woody debris). Pieces that were washed away were replaced by eight trees from the former isthmus; three were full-grown and several tens of metres high, while the rest were between 5 and 15 m high. The total volume of woody debris in this area has not changed significantly except for the removal of part of the Kenická accumulation. Other observed phenomena include the washing away of a tree that had been cut down by a beaver from a gully of the meander near profile No. 11 and its capture in the accumulation with trash as well as a change of orientation of a tree near profile No. 13 that had fallen perpendicular to the basin and had been rotated to the downstream direction due to flow of water.

Further Development of the Locality

The further development of the meander after its rupture is described by Máčka and Krejčí (2006a). The authors based their description on the assumption that the isthmus ruptured at its narrowest point. They expected that a passageway in the isthmus at least 25 m wide would develop rapidly, but this has not happened yet. At the beginning of May, the passageway was only 17.5 m wide. One possible reason is the rapid decline of the flood waters after the rupture, so the river did not have enough energy to widen the passageway. Further erosion is prevented by trees and their roots that have fallen into the river basin. The tree trunks (located in the water perpendicular to the flow) cause deep erosion, prevent lateral erosion, and generate the extreme depth of the former isthmus. This deepening is also supported by the presence of a pool at the bottom of the isthmus, which indicates that the level of the river bottom will decrease due to backward erosion. The rate of erosion will depend on the frequency and strength of future floods. The Kénická accumulation has already started to block the inlet of the meander, where the flow of the water is presently much weaker (in some places, the water is almost stagnant). In the future, it will completely block the inlet. Most of the assumptions of this study were confirmed two months after the rupture of the isthmus, but the future configuration will depend on further development and its speed. The current configuration indicates that a future meander zone that forms from the potential meanders shown on the map will be even longer. It is possible that new meanders will also affect the lower part of the Kénická branch, which would become a basin if it is exposed to the regular flow of water. However, this is only a hypothesis of what may occur in the future.

Conclusions

Floodplain lakes have been studied by ecologists because of their effect on river habitats, conditions and evolution. However, their essential element is the structure and functioning of this part of the fluvial hydrosystem. The morphology of a former channel is a key parameter that affects the frequency of the connection, evolution and specific composition of a floodplain lake. The relationship between the hydrologic connection and processes in a floodplain lake are not linear, but the hydrologic connection model and the changing morphology of a floodplain lake create a different framework of discharge. Our work shows that the water and sedimentary fluxes are the dominant elements that assure the proper functioning of these lakes. If these fluxes are minimised or cut off from a system, the system changes its behaviour and morphological style (reducing the number of channels), and the natural evolution stops.

The processes of sedimentation and erosion affect the evolution of these lakes. These processes are activated during floods and respond to the intensity and frequency of the perturbations. The dynamics of floodplain lakes are directly linked to the dynamics of the fluvial system. The principal hydrologic factors are controlled by anthropic pressure.

The upstream/downstream connections were studied from an ecological perspective in terms of the hydrologic changes, such as the effect of a dam on water level fluctuations. Sedimentary connections are generally studied in the field of geomorphology. The presence of a dynamic floodplain lake system is the sign of a healthy fluvial system. Floodplain lakes are not only places for the diversification of a river corridor but are also an important demonstration of fluvial dynamics.

Floodplain lakes affect both the diversity of the former channels and their importance in flood restoration. The maintenance of fluvial dynamics can be stopped by levees and by the constructions of dams. These factors have affected the incision of channels and decreased the frequency of the connections to the floodplain. The restoration and postrestoration management of floodplain lake sedimentation is a major factor in the proper functioning of a fluvial system.

Based on our study, we conclude that the effects of anthropic influences are still present in the study area. We need to understand the historical and actual anthropic pressures and evaluate the consequences of major human activities. The models of fluvial management are the same even though the local conditions are different. It is difficult to respond to the different expectations of society, but the possibility of restoring the dynamics of the fluvial system should be considered the ideal goal in order to preserve the morphological and ecological values of the system.

(continued)

The consequences of floodplain lake degradation and loss may have direct and indirect effects on people, ecosystems and environments far from the point of impact. Floodplain lakes are an increasing part of international conventions, such as the Ramsar Convention, and transnational documents such as the EU Water Framework Directive.

Acknowledgements The authors gratefully acknowledge the support by the Operational Program Education for Competitiveness—European Social Fund (project CZ.1.07/2.3.00/20.0170) of the Ministry of Education, Youth and Sports of the Czech Republic.
The authors gratefully acknowledge the support by the Visegrad Fund (project No. 31210058).

References

Aber JS (2010) Small-format aerial photography: principles, techniques and geoscience applications, 1st edn. Elsevier Science, Amsterdam

Culek M (1995) Biogeografické členění České republiky. Praha

Grešková A (2002) Dynamika a transformácia nivy rieky Moravy študovaná pomocou historických máp a leteckých snímok. Geomorphologia Slovaca 2:40–44, Bratislava

Holubová K, Hey R, Lisický M (2005) Middle Danube tributaries – constraints and opportunities in lowland river restoration. In: Large Rivers – Rehabilitating large regulated rivers. Hydrobiol Suppl. 155/1-4 (Wageningen, The Netherlands)

Máčka Z, Krejčí L (2006) Prognóza geomorfologického vývoje řeky Moravy v úseku od jezu hynkov po kenickou lávku: NPR Ramena řeky Moravy, CHKO Litovelské Pomoraví.Agentura ochrany přírody a krajiny České republiky, Brno

Máčka Z, Krejčí L (2006) Plavená dřevní hmota (spláví) v korytech vodních tok – případová studie z CHKO Litovelské Pomoraví. In: Měkotová, Jarmila a Štěrbă, O. Říční krajina 4: Sborník příspěvk z konference. Olomouc

Miřijovský J, Vávra A (2012) UAV photogrammetry in fluvial geomorphology. In: Conference proceedings SGEM 2012. 12th international multidisciplinary scientific geoconference STEF92 Technology Ltd., Sofia

Stevens V (1997) The ecological role of coarse woody debris: an overview of the ecological importance of CWD in BC Forests [online]. British Columbia Ministry of Forest

Terrain Analysis for Armed Forces

Vaclav Talhofer, Vladimir Kovarik, Marian Rybansky, Alois Hofmann, Martin Hubacek, and Sarka Hoskova-Mayerova

Introduction

The present theory of using weapon and weapon systems assumes computer models of their behavior in the terrain. It is necessary to consider two basic conditions—the technical parameters of the weapons and weapon systems that are important because of their behavior in the terrain on the one hand, and the content, properties and quality of digital spatial data describing the terrain on the other hand.

If both conditions are fulfilled, it is possible to derive the physical models of the behavior of weapon and weapon systems in a terrain. Rybansky (2009) or STANAG 2999 (2012) could be mentioned as the examples of such physical models. Physical models usually determine conditions of the terrain in which the weapon or weapon system can be used, or, where appropriate, to set limits for these conditions. The conditions laid down then represent the basis for the applications in a computer environment and it is possible to create computer models, often in a form of spatial analysis. The spatial analyses form a part of most of present Command and Control Systems (C2S) in which they support the decision making processes. When limits of physical models evaluation are not considered, the final results of spatial analyses are influenced by content, precision and quality of digital spatial data used in the given model.

V. Talhofer (✉) • V. Kovarik • M. Rybansky • A. Hofmann • M. Hubacek
Faculty of Military Technology, Department of Military Geography and Meteorology,
University of Defence, Kounicova 65, 662 10 Brno, Czech Republic
e-mail: vaclav.talhofer@unob.cz; vladimir.kovarik@unob.cz; marian.rybansky@unob.cz;
alois.hofmann@unob.cz; martin.hubacek@unob.cz

S. Hoskova-Mayerova
Faculty of Military Technology, Department of Mathematics and Physics, University of
Defence, Kounicova 65, 662 10 Brno, Czech Republic
e-mail: sarka.hoskova-mayerova@unob.cz

© Springer International Publishing Switzerland 2015
J. Brus et al. (eds.), *Modern Trends in Cartography*, Lecture Notes in
Geoinformation and Cartography, DOI 10.1007/978-3-319-07926-4_39

Complex models of terrain features can be found in geospatial databases in which all features have given properties (shape, size, location, thematic and time properties, etc.). Additional properties can be derived, for example a slope from digital elevation model. They are two different views on digital features and their properties—with or without consideration data quality and mainly their certainty or uncertainty. For example position of a given feature saved in the geospatial database is determined by its coordinates. But its real position in the terrain may be different depending on its natural properties. Building footprints can be measured with an accuracy of centimeters, but borders of various types of soils are quite indeterminate. If uncertainty of feature properties is not considered in spatial analyses, the final results can be a bit out of reality and using them in decision making process may cause difficulties in the future. To decrease the possibility of a wrong decision, the uncertainty of digital features must be taken into account. Application of fuzzy logic in spatial analyses is one possible and quite frequent way and it is possible to find many examples of using general fuzzy logic (Zadeh 1965; Ahmad and Kharal 2009; Sunila and Hottanainen 2004), or its application in decision making processes (Di Martino and Sessa 2011; D'Amico et al. 2013; Kainz 2007; Talhofer et al. 2012). The following text presents two examples of fuzzy logic application in spatial analyses used in the armed forces. ArcGIS 10 was used for both examples (ESRI 2013).

Analysis of Potential Helicopter Landing Sites

To carry out an analysis of locations suitable for landing of a helicopter in a given area (Helicopter Landing Sites—HLS), both natural and man-made terrain features that can represent obstacles have to be determined. There are features that can act as obstacles due to their height such as forests, power lines, or communication towers. Other features can impede landing by their nature such as vineyards, lakes, or swamps. All these features can be found in geospatial databases and they can be classed to categories as follows (Kovarik 2011):

- *Vegetation:* forest, wood strip, nursery, orchard, vineyard, hop-garden, reed-grass
- *Water:* lake, pond, reservoir, river, stream, canal, ditch, swamp, aqueduct, water tower
- *Transportation:* road, railroad, aerial cableway
- *Utilities:* pipeline, power transmission line
- *Terrain:* rocks, cliffs, crevasse, depression, fault, landslide, karst
- *Built-up areas:* settlement
- *Other objects:* chimney, cooling tower, power station, transformer yard, oil rig, tower, communication tower, windmill, cemetery

The physical model of interaction between helicopter and surrounding landscape of HLS is described in STANAG 2999 (2012). This document does not specify the types or nature of the obstacles but it describes the selecting criteria for locations suitable for tactical or non-permanent landing sites. From a variety of conditions only the following two conditions were selected for further analysis: slope of ground, and obstruction angle on approach and exit paths. These criteria can be formulated as follows:

- Slope should not exceed 7° or 3° in any direction by day or night, respectively.
- Within the selected approach and exit paths, the maximum obstruction angle to obstacle should not exceed 6° to a distance of 500 m by day and 4° to a distance of 3,000 m by night

The analysis was carried out in the ESRI ArcGIS software suite using the NATO standard VMap 1 (Vector Map Level 1) and DTED 1 (Digital Terrain Elevation Data Level 1) databases in the study area which was located in the vicinity of the city of Kosovska Mitrovica in the northern part of Kosovo (Kovarik 2014). The VMap 1 database contains digital geospatial data in a vector format with resolution, accuracy and level of generalization relating to the map scale of 1:250,000. Data is separated into the thematic levels including boundaries, elevation, hydrography, industry, physiography, population, transportation, utilities and vegetation. The DTED 1 database provides a uniform latitude/longitude based matrix of terrain elevation values. These values are recorded approximately every 100 m (for longitude from 0° up to 50° north or south the spacing is 3 by 3 arc seconds, for other areas the spacing is different).

Crisp Set Analysis of Potential HLS

The following features were selected as obstacles for the analysis without uncertainty consideration: roads, railroads, rivers, lakes, power lines, forests, built-up areas, and terrain slope.

- *Roads and railroads.* For the purposes of the analysis, it was assumed that the trees or electricity poles occurred along the features and that their average height was 10 m. According to the STANAG requirements for the obstruction angle for the day operations the buffer zones of 95 m from the feature center lines were generated.
- *Rivers and lakes.* The trees of the height of 20 m were assumed on the banks of these water features therefore the buffer zones of 190 m from the feature boundaries were generated.
- *Power lines.* As the VMap 1 data states neither the pole heights nor the transmitted voltage, the obstacle height of 30 m was assumed and the corresponding buffer zones of 285 m from the feature center lines were generated.

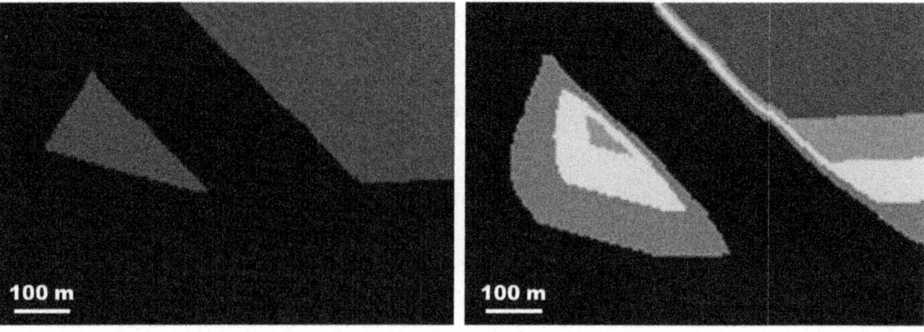

Fig. 1 Example of results of the crisp set analysis (*left*) and the fuzzy set analysis (*right*) (original colour coding of the terrain: *red*—not suitable, *orange*—very limiting, *yellow*—limiting, *light green*—acceptable, *dark green*—suitable. Grayscale version of the image shows *red* as *black*, other colours at corresponding gray tones)

- *Forests*. Similarly to the trees on the banks of rivers and lakes, the obstacle height of 20 m was assumed and the corresponding buffer zones of 190 m from the feature boundaries were generated.
- *Built-up areas*. No height of this feature was assumed therefore no buffer zones were generated and the obstacle was therefore represented by the polygon features themselves.
- *Terrain slope*. The DTED 1 elevation data was used to create the terrain slope and the parts of the terrain having the slope greater than 7° was considered as an obstacle.

All the individual results were finally overlaid to create the output showing parts of the terrain either suitable (green) or not suitable (red) for landing. See the example in Fig. 1.

Fuzzy Set Analysis of Potential HLS

As the geospatial data represent the simplified and schematic reality, the uncertainty of the feature parameters was introduced to the analysis. That led to employing of fuzzy sets.

- *Roads and railroads*. Similarly to the first part of the analysis working with crisp sets it was assumed that the height of trees or electricity poles along the features was 10 m, but the buffer zones were not generated. The uncertainty of 2 m in height was assumed which meant the uncertainty of ±20 m in a position of the boundary defined by the obstruction angle. The linear membership function for roads and railroads (R) can be defined as

$$\mu_R(x) = \begin{cases} 0, & x \le 75 \\ \dfrac{x-75}{40}, & 75 < x \le 115 \\ 1, & x > 115 \end{cases} \tag{1}$$

where x is the distance from the feature center line. The crossover point (i.e. the point with membership value 0.5) is at 95 m from the feature center line.

- *Rivers and lakes.* The trees of the height of 20 m were assumed on the banks with the uncertainty of 3 m which meant the uncertainty of ± 30 m in a position of the boundary defined by the obstruction angle. The linear membership function for these water features (W) can be defined as

$$\mu_W(x) = \begin{cases} 0, & x \le 160 \\ \dfrac{x-160}{60}, & 160 < x \le 220 \\ 1, & x > 220 \end{cases} \tag{2}$$

where x is the distance from the feature boundaries. The crossover point is at 190 m.

- *Power lines.* The height of electricity pylons was assumed to be 30 m with the uncertainty of 2 m which meant the uncertainty of ± 20 m in a position of the boundary defined by the obstruction angle. The linear membership function for power lines (P) can be defined as

$$\mu_p(x) = \begin{cases} 0, & x \le 265 \\ \dfrac{x-265}{40}, & 265 < x \le 305 \\ 1, & x > 305 \end{cases} \tag{3}$$

where x is the distance from the feature center line. The crossover point is at 285 m.

- *Forests.* The height of 20 m was assumed for the trees with the uncertainty of 3 m which meant the uncertainty of ± 30 m in a position of the boundary defined by the obstruction angle. The linear membership function for forests (F) can be defined as

$$\mu_F(x) = \begin{cases} 0, & x \le 160 \\ \dfrac{x-160}{60}, & 160 < x \le 220 \\ 1, & x > 220 \end{cases} \tag{4}$$

where x is the distance from the feature boundaries. The crossover point is at 190 m.

- *Built-up areas.* No height of this feature was assumed however the uncertainty of 100 m in a position of the feature boundary was assumed. The linear membership function for built-up areas (B) can be defined as

$$\mu_B(x) = \begin{cases} 0, & x = 0 \\ \dfrac{x}{100}, & 0 < x \le 100 \\ 1, & x > 100 \end{cases} \tag{5}$$

where x is the distance from the feature boundaries. The crossover point is at 50 m.

- *Terrain slope*. The slope was computed from the elevation data and reclassified. The linear membership function for terrain slope (S) can be defined as

$$\mu_S(x) = \begin{cases} 1, & x \le 6 \\ \dfrac{8 - x}{2}, & 6 < x \le 8 \\ 0, & x > 8 \end{cases} \tag{6}$$

where x is a slope in degrees. The crossover point is at $7°$.

All the individual results were then overlaid. To visualize the level of suitability of particular parts of the terrain for landing in a simple and illustrative way, the military categories for the Cross-Country Mobility—*No Go*, *Slow Go* and *Go*—can be used. As these classes are relatively broad, the more detailed classification was applied. The membership function values of 0.2, 0.4, 0.6, and 0.8 were used to create five classes of suitability for landing which were labelled as *Not Suitable*, *Very Limiting*, *Limiting*, *Acceptable*, and *Suitable* (see Table 1 and the example in Fig. 1).

Results and Discussion

The outputs of overlaying the individual results of both crisp and fuzzy sets are showed in Fig. 1. The example of the crisp set analysis on the left shows parts of the terrain not suitable for landing of helicopters in black (i.e. red in the colour version) and parts suitable for landing in gray (i.e. dark green). The example of the fuzzy set analysis on the right shows the categories of the terrain divided according suitability of landing. The results of the crisp set analysis show simple and clear division of the terrain, but considering the features entering the analysis such as forests or built-up areas we know that in reality they do not have crisp boundaries. Therefore the boundaries between the parts of the terrain labelled as suitable or not suitable for landing might not be fully representative.

Applying the fuzzy sets enables to improve the quality and reliability of the analysis. It enables to identify highly convenient parts of the terrain that may be used as the HLS and to leave the possibility still to use certain areas when accepting a certain risk in relation to a particular feature. It can also eliminate a number of locations that would be necessary to verify in the terrain when searching for the

Table 1 Categories of terrain suitability for landing of helicopters

Membership value	Terrain category	Original colour coding
0.0–0.2	Not suitable	Red
0.2–0.4	Very limiting	Orange
0.4–0.6	Limiting	Yellow
0.6–0.8	Acceptable	Light green
0.8–1.0	Suitable	Dark green

HLS. Needless to say, the ultimate decision whether and where to land or not will always rest with a helicopter commander.

Analysis of Cross Country Movement

The main goal of the Cross Country Movement (CCM) is to evaluate the impact of geographic conditions on movement of vehicles in terrain (Rybansky 2009; Rybansky and Vala 2010). CCM of a vehicle usually depends on numerous geographical, meteorological, technical and personnel factors. There is a great difficulty to identify all such factors and to calculate their impact on a vehicle route and speed across real terrain, due to the complicated expressions of the mathematical and physical relationship between vehicle speed and real environmental factors (Rybansky and Vala 2009). Each of these factors acts on vehicle deceleration individually or collectively with other factors. To resolve these problems it is possible to consider the two following conditions:

- Vehicle route deceleration, which is based on terrain surface configuration (slope), surface roughness, surface materials (soils), vegetation roughness, etc.
- Vehicle route extension that is based on terrain obstacles (mountains, lakes, rivers, settlements, forests, etc.).

The methodology for determining the optimal (least-cost) route of a vehicle depends on individual geographical factors. Each of these factors contributes to the deceleration of vehicle speed relative to the maximum speed that a vehicle can be driven on the road. The resulting speed of a vehicle on terrain can be determined by individual decelerating (cost) factors, as part of the particular set of the above mentioned geographical features.

Four types of possible ways can be determined as a result of CCM analyses that can be the base for a commander's optimal decision making:

- The shortest path
- The fastest path
- The cheapest path
- The safest path

Warfare recognises three basic degrees of CCM:

- Passable terrain (GO)—real speed is approaching maximum speed;
- Terrain passable with restrictions (SLOW GO)—real speed is lower or significantly lower than maximum speed, obstacles can be overcome;
- Impassable terrain (NO GO)—obstacles cannot be overcome.

It is possible to determine the basic degrees of CCM for each transport or fighting vehicle because of their technical parameters (chassis type, power of engine, transmission system etc.), or consider the CCM of the weakest vehicle in the unit.

Specific degree of CCM determination on given part of terrain is possible to express as a complex function in which all impacts of individual geographic factors are evaluated as the coefficients of deceleration 'C_i' and expressed as a number from the interval of 0–1. The individual coefficient of deceleration shows the real (simulated) speed of vehicle v in the landscape in the confrontation with the maximum speed of given vehicle v_{max}. The impact of all the 7 basic geographic factors can be expressed by the formula:

$$v = v_{max} \prod_{i=1}^{7} C_i \tag{7}$$

The main coefficients of deceleration and the theory of their determination are listed in the Table 2 (Rybansky and Vala 2010).

These coefficients are thereinafter indexed and classified into particular discrete factors listed in the Table 3.

According to the theory, each basic coefficient is calculated by the next formula:

$$C_i = \prod_{j=1}^{m} C_j, \quad i = 1, \ldots, 7, \ m \in \{1, \ldots, 6\} \tag{8}$$

If there is the assumption that the route of vehicle movement, including direct segment, consists of various sub-sections, in which values of geographical factors are unchanged, the final degree of CCM can be determined as a cost of the given segment.

For given vehicle (its technical properties) the values of deceleration coefficients are calculated from ascertained properties of geographic objects stored in the spatial geo-database. Using formulas (7) and (8) it is possible to create a cost map in which the value of each pixel is the final (modeled) speed. The *cost map* can be used as a source for calculation of the shortest, fastest, cheapest or safest path.

Similarly to HLS determination it is possible to consider the quality and certainty of digital data describing the geographic features. Crisp sets or fuzzy logic is also applicable for cost map evaluation. The following paragraphs describe shortly both possibilities for which the features from Brno and its vicinity selected from standard spatial database DMU25 produced by the Military Geographic Service of the Army of the Czech Republic was used. The list of used features and their

Table 2 Main coefficients of deceleration

Basic coefficient	Geographic signification and impact
C_1	Terrain relief (gradient of terrain relief and micro relief shapes)
C_2	Vegetation cover
C_3	Soils and soil cover
C_4	Weather and climate
C_5	Hydrology
C_6	Build-up area
C_7	Road network

Table 3 List of particular coefficients

Particular coefficient	Description
C_{11}	Slope gradient
C_{12}	Microrelief
C_{21}	Spacing between stems
C_{22}	Stem diameter
C_{23}	Tree height
C_{24}	Type of tree
C_{25}	Nature of root system
C_{31}	Soil type
C_{32}	Kind of soil
C_{33}	Soil-forming substrate
C_{41}	Dry season
C_{42}	Moist season
C_{43}	Wet season
C_{51}	Kind of waters
C_{52}	Depth
C_{53}	Width
C_{54}	Flow speed
C_{55}	Characteristics of bottom
C_{56}	Characteristics of bank (bank slope)
C_{61}	Block built-up area
C_{62}	Uptown
C_{63}	Cottage built-up area
C_{71}	Highway
C_{72}	1st category road
C_{73}	2nd category road
C_{74}	3rd category road
C_{75}	Hardened way, forest and cart way

properties is in MoD-GeoS (2010). The technical parameters of military heavy vehicle Tatra 815 (Tatra 2010) were used for the cost map evaluation.

Crisp Set Analysis of CCM

The final speed of given vehicle was evaluated according to formulas (7) and (8). Particular obstacles were compared with the given vehicle properties in a pixel of 1 by 1 m. Total of 27 raster layers were created for each particular coefficient C_{ij} where the value of given pixel was the "cost" of the pixel—reclassified value of coefficient to the range of 0–1.

The final cost map was created using map algebra from combination of all particular layers. This cost map enables to find the cheapest way from start point to destination.

Precision, uncertainty, vagueness and similar properties were not taken into account in the calculations of cost map because of crisp set analysis. The final result is easily understandable but the great disadvantage of such analysis is that user has no information about the properties of features entering calculations. While determination of feature borders can be different (for example, the footprint of building can be defined in resolution of centimetres, the border between two types of soil is defined in resolution of approximately 100 m), therefore it is necessary to include other parameters in the calculation. One possible approach is application of a fuzzy logic.

Fuzzy Set Analysis of CCM

Certain processes that differ only in the input conditions are repeated with all coefficients. These are selection processes as well as conversional ones, etc. For solution of vagueness, in-built processes Fuzzy Membership and Fuzzy Overlay are used. While the use of Fuzzy Membership is different, Fuzzy Overlay is the same for all coefficients. Conditions for Fuzzy Membership are dependent on geometric accuracy that shows the types of geographical objects in the database. There is only the example of the coefficient C_1—Terrain relief calculation in the following text.

Coefficient C_1 is very complex, it includes the influence of slope gradient as well as the influence of micro-reliefs. The slope gradient is one of the input layers and it can be—just like in this case—prepared from the actual factor calculation. The calculation is divided into several branches.

In one branch, fuzzification is calculated directly from the values of gradient (expressed in degrees). The limiting value for the chosen vehicle (36°) was taken for the calculation. The conditions for fuzzification are as follows

$$\mu_S(x) = \begin{cases} 0, & x = 0 \\ \dfrac{36 - x}{36}, & 0 < x < 36 \\ 1, & x \geq 36. \end{cases} \qquad (9)$$

Other branches of calculation concern micro-relief. The input layers are point, line, or areal height obstacles. For each of them, Euclidean distance is calculated, and from that also Fuzzy Membership.

For point objects (a_vobj_p) is Fuzzy Membership for Euclidean distance 20 m calculated with the following conditions

$$\mu_{MP}(x) = \begin{cases} 1, & x = 0 \\ \dfrac{20 - x}{20}, & 0 < x < 20 \\ 0, & x \geq 20. \end{cases} \tag{10}$$

Depth and height of an object are the observed parameters. Fuzzification for both parameters is given with these conditions

$$\mu_{MPH}(x) = \begin{cases} 0, & x = 0 \\ \dfrac{0.1 - x}{0.1}, & 0 < x < 0.1 \\ 1, & x \geq 0.1. \end{cases} \tag{11}$$

For line objects (a_vobj_l) Fuzzy Membership is calculated for depth and height as well, besides that, also width of an object is considered. For Euclidean distance 10 m, conditions for fuzzification are:

$$\mu_{HL}(x) = \begin{cases} 1, & x = 0 \\ \dfrac{10 - x}{10}, & 0 < x < 10 \\ 0, & x \geq 10. \end{cases} \tag{12}$$

Fuzzy Membership for height, depth and width

$$\mu_{HLH}(x) = \begin{cases} 0, & x = 0 \\ \dfrac{0.1 - x}{0.1}, & 0 < x < 0.1 \\ 1, & x \geq 0.1. \end{cases} \tag{13}$$

Polygon objects (a_vobj_a) are given by its ground plan; therefore height and depth were used in the calculation. Fuzzification was calculated for Euclidean distance 20 m.

$$\mu_{HP}(x) = \begin{cases} 1, & x = 0 \\ \dfrac{20 - x}{20}, & 0 < x < 20 \\ 0, & x \geq 20. \end{cases} \tag{14}$$

Fuzzy Membership for height and depth with conditions

$$\mu_{HPH}(x) = \begin{cases} 0, & x = 0 \\ \dfrac{0.1 - x}{0.1}, & 0 < x < 0.1 \\ 1, & x \geq 0.1. \end{cases} \tag{15}$$

All the remaining coefficients are calculated similarly with the exception of coefficient C_4, which has not been calculated yet because the system is not connected to the on-line meteorological data.

The final deceleration coefficient is calculated from the individual files cxfuzzy with the help of tool Fuzzy Overlay. The first five coefficients enter the calculation with the help of relation $\max\{C_1, C_2, C_3, C_5, C_6\}$. The resulting value is multiplied by a raster of coefficient C_7 by reason of assigning meaning of the individual communications according to traffic importance (highways, first-class roads, forest roads, etc.). The result is the cost map that can be an input for searching of an optimal route in a decision-making process in CCM.

Results and Discussion

As it was mentioned at the beginning, the aim of the task was to describe thoroughly the creation of process models for calculation of deceleration coefficients within verification of the theory of terrain passability with the use of crisp set and fuzzy theory and also to show the problems we met.

For verification of functionality of the created models, database DMÚ25 was chosen, in the quality of 2010. Brno and its closest surroundings were used as the area of testing. For the chosen area, first, the individual coefficients were calculated, then the resulting cost map. It was followed by the verification of results directly in the terrain.

One of the main problems of creation of the cost map proved to be the length of calculations of the individual coefficients, which was very long, much longer than for crisp set calculation. The complete calculation with fuzzification took place with a basic pixel size of 25 m. Then we came to a pixel size of 5 m. In both given cases, the time consumption of the calculation was acceptable. From the size of mean error in position of the individual object, however, the necessity to use a pixel size of 1 m became apparent. When we used this resolution level for the simplest calculation, the time of calculation prolonged to an unbearable length. Coefficient C_1 was the most time-consuming for calculation. It took 22 min on a desktop PC Dell Precision 490, processor Intel Xeon 5130, 2,6 GHz, RAM 4GB, operation system 32 bit Windows 7, Enterprise. For example calculation of coefficient C_5 took 17 min, C_6 took 15 min. The quickest was the calculation of coefficient C_7, which lasted 3 min. The complete calculation of the cost map together with the already calculated coefficients then took another 45 min. A sample of a part of the calculated cost map is shown in Fig. 2. If it was necessary to calculate in a higher

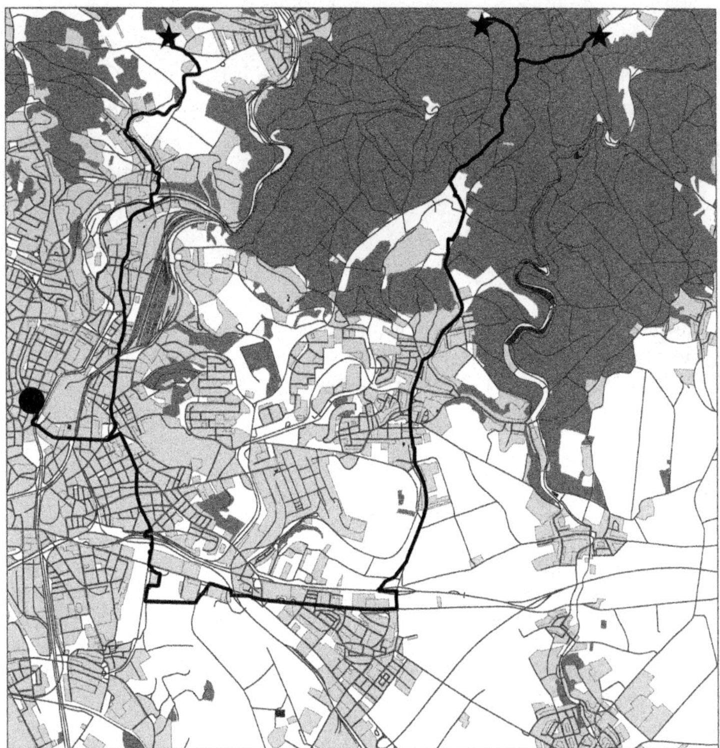

Fig. 2 Cost map and path example

resolution level, it would have to be necessary to significantly reinforce the performance of computer equipment or optimize the computing procedures.

When verifying calculated cost maps directly in the terrain, we discovered certain drawbacks caused mainly by absence of some of the declared characteristics of objects in the DMU25 database. However, these were mostly insignificant faults which did not affect the total calculation result. When consulting these problems with the provider, we were assured that the current new database edition, which we could not use at the time of calculations, would be more complete.

On the other hand the differences between crisp set and fuzzy set analyses were not very visible in build-up area because there were mostly objects with certain position (building, streets etc.). Therefore the calculated coefficient will be verified in an area outside build-up areas (in military training area) in the next phase of the project, and the authors expect verification also with the use of various types of military vehicles.

Conclusion

Both examples showed the differences between crisp set and fuzzy logic in spatial analyses and demonstrated the appropriateness of the use of fuzzy logic for the inclusion of the various characteristics of spatial objects in the calculations. They also showed the disadvantages at the same time, mainly worst interpretation of results from user point of view and time for calculation.

Acknowledgement The work presented in this paper was supported within the project for Development of Military geography and meteorology and for Support for Mathematical and Physical Research supported by the Ministry of Defence of the Czech Republic.

References

Ahmad B, Kharal K (2009) FuzzySets, FuzzyS-OpenandS-closed mappings. Adv Fuzzy Syst 5

D'Amico P, Di Martino F, Sessa S (2013) A GIS as a decision support system for planning sustainable mobility in a case-study. In Ventre A, Maturo A, Hoskova-Mayerova S, Kacprzyk J (eds) Multicriteria and multiagend decision making with applications to economics and social sciences (Studies in fuzziness and soft computing). Springer, Heidelberg, pp 115–128

Di Martino F, Sessa S (2011) Spatial analysis and fuzzy relation equations. Adv Fuzzy Syst 2011:14

ESRI (2013) User documentation. Copyright © 1995–2013 ESRI

Kainz W (2007) Fuzzy logic and GIS. University of Vienna, Vienna

Kovarik V (2011) Possibilities of geospatial data analysis using spatial modeling in ERDAS IMAGINE. In: Proceedings of the international conference on military technologies (ICMT'11). University of Defence, Brno, pp 1307–1313

Kovarik V (2014) Use of spatial modelling to select the helicopter landing sites. Adv Mil Technol 89(1):10

MoD-GeoS (2010) Catalogue of the topographic objects DMU25, 7.3 edn. Ministry of Defence of the Czech Republic, Geographic Service, Dobruska

Rybansky M (2009) Cross-country movement, the impact and evaluation of geographic factors, 1st edn. Akademické nakladatelství CERM, s.r.o. Brno, Brno

Rybansky M, Vala M (2009) Analysis of relief impact on transport during crisis situations. Moravian Geogr Rep 17(3):19–26

Rybansky M, Vala M (2010) Relief impact on transport. In: International conference on military technologies (ICMT'09). University of Defence, Brno, pp 551–559

STANAG 2999 (2012) Use of Helicopters in land operations doctrine, 9th edn. NATO Standardization Agency (MAS), Brussels

Sunila R, Hottanainen P (2004) Fuzzy model for soil polygons for mapping the imprecision. In: Proceedings of the StartGIS 2003, Pörtschach, p 8

Talhofer V, Hoskova-Mayerova S, Hofmann A (2012) Improvement of digital geographic data quality. Int J Prod Res 50(17):4846–4859

Tatra (2010) Tatra is the solution. (Tatra, a.s.) Retrieved 5 17, 2010, from TATRA: http://partners.tatra.cz/exter_pr/vp/new/typovy_listprospekt.asp?kod=341&jazyk=CZ

Zadeh I (1965) Fuzzy sets. Inf Control 8:338–353

Index